AF247448

Reviews of Accelerator Science and Technology

Volume 4

Reviews of Accelerator Science and Technology

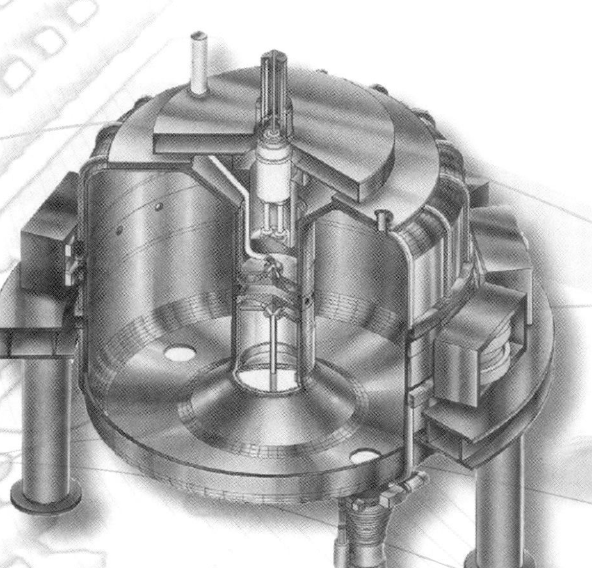

Volume 4

Accelerator Applications in Industry and the Environment

Editors

Alexander W. Chao
SLAC National Accelerator Laboratory, USA

Weiren Chou
Fermi National Accelerator Laboratory, USA

World Scientific

NEW JERSEY · LONDON · SINGAPORE · BEIJING · SHANGHAI · HONG KONG · TAIPEI · CHENNAI

Published by

World Scientific Publishing Co. Pte. Ltd.

5 Toh Tuck Link, Singapore 596224

USA office: 27 Warren Street, Suite 401-402, Hackensack, NJ 07601

UK office: 57 Shelton Street, Covent Garden, London WC2H 9HE

British Library Cataloguing-in-Publication Data
A catalogue record for this book is available from the British Library.

REVIEWS OF ACCELERATOR SCIENCE AND TECHNOLOGY
Volume 4: Accelerator Applications in Industry and the Environment

ISBN-13 978-981-4383-98-1
ISBN-10 981-4383-98-8

Printed in Singapore by Mainland Press.

Contents

Reviews of Accelerator Science and Technology
Vol. 4 (2011) vii–viii
© World Scientific Publishing Company
DOI: 10.1142/S1793626811000628

Editorial Preface

Since their debut in the late 1920s, particle accelerators have grown to become a backbone for the development of science and technology in modern society. These versatile and powerful instruments contribute in many areas, and exist in a large variety of sizes and capacities, each type fitting a particular application. Particle accelerators are used in the basic research of discovery sciences, as well as for medical treatment and research. An increasingly important role is played by photon sources. We have elaborated on many of these topics in previous RAST volumes. The largest number of accelerators is used in industry. Of about 30,000 accelerators at work in the world today, less than a fraction of a percent is used for basic scientific research. The vast majority — more than 99% — is for applications in industry (about 20,000 systems worldwide) and medicine (about 10,000 systems worldwide). Volume 2 of this journal (2009) was dedicated to the medical applications. The present volume, Volume 4, is focused on *Accelerator Applications in Industry and the Environment.*

There are two major categories of industrial applications: materials processing and treatment, and materials analysis. Materials processing and treatment includes ion implantation (semi-conductor materials, metals, ceramics, etc.) and electron beam irradiation (sterilization of medical devices, food pasteurization, treatment of carcasses and tires, cross-linking of polymers, cutting and welding, curing of composites, etc.). Materials analysis covers ion beam analysis (IBA), non-destructive detection using photons and neutrons, as well as accelerator mass spectrometry (AMS). All the products that are processed, treated and inspected using beams from particle accelerators are estimated to have a collective value of US $500 billion per annum worldwide. Accelerators are also applied for environment protection, such as purifying drinking water, treating waste water, disinfecting sewage sludge and removing pollutants from flue gases.

Industrial accelerators continue to evolve, in terms of new applications, qualities and capabilities, and reduction of their costs. Breakthroughs are encountered whenever a new product is made, or an existing product becomes more cost effective. Their impact on our society continues to grow with the potential to address key issues in economics or society of today.

This volume contains fourteen articles, all authored by renowned scientists in their respective fields. The first eight articles are reviews of various industrial accelerator applications. Sueo Machi, a pioneer of electron accelerators in industry, gives a comprehensive review of electron beam accelerator applications. The article by Larson, Williams, and Current is on ion implantation of semi-conductor materials, the largest industrial application of accelerators. Following are six articles: Jeynes, Webb, and Lohstroh on IBA; Harmon, Wells, and Hunt on non-destructive detection using neutrons and photons; Schmor on radioisotope production; Chen, Guo, Liu, and Zhou on AMS; Chmielewski on environment protection; and Hosemann on radiation damage.

The next three articles discuss accelerator technology developed specifically for industry: Hellborg and Whitlow on direct current accelerators; Cleland on radio-frequency electron accelerators; and Mank, Bauer, and Mulhauser on accelerators for neutron generation.

It is a challenging subject to write about future prospects in this rapidly evolving branch of technology. Alan Todd offers his views in his article.

The European Organization for Nuclear Research (CERN) is a major science institution renowned for its high-quality and innovative research since its foundation in 1954. With the successful construction and operation of the world's largest accelerator, the Large Hadron Collider (LHC), CERN has become the energy-frontier of fundamental scientific research. In this volume, we invite Herwig Schopper, a former Director General of CERN, to write a review of CERN's history and principles, and its future. Vinod Chohan writes about Simon van der Meer, a Nobel laureate and a giant in the accelerator field who passed away this year. He is our choice for "Person of the Issue." His ingenuity and dedication to perfection can be seen vividly in Chohan's article.

Alexander W. Chao
SLAC National Accelerator Laboratory, USA
achao@slac.stanford.edu

Weiren Chou
Fermi National Accelerator Laboratory, USA
chou@fnal.gov

Reviews of Accelerator Science and Technology
Vol. 4 (2011) 1–10
© World Scientific Publishing Company
DOI: 10.1142/S1793626811000562

Trends for Electron Beam Accelerator Applications in Industry

Sueo Machi

Japan Atomic Energy Agency,
1233 Watanuki, Takasaki, 370-1292, Japan
machi.sueo@jaea.go.jp

Electron beam (EB) accelerators are major pieces of industrial equipment used for many commercial radiation processing applications. The industrial use of EB accelerators has a history of more than 50 years and is still growing in terms of both its economic scale and new applications. Major applications involve the modification of polymeric materials to create value-added products, such as heat-resistant wires, heat-shrinkable sheets, automobile tires, foamed plastics, battery separators and hydrogel wound dressing. The surface curing of coatings and printing inks is a growing application for low energy electron accelerators, resulting in an environmentally friendly and an energy-saving process. Recently there has been the acceptance of the use of EB accelerators in lieu of the radioactive isotope cobalt-60 as a source for sterilizing disposable medical products. Environmental protection by the use of EB accelerators is a new and important field of application. A commercial plant for the cleaning flue gases from a coal-burning power plant is in operation in Poland, employing high power EB accelerators. In Korea, a commercial plant uses EB to clean waste water from a dye factory.

Keywords: Electron accelerator; industrial application; polymeric material; environmental protection.

1. Introduction: History of Electron Accelerator Applications

The International Atomic Energy Agency (IAEA) was established in 1957 by an initiative of the US President Dwight D. Eisenhower to promote the peaceful application of atomic energy and to verify the nonproliferation of nuclear weapons.

Two years later, in 1959, the IAEA Conference on Application of Large Radiation Sources (including EB accelerator applications) was held in Poland (Fig. 1).

The Raytherm Corporation [1] was established by Mr. Paul Cook in 1957 in the US. It was the world's first company to commercially produce radiation cross-linked polyethylene by the use of EB accelerators. EB cross-linked polyethylene was used for heat-shrinkable tubing and heat-resistant insulation of wires.

From the 1940s to the 1960s, Robert Van de Graaff, John Trump and Roy Emanuelson developed an electron accelerator based on an insulating core transformer (ICT) for industrial purposes, and they established High Voltage Engineering Co. Ltd. to produce commercial accelerators.

In Japan, the government-funded institute Takasaki Radiation Chemistry Research Establishment (TRCRE), under the Japan Atomic Energy Research Institute, was established in 1963 to promote the industrial application of radiation processing by the use of both EB accelerators and large cobalt-60 sources. The first large electron accelerator used at TRCRE was imported from Radiation Dynamics Inc. of the US.

In 1971, Sumitomo Electric Industry produced its first commercial radiation cross-linked polyethylene using a large capacity electron accelerator in Japan.

Since then, applications of radiation processing in industry have been growing rapidly in both the US and Japan, followed by Germany, France, Korea and developing countries such as China and Brazil.

2. Trends of Accelerators for Radiation Processing Applications [2]

Radiation processing in industry is carried out by the use of cobalt-60 irradiators and/or EB accelerators. Cobalt-60 has the advantage of a high

Fig. 1. The first IAEA International Conference on Application of Large Radiation Sources, in 1959 in Warsaw, Poland.

for medical purposes in radiation therapy of cancer patients, a medical, nonindustrial use.

More recently, low energy ion beam accelerators have been employed extensively for ion implantation in silicon for semiconductor production.

It should be noted that in the past several years EB accelerators have been replacing cobalt-60 for the sterilization of medical products because of the greater commercial acceptance of nonradioactive sources and their economic competitiveness. For achieving higher penetration, 10 MeV EB accelerators are produced by a few companies. X-rays, which have high penetration comparable to that of gamma rays, are derived from high energy EB accelerators and are being used to replace isotope sources for medical product sterilization.

penetration range of its gamma rays and has been found to be easy to operate and maintain. It has been widely used for radiation sterilization of disposable medical supplies, such as syringes, bandages and dialyzers, as well as food irradiation.

EB accelerators have been employed for radiation processing of polymeric materials in industries, such as wire/cable, automobile tire manufacture, food packaging and heat-shrinkable materials. A large number of linear accelerators are used

3. Economic Scale of Radiation Applications in Comparison with the Use of Nuclear Energy to Produce Electrical Power

Japan's Cabinet Office has studied the economic scale of radiation applications in industry, medicine and agriculture, and found it to be nearly comparable with the nuclear power application in Japan, as shown in Fig. 2. A major contributor to the economic size of the industrial sector is the

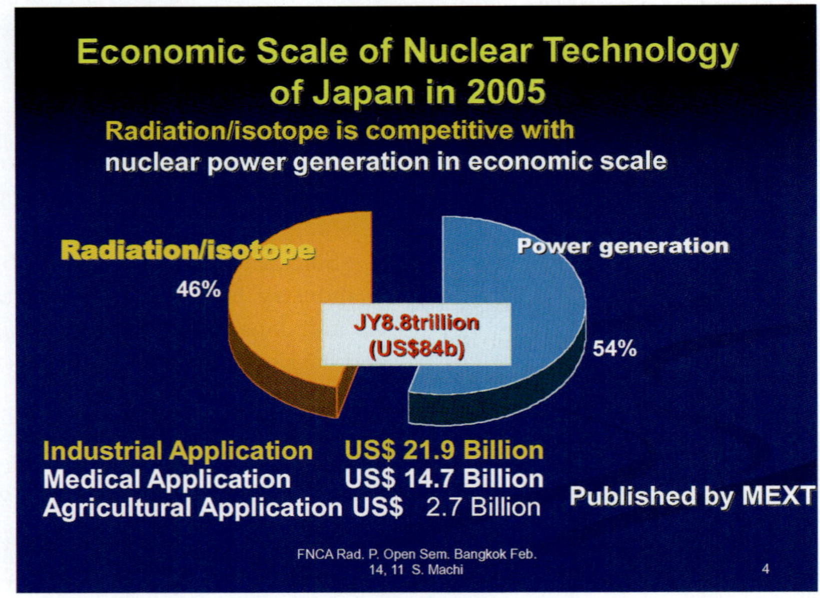

Fig. 2. Economic scale of radiation applications in Japan.

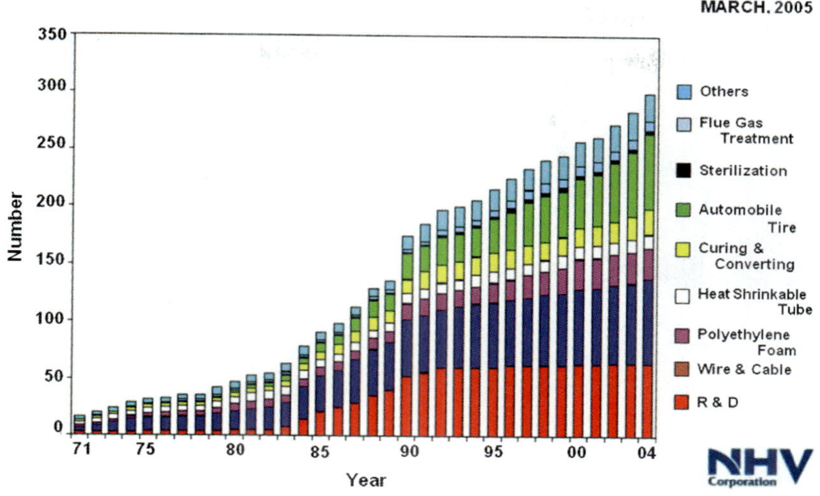

Fig. 3.   Growth in the number of electron accelerators for industrial applications and research and development in Japan.

scale of ion implantation into silicon to produce semiconductors.

There has been a remarkable rate of growth in EB accelerator production for industrial and research purposes in Japan for more than 30 years. This is clearly shown in Fig. 3.

4. Application of Electron Accelerators in the Polymer Industry

Since the 1950s, the polymer industry has become one of the largest segments of the entire chemical industry. EB accelerators are used for the modification of a variety of polymeric products, as shown in Table 1.

Recently, the IAEA reported the ratios of the different end-use applications of EB accelerators, as shown in Fig. 4.

4.1. *Cross-linking of polymers to improve properties*

Radiation-induced cross-linking of polymers was discovered by Prof. Arthur Charlesby in the UK in the 1950s. It is widely used in industry for a variety of

Table 1.   List of commercial products made through EB processing.

Cross-linked or grafted polymers commercially produced through radiation processing	
Products	Applications
Cross-linked polyethylene and PVC	Wire insulation resistant to heat and chemicals; pipes for hot water of heating systems
Cross-linked foamed polyethylene	Insulation, packing, floating materials
Heat-shrinkable tubes and sheets	Food packaging, wire insulation, protection of welding for corrosion
Cross-linked rubber sheets	Automobile tires (high quality)
AA-grafted PE film	Battery separators
Cross-linked polyurethane	Cable insulation for antilock brake sensors
Cross-linked nylon	Automobile parts resistant to heat and chemicals
Super-heat-resistant SiC fiber	Metal and ceramic composites
Cross-linked hydrogel	Wound dressing (Japan, Korea, Poland), face mask (Malaysia)
Curing of paints and inks	Surface coating and printing
Grafted PE fiber	Deodorants

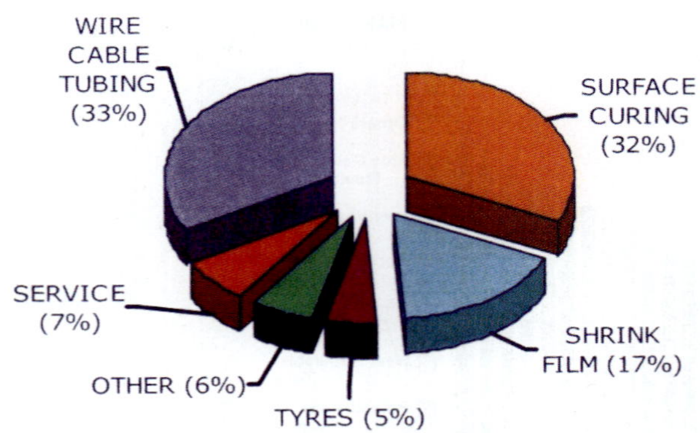

Fig. 4. Ratios of the commercial use of EB accelerators by applications [3].

Fig. 5. Cross-linking of wires by an EB accelerator (Nuclear Malaysia).

applications. Several major applications are:

4.1.1. *Automobile tires*

The production of automobile tires using EB accelerators was first commercialized by Firestone Co. Ltd. and by Goodyear Co. Ltd. in the US in 1960s. Currently, most of the automobile tires in Japan are manufactured by the use of EB accelerators to partially cross-link rubber extrusions before the final tire molding. This process is used in many countries, such as in France by Michelin Co. Ltd., in Japan by Bridgestone Co. Ltd., in Brazil, Korea and the US, etc. Electron beams quickly cross-link both natural rubber and synthetic rubber at room temperature without any chemical additives. They are employed to irradiate rubber extrusions continuously by using a conveyor system at room temperature running at high production rates. This application is growing in a variety of countries.

4.1.2. *Wire and cable*

Another large commercial application of EB accelerators is cross-linking of the insulation of wire and cable. Major manufacturers of wire and cable are using EB accelerators for the production of heat-resistant wire and cable. The wire insulation of polyethylene and polyvinyl chloride is made heat- and chemical-resistant by EB cross-linking. The cross-linking process is very simple, as shown in Fig. 5. Extruded polymer-insulated wire or cable is run under an EB accelerator at room temperature.

Electron-induced cross-linking of the insulating polymers not only renders them more heat-resistant but also enhances chemical resistance and improves some of the mechanical properties to the wire insulation.

The heat and chemical resistance of EB cross-linked wire and cable is of benefit in widely used areas, such as for automobiles, electrical industrial equipment and home appliances where these properties are needed.

4.1.3. *Heat-shrinkable tubes and films*

Extruded plastic films and tubing are EB cross-linked and then expanded at temperatures above the melt transition of the plastic, chilled, and sold as heat-recoverable materials. Upon subsequent heating, these materials shrink back to their dimensions before being expanded. The creates airtight sealing of food packaging films and of tubing placed around such items as electrical connectors.

This is the widely employed process for the production of heat-shrinkable tubing and films. Heat-shrinkable tubing is used for electrical insulation around the connections of wires and cables. Heat-shrinkable films and sheets are used for a variety of applications, including food packaging and protection of welded joints on pipelines.

4.1.4. *Hydrogel wound dressings and face masks*

Hydrogel wound dressings have been developed by Japan Atomic Energy Agency (JAEA) [4] and by research groups in Poland [5], and are commercially produced in both countries using electron-induced cross-linking of polyvinyl alcohol. These products

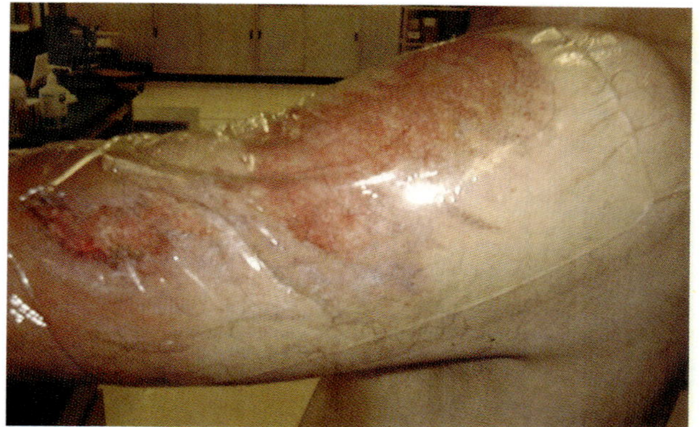

Fig. 6. Application of a transparent hydrogel dressing produced by an EB accelerator.

Fig. 7. Industrial plant for curing of the surface coating of printed paper by a low energy EB accelerator.

have no toxic impurities, because the cross-linking process does not require a catalyst, which is very desirable for medical applications (Fig. 6). Hydrogel wound dressings are much better than gauze dressings in terms of the healing speed of wounds and friendliness to patients (no pain when the dressing is peeled off).

Face masks made of hydrogels were developed by a group at Nuclear Malaysia. The hydrogel sheets were produced by EB-induced cross-linking of sago starch using an accelerator. The hydrogel face mask is commercially used for reducing face wrinkles and for restoring skin texture by increasing skin moisture.

4.2. *Curing of surface coatings and printing inks by electron beam accelerators*

Surface coating and printing are large areas of use in the polymer industry.

Conventional curing processes for coatings and printing inks are carried out through the evaporation of an organic solvent using a thermal process. This process causes air pollution due to the emission of the evaporated organic solvent into the atmosphere, and also consumes a significant amount energy for the thermal evaporation of the solvent.

Electron-induced curing of coatings and printing inks is based on a mechanism different from that of thermal curing. EB coating and printing ink materials are composed of color pigments and reactive liquid monomers and oligomers without solvents. The coating process is followed by irradiation from EBs from accelerators, which polymerize the liquid monomers and oligomers to form solid cross-linked

films and printing. This innovative curing process needs little energy and does not emit organic solvent, because monomers and oligomers convert to solid polymers using EBs. The energy consumption of this electron curing process is estimated to be about 370 times lower than that of a conventional thermal process. Cured coatings and printing have higher solid contents than those of the conventional process, and are therefore of better quality in terms of surface coating hardness (Fig. 7).

4.3. *Radiation-induced grafting of polymers*

Polymeric materials such as polyethylene can be modified by a radiation-induced grafting technique. Carbon hydrogen bonds of polyethylene chains are broken by high energy radiation to form radicals which are chemically active. Graft polymerization of monomers can be initiated on these radicals to form the grafted polymer chains which modify the properties of the backbone polymers.

The first commercial application of radiation grafting was the production of membrane separators for silver oxide batteries by grafting acrylic acid onto polyethylene film, in Japan. The research for this was carried out by a group from Japan Atomic Energy Agency and development work was implemented in collaboration with Yuasa Battery Co. in the 1970s [6].

Polyethylene film is first irradiated by an EB accelerator at room temperature, followed by contact with an acrylic acid solution for a few hours to initiate graft polymerization of the acrylic acid. Since the lifetime of polyethylene radicals is in the range of days, graft polymerization reaction is a separate process from irradiation of polyethylene film.

The EB-produced membrane separator has excellent chemical resistance compared to more conventional membranes. As a result, batteries with EB-grafted membranes have a life that is up to five years longer.

5. Food Irradiation

The commercial acceptance of food irradiation is expanding worldwide because of better food safety, disinfestations that eliminate quarantines, and decreased postharvest loss by sprout inhibition and disinfestations.

Postharvest loss in developing countries is higher than 40% — a very serious problem that can be partly addressed by food irradiation integrated through improved storage and transport infrastructure.

The IAEA reported that about 500,000 tons of a variety of foods are irradiated worldwide (as shown in Table 2) by the use of 180 gamma irradiation facilities and 12 EB or X-ray accelerator facilities.

The wholesomeness of irradiated foods has been officially announced by the World Health Organization (WHO), the Food and Agriculture Organization (FAO), and in an IAEA based on extensive study. The international food safety standard CODEX has also affirmed the wholesomeness of irradiated foods.

Food irradiation is an important growing potential market for industrial EB accelerators. For example, in Odessa, Ukraine, an EB accelerator has been commercially used for the irradiation of wheat grains to kill fungi and insects for more than 20 years.

Table 2. List of commercially irradiated foods in major countries.

Irradiated foods increasing permitted in 60 countries	
World total of irradiated foods	*ca. 500,000 ton*
China: Garlic, dried vegetables, etc.	146,000 ton
Vietnam: Frozen shrimps, etc.	14,000 ton
Japan: Potatoes	8,000 ton
USA: Spices, ground meats, fruits	92,000 ton
Ukraine: Wheat grain	70,000 ton
Brazil: Spices, fruits	23,000 ton
South Africa: Spices	18,000 ton
Belgium: Spices, frozen chicken	7,000 ton
Others:	ca. 120,000 ton

6. Sterilization of Medical Products and Food and Beverage Packaging

Disposable medical products, such as syringes, catheters, dialysis equipment, sutures and bandages, are sterilized by irradiation or fumigation with ethylene oxide. Radiation sterilization is increasing to replace ethylene oxide (EtO) worldwide because of its better quality assurance.

Cobalt-60 has been major source of radiation for applications because of the greater penetration of gamma rays. However, there has recently been a trend toward using high energy EB accelerators and/or X-rays derived from such accelerators in lieu of cobalt-60, owing to difficulties in isotope availability and the necessity to replenish it every few years.

In Japan, the sterilization of PET (polyethylene terephthalate) bottles for beverages by an EB accelerator is a new application recently introduced using a commercial in-line 300 keV EB unit, as shown in Fig. 8.

The advantages of the irradiation process over the conventional bottle sterilization system which uses hydrogen peroxide are:

- Sterilization in-line with the filling process;
- No chemical residues in bottles;
- Cost reduction by 10–25%;
- Compact system requiring less space;
- High capacity for sterilization, such as 600 500 ml bottles per minute.

Plastic food packaging has also been sterilized by EB accelerators very efficiently for commercial purposes in Japan in the past several years.

Fig. 8. Industrial scale PET bottle sterilization system.

7. Application for Environmental Conservation

Environmental protection to reduce pollutants released from industrial plants is an important technical challenge in both developing and developed countries. For instance, the emission of SO_2 and NO_x mostly from coal- and/or oil-burning power stations has been increasing. This in turn causes acid rain, which damages woods, forests and lakes, particularly in eastern Europe, China and India. World Bank statistics reported that the emission of SO_2 in 2020 will be 80,000,000 tons, mainly from coal-burning power stations.

There have been two successful technical developments using EB accelerators for environmental purposes: removing SO_2 and NO_x from the exhaust gases of fossil-fueled power plants, and cleaning waste water from dyeing textile factories.

7.1. *Cleaning industrial flue gases by the use of electron beams*

The first pioneering work in using EBs to clean flue gases was carried out by a group from JAEA in collaboration with Ebara Co., in 1971 [7, 8]. This was followed by work done by research groups in the USA, Poland, Germany, China and Brazil.

7.1.1. *Technology of cleaning flue gas and reaction mechanism*

This innovative technology provides the only method that is able to remove SO_2, NO_x and volatile organic compounds (VOCs) simultaneously. As shown in the schematic flow diagram in Fig. 9, flue gas from a power station is first cooled down by a water spray to a temperature ranging between 65°C and 70°C, and then introduced into a process vessel where the flue gas is irradiated by EBs through thin windows of the accelerators process vessel.

The reaction mechanism in Fig. 10 shows that hydro-oxy radicals (OH·) are first formed by the radiolysis of water contained in the flue gases, which then reacts with SO_2 and NO_x to convert these gases into H_2SO_4 and HNO_3, respectively. These acids are next converted to the solids ammonium sulfate and ammonium nitrate through reactions with ammonia added to the process vessel.

Particulates of ammonium sulfates and ammonium nitrates are collected by an electrostatic precipitator. The collected byproducts can be used as an agricultural fertilizer.

7.1.2. *Industrial application of the technology*

Based on laboratory scale experiments, pilot scale plants to clean flue gases at rates ranging from

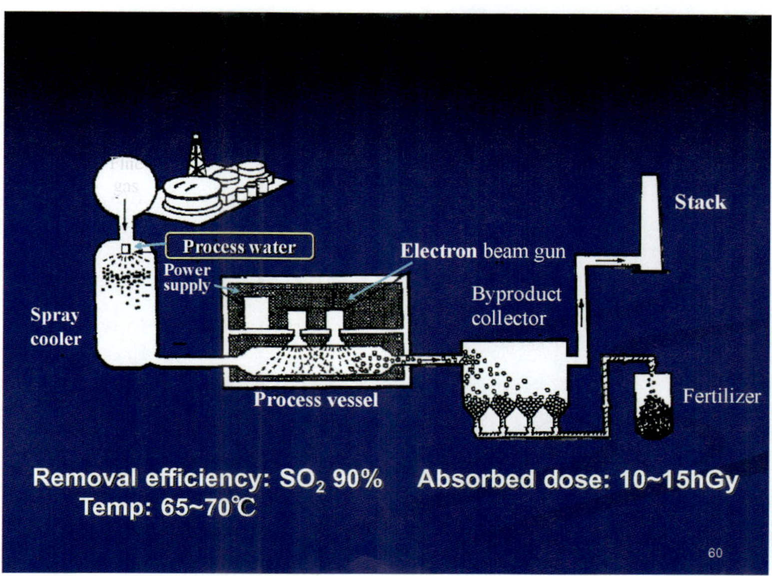

Fig. 9.   Flow diagram of flue gas cleaning by EB accelerators.

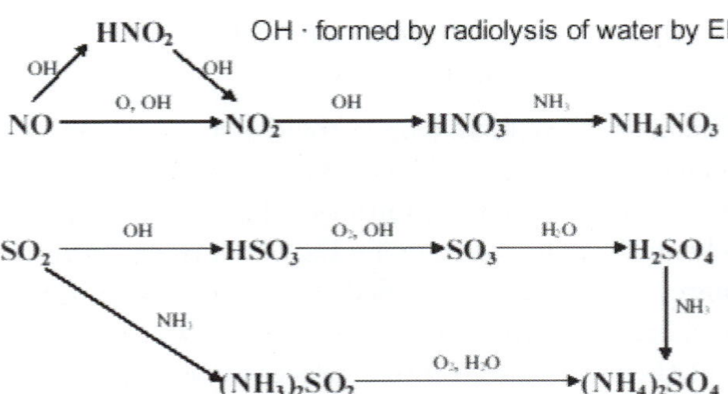

Fig. 10. Reaction mechanisms for the removal of SO_2 and NO_x by EBs.

1000 to 10,000 Nm^3/h were installed and successfully operated to obtain engineering data in Japan [9], the USA, Germany, Poland [10] and China.

Based on the promising results of these pilot plant experiments and engineering data, the first industrial scale plant with a 300,000 Nm^3/h capacity was installed in Chengdu, China by Japanese firm Ebara Co., in 1997 [11] (Fig. 11). The plant had been operated successfully to clean flue gas from a 90 MWe coal-fired power plant by using two electron accelerators of 0.8 MeV with a 160 kW capacity. The removal rates of SO_2 and NO_x are 80% and 20%, respectively. Byproducts of ammonium sulfates and ammonium nitrates are being sold as agricultural fertilizer.

A second industrial plant was installed in Poland on the basis of extensive laboratory and pilot plant scale work at the Institute of Nuclear Chemistry and Technology (INCT). It is based on a Polish design [12] developed in collaboration with the IAEA and the Government of Japan and their experts and funding (Fig. 12).

Fig. 11. Industrial plant to clean flue gas from a coal-burning power station in Chengdu, China.

Fig. 12. Process vessel and electron accelerator for removal of flue gas from an industrial coal-burning power plant in Poland.

The treatment capacity of this plant is 270,000 Nm^3/h, which is achieved by the use of 4 EB units of 700 keV and 375 mA. The plant started operation in 2001. There were some difficulties at the early stage of operation, involving the continuous operation of such a large power supply for accelerators. Otherwise the plant has been successfully operated and the byproducts are being used as fertilizer.

7.1.3. *Advantages of the technology*

The innovative technology for cleaning flue gases by using EB accelerators has the following advantages over the conventional gypsum process:

- Simultaneous removals of SO_2 and NO_x by a single process;
- Zero waste water;
- Useful byproduct of ammonium sulfate and ammonium nitrate, which can be used as fertilizer;
- Smaller space for installation of the EB plant (better for retrofitting existing plants);
- Simultaneous removal of pollutants of polyaromatic hydrocarbons;
- Low cost of plant construction and operation.

7.2. *Cleaning of industrial waste water by the electron accelerator*

Biological treatment is the most common technology for removing organic contaminants in waste water. Some pollutants, however, are not easily degraded by such treatment.

In Korea, an innovative technology is used to clean waste water from factories that dye textiles. This has been successfully developed for industrial use. The total system consists of EB irradiation of water combined with biological treatment. Waste water from the dyeing factories is heavily colored and has a high BOD (biological oxygen demand). It is difficult to remove the color of waste water by biological treatment alone.

The irradiation of waste water produces hydro-oxy radicals (OH·) which decompose or convert organic pollutants or dye molecules in the water to compounds which are more readily degraded by biological treatment.

Pilot scale experiments were conducted by EB Tech Co. in Korea, starting in 1998, and they demonstrated the cleaning of $1000\,\mathrm{m}^3$/day waste water from a dyeing factory using a 1 MeV, 40 kW EB accelerator in Tegu city, Korea. Following successful results on the pilot plant scale, an industrial plant for waste water treatment was installed to clean $10{,}000\,\mathrm{m}^3$/day of waste water by the use of a large electron accelerator of 1 MeV, 400 kW with three beam scanners, as shown in Fig. 13 [13]. This project was supported by the IAEA.

Fig. 13. Industrial plant for the cleaning of waste water by an electron accelerator in Korea: capacity $10{,}000\,\mathrm{m}^3$/day; electron accelerator 1 MeV, 400 kW.

8. Conclusion: Challenges of Electron Beam Accelerator Applications

Industrial applications of electron beam accelerators are growing to meet industrial demands for advanced materials and energy-saving processes. The rate of development of radiation processing is even faster in developing countries such as China, Brazil and India. The IAEA is supporting the development of radiation technology and safety in its member states.

One obstacle to further growth of radiation processing in diverse applications is public acceptance of radiation. A good example is food irradiation. In some countries, including Japan, the public has fears concerning irradiated foods because of misinformation. A second obstacle to expanding industrial applications is the lack of technical information and the lack of knowledge within the management of commercial firms with regard to the advantages of radiation processing.

Electron beam accelerators are essential facilities for industrial radiation processing applications. There are still challenges for engineers and manufacturers of accelerators. One important challenge is the improvement of accelerator reliability for continuous operation for more than one year. A second challenge is to develop a reliable large capacity accelerator higher than 500 kW for applications in the field of environmental protection, which could be the largest potential market for accelerators. A third challenge is the development of large capacity X-ray accelerators with high energy efficiency, which would have large potential applications in irradiation of foods and medical products.

Last but not least, there is the challenge of cost reduction. The costs of large capacity electron beam accelerators are quite high for developing countries, including radiation-shielding construction.

Human resource development and information dissemination regarding the advantages and safety of radiation technology should be better implemented by governments, the IAEA, universities, and in the private sector to enhance the opportunities for radiation applications in view of the sustainable development of society.

References

[1] P. Cook, *Rad. Phys. Chem.* **35**, 7 (1990).
[2] S. Machi, *Rad. Phys. Chem.* **42**, 13 (1993).
[3] *Nuclear Technology Review* (IAEA, 2009).

[4] F. Yoshii *et al.*, *Rad. Phys. Chem.* **46**, 169 (1995).

[5] M. Rosiak, Hydro-gel dressings, *ACS Symposium Series 475* (Washington D.C., 1991), p. 271.

[6] S. Machi *et al.*, Ion exchanges membrane by radiation cross-linking, Second International Meeting on Radiation Processing (Miami, Florida, 22–26 Oct. 1978).

[7] S. Machi *et al.*, *Rad. Phys. Chem.* **9**, 371 (1977).

[8] S. Machi, *Rad. Phys. Chem.* **22**, 91 (1983).

[9] H. Namba *et al.*, *Rad. Phys. Chem.* **46**, 1103 (1995).

[10] A. G. Chmielewski *et al.*, *Rad. Phys. Chem.* **46**, 1067 (1995).

[11] Y. Doi *et al.*, *Rad. Phys. Chem.* **57**, 495 (2000).

[12] A. G. Chmielewski *et al.*, *Rad. Phys. Chem.* **71**, 441 (2004).

[13] B. Han *et al.*, *Water Sci. Technol.* **52**, 317 (2005).

Sueo Machi served as Director General of Takasaki Radiation Chemistry Research Establishment at Japan Atomic Energy Research Institute (1989–1991), Deputy Director General of the IAEA (1991–2000), Commissioner of Japan Atomic Energy Commission (2004–2007) and Coordinator of Japan, Forum for Nuclear Cooperation in Asia (2000–present). He has been doing research on radiation polymerization of ethylene, radiation grafting of acrylic acid onto polyethylene to produce battery separators, radiation degradation of polymers for nuclear power safety, and cleaning of flue gas by electron beams for environmental protection.

Reviews of Accelerator Science and Technology
Vol. 4 (2011) 11–40
© World Scientific Publishing Company
DOI: 10.1142/S1793626811000616

Ion Implantation for Semiconductor Doping and Materials Modification

Lawrence A. Larson* and Justin M. Williams

*Ingram School of Engineering, Texas State University at San Marcos
RFM 5210, 601 University Drive, San Marcos, TX 78666, USA
Larry.Larson@txstate.edu

Michael I. Current

*Current Scientific
1729 Comstock Way, San Jose, CA 95124, USA
currentsci@aol.com*

In the 50-plus years since the patent was issued to William Shockley in 1957, ion implantation has become a key process in the commercial production of semiconductor devices, advanced engineering materials and photonic devices. This article reviews the fundamental concepts of production ion implanters for both the processes used in manufacturing and also in the design of the tools themselves. Recent publications in the application areas of semiconductors and materials modification are summarized, focusing on the attendant process effects. These results demonstrate that ion implantation is a well understood technology with abundant and evolving applications.

Keywords: Ion implantation; semiconductor processing; materials modification.

1. Introduction

This article aims to review the technology of ion beam implantation and its many applications in materials processing. The majority of the commercial applications of ion implantation are in semiconductor processing and we will review recent advances in process space, machine types, semiconductor applications and process effects including shallow junction source/drain extension [SDE], junction leakage, source/drain contact [SDC] and SDE strain, photoresist, photovoltaic [PV] doping, and semiconductor process issues. The other major area of commercial application of ion processing is materials modification applications and processes; major sections within that area include tool hardening, H-cut for silicon-on-insulator [SOI] and PV, silicide stability, and a collection of similar applications that can be considered materials modification using an ion beam.

2. Dose and Energy Ranges

Ions are accelerated for implantation into integrated circuit (IC) and photovoltaic (PV) devices over an energy range from $\approx 100\,\mathrm{eV}$ to nearly $10\,\mathrm{MeV}$ and a dose range from 10^{10} to over $10^{18}\,\mathrm{ions/cm^2}$. The specific applications include doping of semiconductor junctions and a rapidly increasing number of materials modification uses throughout the manufacture of ICs and PV systems. The key application areas are shown Fig. 1.

In addition to IC and PV uses, ion implantation is used in the fabrication of hardened metals, polycarbonate foils, nuclear reactor containment structures, catalysts and micro-electro-mechanical (MEMS) devices, discussed later in this article.

3. Implanter Machine Types

The acceleration and distribution of ions over target surfaces over the wide dose and energy ranges shown in Fig. 1 has led to the development of many and diverse system designs for operation in specific areas of the ion implantation "phase space." The majority of implantation machines operate in the ion energy regime from a fraction of a keV to $\approx 200\,\mathrm{keV}$, with "medium current" tools spanning this energy range with beam currents of a few mA or less and "high current" tools operating at maximum beam currents

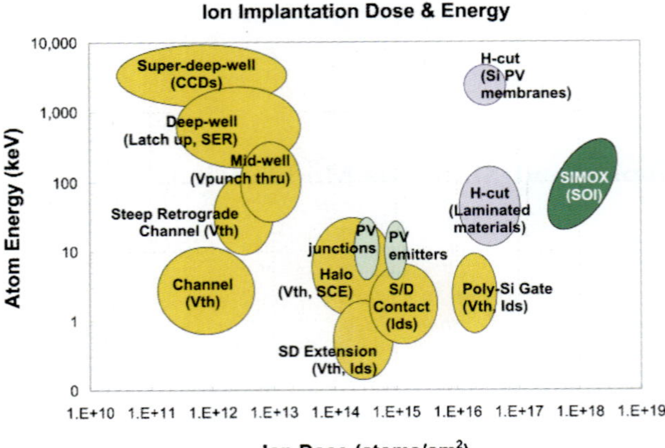

Fig. 1. Ion energy and dose for key applications in IC and PV device manufacturing.

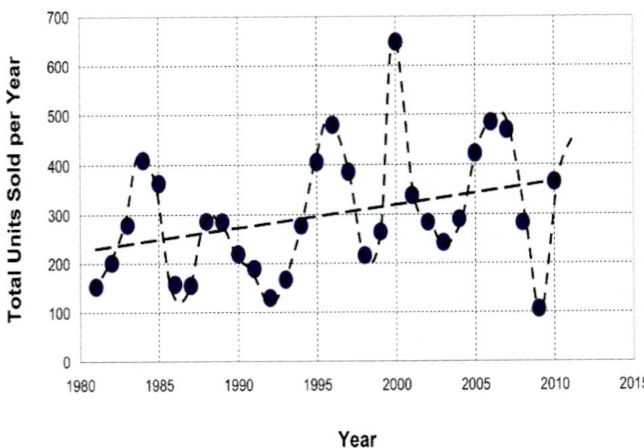

Fig. 2. Yearly unit sales of ion implantation systems, mostly for IC device fabrication.

of several 10 s of mA. The class of "high energy" tools provide energetic ions in the range from $\approx$20 keV to several MeV for doping of deep junctions for Flash memories and CMOS-based optical sensors. Versions of high energy implanters have been developed to provide mA beam currents of protons at energies up to 4 MeV for use in splitting of $\approx$100 μm thick freestanding Si membranes. A recent addition to the ion implant tool set is a class of "high dose" tools which either have a simplified beamline structure to provide an ion "shower" at high ion flux rates or forgo the use of accelerating columns entirely and utilize wafer bias to draw ions out of nearby plasma in "plasma immersion" ion implant (PIII) systems. This diversity of machine types can be expected to increase in the near future, with several "hybrid" designs in development dedicated for high throughput doping of Si-based PV junctions. The design principles and operational characteristics of these various types of ion implantation systems are discussed in the following section.

The industrial demand for ion implantation systems, almost entirely for doping of IC transistors, has led to the development of a set of tool vendors with many members active in this field since the 1970s. The yearly unit sales of commercial implanters is shown in Fig. 2. The yearly number of units sold averages around 300, with large fluctuations year to year, usually with a six-year cycle, due to variations in facility construction and expansion schedules, transitions to new wafer sizes and the opening of new markets for IC devices. With a typical selling price of US$4 million for a fully-featured production tool,

the primary implanter vendors sell over US$1 billion in new machines in a year with active plant expansion. In addition, there is a well-developed group of secondary vendors who provide spare parts, refurbishment and upgrade kits, implant services and the dopant and other materials to support continuous operations as well as a vigorous market for resale of older machines to developing markets.

4. Doping Planar CMOS Transistors

The principal doped regions of a transistor are shown in Fig. 3, where the full CMOS device has approximately equal numbers of n- and p-type channel regions (the area under the gate oxide between the source and drain extension junctions) with opposite doping in each type, hence the "C" (complementary) in CMOS. The most important junction in the

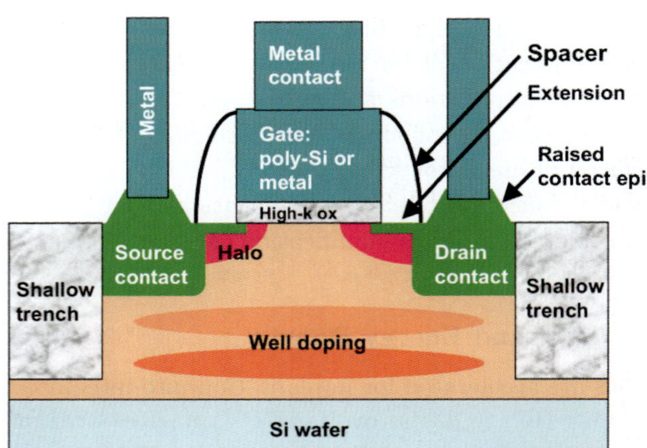

Fig. 3. Principal doped regions of a planar CMOS transistor.

transistor is the extension region under the spacers because the location and doping levels in the extensions determine the threshold voltage and drive current. The extension junctions in advanced logic devices are "ultra-shallow," approximately 10 nm, with desired doping levels above 10^{20} carriers/cm^3. The deeper contact junctions connected to the source/drain extensions often also contain additional elements to provide carrier mobility enhancing strain to the channel region; doped SiGe alloy epi in pMOS transistors (providing compressive channel strain) and C, added by implant or selective epi (providing tensile strain) in the nMOS contacts. The contact junctions are often capped by "raised" regions, grown by selective doped epi CVD, so that the metal contacts to the source/drain junctions are sufficiently far away from the extension and channel to prevent metal shorts and reduce current flow crowding heating during transistor operations. The doping in the region just under the metal contacts is often enhanced by additional shallow implants to reduce contact resistance and thereby increase transistor drive currents.

The region below the source/drain contacts and extensions, the CMOS "well," has carefully engineered doping in a series of opposite type implants compared to the source/drain junctions. The source/drain extensions are bounded by an extra well doping "halo," usually formed by a set of angled implants before the gate spacer cladding is put in place. The extra doping level in the halos is intended to suppress the lateral encroachment of the source depletion layer into the channel area and thereby stabilize the transistor threshold voltage characteristics and reduce off-state currents. As we will see later, in the discussion of junction leakage currents, halo doping can have deleterious effects by increasing the reverse bias junction leakage through enhanced band-to-band tunneling, and so the halo doping levels are one of the many aspects of the transistor doping that must be balanced in process of "integration" of the final IC design and production flow.

The doping profile of CMOS wells are often formed with a series of multiple energy implants. The location of shallow local doping peaks near the channel sets the threshold voltage. There is also a mid-well local peak to set the well punch through characteristics and a deep well region peak to establish lateral isolation between adjacent wells.

This deep peak suppresses "soft errors" arising from local charge upsets from ionizing radiation from cosmic rays and other sources. In Flash memory devices, some wells are themselves isolated by surrounding deeper doped regions of an opposite type, forming "triple well" structures so that opposite polarity biases can be sustained during cell programming which are independent of the background Si wafer ground reference. Even deeper well structures, extending to several microns deep, are formed in CMOS imager devices by implants of dopants at energies above 5 MeV in some cases.

5. System on a Chip (SOC) Doping

Complex control and logic circuits can require more than 40 implant steps per finished device. The balance of chip speed and power consumption is often accomplished by operating various regions of complex "system on a chip" (SOC) devices with different threshold voltages and drive current characteristics. This is achieved by forming transistors with locally specific implant conditions and by adjustments in other factors such as the gate dimensions and gate oxide thickness. The dose and energy distribution of a modern 28 nm SOC process is shown in Fig. 4, where over 40 implant steps are used in the process [1]. The implants near and above 100 keV are the well doping, the lower dose implants between 10 and 100 keV are mainly the various halo and shallow

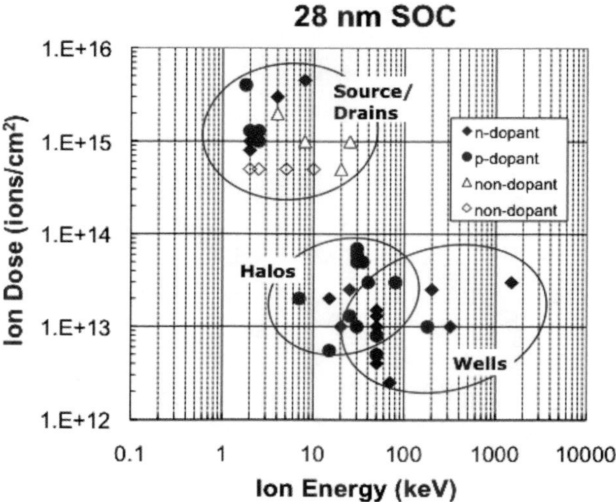

Fig. 4. Dose and energy distribution of ions used to fabricate 28 nm CMOS transistors. This comprises over 40 implant steps [1].

well doping and the low energy, high dose implants are various depth S/D extensions and contacts and the poly-Si gate doping. The non-dopant implants are primarily C profiles implanted as "cocktails" to suppress deep diffusion of B and P doped shallow junctions during annealing [2].

6. Doping of Non-Planar CMOS Transistors

Advanced logic and SRAM memory devices are beginning to use non-planar CMOS transistors, where the channel regions are formed from etched "fins" of Si with source/drain regions doped on either side of fully depleted, undoped channels (Fig. 5). The principal advantage of "finFET"-type transistors is that the gate region surrounds the channel on three sides (leading to the description of some types as "tri-gates") so that the channel can be fully depleted by the gate bias, leading to excellent off current levels.

A limitation of narrow fin-shaped channels is the relatively low levels of drive currents that can

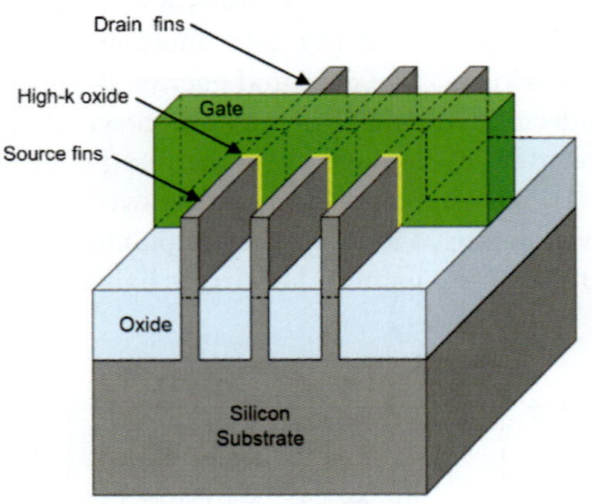

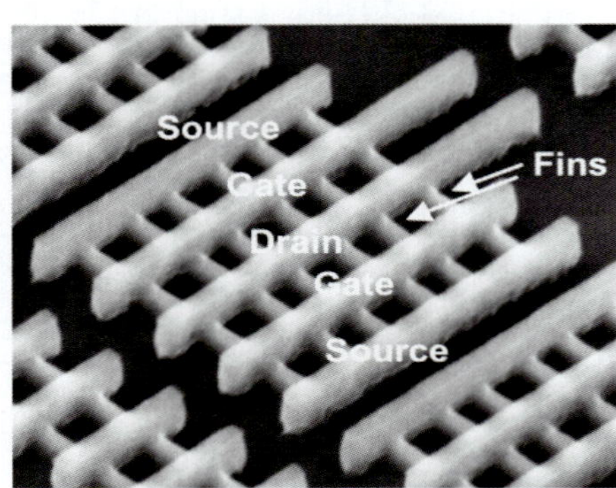

Fig. 5. Sketch of a 3-fin non-planar transistor (top) and SEM of a 6-fin, multi-gate inverter (bottom). Freely adapted from Intel images of 22 nm tri-gate CMOS transistors [3].

be supported by a single fin, so finFET transistors which need to supply high drive currents for high speed circuit operation are constructed with multiple fin channels tied to source and drain contact "bars." The doping of the source and drain regions of finFET transistors presents several challenges not present in planar CMOS designs. Most important is the need to provide uniform doping throughout the full exposed height of the fins at the source and drain ends to ensure uniform threshold characteristics in the undoped channel region under the gate. This requires that implanted ions impinge on the sides of the fin in a controlled fashion. When using beamline accelerators with high tilt angle end stations, shadowing by adjacent fins limit the tilt angles that can be used. An alternative is to employ plasma immersion-type doping systems under conditions that favor side wall "conformal" doping [4, 5].

7. PV Doping

Doping of junctions and contact regions in PV cells is a relatively new and rapidly growing application of ion implantation techniques [6]. The main advantage of implant doping is that the doping levels in the front surface junction and the contact regions under the metal grid lines can be separately and accurately set to provide optimal PV efficiency. Advanced PV cell designs, with p- and n-type junctions on the backside of the PV cell and a fully exposed (no metal line) frontside, require the doping of densely patterned interdigitated junctions of opposite doping types on cell backside, another situation where doped junctions can be efficiently formed by implantation techniques. This is discussed in detail in a later section of the paper.

8. Ion Types

Although singly ionized, atomic ions, such as B^+, P^+, As^+, etc., are widely used for ion implantation purposes, many other types of ions are used as well. Higher charge states are often used to obtain higher energy ions for deep implant profiles. With tandem type accelerators (discussed in the next section), the charge state is switched one or more times as the ions pass through the accelerator to maximize the acceleration effect of the high voltage terminal, obtaining ions at 2 to 5 times higher kinetic energy than the high voltage.

Using large molecules or clusters containing many dopant atoms instead of mono-dopant ions, such as B^+ or BF_2^+, can overcome two fundamental limitations to providing sub-keV dopant beams at commercially viable beam currents; (1) limitations on the extraction currents from the ion source (Child–Langmiur law) and (2) beam transport limitations due to ion repulsion which results in increases in the beam diameter as the ions are transported along a beamline (Coulomb repulsion effects).

In addition to improvements to the dopant flux rates for shallow junction doping, the use of molecular ions increases the damage accumulation rate compared to single ion implants at the same equivalent energy. Since higher damage accumulation rates result in thicker amorphous layers, a higher fraction of the implanted dopant profile can be included within the amorphous layer for molecular ions, resulting in a higher fraction of dopants which are electrically active following recrystalization of the Si region during the thermal anneal. Since more of the implant damage is incorporated in creating the amorphous layer, which can be regrown to a high level of crystalline quality during the anneal, molecular ion implants also result in lower levels of residual lattice damage in the deep portion of the profile, the "end-of-range" (EOR), resulting in lower defect levels in the junction depletion layer and significantly lower junction leakage currents. This will be discussed in more detail in a later section of this article.

9. Ion Implantation Machines

Implanters have been classically separated into medium and high current as the major sub-types of machines. This designation is driven by the ability to handle the power in the beam differently for each. Most implantation machines operate in the ion energy regime from a fraction of a keV to ≈ 200 keV, with "medium current" tools spanning this energy range with beam currents of a few mA or less and "high current" tools operating at maximum beam currents of several 10s of mA (at ≈ 80 keV) with a focus on high ion flux rates, for high wafer throughput, for low energy (<2 keV) and higher dose (5×10^{14} to 5×10^{15} ions/cm^2) doping for "ultra-shallow" transistor junctions.

More recently, advanced implant tools have been introduced which extend the useable energy scale in both directions. MeV ion implanters enable significantly higher energies than the typical commercial ion implantation system. These "high energy" tools provide energetic ions in the range from ≈ 20 keV to several MeV. Another group of high energy implanters have been developed to provide mA beam currents of protons at energies up to 4 MeV for use in splitting of $\approx 100 \, \mu$m thick free-standing Si membranes. A recent addition to the ion implant tool set is a class of "high dose" tools which utilize wafer bias to draw ions out of nearby plasma in plasma immersion ion implantation (PIII) systems operating with ion energies of a few keV or less and doses in the high 10^{15} to mid-10^{16} ions/cm^2 range.

Figure 6 illustrates the capabilities of these tool sub-types in a figure with the same scaling of dose and energy as processes were overviewed in Fig. 1. A good overview of these concepts is reviewed by Glawischnig *et al.* in Ref. 7.

Medium current implanters are characterized by relatively low currents so that the power deposition is relatively low and single wafer processing can be used. Standard energy ranges are 10 to 200 keV and the beam current ranges from nanoamps to a few hundred microamps. Recent tool improvements have extended the effective energy to nearly 1 MeV and the current capability to a few milliamps. Suppliers marketing this type of tool have included Axcelis Technologies Incorporated [8], Nissin Ion Equipment Company [9], SEN Corporation [10], Tokyo Electron Limited [11], ULVAC Technologies, Inc. [12], and Varian Semiconductor Equipment Associates [13].

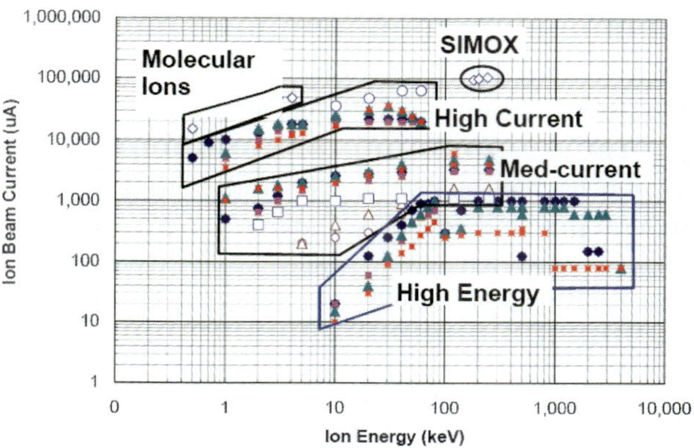

Fig. 6. Beam currents (in particle-μA) vs. energy (keV) for various ions in contemporary ion implantation systems. (Data courtesy of Axcelis and the literature.)

High current implanters are characterized by relatively high currents so that the power deposition is high. Because of this, schemes must be developed to reduce the duty cycle of the beam so that wafers do not excessively heat the wafer and/or the photoresist. Standard energy ranges are 10 to 80 or 160 keV with beam currents on the order of 10 mA. Recent tool improvements have extended the lower end of the energy below 500 eV and have produced beam currents up to 30 mA. Suppliers marketing this type of tool have included Advanced Ion Beam Technology Inc. [14], Applied Materials [15], Axcelis Technologies Incorporated [8], Nissin Ion Equipment Company [9], SEN Corporation [10], Tokyo Electron Limited [11], ULVAC Technologies, Inc. [12], and Varian Semiconductor Equipment Associates [13]:

Plasma immersion ion implanters do not use a classic DC acceleration and instead the target is immersed in a plasma of the dopant and a high energy pulse to the target pulls the dopant to it. Standard energy ranges are a few eV to a few tens of keV. Effective doses are relatively high. This approach has found utility in the materials modification market and low doping energy processes for ICs. Suppliers marketing PIII and ion shower tools include Applied Materials [15], Ion Beam Services [16], Nissin Ion Equipment Company [9], SEN [10], and Varian Semiconductor Equipment Associates [13].

MeV ion implanters are characterized by innovative accelerators that can accelerate beams to a few MeV. The amount of current available depends strongly on the accelerator; however, even with low beam currents, wafer handling for high power deposition needs to be used. Suppliers marketing this type of tool include Axcelis Technologies Incorporated [8], NEC Corporation [17], SEN Corporation [10], and Varian Semiconductor Equipment Associates [13].

The standard DC acceleration scheme involves an ion source, extraction electrodes, magnetic ion separation, and some amount of post separation focusing and deflection. A schematic illustration of this scheme is shown in Fig. 7 and has been well described in Ref. 18.

PIII has the simplest acceleration scheme. The wafer/target is placed in the center of a plasma of dopant ions. When the target is pulsed to a negative potential, the positive dopant ions are accelerated directly to the target. A schematic illustration of this

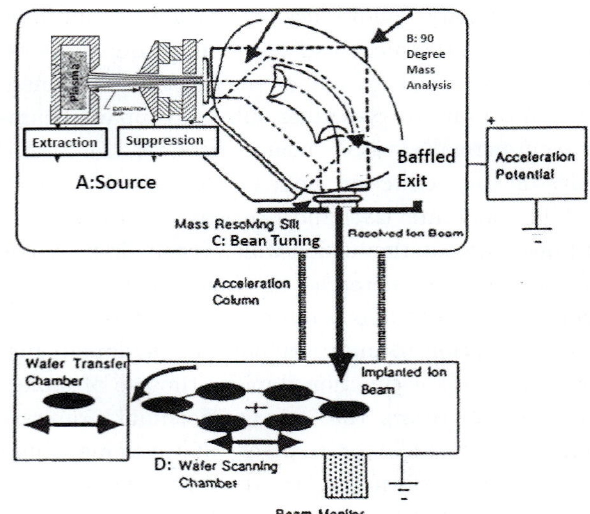

Fig. 7. Standard DC acceleration scheme. (A) source and extraction, (B) magnetic ion separation, (C) post separation beam tuning, (D) wafer target.

technology is shown in Fig. 8 and early descriptions were published by Conrad *et al.* [19] and Mizuno *et al.* [20]. This technology favors materials modification processes in that relatively high doses can be easily implanted and the field lines direct the dopants evenly to irregular shapes.

MeV implanters have more complex acceleration schemes. The major types are charge exchange and linear accelerator (linac). The charge exchange scheme is used by NEC and Varian. In this method, a positive ion is extracted from an ion source and converted by charge exchange to a negative ion by passing through a metal vapor, then accelerated as

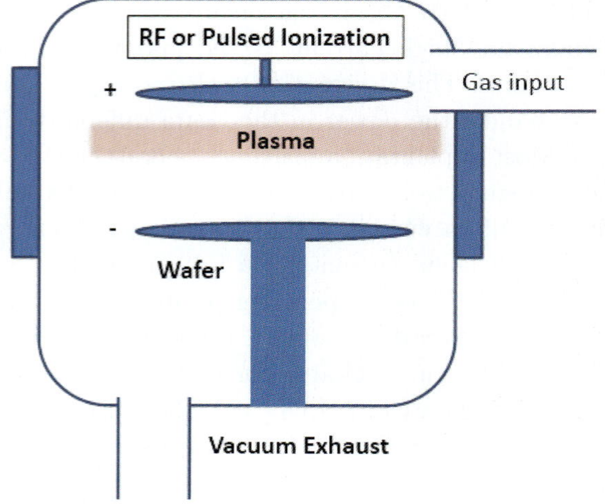

Fig. 8. Plasma immersion ion implantation scheme.

a negative ion to the center of a high potential node where a second charge exchange cell is located. Two or more electrons are removed by collisions in this charge exchange cell and the resulting positive ion is further accelerated on its return from the high voltage node to ground [7]. Both beam currents and energies are dependent on the specific accelerator, however they are on the order of hundreds of microamps at a few MeV. The linac acceleration scheme is used by Axcelis. In this scheme the beamline is a series of klystron accelerators where each individual node is tuned in both phase and frequency to accelerate a bunch of the desired ions into the klystron node and further accelerate that bunch as it exits the node. Beam currents and energies are even more dependent on the tool and recipe; however, effective currents near to a milliamp at a few MeV may be achieved [7].

As described above, several orders of magnitude of beam current and power deposition are possible through the different tool types. This has generated several approaches to beam scanning and wafer motion to most effectively expose the wafers to a uniform dose at a minimum time while keeping the power deposition within limits. Early medium current machines had relatively low power deposition and could therefore continuously expose the wafer to the beam. Electrostatic scanning at non-commensurate X and Y frequencies produced uniform deposition in about 10 s [21]. Higher current machines reduced the duty cycle of the beam on wafer by mounting several wafers on a heat-sinking disk. The beam is held stationary and the disk is rotated and translated so that the dose is uniform across the wafers. Nominal implant times would be one to five minutes, however up to 25 wafers could receive a relatively high dose in this time.

Early processes had an average beam incidence angle of 7° which was chosen to minimize channeling effects and non-uniformities due to angular differences. Later processes determined that the angle of the beam to the wafer was a process variable that could be used to optimize device performance. Tools were then developed so that both tilt and twist angles could be controlled to the best effect. For disk implanters, this took the form of capability to modify the angle of the disk in reference to the beam. For medium current implanters, the approach used was to mount the wafer on a platform that had a

controllable angle with respect to the beam. Many medium current designs produce a flat beam path through scanning in one direction while the platform with the wafer at the appropriate angles would translate in an approximately perpendicular direction.

10. Advanced Semiconductor Processes

10.1. *Shallow junction SDE*

Producing shallow junction depths is one of the more important development goals for the further shrinking of CMOS technology. In the International Technology Roadmap for Semiconductors [22] junctions have the goal of 12 nm in 2010 shrinking to 7.5 nm in 2015. The roadmap text gave the doping related difficult challenge: "In the near term, while maintaining the bulk planar architecture, the difficult challenge for doping of CMOS transistors is achieving doping profiles in the source/drain extension regions to attain progressively shallower junction depths (∼10 nm), while concomitantly optimizing the sheet resistance (∼500 ohms/sq), doping abruptness at the extension-channel junction, and extension-gate overlap."

The formation of these shallow S/D regions is one of the most important techniques for reducing MOSFET characteristics called "short-channel effects." Several of these are threshold voltage (V_t) reduction, high subthreshold leakage current, hot electron effects and source-to-drain punch-through breakdown. These effects cause increased leakage of the devices as the length of the device becomes small. As MOSFET devices are scaled below 100 nm in channel length and power supply voltages are reduced to below 2.5 V, short channel effects like threshold voltage lowering and bulk source-to-drain punch-through are more critical ULSI scaling limits than hot electron effects [1, 2]. Suppression of these short channel effects in scaled-down devices requires close attention to doping profiles. These device characteristics and the effect which junction technologies can have in engineering them are discussed more fully by Jones and Ishida [23] and also by Chason *et al.* [24].

A useful figure of merit for evaluating the effectiveness of shallow junction development activities has been the junction depth vs. sheet resistance plot shown in Fig. 9. In this plot, junction depth is one of

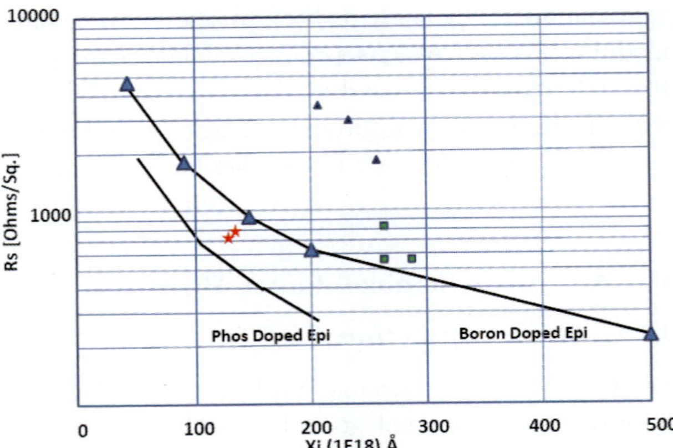

Fig. 9.   A junction depth (X_j) vs. sheet resistance plot for Boron implantations in silicon. Data represent three different advanced anneal techniques.

the axes, making evaluation of reduction of junction depth by itself trivial. However, the other parameter of interest is how well the layer is activated. To a first approximation, simple reduction on the resistivity axis indicates improvement. A useful comparator is to calculate and plot the hyperbolic for the resistance of a layer of constant resistivity [the solid solubility for that dopant at a nominal temperature] on the same plot.

Early development efforts centered on reducing implant acceleration energy [24, 25] and annealing times at increased temperature [7, 21]. The result of these trends was the development of implant tools that would have higher throughput at relatively low energies [21] and the development of Rapid Thermal Annealing tools [RTAs] that had increasing capability to anneal for consistent short times at higher temperatures [19, 26]. The annealing tool development has continued to the development of Laser Annealers [26, 27] and to Flash Thermal Annealers [28, 29] which represent the state of the art of this technology at this time.

The throughput of implant tools is directly related to the acceleration energy that is used to extract the beam and propagate that through the beamline since it is more difficult to transport high current ion beams as the energy is reduced. The trend to lower energies is then counter to the need for higher throughput. The balance of these two effects seems to saturate on the order of 500 eV [30–32]. In addition to the pure productivity issue, it was found that the junction depth reduction as a function of

beam energy was limited by transient enhanced diffusion (TED) [7]. This effect was discovered to result from the annealing of implant damage to ⟨110⟩ defect clusters, which then would slowly release the trapped silicon interstitials resulting in enhanced diffusion for extended times [7].

Higher effective doses and throughput may be achieved by implanting molecules or clusters. Using large molecules or clusters containing many dopant atoms instead of mono-dopant ions, such as B^+, can overcome two fundamental limitations to providing sub-keV dopant beams at commercially viable beam currents; (1) limitations on the extraction currents from the ion source (Child–Langmiur law) and (2) beam transport limitations due to ion repulsion which results in increases in the beam diameter as the ions are transported along a beamline (Coulomb repulsion effects). The Child–Langmuir limit to the ion current that can be extracted from an ideal, planar plasma boundary with an extraction voltage of V, is

$$J_{\text{ion}}(\text{A/cm}^2) = (4\varepsilon/9)(Ze/M)^{1/2}(1/d^2)(V)^{2/3} \quad (1)$$

where: Ze is the ion charge, M is the ion mass, d is the sheath distance and ε is the permittivity of free space. The sharp decrease in the extractable ion current as the extraction voltage, V, is reduced is a serious limit to the use of beamline implanters in "drift" mode. This limitation for beamline systems at low energies has favored the development of PIII systems, where the plasma-to-target transport distance is the plasma sheath distance, of the order of cm or less, with correspondingly high delivered ion current density per pulse.

If multi-atom ions are used instead of single-atom ions, the Child–Langmiur limitations can be relaxed by allowing extraction and drift mode operation at energies that are multiples of the atom number per ion higher than the single ion case. This concept has been used for many years through the use of BF_2 ions to allow use of extraction voltages that are 49/11 (the ion mass ratio) times higher than the extraction needed for single boron ions at the same impact velocity. The use of molecular ions, such as $B_{10}H_{14}$, $B_{10}C_2H_{12}$ and $B_{18}H_{22}$, allows for extraction at energies that are up to 20 times larger than the single-ion case (Fig. 10), with corresponding increases in the available ion current. The use of massive clusters, with up to $\sim 10^4$ atoms/ion, can

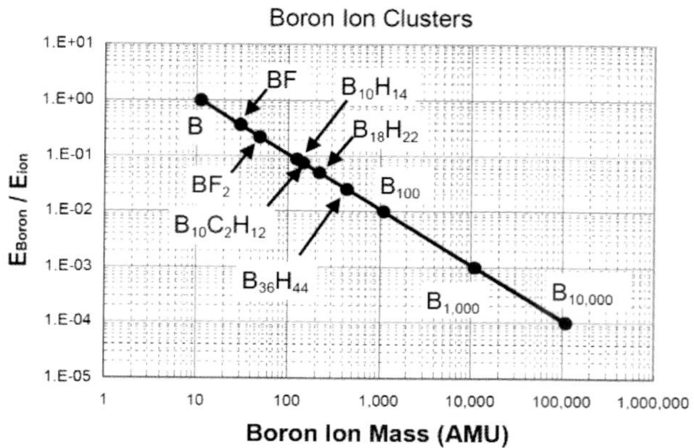

Fig. 10. The ratio of atomic ("Boron equivalent") and ion energies for various Boron-containing atomic, molecular and "massive" gas-cluster ions. The larger (100 and above) Boron clusters are formed by adiabatic expansion cooling of gas mixtures of dopants and other elements, such as Ar.

facilitate the use of extracted dopant fluxes that are correspondingly many times higher before Child–Langmiur limits become relevant.

In addition to improvements to the available dopant flux rates for shallow junction doping, the use of high-atom count molecular ions increases the damage accumulation rate compared to single ion implants at the same equivalent energy, which is of benefit to USJ formation. As more of the implant damage is incorporated in creating the amorphous layer, which can be regrown to a high level of crystalline quality during anneal, molecular ion implants also result in lower levels of residual lattice damage in the deep portion of the profile, resulting in higher quality, lower leakage junctions.

The development of the advanced annealing techniques mentioned above is one effort to move the temperature profile past the region where defect cluster formation is preferred. Another research direction is to implant mixtures of ions where additional elements suppress defect enhanced diffusion. Table 1 shows a summary of research on novel directions in shallow junction research over the last decade.

Early work looked at simple modifications that could be made to annealing so that the effect of "transient enhanced diffuson" (TED) could be minimized. A similar approach is called "defect engineering" where an attempt is carried out to make a mix of implants such that the placement of interstitials and vacancies is engineered such that the resulting defect distribution minimizes the subsequent diffusion. Another similar approach, perhaps more effective, is to co-implant carbon or fluorine. The effect of these species is to embed strain fields which capture or delay the defects that cause the various varieties of defect enhanced diffusion. Following the directions described in the discussion of principles of molecular and cluster ion implantation, there have been a number of works aimed at describing those systems. Among the molecules are octadecaborane, decaborane, carborane, and molecular ions such as borides. Commercial approaches have been developed for octadecaborane, decaborane [51] and cluster implantation [52].

10.2. *Junction leakage*

Junction leakage current is a critical factor to be limited as much as possible for high-speed server

Table 1.   A summary of research on shallow junction technologies.

Shallow junction technology	Additional processing	Depth/Improvement comment	Reference
B	—	2-step anneal	33
B	B atoms, such as SiB, SiB$_2$ GeB, and Bn	molecular ions	34
B	—	Diffused from SiGe	35
B	recoil implantation	recoil implantation	34
B	defects	Defect Engineering	34, 36
B	He	Defect Engineering	37, 38
B	Boride enhanced Diffusion	Defect Engineering	39
B	C/F	and Spike Anneal	40, 41, 42
B	—	from SOD diffusion and RTP	43
B$_2$H$_6$	—	Plasma Doping, Spike RTA	4
Decaborane	—	—	44, 45
Octodecaborane	—	—	46, 47
Cluster Ion Implantation	—	—	48, 49
B	C cluster co-implant	—	50

circuits (where device cooling is a strong environmental and cost issue) and portable electronics (where long battery lifetime is a premium). Design of low-leakage CMOS processes involves nearly every facet of transistor structure, including gate oxide thickness and composition, implant and anneal conditions, local strain and control of metal contamination. Regarding transistor doping, the key factors are the junction carrier activation, background or halo profile doping levels and the density and position of any residual implant defects present after completion of the thermal annealing process. The key mechanisms for junction leakage are (1) band-to-band tunneling, active under reverse bias, (2) carrier generation and recombination and (3) trap-assisted tunneling, where the last two mechanisms being active for both bias directions (Fig. 11). Defects that are located within the junction depletion layer are particularly problematic because they can greatly magnify leakage effects from carrier recombination and generation and trap-assisted tunneling. The thickness of the junction depletion layer, which is determined by the bias amplitude and the relative doping levels near the metallurgical junction depth location, is another prime factor. For high sub-junction doping, such as a halo profile surrounding a shallow source/drain extension, the depletion layer thickness can be a few 10s of nanometers, greatly increasing the effects of carrier tunneling and decreasing the diffusion path of carriers before

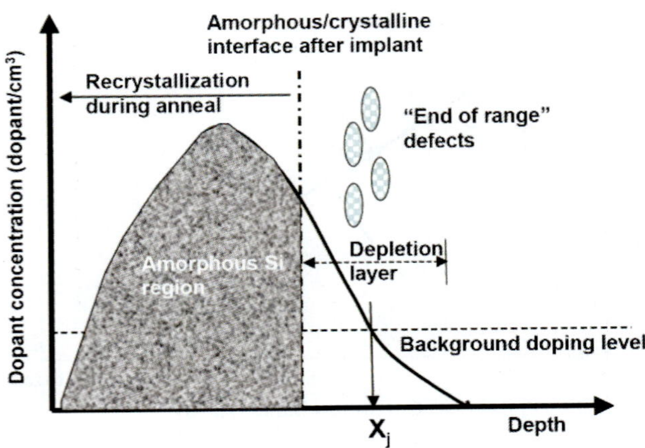

Fig. 12. Sketch of damage effects during implantation and annealing. For a high dose implant, an amorphous layer is formed from the surface to below the depth of the implant profile peak, which can be recrystallized during annealing. Implant damage deeper than the amorphous layer often results in "end of range" defects present after annealing.

they interact with local defects (by recombination or generation).

The general characteristics of residual damage formation for high-dose implants are sketched in Fig. 12. Each energetic ion that strikes a crystalline target, such as Si, creates a large number of point and cluster defects, vacancies and interstitial target atoms, as it slows down and stops in the target. Most of the vacancies and interstitials rapidly recombine within a few milliseconds of the ion impact. However, if enough ions are implanted and other factors apply, the target lattice is converted to an amorphous structure in the region around the peak of the implant profile. The depth of the amorphous region depends not only on the ion mass and energy and the implanted dose but also such factors as the ion beam current density, beam scanning rates and the target temperature during implant. The ion type, a single atom or a multi-atom molecule, also greatly influences the amount and location of damage that accumulates during the implant.

During thermal annealing, the amorphous layer can be recrystallized to a high degree of perfection and the near-surface dopants are electrically activated. However the implant damage below the amorphous/crystalline interface is often not effectively removed, leaving "end of range" (EOR) defects. When EOR defects are located within the junction depletion layer formed during transistor operation, high levels of leakage can occur.

Reverse Bias

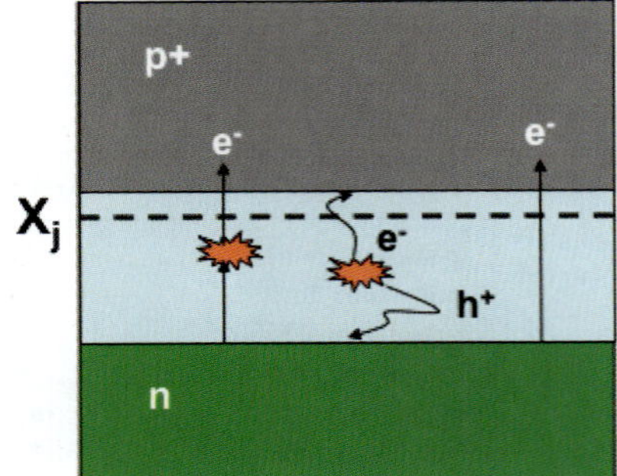

Trap-assisted tunneling   Band-to-band tunneling
Carrier generation

Fig. 11. The three key junction leakage mechanisms for a *p/n*-junction under reverse bias, with residual implant damage defects in the depletion layer, in a lighter doped *n*-substrate.

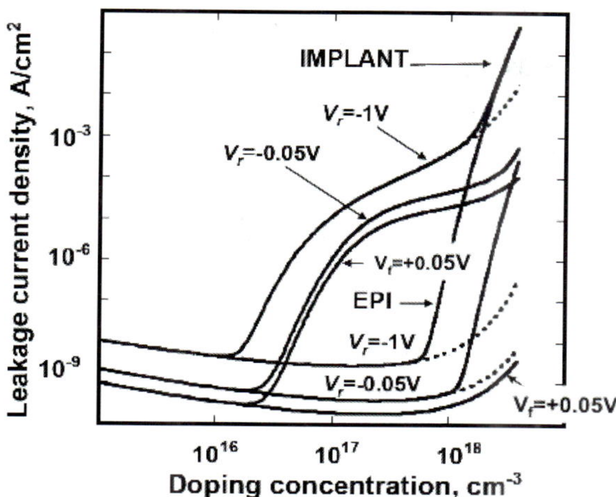

Fig. 13.   Leakage current density vs. sub-junction doping concentration and junction bias for an implanted junction, with a peak damage density of 10^{17} defects/cm^3 located 20 nm below the junction, and an epi-doped junction, with traps located at the epi/substrate interface. The dotted lines show the effects of carrier generation and trap-assisted tunneling alone, without the contributions from band-to-band tunneling [53].

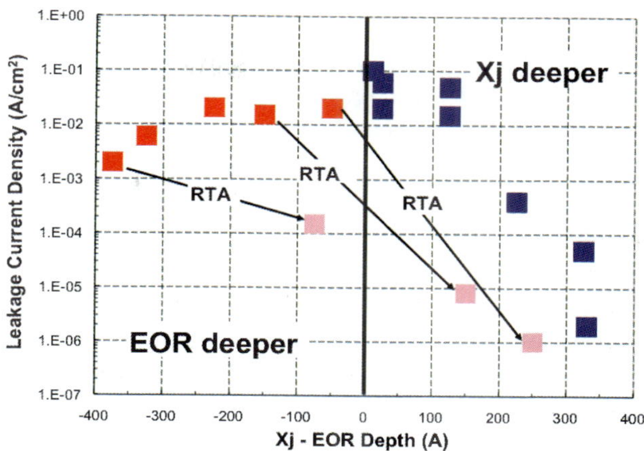

Fig. 14.   Leakage current levels for a variety of implanted junctions where the EOR damage is deeper than the junction (on the left) and where the junction is deeper than the damage layer (on the right), resulting in much lower leakage [54]. The anneal in this study was a limited-diffusion "solid-phase epitaxy" process at temperatures between 560°C and 650°C. Three junctions were also annealed by "rapid thermal annealing" (RTA) at 900°C for 30 s, diffusing the dopant junction deeper and strongly reducing the observed leakage levels.

The combined impact of damage and subjunction doping on leakage current is shown in Fig. 13. For the implanted junction, with a layer of EOR damage located 20 nm below the junction depth, leakage currents increase strongly for all bias conditions as the depletion layer thickness is reduced for sub-junction doping levels above 10^{17} dopants/cm^3. While an epi-doped junction, with no sub-junction damage but a set of traps at the epi/substrate interface shows much lower leakage levels. However, for substrate doping levels above 10^{18} dopants/cm^3, band-to-band tunneling (under reverse bias) presents strong leakage effects for both junctions.

"Damage engineering" involves adjusting transistor doping levels and the implant and annealing conditions so that the desired junctions are formed while damage in the depletion region is reduced or eliminated. For instance, high halo profile doping levels can be beneficial for controlling subthreshold leakage, threshold voltage rolloff and other "short channel" effects. However, the strong junction leakage current effects, shown in Fig. 13, force tradeoffs in the process design that limit the practical halo doping levels.

The relative location and density of any regions containing residual defects remaining from the implant damage and the position of the junction and the surrounding depletion layer are of primary importance for control of defect-driven leakage currents. In the leakage current data shown in Fig. 14, by adjusting the relative location of the EOR damage layer and the dopant junction, which can be done by adjusting the ion energies used for boron doping and Ge ion "preamorphization" implants, the leakage current levels can be varied over a range of five orders of magnitude [54]. If the junctions are diffused deeper with a second anneal, a similar shift in leakage currents can be obtained.

Although diffusing dopant junctions deeper that implant damage layers was a routine process choice through the production of 65 nm CMOS [55], later dimension shrinks relied on "diffusion-less" anneal steps using short thermal pulses from flash lamps or scanned laser beams. Under these thermal cycle conditions, where dopant diffusion is minimized, annealing of implant damage is also strongly curtailed so the details of the implant conditions and the net damage accumulation process become critical choices. As discussed earlier (see Fig. 12) a key factor is the thickness of the amorphous layers created by high dose implants. If the damage accumulation rate for a particular implant is increased so that more of the implanted dopant profile is contained within the amorphous layer, a higher level of dopant activation is often achieved during the re-crystallization process during annealing. If the methods used to increase the main peak of the damage accumulation do not also increase the damage levels in the EOR region, then fewer defects remain to contribute to leakage currents.

For the same ion mass, energy and dose, the damage accumulation during implantation can be increased by (1) increasing the ion flux rate with higher beam current densities and slower beam scanning rates, (2) decreasing the rate of recombination of point defects during implantation by lowering the target temperature and (3) increasing the immediate damage rate by using ions which contain multiple atoms, such as molecules or clusters. Present day process development programs are exploring all of these methods, with the construction of production implantation systems fitted with abilities to cool wafers to temperatures of $-30°C$ to $-100°C$ at the start of the implant cycle [56, 57]. Slow ion beam scan rates have been combined with wafer cooling to $-80°C$, also achieving thicker amorphous layers [58].

In addition to the machine controlled factors (beam current, scan rates and temperature), the form of the ions used for implantation has a major impact on damage accumulation effects. When multiple atoms hit a target surface during the picosecond-timeframe of ion collisions and stopping, the net damage creation rate increases strongly as the number of atoms in the ion increases [59]. Source materials and implantation systems have been developed for the use of multi-atom molecular ions, such as $B_{36}H_x$, where 98% of the implanted dopant is contained within the amorphous layer for a 300 eV implant (at room temperature) and activated [60]. A comparison of the effects of reduced target temperature and the use of molecular ions for 3 kV implants is shown in Fig. 15 [61]. When atomic C ions are used, a 10 nm thick amorphous layer is formed for a dose of 10^{15} C/cm^2 at an implant temperature of $-30°C$, more than twice as thick as the amorphous layer formed by C ions at 25°C. However, when molecular carbon ions are used, C_7H_7 and $C_{16}H_{16}$, a $\approx$10 nm amorphous layer is formed at a 10× lower C dose even at room temperature, indicating that the use of molecular ions is intrinsically more efficient than cryo-conditions for enhancing damage accumulation rates. The "payoff" demonstration of the use of enhanced damage accumulation techniques for reduction of device leakage currents is being

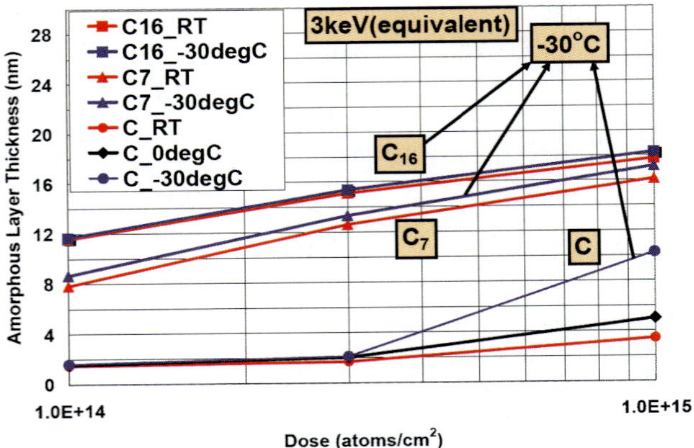

Fig. 15. Amorphous layers formed with 3 keV equivalent ions using atomic and molecular ions at implant temperatures between 25°C to $-30°C$, showing the strong effect of multiatom impacts on damage accumulation [61].

vigorously explored by all advanced CMOS development groups.

10.3. *Mobility enhancing strain*

For many decades the key to increasing CMOS transistor speeds, drive currents and power requirements was to shrink the lateral and vertical dimensions of the transistor structure in proportion [62]. By 2000, with the introduction of 90 nm features, several physical limitations had halted this "scaling" process. The principal problem was that leakage currents through SiO_2 and $SiON_x$ gate were too high to allow the use of oxide thinner than about 1.2 nm. Without the ability to scale the gate dielectric, shrinking the gate length alone did not result in a net performance gain. In order to obtain strong increases in transistor performance with only modest shrinks in gate size, Intel, soon followed by others, introduced a set of intentional stressors in order to increase CMOS drive current by increasing carrier mobility [63].

Carrier mobility in Si depends not only on local stress levels but also on the crystalline orientation of the channel conduction (Fig. 16). In pMOS transistors, where holes are the channel carriers, compressive stress is favored. In nMOS transistors, where electrons are the channel carriers, tensile stress is desired. An additional mobility benefit can be gained by orienting pMOS channels along a (110) direction and nMOS channels along (100), although this greatly complicates circuit layout for complex devices and is not widely used.

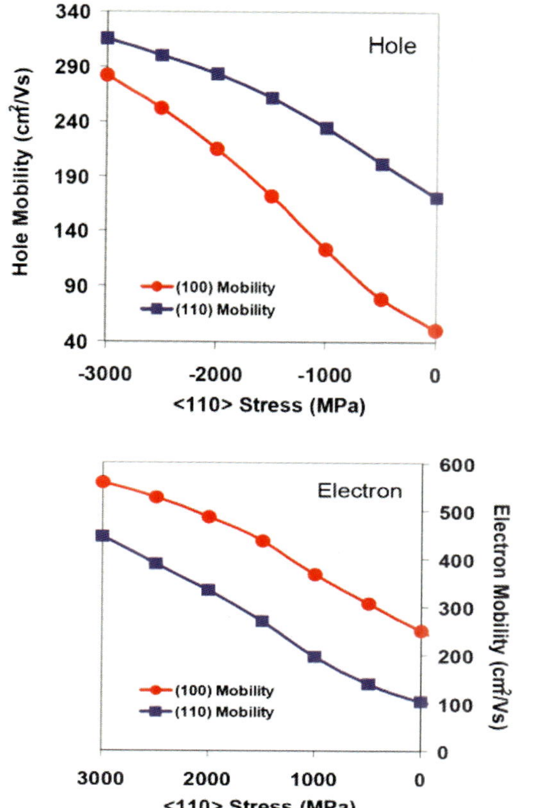

Fig. 16. Hole (top) and electron (bottom) mobilities in (110) and (110) Si directions as a function of uniaxial stress [63].

In the decade since the initial use of strain sources in CMOS, an extensive suite of stress enhancing processes have been developed (Fig. 17). The earliest implementation of pMOS strain was the use of Si_xGe_{1-x} selective epi to form compressive stress sources in S/D contact regions. 90 nm pMOS used

Ge fractions of $\approx$17%, which is now increased to $\approx$40% for 32 nm processes. In addition, compressive strain is impressed on the Si regions at the boundaries of shallow trench isolation (STI), and stressed nitride cladding layers can be formed over the gate stack with compressive properties for pMOS and tensile effects for nMOS (locally adjusted by nitride deposition conditions and implants). Highly doped junctions contain additional strains arising from the lattice mismatch of dopant and Si atoms, leading to modification of carrier mobility within the junction, among other effects.

For nMOS transistors, intentional stacking fault defects formed in the S/D contact and extension regions can be used to provide tensile strain in the channel. The process involves forming an amorphous layer, with a Ge implant, in the region of the S/D contact junction (Fig. 18). During the anneal process, an angled stacking fault defect is formed at the intersection of the lateral and vertical regrowth fronts, providing a source for tensile strain on the nMOS channel region [64, 65]. The doped S/D contact is placed so that the stacking fault is entirely contained within the heavily doped junction and does not result in a leakage current path through the junction. Additional tensile strain can be obtained by repeating this process with increasingly thick spacer linings [66].

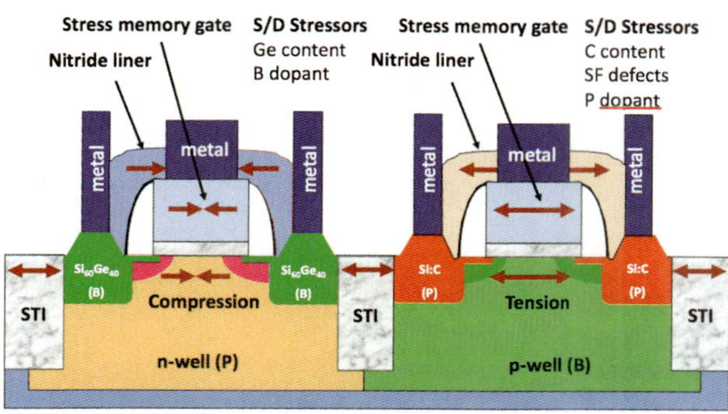

Fig. 17. Sketch of the main stressor sources in planar pMOS (left) and nMOS (right) CMOS transistors. The principal doping elements in the SD and well regions are shown in parenthesis. The stress directions are shown as red arrows.

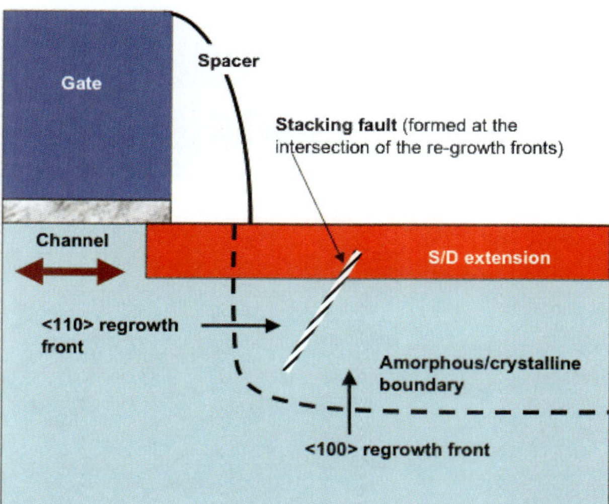

Fig. 18. Sketch of tensile strain effect in nMOS transistors of stacking fault formation along the intersection of the re-growth regrowth fronts of a deep amorphous layer formed by Ge implantation.

Higher levels of tensile strain can be obtained by forming C-rich regions in the S/D contact and extension junctions. Optimal strain is provided by an $\approx 2\%$ population of substitutional C embedded within a Phosphorus-doped nMOS S/D. Efforts to adapt the SiGe selective epi deposition process used in pMOS transistors are limited by the slow and difficult epi process needed to grow Si:C epi with high substitutional concentrations of C. A direct implant process uses a Ge amorphization implant coupled with a C^+ ion implant into the MOS S/D contact region followed by annealing [67].

A more efficient process uses molecular ions, such as $C_7H_7^+$, to simultaneously add the required C concentration profile and directly form a deep, dense amorphous layer, without the need for the separate Ge implant [68].

A comparison of damage layers created by 6.6 keV C^+ and 49.5 keV $C_7H_7^+$ ions (Fig. 19) shows that atomic C^+ ions do not form a continuous amorphous layer [69]. After anneal, only a small fraction of the implanted C is substitutional and the result is a low level of strain. Molecular ions form a thick and dense amorphous layer containing a large fraction of the implanted C. After anneal, $\approx 2\%$ substitutional

fraction of C are present, resulting in strong tensile strain effects, as seen in the XRD spectra. When a heavier molecular ion, $C_{14}H_{14}^+$, is used, a thicker amorphous layer is formed containing a larger fraction of the implanted C, resulting in higher strain after millisecond anneal.

10.4. *Photoresist "freezing" for lithography*

Ion implantation has long been used to pre-condition photoresist (PR) patterns for various processes, such as adding a deep cross-linked and C-rich skin to PR for enhanced durability during exposure to reactive ion etch plasmas and to reduce the amount of PR outgassing for sensitive dopant implants in order to avoid dose errors.

As ions are stopped in organic PR masks, many local bonds are broken by the ion and secondary recoil atom impacts. This forms a C-rich, heavily cross-linked "crust," is formed with a thickness roughly equal to the full extent of the implanted ion profile [70, 71]. As the C-rich crust is formed, the PR line shrinks in both the vertical and lateral directions. This effect as a process issue will be discussed later in this article.

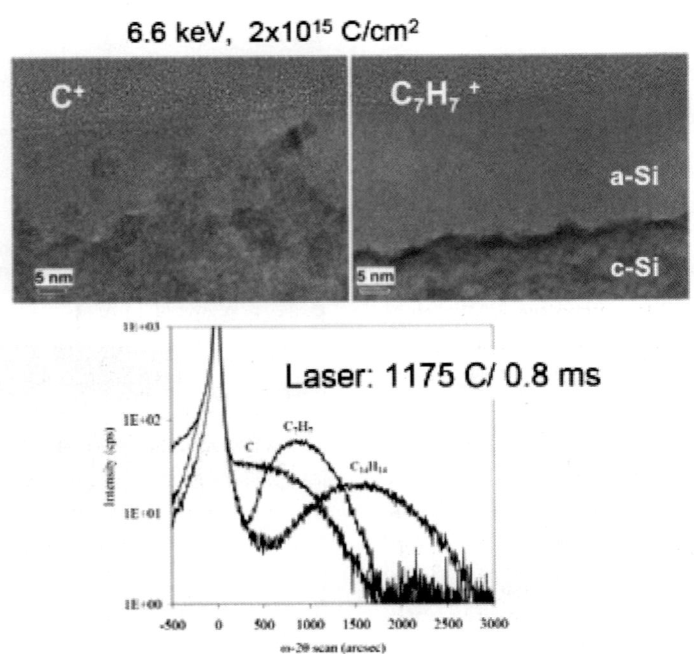

Fig. 19. TEM images (top) of atomic C^+ (left) and molecular $C_7H_7^+$ (right) ion implants in Si, forming a mixed damage-amorphous region with C^+ and a thick, uniform amorphous layers with C_7^+. After laser annealing, XRD (bottom) shows high strain for C_7 implanted layers and a higher level of strain following the use of a larger molecular ions, $C_{14}H_{14}^+$ [69].

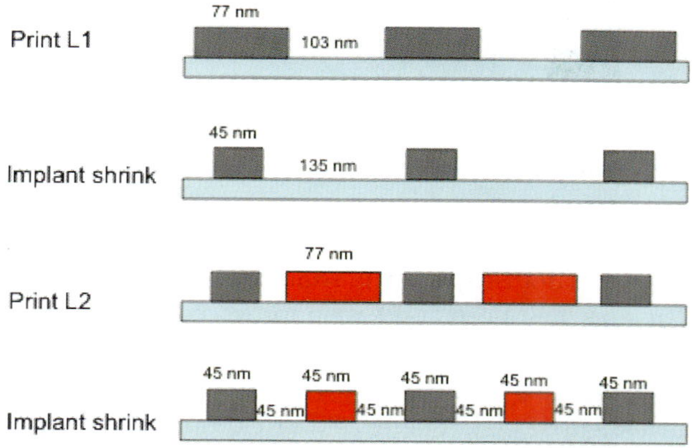

Fig. 20. Sketch of PR lines for a dual-exposure process utilizing ion implant line shrinks.

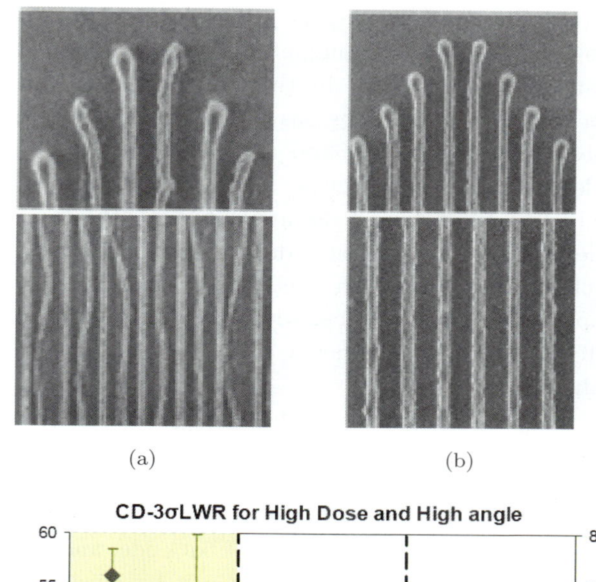

Fig. 21. PR lines implanted with 2 keV (left (a)) and 5 keV (right (b)) Ar$^+$ ions and shrinkage of PR line width, CD, and reduction of line width roughness, LWR, with 2 and 8 keV Ar$^+$ ion implants into PR [74].

With the extended use of subwavelength optical patterning, ion implant conditioning and the resulting PR line shrinks have become a routine process factor. With the elaboration of litho patterning to include dual and quad exposures, the effects of ion implant conditioning have been extended to include enhancement of the adhesion of multiple litho patterns of PR to the underlying wafer coatings and reduction of line edge and line width roughness [72].

PR line shrinkage with ion implantation is illustrated in Fig. 20 for a dual-exposure process. The usual ion for PR conditioning is Ar$^+$, mainly for its ease of use and relative lack of process complications. A challenge for dual-exposure litho process is that the first level PR layer needs to stay in place throughout the spin coat, exposure, develop and curing of the second layer. If the first level PR adhesion to the wafer is not sufficient, the solvents in the develop cycle of the second layer can delaminate the PR lines. Here again, the details of the implant conditions can vary the PR adhesion strength [72].

As the C-rich crust forms on the surface of the PR line, the dimensional integrity of the printed line increases as the surface "stiffens" and solvent absorption decreases. These effects improve both the line edge and line width roughness of fine lines, especially if the implant included exposures with a high beam tilt angle so that the sides as well as the tops of the PR line are encrusted [73]. The effects of both adhesion and line roughness improvements with ion implantation are shown in Fig. 21.

11. Photovoltaics Doping

Makers of photovoltaic cells have historically avoided the more complex and expensive semiconductor manufacturing techniques as the solar industry has been driven mainly by cost. The surface junctions in solar devices are optimized by balancing their depth and that of the intrinsically doped silicon to the penetration depth of the light. This characteristic depth is a few microns, so the surface junction optimizes at a depth which is large in comparison with the shallow junctions used in CMOS devices.

Similarly, the doping level is optimized high enough to give good resistivity and field penetration, but at the cost of a reduction in the lifetime of the carriers being collected in the front junction. Standard production thus favors the use of polycrystalline silicon and deposited doped layers. However, advanced development favors implanted doped layers where control in placement and doping level is desired for higher photoconversion efficiency. The main advantage of implant doping is that the doping levels in the front surface junction and the contact regions under the metal grid lines can be separately and accurately set to provide optimal PV efficiency. In addition, by using a shadow mask during the implant, both the front side junction and the extra doping desired for good contact resistance with metal grid lines can be implanted in a single step, without the need for on-cell patterning, a significant process time and cost reduction [6].

Advanced PV cell designs, with p- and n-type junctions on the backside of the PV cell and a fully exposed (no metal line) frontside, require the doping of densely patterned interdigitated junctions of opposite doping types on the cell backside, another situation where doped junctions can be efficiently formed by implantation techniques. This is discussed in detail in a theory and simulation paper by Aleman *et al.* [75]. This paper discusses interdigitated back contacted cells (IBC) as an option for the achievement of high-efficiency on silicon material. Implementation of an electrical field on the front side, also called the Front Surface Field (FSF), is included in this analysis and is considered beneficial. The back surface field was modeled and experimentally tested resulting in an analysis of recombination parameters by Bordin *et al.* [76].

Doping can be used to modify the solar cell materials to improve the solar efficiency of the materials. Impurity traps (IPV effect) were tested by both simulation and experiment for boron in silicon by Kasai and Matsumura of KAIST [77]. They assert that impurity-traps could possibly act as stepping stones in two-step excitation of electrons from the valence band to the conduction band, and thus, sunlight of longer wavelength than that equivalent to the band gap could be utilized to improve cell efficiency.

Hydrogen implantation to form shallow donor levels has been tested to make deep n–p junctions in p-type silicon by hydrogen ion implantation [78]. To use less silicon (a cost improvement), H-cut methods have been used to make thin silicon layers at an encouraging efficiency level — Microsystems enabled photovoltaics: 14.9% efficient 14 micron thick crystalline silicon solar cell [79]. (This is also discussed in the H-cut section).

Implantation may be enabling for more novel technologies. The surface morphology is an area where increased absorption of incoming photons can be produced by appropriate scale surface roughening. Tesfamichael, Will, and Bell have implemented this in titania films used in dye-sensitized solar cells [80]. Doping of more complex PV systems is a topic of greater interest. Metal plasma ion implantation of Ru ions has been used to improve conversion efficiency of dye-sensitized solar cells [81]. Phosphorus implantation of C_{60} for PV applications has been investigated [82]. The effects of Cl ion implantation on the properties of $CuInSe_2$ epitaxial thin films have been reported [83]. Polycrystalline CdTe layers were implanted with Phosphorus in order to obtain an enhanced p-type doping close to the back contact of CdTe solar cells [84].

As standard implantation tools are considered expensive and the layers needed are in an inefficient part of their process space, plasma implant and ion shower tools and techniques have been investigated as a more appropriate technology. Torregrossa *et al.* used PIII to implant BF_3 to form boron junctions and illustrate the use of Plasma Immersion Ion Implantation [85]. Moon *et al.* investigated amorphous silicon p-i-n solar cells using the ion shower doping technique [86].

12. Semiconductor Process Issues

Ion implantation tools, which continue to be the most complex combination of technologies operating in an IC fabrication line, also need to be the cleanest and most controlled in order to function as expected and needed. For example, the balance of doping levels and the precise location of junction location in both vertical and lateral dimensions required for high performance CMOS transistors translates into a need for ion dose control often to better than 1% over a dose range scanning six orders of magnitude. Also the energy and angle of incidence of the ion beam is specified to within a few percent (for energy) and fractions of a degree (for angle of incidence). The angular control requires not only a tight control on mechanical tolerances for wafer positioning but also the need to minimize beam divergence effects arising from electrostatic repulsion of the ions in the beam. Methods to reduce the effects of beam divergence include the introduction of dense local plasmas at critical points in the beam path to the wafer and the use of high-mass molecular ions for low energy beams [87].

Once the ion beam and wafers are arranged so that the desired ion type, with the correct energy, incident angle and dose is delivered across the wafer with sufficient uniformity, a significant number of additional process issues must be controlled by proper system design and operational procedures.

12.1. *Particles*

Since the ion energies used for implantation are usually quite modest, a fraction to a few tens of keV, particles and thin films on the wafer can result in unwanted blocking of the incoming ions. In addition, quite small particles that do not completely block the ion flux can result in distortions of the junction edge location and small particles and thin films that are penetrated by the ion beam contribute surface contamination by recoil implantation of atoms from the particles and films. A schematic of these effects is shown in Fig. 22.

The larger particle on the left in Fig. 22 results in a locally undoped junction that will provide a direct short to the substrate when a metal contact is formed on the implanted region. For shallow junctions, aimed to be 10 nm and less in depth, the presence of small particles and absorbed thin films and native oxides as thin as a few nanometers can block a large fraction of the incoming low energy ion beam.

The mid-sized particle, small enough to be penetrated by the ion beam, will still cause problems because the region under the particle will be

under-doped and the junction depth will in most cases be locally shallower, leading to increased risk of junction breakdown events and changes to transistor functional properties. Even quite small particles and absorbed thin films result in junction contamination from recoil implantation driven by the passing of the beam ions through these small particles and films. Since particles and absorbed films are largely made up of sputtered and vaporized materials from beam-line apertures, beam stops and wafer holding fixtures as well as outgassed and sputtered debris from previously implanted wafers, the recoil implanted atoms can be a complex mix of dopants, metals, graphitic compounds and hydrocarbons. In addition, any particles which are not removed by the post-implant resist stripping cleans and remain on the wafer during the subsequent anneal cycles are dangerous sources of diffused contaminants and present further problems for the metal contact formation step.

Particles that are trapped in the positive beam potential of the ion beam and are transported to the wafer can cause an additional problem through kinetic impact with rapidly moving wafers being scanned in front of the beam. For wafers mounted on spinning wheels and moving at scan velocities of up to 90 m/s, the impact of a relatively slow moving beam-transported particle with thin gate structures in devices in the rapidly moving wafers is sufficient to shatter stacks of nearby poly-silicon gates for CMOS scales of 0.25 μm and less [88]. This problem has led to the nearly complete abandonment of spinning wheel scanning designs for advanced CMOS manufacturing and the use of much slower mechanical scanning of single wafers.

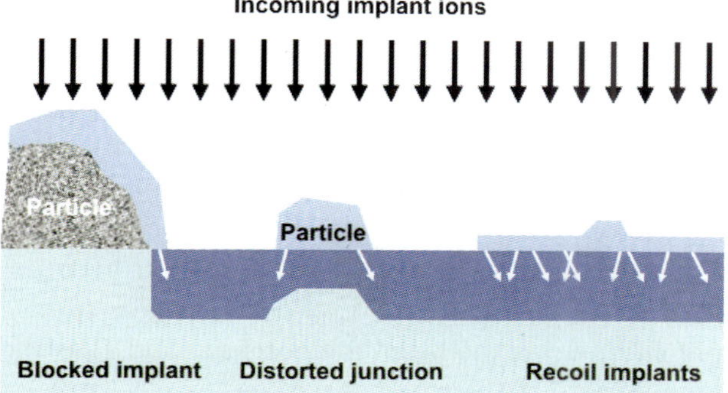

Fig. 22.   Schematic of particle effects for a particle larger than the ion range (left), a particle that blocks a portion of the ion flux (middle) and a combination of a small particle and a thin film (right).

12.2. *Charging*

Since ion implantation operates with charged ions, the balance of charge flows between the ion beam and the wafer must be carefully controlled to minimize the net flow of charge through fragile device structures, buildup of dangerous local potentials and distortions of ion beam characteristics. In early ion implantation systems charge balance on the wafer was attempted, with varying degrees of success, by direct electron beams aimed at the intersection of the ion beam and wafer, by mixes of primary and secondary electrons emitted by an electron-bombarded metal target placed near the implanted wafer and an increasing variety of plasma flows designed to mix with incoming ion beam plasma and flow to the wafer surface.

The modern understanding of beam-wafer interactions includes control of the stability of the ion beam plasma, consisting of a central column of rapidly moving positive ions and an approximately equal charge of slow moving electrons trapped in the ion potential. Maintaining space-charged neutral conditions throughout most of the ion beam path from the ion source to the wafer is usually essential to avoid significant loss in beam transmission through beam size expansion driven by Coulomb repulsion of the energetic ions, particularly for low-energy, high-current beams.

In the region near the implanted wafer surface, additional methods are used to ensure a quasi-neutral wafer surface and to minimize the net current flow to the wafer. Early efforts to maintain a neutral wafer surface during ion implantation used beams of 300 eV electrons directed toward the intersection of the ion beam with the wafer surface. However, if the beam alignments were not accurate and the various ion and electron currents not well controlled, the high-energy electron beam could cause as much disruption to device structures as the un-neutralized ion beam. An improved scheme directed the energetic electron beam towards a grounded metallic surface placed near the wafer location, expecting that low-energy secondary electrons would be a gentler means of neutralizing the ion beam on the wafer surface. However this method also failed in practice since organic material outgassed from photoresist on wafers quickly coated the metallic target surface. As a result, no secondary electrons were emitted and the energetic primary electron beam was reflected by the charged organic film back towards the wafer surface, causing similar damage as the older direct electron beam systems.

A better way to control ion beam charging was to add more ions and electrons to the wafer vicinity in the form of neutral, low-energy plasmas. When the plasma ion density, and corresponding electron population, is much higher than the incoming ion beam density, the charge and current flows on the wafer surface is dominated by the characteristics of the added plasma, not the implanted ion beam. A sketch of the basic charge flows with an ion

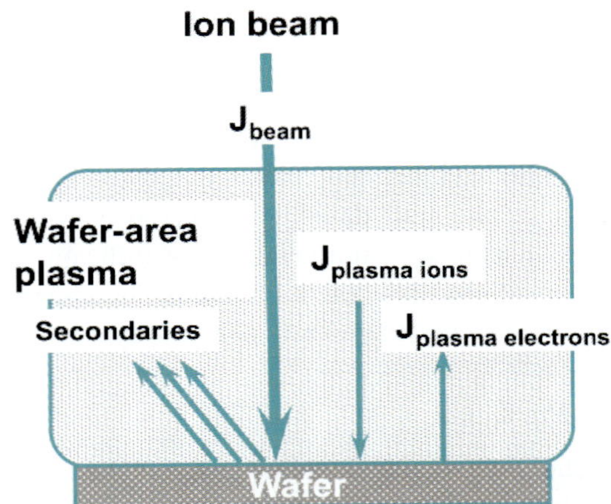

Fig. 23.   Charge flows with an ion beam and local plasma in contact with an implanted wafer.

beam and local plasma at a wafer surface is shown in Fig. 23.

The net current flow in Fig. 23 is a sum of the incoming positive ion beam and accompanying beam plasma electrons and the ion and electrons from the local plasma, mixed with the secondary electrons and ions generated on the wafer surface by the energetic ion beam impact.

The actual current components at the wafer surface are somewhat more complex than those shown in Fig. 23. For example, gas atoms in the ion beam path are ionized by collisions with the energetic ions, increasing the local plasma density and generating a population of "slow ions" that are repelled by the beam potential to the sides of the ion beam path (Fig. 24). Measurements of the energy of the slow ions emerging from the ion beam region can be used to probe the beam potential characteristics of various ion beams [89].

A reasonably accurate and compact description of the net current flow on the wafer surface can be developed for the case where the local plasma is space-charge neutral, i.e. the total electron density is balanced by the beam and plasma ions, $n_{\text{e}} = n_{\text{ib}} + n_{\text{ip}}$. For a thermal equilibrium local plasma, the electron population can be described by $n_{\text{e}} = n_{\text{eo}}e^{([V_{\text{surf}} - \Phi_p]/T_{\text{e}})}$, where V_{surf} is the wafer surface potential, Φ_{p} is the beam ion potential and T_{e} is the temperature of the thermalized plasma electrons.

The net ion current, where γ is the secondary charge generation coefficient, is:

$$J_{\text{net}} = j_{\text{ib}}(1 + \gamma) + j_{\text{ip}} - j_{\text{e}}. \qquad (2)$$

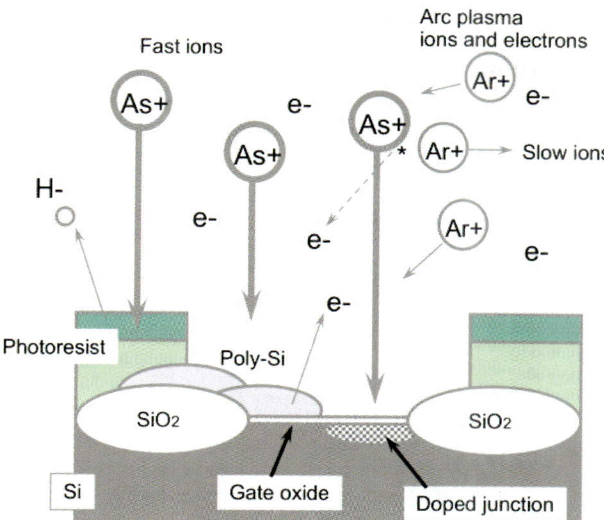

Fig. 24. Sketch of the charges generated near and at the wafer surface by the energetic ("fast") ion beam, including ionization of Ar gas atoms in the beam path, secondary electrons from conductive regions on the wafer and outgassing of charged ions, such as H^-, from the photoresist mask areas. Also shown are Ar^+ ions and electrons from the local plasma source.

For a local plasma made up of ions of mass, A_p, the net current density is:

$$J_{net} = j_{ib}(1 + \gamma) + j_{ip} * [1 - (1 + n_{ib}/n_{ip})$$
$$* 34.2 * A_p^{1/2} * e^{([V_{surf} - \Phi_p]/T_e)]}. \quad (3)$$

This compact description of the net current matches well with detailed measurements of surface potentials and positive and negative current flows obtained by on-wafer sensing structures connected to EEPROM-like data storage transistors [90, 91]. Two aspects of the local plasma characteristics are illustrated in Fig. 25, the effect of ion mass and electron temperature in the local plasma. For an Ar^+ ion beam with a current density of $2\,mA/cm^2$ and a plasma electron temperature of $4\,eV$, the net positive current flow into the wafer ($+V$) as well as the maximum surface potential decreases substantially when a heavier ion is used in the local plasma. This accounts for the choice of Xe as the local charge control plasma ion in many commercial ion implanters. The net negative current flows ($-V$) are mainly controlled by the plasma electron temperature. If the local plasma source is biased relative to the wafer and ion beam, the current flow from the source into the region of the beam and wafer can be greatly increased, however the higher electron temperature of the biased

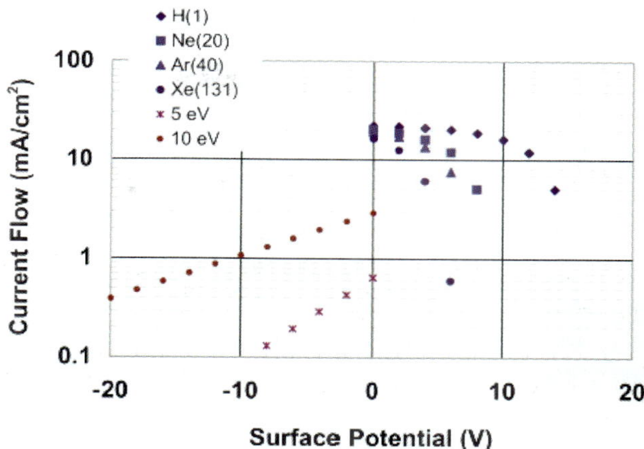

Fig. 25. Positive and negative current flows for an Ar^+ ion beam with a current density of $2\,mA/cm^2$ and a beam potential of $15\,V$.

plasma also greatly increases the net negative current flow from the local plasma, primarily from outside of the ion beam area. This can lead to additional IC device yield problems in sensitive device thin films and structures.

In Fig. 25, the local plasma density is 10 times higher than the ion beam density with an electron temperature of $4\,eV$. The average wafer surface secondary coefficient is 10. The positive current flows show the effects of using various ions for the local plasma and the negative current flows are calculated for an Ar plasma with electron temperatures higher than $4\,eV$.

12.3. *Photoresists*

The ability of ion implantation to use the patterning photoresist as a blocking mask at temperature close to room temperature has long been a major factor in the economic superiority of implant to thermal-driven CVD methods for doping. However the damage and degradation of organic photoresist films as the energetic ions are stopped in the mask layer leads to a number of process effects (Fig. 26).

The principal effect is that, as ions are stopped in the resist, bonds in the organic material are broken by ion-target collisions and large amounts of volatile materials, mostly H_2, H, N_2 and O_2, are released during the initial stages of the implant. The high flux of outgassed atoms raises the pressure in the region of the wafer and leads to dose and energy errors in certain beamline and dosimetry designs

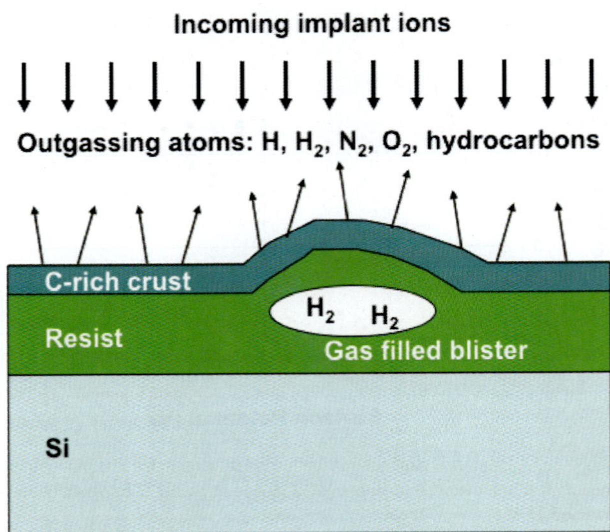

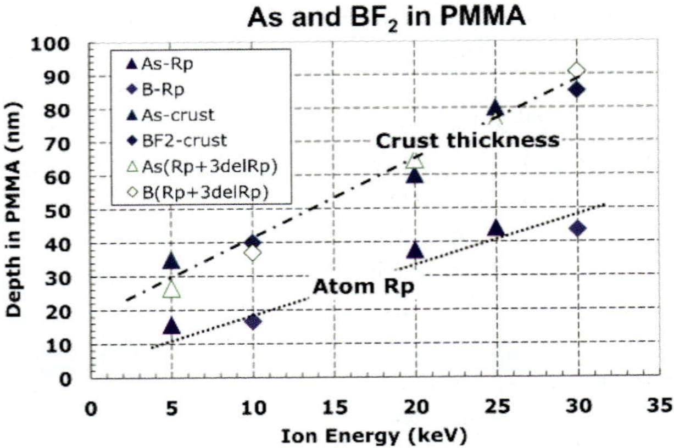

Fig. 27. Dopant atom ranges and C-rich crust thickness for As and BF_2 ions implanted into PMMA photoresist for doses of $>5 \times 10^{15}$ ions/cm^2. The crust thickness data is from [70].

Fig. 26. Sketch of process effects occurring in organic photoresists during ion implantation, including outgassing of volatile atoms released from the implanted mask layer, formation of a top-surface C-rich crust layer and formation of gas-filled blisters if the resist film is heated too much by a high-power ion beam with insufficient wafer cooling.

high-dose implants, the C-rich crust thickness closely follows the sum of the mean ion range plus three times the range straggling.

through neutralization and other charge exchanges during collisions with the incoming ion beam.

As the dose accumulates in the resist film, the implanted layer becomes depleted of many volatile components of the organic material and becomes a dense, C-rich "crust" [71]. If the ion beam heating of the wafer raises the temperature in the resist film higher than the boiling point of some of the un-crosslinked short-chain resist components, internal gas-filled blisters are formed. If the heating continues, these blisters grow and can burst through the upper crust layer, causing breakdown of the mask integrity. To avoid blister formation, ion implanters are designed with sufficient heat sinking and coolant flows under the wafer such that the wafer temperature is kept below $\approx 120°C$ for high-current and high-energy implants.

At the conclusion of the implant cycle, the resist layer is removed to clear the way for subsequent processing and patterning steps. The C-rich crust layer usually requires plasma exposure in an "asher," followed by a wet-chemical clean to remove the residual lower resist layers [70]. The thickness of the C-rich crust, which determines the power settings and process time in the plasma asher tool, depends on the range of the implant ions in the resist (Fig. 27). For

13. Materials Modification Applications and Process Effects

Ion implantation has been used for its capability to beneficially modify the chemical and mechanical surface properties of materials since the early 1970s. This is well described by Nastasi, Mayer and Hiroven [18]. An early paper describing the technique of plasma immersion ion implantation by Conrad *et al.* also has a discussion of the many applications of these techniques [19]. Similarly, Ueda, Berni and Castro published "Application of Plasma Immersion Ion Implantation for improved performance of tools and industrial components" which specifically focuses on materials modification for surface and mechanical property improvement [92].

Advantages of ion implantation in comparison to other surface treatments include the following: the solid solubility limit can be exceeded, that the preparation is independent of diffusion, and that the depth distribution is controllable. At the same time there is no sacrifice of the bulk material properties, no significant dimensional changes and adhesion is expected as there is no clear interface.

A limitation of beam-line implantation is that it is a line-of-sight process. Samples having complicated re-entrant surfaces cannot be treated effectively. However, one of the benefits of PIII is that the ions follow the field lines, which trace the surface geometry of the sample. This provides a much more even treatment of the material. A paper that describes the theoretical background to this statement is "PIII Implantation Inside a Small Bore" [93].

Tools for materials modification are described in Chapter 15 of the book by Nastasi, Mayer and Hiroven [18]. Directed ion beam implanters have followed closely the same evolution and features as described earlier for tools developed for semiconductor technology. Ion sources capable of larger currents have been developed for their use in these applications. These include CHORDIS (cold and hot reflex discharge ion source) [94] and MEVVA (metal vapor vacuum arc) [95, 96]. PIII tools were developed in the 1990s [97] and found quick application due to the field-line implantation attribute and the promise of high doses in a simple, fast, cost-effective manner [98]. Most PIII tools and applications use relatively low energies; however, Rossi *et al.* [99] have developed a high energy immersion approach using short repetitive pulses for aerospace materials. An associated technology is IBAD (ion beam assisted deposition). Nastasi *et al.* [18] and Kaufman and Robinson [100] discussed source technology and system for this tool family.

Table 2 summarizes the results of an active area of the literature, where researchers have published a large number of results on the effect of several ions on metallurgical systems. Two areas in particular have found multiple applications. The first three rows of Table 2 show variants of Ti-Al alloys. These materials are favored for medical implants, such as knee joints and hip replacement. Reviewing the effects of the treatments show that nitrogen and several other elements can increase the hardness, decrease

Table 2. A summary of the literature materials treatment on metal, ion and a short comment on the results are listed.

Substrate	Ion	Improvements observed	Reference ID
Titanium Alloys			
Ti-6Al–4V	N	Hardness, coefficient of friction, wear and adhesion	101, 102, 103, 104, 99
Ti-Al-Si-N	C	Frictional coefficient; wear	105
Ti-Al-Zr	Ta, N	Anodic current density; corrosion potentials	106
Chromium Alloys			
CrCoMo, CrNiMo	N, H	Hardness improvement	92
Aluminum Alloys			
A17475, A15052	N	Corrosion reduction, hardness improvement	99
Steels			
5160, SKD11, SKS3, M2, SI, E-52	N, H, Cr, Ar, Si	Hardness, wear rate, corrosion potential	19, 107, 102, 103, 92, 108, 109, 110, 111, 112
H13, 304, 420, 430, 304	N, Cr, N and Cr		113, 114, 115, 116
Polyethylene			
Polyethylene (UHMWPE)	N	Hardness; elastic modulus; coefficient of wear	117, 99, 92
Nanoparticles			
Silicon with silicon dioxide layer	Ni, Fe, Si	Created nanoparticles with controllable size and density	118, 119, 120
Silica microbeads, soda-lime glass beads	Ar and C ions, + and −	Tested threshold charging voltage and particle	121
Photo			
Silicon, H-implanted Si	H_2^+, Acetylene and Ar	Photoluminescence, roughness, isolation	122
Rutile TiO$_2$	Cu and Ag	Photocatalytic efficiency	123
PbTe	Tin	Thermopower and electrical conductivity	124

the wear coefficient and reduce the potential for corrosion. Similarly, much work has been done on steels; the majority of the references address this topic. For multiple alloys, results indicate that nitrogen implantation along with several other elements can increase the hardness and decrease the wear rate. Notably, several specific applications are quoted as the target of this research with a beneficial effect.

Increasing the hardness of metal alloys is not the only application of these surface treatments, but may be the largest. Several publications detailed work on plastics, where the irradiation created surface layers that increased their wear resistance. Kapton was found to create a metal oxide layer that increases its lifetime in the space atmosphere of atomic oxygen [125]. Similarly, Shi, Li and Dong of Birmingham found that PIII of nitrogen would improve the hardness, elastic modulus, and coefficient of wear reduction of ultra-high molecular weight polyethylene [117]. Tin implantation has been studied to improve the activity of PbTe as a thermoelectric material [124]. Several materials segregate to form particles when present in sufficient amounts. It has been found that these segregations naturally form into nanoparticles. Silicon particles have been formed in SiO_2 [120] and nanotubes have been grown from both nickel [118] and iron [119] implantations.

In the area of catalysis, metal ions have been used to improve the catalytic properties of rutile TiO_2 [123] and a machine process paper has described how the implantation may scatter catalytic particles [121]. The use of PIII techniques for formation of bioactive materials and devices has been reviewed by Chu [126].

13.1. *H-cut wafer splitting*

Hydrogen implantation at high enough doses ($\approx 5 \times 10^{16}\,H/cm^2$) to induce planar lateral splitting in crystalline Si, Ge and GaAs, etc., under proper conditions has been used for the last decade to manufacture silicon-on-insulator (SOI) wafers as well as for lamination of diverse types of photonic and photovoltaic materials and structures [127–129].

The effects of high-dose H implant into Si are sketched in Fig. 28. For relatively shallow H implants, a few hundred nm, into an exposed Si wafer, H_2-filled surface blisters form during the implant and during subsequent thermal processing. The thickness of the blister skin is equal to

the H range in Si, increasing with energy. If the H-implanted Si wafer is bonded to an oxidized "handle" wafer before thermal processing, the chemical etching of Si by the accumulated H atoms results in the formation of thin "platelet" voids aligned along the wafer surface plane in Si(100) at the depth of the H implant peak. If this bonded pair of wafers is heated to $\approx 450°C$ or subjected to appropriate mechanical force, a planar "cleave plane" will form along the platelets in the H-rich region and the wafers can be separated. After removal of the H-rich damaged layer, the handle wafer, oxide, and transferred layer become an SOI wafer ready for CMOS and photonic device processing.

If a higher energy proton beam is used to implant H at a deep enough location, the overlying Si is stiff enough to drive the formation of planar platelets rather than surface blisters, without the need for the bonded handle wafer. Free-standing Si membranes as thin as $20\,\mu m$, using an $\approx 1.2\,MeV$ proton beam, have been reported [129].

Monte Carlo calculations [130] of H and Si recoil profiles for high-dose, low and high-energy proton profiles are shown in Fig. 29. The residual damage associated with the Si recoils during the ion stopping is important for H implants because H diffuses easily in Si and migrates away from the initial stopped profile unless it is trapped by local Si defects. The

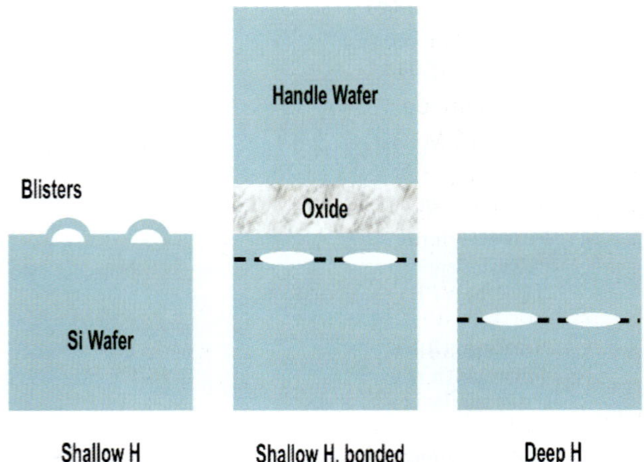

Fig. 28.　Schematic for high dose ($\approx 5 \times 10^{16}\,H/cm^2$) implants into Si after thermal treatment for (left) a shallow H profile with an open Si wafer surface, resulting in surface blisters, (middle) a shallow H profile with an oxide covered Si handle wafer bonded to the implanted wafer before thermal treatment, resulting in Si layer splitting, and (right) a deep H profile into a Si wafer, also resulting in Si layer splitting.

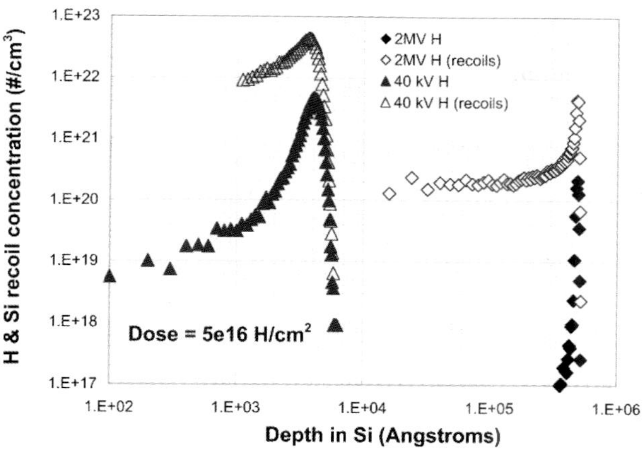

Fig. 29. Monte Carlo calculations of initial H and Si primary recoil distributions for high dose 40 keV and 2 MeV protons [130].

resulting H distribution follows the damage profile, peaking near the H range but slightly shallower than the initial H depth profile. For a 40 keV H profile, typical of SOI wafer and photonic materials fabrication, the H depth and transferred layer thickness is $\approx 0.4\,\mu$m. For a 2 MeV proton beam, used for splitting of free-standing Si membranes for low mass PV cells, the range and thickness is $\approx 50\,\mu$m.

H-cut splitting can be accomplished in a number of ways, such as thermal splitting at $\approx 450°$C for ≈ 30 min [127], mechanical separation at room temperature [131], by exposure to microwave radiation for short periods of time [132] and thermal stress from high power laser scans [133]. The heavily damaged and H-rich layer that surrounds the cleave plane can be removed by chemical mechanical polishing (CMP) [127], exposure to H_2 at temperatures above $\approx 1100°$C [134], or by specialized etching procedures, leaving relatively undamaged crystalline material in the transferred layer. Measurements of carrier recombination rates in free-standing Si membranes, an important consideration for PV cell materials, indicate that a high quality Si material survives the passage of high dose MeV proton beams [129].

The accelerators used for implantation of H for creation of free-standing Si membranes of several to $150\,\mu$m thick require proton beams of sufficient beam current to be economically viable for doses of $\approx 5 \times 10^{16}$ H/cm^2 per membrane at energies up to 4 MeV. Single-ended 4MV Van de Graaff-type accelerators provide a useful high-current proton flux over

a range from 1 to 4 MeV for fabrication of thin Si membranes for use in high efficiency PV cell manufacturing [135]. Multichannel linac designs are also in use to provide multi-mA-level proton currents for Si membrane separation [136]. In addition to providing low weight solar modules, H-cut methods also are highly efficient in their utilization of Si material with minimal waste of Si in the cutting and removal of the heavily damaged cleave plane region. This is in strong contrast to wire-saw methods which consume 30–50% of the Si block in the cutting and saw damage etching process for thin Si sections.

H-cut techniques are envisioned to play a central role in formation of 3D ICs [137]. In the methods proposed by MonolithIC 3D, a completed CMOS circuit, stopping at the S/D contact formation, is implanted with H at a depth sufficient to separate the device level from the fabrication wafer. Then a temporary bond is made to a transparent handle wafer and the CMOS device layer is separated from the wafer, aligned with a fully interconnected second CMOS device layer and bonded to it. The 3D structure is completed with the fabrication of shallow vias and interconnects linking the two device layers. Many similar 3D integration methods have been explored, including bonding of CMOS device layers formed on SOI wafers thinned by removal of the handle wafer material by grind and etching methods using the buried oxide layer as an etch stop [138]. It is estimated that proton acceleration to less than 50 keV would be sufficient to implement the process shown in Fig. 30, with a cleave plane depth of less than 500 nm and the use of a temporary bonded carrier wafer.

13.2. *Phase stabilization implants for NiSi contacts*

The effect of source/drain junction contact resistance on the transistor total series resistance increases as the contact pitch and contact area decrease with more strongly scaled devices. Ni silicide has been widely adopted for a junction contact metal, with very good results (low Schottky barrier height (SBH) and low contact resistivity for contact to p-type junctions). However the NiSi form of Ni silicide has a high SBH and high contact resistivity on n-type junctions due to the contribution of the Si band gap.

Implants of rare-earth ions (Dy, Pr, La) into the surface regions of n-type contacts prior to Ni

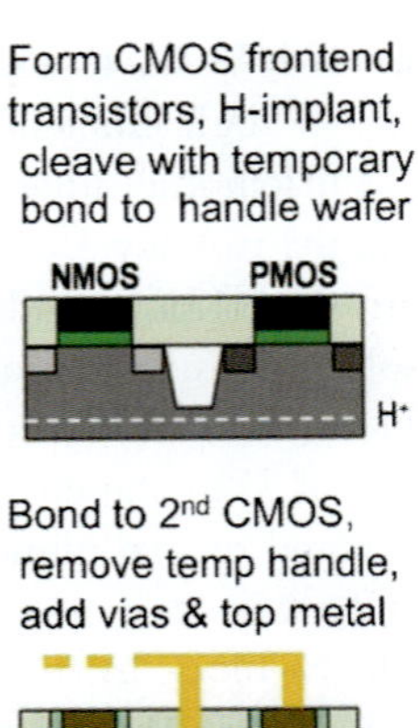

Fig. 30. Schematic of the use of H-cut techniques to split away a completed CMOS front end device (top) and laminate it to a fully metalized 2nd CMOS layer, to be connected by vias and final metal layers (bottom) [137].

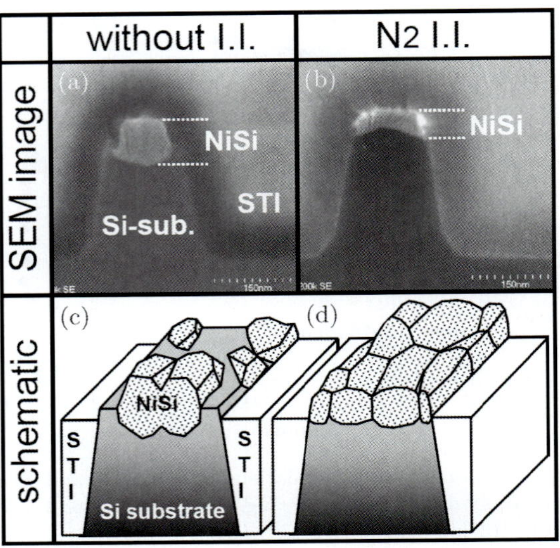

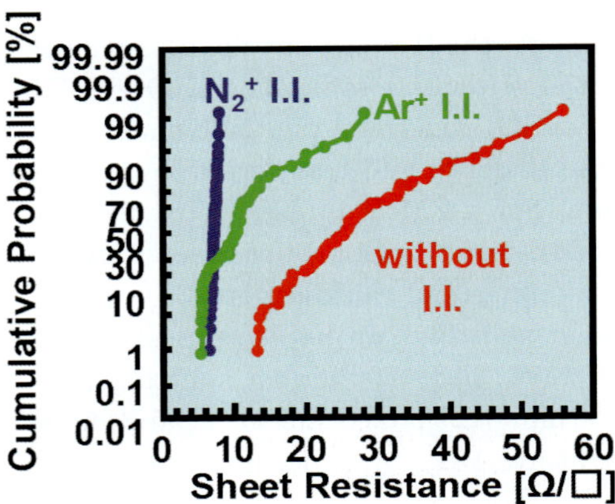

Fig. 31. SEM cross-sections and sketches (top) and sheet resistance distributions for 150 nm thick Ni-silicide layers (bottom) showing the positive effect of N_2^+ ion implantation into NiSi layers on n-type junctions [1].

deposition and silicide formation have shown lower SBH. However the contact resistance to heavily n-type doped junctions is still limited by the formation of local and discontinuous NiSi grains (Fig. 31). The implantation of N at doses of $\approx 3 \times 10^{15}\,\mathrm{N/cm^2}$ results in a stable phase of NiSi for anneal temperatures of 300°C–400°C with lower SBH values, similar to $\mathrm{NiSi_2}$ but without the need for high anneal temperatures (>750°C for formation of $\mathrm{NiSi_2}$ phases) and at a lower metal layer resistivity than $\mathrm{NiSi_2}$. The phase-stabilized N-implanted NiSi also has a more uniform grain distribution, allowing the use of thinner silicide contact layers, and a much narrower and lower value distribution of sheet resistance (Fig. 31). The use of N-implants into NiSi contacts reduces the total series resistance in a modern nMOS transistor by 22% (Fig. 32).

Additional studies have shown that C doping of NiSi films and substrates, combined with dopants, also contributes to the phase stability of NiSi contacts. When combined with dopants that diffuse by interstitial mechanisms, mainly B or P, the C implants can serve the dual functions of reducing dopant diffusion (the "cocktail" implant effect) and allowing the NiSi contact film to withstand higher anneal temperatures without agglomerating into local islands.

The enhanced damage rates produced by the molecular ions of B_{18} and C_{16} also seem to increase the stability of NiSi films during post-deposition anneals (Fig. 33). The C in combination here with B dopants results in shallower junctions after anneal.

13.3. *Optical constant modification*

A well known property of ion implanted materials is that the fluence of ions causes damage in the material exposed to the beam. This damage often changes the optical properties of the material. An active application of ion implantation is to exploit

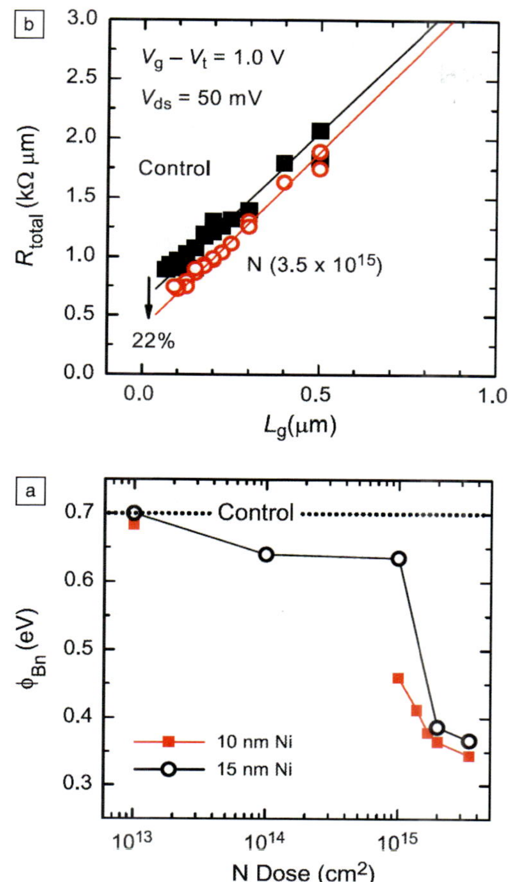

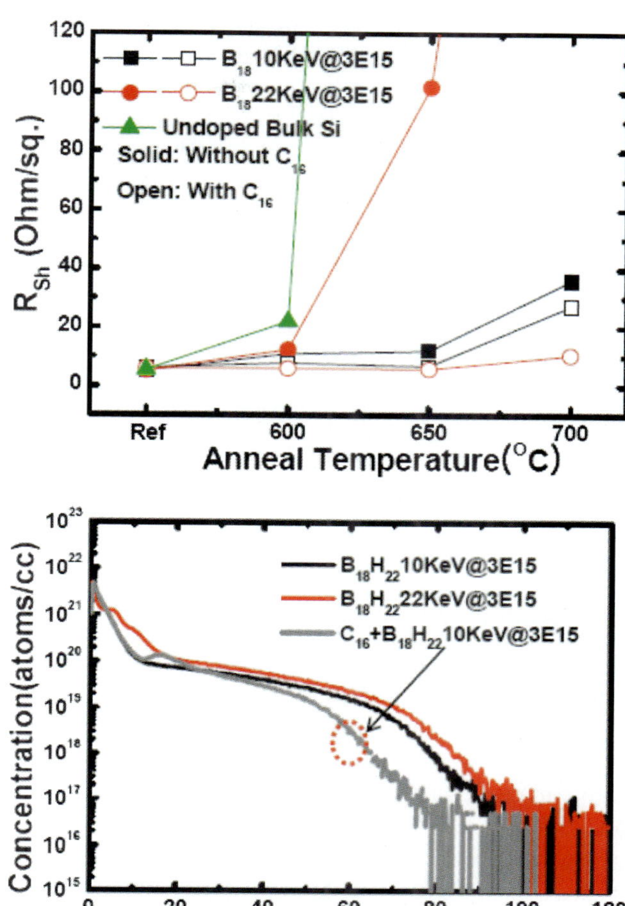

Fig. 32. Reduction of SBH for NiSi on n-type junctions by high dose N^+ implant (top) and reduction of total series resistance in an nMOS transistor (bottom) [139].

Fig. 33. NiSi sheet resistance for 15 nm films with additional B_{18} and C_{16} molecular ion implants (top) and B diffusion profiles (bottom) [140].

this effect by modifying the optical properties of optical materials using implantation. In particular, ion implantation is a powerful and promising technique to use to fabricate optical waveguides [141]. Accurate control of both the depth and lateral concentrations of the dopant at low temperature make the implantation process an extremely attractive technique for this purpose. In the waveguides formed by light-ion implantation, such as He or H, ion doses of the order of 10^{16} ions/cm^2 or higher are needed, and an optical barrier is built up at the end of track due to the damage induced by nuclear energy loss [142, 143].

This damage in the optical material primarily causes changes in the index of refraction, n. Several other physical and optical properties may also be affected. Table 3 surveys the literature on this area and although the majority of the work

focuses on changes in n and associated optical constants, there are several other effects that have been explored.

Implantation has been used to modify luminescent properties of a laser crystal [154] and of silicon [122]. The silicon work is of particular interest as this is a link coupling optical materials for possible use with silicon microelectronics. Another effect revealed is the formation of copper nano-particles in polymer substrates [157] and in ZnO [158]. Zinc oxide was chosen, because of its utility for photonic applications, as a semiconductor with high radiation resistance [158]. In the polymers, the copper nano-particles are the agent that affects the optical properties of the material [157].

Table 3.   Survey of the literature regarding implantation on optical material systems.

Substrate	Ion	Modification	Reference
Chalcogenide glasses	Ar, N	Waveguides, step-like n distribution	144
α-SiO$_2$	F, Cl	Waveguides, step-like n distribution	145
LiNbO$_3$	O	n enhancement in the waveguide layer	118
PMMA, PVC, PI, PC	Li, N	Waveguides fabricated with n	146
PMMA, Polystyrene	H, H$_2$, N, B	up to 30% change in index of refraction (n)	147
PMMA	Protons	refractive index depth profile	148
Polycarbonate	O	n, loss factor, hardness	149
Polycarbonate	C	optical, chemical and thermal properties	150
polycarbonate	Cu, Ni	structural degredation characterized	151
Sapphire, a-Al$_2$O$_3$	Au	optical absorption shifts	152
Cd$_2$SnO$_4$	Ag	optical and electrical properties	153
Nd:MgO:LiNbO$_3$	C	fluorescence modifications	154
polycarbonate	O	structural degredation characterized	155
Sapphire, a-Al$_2$O$_3$	Co	structural and optical changes	156
Si	H, by PIII	Photoluminescence	122
HDPE, PS, PO	Cu	Cu nanoparticle formation	157
ZnO	Cu	Cu nanoparticle formation	158

14. Summary

Ion implantation in its many forms continues to provide the principal method for delivering dopants and other ions for the manufacture of IC devices. In addition, new applications for doping of PV materials, materials modification of metals, plastics, and other forms, and bioactive surfaces and devices, continue to develop. The diversity of implant machine types continues to grow, in many cases still nurtured by the deep roots of accelerator physics developed through the 1930s to 60s. The high expectations for the ion implantation processes for doping accuracy in amount and placement as well as in elemental purity mean a wide variety of process control issues that must be resolved on a daily basis by machine builders, engineers, maintenance workers, and operators throughout the world. A sample of these implantation tools, applications and issues have been reviewed in this paper.

Acknowledgments

The authors would like to thank Dr. Harold Stern and the Ingram School of Engineering for partial support of this effort.

References

[1] K. Tsukamoto, T. Kuroi and Y. Kawasaki, Evolution of ion implantation technology and its contribution to semiconductor industry, in *Proc. Int. Conf. Ion Implantation Technology* (2010), pp. 9–16.

[2] S. B. Felch *et al.*, *J. Vac. Sci. Technol. B* **26**, 281 (2008).

[3] http://download.intel.com/newsroom/kits/22nm/pdfs/22nm-Announcement_Presentation.pdf

[4] Y. Sasaki *et al.*, Conformal doping for finFETs and precise controllable shallow doping for planar FET manufacturing by a novel B$_2$H$_6$/helium self-regulatory plasma doping process, in *Proc. Int. Conf. Electron Devices Meeting IEDM* (2008), pp. 917–920.

[5] J. Mody *et al.*, *J. Vac. Sci. Technol. B* **28**, C1H5–C1H13 (2010).

[6] A. Rohatgi and D. Meier, *Photovolt. Int.* **10**, 87 (2010).

[7] N. Cheung *et al.*, *Ion Implantation Science and Technology*, ed. J. F. Ziegler (Ion Implantation Technology Corp., Yorktown, New York, 1996).

[8] http://www.axcelis.com

[9] http://www.nissin-ion.co.jp

[10] http://www.senova.co.jp/english/index.html

[11] http://www.tel.com/eng/about/index.htm

[12] http://www.ulvac.com

[13] http://www.vsea.com

[14] http://www.aibt-inc.com

[15] http://www.appliedmaterials.com

[16] http://www.ion-beam-services.com

[17] http://www.pelletron.com/pellet.htm

[18] M. Nastasi, J. W. Mayer and J. K. Hirvonen, *Ion–Solid Interactions: Fundamentals and Applications* (Cambridge University Press, 1996).

[19] J. R. Conrad, J. L. Radtke, R. A. Dodd, F. J. Worzala and N. C. Tran, *J. Appl. Phys.* **62**, 4591 (1987).

[20] B. Mizuno, I. Nakayama, N. Aoi, M. Kubota and T. Komeda, *Appl. Phys. Lett.* **53**, 2059 (1988).

[21] S. Wolf and R. N. Tauber, *Silicon Processing for the VLSI Era* (Lattice Press, Sunset Beach, California, 1986).

[22] http://www.itrs.net

[23] E. C. Jones and E. Ishida, *Mater. Sci. Eng.* **R24**, 1 (1998).

[24] E. Chason *et al.*, *J. Appl. Phys.* **81**, 6513 (1997).

[25] D. H. Lee and J. W. Mayer, *Proc. IEEE* **62**, 1241 (1974).

[26] K. S. Jones, E. Kuryliw, R. Murto, M. Rendon and S. Talwar, Boron diffusion upon annealing of laser thermal processed silicon, in *Proc. Int. Conf. Ion Implantation Technology* (2000), pp. 111–114.

[27] F. Torregrosa *et al.*, *AIP Conf. Proc.* **1066**, 473 (2008).

[28] http://www.mattson.com/index.asp

[29] http://www.screen.co.jp/spe/eng/products/LA-3000-F/index.html

[30] B. L. Yang *et al.*, *Microelectron. Reliab.* **38**, 1489 (1998).

[31] N. W. Cheung, *Nucl. Instrum. Methods Phys. Res. B* **55**, 811 (1991).

[32] J. V. Mantese, I. G. Brown, N. W. Cheung and G. A. Collins, *MRS Bull.* **21**, 52 (1996).

[33] A. Sultan *et al.*, The effect of implant dose rates and two step anneals on p^+-n ultra-shallow junctions, in *Proc. University/Government/Industry Microelectronics Symp.* (1995), pp. 108–112.

[34] W.-K. Chu *et al.*, The alternative ion implantation approaches for ultra-shallow junction, in *16th Int. Conf. Applications of Acceleraors in Research and Industry* (2001), pp. 891–895.

[35] T. Uchino, A. Miyauchi and T. Shiba, *IEEE Trans. Electron Devices* **48**, 1406 (2001).

[36] M. Canino, G. Regula, M. Xu, E. Ntzoenzok and B. Pichaud, *Mat. Sci. Eng. B* **159–160**, 338 (2009).

[37] E. Bruno *et al.*, *Nucl. Instrum. Methods B* **257**, 181 (2007).

[38] E. Bruno *et al.*, *J. Vac. Sci. Technol. B* **26**, 386 (2008).

[39] L. Shao, J.-R. Liu, X.-M. Wang, H. Chen, P. E. Thompson and W.-K. Chu, *Nucl. Instrum. Methods B* **206**, 413 (2003).

[40] C. I. Li *et al.*, Integration of advanced source and drain extension process using carbon/fluorine co-implants and spike anneal in 65 nm PMOS devices, in *Proc. 16th Int. Conf. Ion Implantation Technology* (2006), pp. 46–49.

[41] S. B. Felch *et al.*, *J. Vac. Sci. Technol. B* **26**, 281 (2008).

[42] N. Ohno, T. Hara, Y. Matsunaga and M. I. Current, Dual ion implantation of non-dopant and dopant ions into Si for defect engineering of shallow p^+ junctions, in *Proc. Int. Conf. Ion Implantation Technology* (1998), pp. 1047–1050.

[43] J. Plaza Castillo, A. Torres Jacome, O. Malik and N. Torres Lopez, *J. Microelectron* **39**, 678 (2008).

[44] K. Goto, J. Matsuo, Y. Tada, T. Sugii and I. Yamada, *IEEE Trans. Electron Devices* **46**, 683 (1999).

[45] K. Goto *et al.*, Novel shallow junction technology using decaborane ($B_{10}H_{14}$), in *Proc. Conf. Int. Electron Devices Meeting, IEDM* (1996), pp. 435–438.

[46] D. S. Chao, D. Y. Shu, S. B. Hung, W. Y. Hsieh and M.-J. Tsai, *Nucl. Instrum. Methods B* **237**, 197 (2005).

[47] M. S. Ameen, L. M. Rubin, M. A. Harris and C. Huynh, *J. Vac. Sci. Technol. B* **26**, 373 (2008).

[48] J. Matsuo, D. Takeuchi, T. Aoki and I. Yamada, Cluster ion implantation for shallow junction formation, in *Int. Conf. Ion Implantation Technology* (1996), pp. 768–771.

[49] D. Takeuchi, N, Shimada, J. Matsuo and I. Yamada, Shallow junction formation by poly-atomic cluster ion implantation, in *Int. Conf. Ion Implantation Technology* (1996), pp. 772–775.

[50] K. Yako *et al.*, 26 nm gate length MOSFETs with aggressively reduced silicide position by using carbon cluster implanted raised source/drain extension structure, in *IEEE Symp. VLSI Technology* (2009), pp. 160–161.

[51] http://www.semequip.com

[52] http://www.tel.com/eng/about/us/telepion.htm

[53] V. N. Faifer *et al.*, Leakage current and dopant activation characterization of SDE/halo CMOS junctions with non-contact junction photo-voltage metrology, in *Proc. Int. Workshop Junction Technology* (2007), pp. 43–46.

[54] S. B. Felch *et al.*, Optimized BF3 P2LAD implantation with Si-PAI for shallow, abrupt and high quality p^+/n junctions formed using low temperature SPE annealing, in *Proc. 14th Int. Conf. Ion Implantation Technology* (2002), pp. 52–55.

[55] C.-H. Jan *et al.*, A 65 nm ultra low power logic platform technology using uni-axial strained silicon transistors, in *Proc. Conf. Int. Electron Devices Meeting, IEDM* (2005), pp. 60–63.

[56] C. I. Li *et al.*, Enabling solutions for 28 nm CMOS advanced junction formation, in *Proc. Int. Conf. Ion Implantation Technology* (2010), pp. 45–48.

[57] E. J. H. Collart *et al.*, Process characterization of low temperature ion implantation using ribbon beam and spot beam on the AIBT iPulsar high current, in *Proc. Int. Conf. Ion Implantation Technology* (2010), pp. 49–52.

[58] S. Ninomiya *et al.*, New combination of damage control techniques using SEN's single-wafer implanters, in *Proc. 11th Int. Workshop Junction Technology* (2011), pp. 88–91.

[59] T. Aoki, T. Seki and J. Matsuo, in *Proc. 7th Int. Workshop Junction Technology* (2007), pp. 23–24.

[60] B. Sekar, W. Krull, K. Huet, C. Boniface and J. Venturini, Larger clusterboron (B36Hx) implant

for USJ applications, *Int. Conf. Ion Implantation Technology* (2010), pp. 101–104.

[61] K. Sekar *et al.*, Cluster carbon implants — cluster size and implant temperature effect, in *Proc. Int. Workshop Junction Technology* (2011).

[62] R. Dennard *et al.*, *IEEE J. Solid State Circuits* **9**, 256 (1974).

[63] P. Packan *et al.*, High performance Hi-K + metal gate strain enhanced transistors on (110) silicon, in *Proc. Int. Electron Devices Meeting, IEDM* (2008), pp. 1–4.

[64] K. L. Saenger, K. E. Fogel, J. A. Ott, D. K. Sadana and H. Jin, *J. Appl. Phys.* **101**, 104908 (2007).

[65] K.-Y. Lim *et al.*, Novel stress-memorization-technology (SMT) for high electron mobility enhancements of gate last high-k/metal gate devices, in *Proc. Int. Electron Devices Meeting, IEDM* (2010), pp. 10.1.1–10.1.4.

[66] A. Wei *et al.*, Multiple stress memorization in advanced SOI CMOS technologies, in *IEEE Symp. VLSI Technology* (2007), pp. 216–217.

[67] Y.-C. Liu *et al.*, Strained channel MOSFETs with embedded silicon carbon formed by solid phase epitaxy, in *IEEE Symp. VLSI Technology* (2007), pp. 44–45.

[68] K. Sekar, W. Krull, J. Gelpey and S. McCoy, Optimization of Si:C stress retention and junction quality with ClusterCarbon implantation, in *Proc. Int. Conf. Ion Implantation Technology* (2010), p. 93.

[69] A. Li-Fatou *et al.*, *Electrochem. Soc. Trans.* **11**, 125 (2007).

[70] T. Hattori, Y.-J. Kim, C. Yoon and J.-K. Cho, *IEEE Trans. Semicond. Manuf.* **22**, 468 (2009).

[71] G. Oehrlein, R. Phaneuf and D. Graves, *J. Vac. Sci. Technol. B* **29**, 010801-1 (2011).

[72] N. Samarakone, P. Yick, M. Zawadzki and S.-J. Choi, Double printing through the use of ion implantation, in *Optical Microlithography 6924th Proc. SPIE Int. Soc. for Optical Engineering* (2011), pp. 69242B-1–69242B-13.

[73] Y. Kikuchi, D. Kawamura and H. Mizuno, Study of ion implantation into EUV resist for LWR improvement, in *7969th Proc. SPIE Int. Soc. for Optical Engineering* (2011), pp. 7969–7988.

[74] P. M. Martin, L. Godet, A. Cheung, G. de Cock and C. Hatem, Ion implant enabled 2× lithography, in *Proc. Int. Conf. Ion Implantation Technology* (2010), pp. 171–175.

[75] M. Aleman *et al.*, *IEEE PVSC* **35**, 1291 (2010).

[76] N. Bordin *et al.*, Recombination parameters of Si solar cells with back surface field formed by ion implantation, in *World Conf. Photovoltaic Energy Conversion* (2006), pp. 1171–1174.

[77] H. Kasai and H. Matsumura, *Sol. Energy Mater. Sol. Cells* **48**, 93 (1997).

[78] D. Barakel, A. Ulyashin, I. Perichaud and S. Martinuzzi, *Sol. Energy Mater. Sol. Cells* **72**, 285 (2002).

[79] J. L. Cruz-Campa *et al.*, *Sol. Energy Mater. Sol. Cells* **95**, 551 (2011).

[80] T. Tesfamichael, G. Will and J. Bell, *Appl. Surf. Sci.* **245**, 172 (2005).

[81] C.-C. Yen *et al.*, *Thin Solid Films* **519**, 4717 (2011).

[82] K. L. Narayanan and M. Yamaguchi, *J. Appl. Phys.* **89**, 8331 (2001).

[83] T. Tanaka *et al.*, *Sol. Energy Mater. Sol. Cells* **75**, 109 (2003).

[84] C. Kraft *et al.*, *Thin Solid Films* (2011), doi:10.1016/j.tsf.2011.01.389.

[85] F. Torregrosa *et al.*, *Surf. Coat. Technol.* **186**, 93 (2004).

[86] B. Y. Moon *et al.*, *Sol. Energy Mater. Sol. Cells* **75**, 113 (1997).

[87] T. N. Horsky *et al.*, Beam angular divergence effects in ion implantation, in *Proc. Int. Conf. Ion Implantation Technology* (2008), pp. 403–406.

[88] L. Pipes *et al.*, *Nucl. Instrum. Methods Phys. Res. B* **237**, 330 (2005).

[89] J. England *et al.*, *Nucl. Instrum. Methods Phys. Res. B* **74**, 613 (1993).

[90] M. I. Current *et al.*, A study of wafer charging with CHARM and SPIDER monitors, in *Proc. Int. Conf. Ion Implantation Technology* (1996), pp. 61–64.

[91] M. I. Current, W. Lukaszek and M. C. Vella, Control of wafer charging during ion implantation: issues, monitors and models, in *Proc. 5th Int. Symp. Plasma Process-Induced Damage* (2000), pp. 137–140.

[92] M. Ueda, L. A. Berni and R. M. Castro, *Surf. Coat. Technol.* **200**, 517 (2005).

[93] X.-C. Zeng *et al.*, *IEEE Trans. Plasma Sci.* **26**, 175 (1998).

[94] R. Keller, *Nucl. Instrum. Methods Phys. Res. B* **298**, 247 (1991).

[95] I. G. Brown, M. R. Dickinson, J. E. Galvin, X. Godechot and R. A. MacGill, *J. Mater. Eng.* **13**, 217 (1991).

[96] I. G. Brown, *J. Vac. Sci. Technol. A* **11**, 1480 (1993).

[97] M. M. Shamim, J. T. Scheuer, R. P. Fetherston and J. R. Conrad, *J. Appl. Phys.* **70**, 4756 (1991).

[98] D. J. Rej and R. B. Alexander, *J. Vac. Sci. Technol. B* **12**, 2380 (1994).

[99] J. O. Rossi *et al.*, *IEEE J. Trans. Plasma Sci.* **37**, 204 (2009).

[100] H. R. Kaufman and R. S. Robinson, *Operation of Broad-Beam Sources* (Commonwealth Scientific Corp., Alexandria, Virginia, 1987).

[101] A. Fabre, L. Barrallier, F. Torregrosa and L. Roux, *Microsc. Microanal. Microstruct.* **8**, 413 (1997).

[102] M. Ueda *et al.*, *Surf. Coat. Technol.* **156**, 71 (2002).

[103] M. Ueda *et al.*, *Surf. Coat. Technol.* **169–170**, 408 (2003).

[104] M. Ueda *et al.*, *Surf. Coat. Technol.* **201**, 4953 (2007).

[105] Y. F. Xu, P. W. Shum, Z. F. Zhou and K. Y. Li, *Surf. Coat. Technol.* **204**, 1914 (2010).

[106] Y. Z. Liu *et al.*, *Surf. Coat. Technol.* **201**, 7538 (2007).

[107] L. D. Yu *et al.*, *Surf. Coat. Technol.* **103–104**, 328 (1998).

[108] R. M. Oliveira, J. A. N. Gonçalves, M. Ueda, S. Oswald and S. C. Baldissera, *Surf. Coat. Technol.* **204**, 2981 (2010).

[109] L. L. G. da Silva, M. Ueda and R. Z. Nakazato, *Surf. Coat. Technol.* **201**, 8291 (2007).

[110] E. Cano *et al.*, *Surf. Coat. Technol.* **200**, 5123 (2006).

[111] R. K. Y. Fu, D. L. Tang, G. J. Wan and P. K. Chu, *Surf. Coat. Technol.* **201**, 4879 (2007).

[112] S. Mandl, R. Gunzel, B. Rauschenbach, R. Hilke, E. Knosel and K. Kunanz, *Surf. Coat. Technol.* **103–104**, 161 (1998).

[113] A. A. Youssef *et al.*, *Vacuum* **77**, 37 (2004).

[114] H.-X. Liu, B.-Y. Tang, L.-P. Wang, X.-F. Wang and B. Jiang, *Surf. Coat. Technol.* **201**, 5273 (2007).

[115] C. B. Mello, M. Ueda, R. M. Oliveira and J. A. Garcia, Corrosion effects of plasma immersion ion implantation–enhanced Cr deposition on SAE 1070 carbon steel, *Surf. Coat. Technol.* (2011), doi:10.1016/j.surfcoat.2011.01.030.

[116] C. B. Mello, M. Ueda, R. M. Oliveira, C. M. Lepienski and J. A. Garcia, *Surf. Coat. Technol.* **204**, 2971 (2010).

[117] W. Shi, X.Y. Li and H. Dong, *Wear* **250**, 544 (2001).

[118] A. R. Adhikari *et al.*, *Appl. Phys. Lett.* **86**, 053104-0 (2005).

[119] Y.-H. Choi, J. Sippel-Oakley and A. Ural, *Appl. Phys. Lett.* **89**, 153130-1 (2006).

[120] J. Grisolia *et al.*, *Nanotechnology* **16**, 2987 (2005).

[121] J. Ishikawa, H. Tsuji and Y. Gotoh, Particle-scattering phenomenon of powders caused by charging voltage of the surface during ion implantation, in *Proc. 11th Int. Conf. Ion Implantation Technology* (1996), pp. 46–49.

[122] P. Chu, Recent advances and applications of plasma immersion ion implantation, in *Proc. 7th Int. Conf. Solid-State and Integrated Circuits Technology* (2004), pp. 2172–2177.

[123] H. Tsuji, T. Sagimori, K. Kurita, Y. Gotoh and J. Ishikawa, *Surf. Coat. Technol.* **158–159**, 208 (2002).

[124] Q. Shen, J.-G. Li and L. M. Zhang, *Sol. Energy Mater. Sol. Cells* **62**, 167 (2000).

[125] I. H. Tan *et al.*, *Surf. Coat. Technol.* **186**, 234 (2004).

[126] P. K. Chu, *Biomaterials Fabrication and Processing Handbook*, eds. P. K. Chu and X. Liu (CRC, Florida, 2008).

[127] http://www.soitec.com/pdf/SmartCut_WP.pdf

[128] H. A. Atwater, M. I. Current, M. Levy and T. Sands, Integration of heterogeneous thin-film materials and devices, in *768th Symp. Proc. Materials Research Society* (2003).

[129] F. Henley, A. Lamm, S. Kang, Z. Liu and L. Tian, Direct film transfer (DFT) technology for kerf-free silicon wafering, in *23rd European PV Solar Energy Conf.* (2008).

[130] J. F. Ziegler, J. P. Biersack and M. D. Ziegler, SRIM: Stopping and range of ions in matter (2008), www.srim.org

[131] M. I. Current, S. W. Bedell, I. J. Malik, L. M. Feng and F. J. Henley, What is the future of sub-100 nm CMOS: ultrashallow junctions or ultrathin SOI?, *Solid State Technol.*, Pennwell (2000).

[132] D. C. Thompson *et al.*, *Appl. Phys. Lett* **87**, 224103-1 (2005).

[133] F. J. Henley and N. W. Cheung, Controlled process and resulting device, U.S. patent #US2007/0122997 A1 (2007).

[134] A.-L. Thilderkvist, S. Kang, M. Fuerfanger and I. J. Malik, Surface finishing of cleaved SOI films using epi techniques, in *Proc. IEEE Int. SOI Conference* (2000), pp. 12–13.

[135] A. A. Brailove, Kerf-free wafering: Technology overview, (2011), http://www.avsusergroups.org/joint_pdfs/2011-2brailove.pdf

[136] Z. Or-Bach, D. C. Sekar, B. Cronquist and I. Beinglass, *Future Fab. Int.* **37** (2011).

[137] T. Parrill and V. Benveniste, Hydrogen ion implanter using a broad beam source, U.S. patent #US 7,897,945 B2 (2011).

[138] K. W. Guanari *et al.*, Electrical integrity of state-of-the-art 0.13 um SOI CMOS devices and circuits transferred for three-dimensional (3D) integrated circuit (IC) fabrication, in *Proc. Int. Electron Devices Meeting, IEDM* (2002), pp. 943–945.

[139] W.-Y. Loh and B. Cross, *MRS Bull.* **36**, 97 (2011).

[140] K. Sekar, SemEquip, private communication (2011).

[141] P. D. Townsend, P. J. Chandler and L. Zhang, *Optical Effects of Ion Implantation* (Cambridge University Press, 1994).

[142] G. L. Destefanis, P. D. Townsend and J. P. Gailliard, *Appl. Phys. Lett.* **32**, 293 (1978).

[143] P. J. Chandler, L. Zhang and P. D. Townsend, *Appl. Phys. Lett.* **55**, 1710 (1989).

[144] F. Qiu, T. Narusawa and J. Zheng, *Appl. Opt.* **50**, 733 (2011).

[145] J. Manzano *et al.*, *Nucl. Instrum. Methods Phys. Res. B* **268**, 3147 (2010).

[146] J. R. Kulish, H. Franke, A. Singh, R. A. Lessard and E. J. Knystautas, *J. Appl. Phys.* **63**, 2517 (1988).

[147] A. V. Leont'ev, *Russ. Microelectron.* **30**, 324 (2001).

[148] W. Hong, H.-J. Woo, H.-W. Choi, Y.-S. Kim and G.-D. Kim, *Appl. Surf. Sci.* **169–170**, 428 (2000).

[149] N. L. Singh, A. Qureshi, F. Singh and D. K. Avasthi, *Mater. Sci. Eng. A* **457**, 195 (2007).

[150] R. Kumar *et al.*, *Nucl. Instrum. Methods B* **251**, 163 (2006).

[151] L. Singh and K. S. Samra, *Nucl. Instrum. Methods B* **263**, 458 (2007).

[152] C. Marques, E. Alves, R. C. da Silva, M. R. Silva and A. L. Stepanov, *Nucl. Instrum. Methods B* **218**, 139 (2004).

[153] R. Kumaravel *et al.*, *Nucl. Instrum. Methods B* **268**, 2391 (2010).

[154] N.-N. Dong, F. Chen and D. Jaque, *Opt. Express* **18**, 5951 (2010).

[155] M. Mujahid, D. S. Srivastava and D. K. Avasthi, *Rad. Phys. Chem.* **80**, 582 (2011).

[156] C. Marques, M. M. Cruz, R. C. da Silva and E. Alves, *Surf. Coat. Technol.* **158–159**, 54 (2002).

[157] N. Umeda, V. V. Bandourko, V. N. Vasilets and N. Kishimoto, *Nucl. Instrum. Methods B* **206**, 657 (2003).

[158] K. Kono, S. K. Arora and N. Kishimoto, *Nucl. Instrum. Methods B* **206**, 291 (2003).

Lawrence A. Larson is a Senior Lecturer and Research Professor in the Ingram School of Engineering at Texas State University at San Marcos. At Texas State, he has taught classes on the ITRS in addition to core engineering classes. Previously, he was Associate Director of the FEP Division of SEMATECH. This program targeted the development of processes and materials needed to manufacture the most advanced semiconductor devices. He established the project and funding for the SEMATECH/SRC Front End Processes Research Center, and worked on programs in installed base implant, ultrashallow junction development, rapid thermal processing, implant contamination, and the conversion of SEMATECH's fab to 200 mm. Larson received a B.A. in Physics from Whitman College and a Ph.D. in Solid-State Physics from Washington State University. He is the author of over 130 publications.

Justin M. Williams attended the University of Hawai'i at Manoa for his freshman year, after graduating from James E. Taylor High School in Katy, Texas. He is currently studying electrical engineering with a specification toward networks and telecommunications systems at Texas State University at San Marcos, and expects to graduate in December 2012 *summa cum laude*.

Michael I. Current's Research interests range over three decades of the applications and process controls of ion beam processing of Si and related materials for fabrication of advanced IC and photonic devices. He has worked at Current Scientific, Frontier Semiconductor, Silicon Genesis, Applied Materials, Xerox Palo Alto Research Center, Trilogy Systems, Signetics/Philips, and Kyoto and Cornell Universities. He earned his Ph.D. in Physics and worked at Rensselaer Polytechnic Institute. He has written over 200 papers and book chapters and co-chaired 6 international conferences, was a co-founder of the Silicon Valley Implant Users Group, and is an active member of the MRS and AVS.

Reviews of Accelerator Science and Technology
Vol. 4 (2011) 41–82
© World Scientific Publishing Company
DOI: 10.1142/S1793626811000483

Ion Beam Analysis: A Century of Exploiting the Electronic and Nuclear Structure of the Atom for Materials Characterisation

Chris Jeynes* and Roger P. Webb[†]

*University of Surrey Ion Beam Centre,
Guildford GU2 7XH, England*
*c.jeynes@surrey.ac.uk
[†]r.webb@surrey.ac.uk

Annika Lohstroh

*Department of Physics, University of Surrey,
a.lohstroh@surrey.ac.uk*

Analysis using MeV ion beams is a thin film characterisation technique invented some 50 years ago which has recently had the benefit of a number of important advances. This review will cover damage profiling in crystals including studies of defects in semiconductors, surface studies, and depth profiling with sputtering. But it will concentrate on *thin film depth profiling* using Rutherford backscattering, particle induced X-ray emission and related techniques in the deliberately synergistic way that has only recently become possible. In this review of these new developments, we will show how this integrated approach, which we might call "*total IBA*", has given the technique great analytical power.

Keywords: RBS; EBS; PIXE; ERD; NRA; MEIS; LEIS; SIMS; IBIC.

1. Introduction

Ion beam analysis is a very diverse group of characterisation techniques which have been applied to every class of material where the interest is in the surface or near-surface region up to a fraction of a millimetre in thickness. Such a field is far too broad to be reasonably covered by such a review as this; we will concentrate on thin film elemental depth profiling methods using ions with energies of order 1 MeV/nucleon. We will emphasise complementary techniques, including some other closely related IBA[a] methods.

Thin film elemental depth profiling is of critical importance to a wide variety of modern technologies, including the semiconductor, sensor, magnetics, and coatings industries (including both tribology and optics), among others. It is also valuable in many other disparate applications such as cultural heritage, environmental monitoring and forensics.

We will be describing examples in many of these areas in our review of MeV IBA.

Historically, IBA labs have tended to split into (at least) two "traditions": on the one hand nuclear methods (RBS, ERD, NRA), and on the other atomic methods (PIXE). We will outline various reasons for this, but will show that recent advances have facilitated the integration of these two traditions giving us what is effectively a new and much more powerful technique.

Curiously, this review seems to touch on all the main breakthroughs in 20th century physics, and all IBA techniques hinge on spectroscopy. So perhaps we should start by acknowledging our debt to Isaac Newton, who in 1671 was the first to use the word "spectrum" with the modern connotation of *quantum phænomenon observed with a dispersive mechanism* [1].[b] Previously (and subsequently)

---

[a]See the end of the article for expansion and explanation of acronyms. Appendix A is a glossary of IBA techniques, and Appendix B is a glossary of related techniques.
[b]Using Newton's spelling of "phænomenon" which transliterates the Greek $\phi\alpha\iota\nu\circ\mu\varepsilon\nu\circ\nu$. In his use of "spectrum" Newton had in mind the Latin etymology of *specere*, "to see".

"spectrum" had the connotation of "spectre" or "apparition". But philosophers see reality.

1.1. *Scope of chapter*

Although concentrating on MeV IBA depth profiling, we will mention methods for depth profiling crystallographic defects using channelling and MEIS; characterising defects in semiconductors, including their spatial distribution, by IBIC; and depth profiling using sputtering by SIMS (including MeV SIMS), although we will not discuss beam damage extensively. We will also mention ion beam methods sensitive to the true surface (MEIS, LEIS) since they also use RBS.

We will not specifically review microbeam applications, taking microbeam technology for granted throughout the text. But it will be clear that we think that 3D spatial resolution is central to the general usefulness of IBA.

1.2. *Complementary techniques*

Throughout, we will also mention complementary techniques wherever appropriate. Materials analysis must be a strongly interdisciplinary field, and the characterisation problems of modern materials almost invariably require the use of a variety of techniques for their solution. Any discussion of a technique without the context of complementary techniques is likely to be strongly misleading.

In all these fields the analyst has various standard tools: the electron microscopies and spectroscopies[c] (SEM, TEM, XPS, AES and their variants),[d] the scanning probe microscopies (AFM and variants including the new optical near-field methods), X-ray techniques like XRF and XRD (also with many variants) and optical methods like ellipsometry, Raman, FTIR and other spectroscopies. Elemental depth profiling can be done destructively using sputtering techniques with SIMS (or, frequently, AES). If destructive techniques are considered then bulk methods like ICPMS and AMS should be mentioned, and of course there are a wide variety of wet chemical analytical methods. XRF and

XRD are frequently applied to "bulk" as well as thin film samples, and other comparable fluorescence techniques are cathodoluminescence or photoluminescence. Molecular imaging can already be done in air by MALDI, DESI and DART.

Where does IBA fit in this kaleidoscope of techniques? IBA typically uses an accelerator which needs a hall of at least $200\,\mathrm{m}^2$, a footprint well over an order of magnitude larger than any of the other techniques mentioned — it is necessarily a technique with high running costs. What can it do which cannot be done reasonably easily by other techniques? If a materials research organisation (for example, a university) were to set up a central analytical laboratory to service the needs of all its research groups and other collaborators, would IBA be one of the techniques considered "essential"?

We believe that modern integrated IBA methods are exceptionally powerful for a wide range of materials problems, and we will show a number of significant examples that illustrate this.

1.3. *Overview of article*

In Secs. 2 and 4, we will summarise the nuclear and atomic IBA techniques we will be concentrating on, and in Sec. 6, we will address their integration. Section 3 will briefly describe the other IBA techniques, for a more complete overview of the field. In Sec. 5, we will discuss the important issue of computer codes used in IBA, together with a further discussion on the high accuracy available with these methods.

A general introduction to the field must mention the three IBA handbooks published over the last 35 years. The 1977 "green book" [2] includes a section on PIXE which was dropped from the more extensive "black book" from 1995 [3]. PIXE has been restored in the recent 2-volume Handbook [4]. There have been an almost unbroken series of biennial IBA Conferences, the first of which was held in 1973 [5]. Currently the most recent published proceedings are from the Cambridge conference in September 2009 [6]: the Proceedings are not yet available from the Brazil conference of April 2011. There is also a triennial PIXE conference series, the

---

[c] "spectrometry" or "spectroscopy"? A spectrum is the object that results from some dispersive process: we do *spectrometry* where we measure the spectrum and *spectroscopy* where we look at the spectrum. The Hubble red-shift for example is a *spectroscopic* effect since the recognisable pattern of atomic absorption lines has shifted to different frequencies. However, the boundary between spectroscopy (as in XPS) and spectrometry (as in RBS) is ill-defined, and it is mostly a conventional distinction.

[d] See Appendix B for a glossary expanding and explaining acronyms for complementary characterisation techniques.

latest of which was in Surrey in 2010 [7]. Other useful recent reviews include Giuntini (2011) on the use of external beams [8].

2. Nuclear IBA Depth Profiling Methods

Large angle ion scattering was first observed in Geiger & Marsden's experiments in 1909 [9], which were interpreted by Ernest Rutherford in 1911 to demonstrate the existence of the positively charged atomic nucleus [10]. The transition from RBS to EBS as the Coulomb barrier is exceeded was first demonstrated by Chadwick & Bieler in 1921 (for alphas on H [11]). The wave-mechanical interference between identical scattered and recoil nuclei due to their indistinguishability was pointed out (for electrons) by Mott in 1930 [12] and immediately verified using magnetic spectrometers for proton-proton RBS [13] and EBS (with measurements [14] and theory [15]). Explicit energy spectra were not published until the 1950s (in papers on the quantum mechanical calculation [16] and analytical chemistry [17]). Davies, Amsel & Mayer point out in their nice 1992 "reminiscences" paper [18] that J. O. Nielsen observed a beautifully Gaussian implantation range profile of 40 keV Gd in Al in 1956. But the technique did not become useful for materials analysis until more convenient silicon diode detectors were available, with the first paper by Georges Amsel on Si diode detectors in 1960 [19] and Turkevich's immediate proposal for the Surveyor Moon mission in 1961 [20] with the report in 1967 [21]. Explicit depth profiles were not published until 1970 [22].

2.1. *Energy loss*

It was obvious to all the early workers that the *energy loss* of scattered particles represented *depth* in the samples, and the famous *Bragg rule* (1905) [23] obtained the compound stopping power for a fast particle from a linear combination of elemental stopping powers.

To interpret IBA spectra it is in general essential to have energy loss ("stopping power") data for the whole periodic table and all the ion beams of interest. This is a massive task both of measurement and of evaluation against a theoretical model. The measurements are difficult to make and the model enables a valid comparison between different

sets and also extrapolation to materials or beams for which measurements are not available. Happily this has been done, with comprehensive stopping power databases now available from Jim Ziegler's SRIM website [24–26]. Helmut Paul has also recently reviewed this field with references to other compilations (H. Paul [27], MSTAR, ICRU...) [28].

The Bragg rule is an approximation that clearly implies that the inelastic energy loss of an energetic particle is largely due to inner-shell (strongly bound) electrons: otherwise there would be more noticeable chemical effects, which have long been observed (Bourland & Powers, 1971, studied alphas in gases [29]) but are not large. For example, Bragg's rule applies even for heavy ions in ZrO_2 [30] and TiO_2 [31], but ~5% deviations were measured for light ions in polyvinyl formal [32]. Up to 20% deviations can be seen in some cases, and these are discussed in detail in the SRIM 2010 paper [33].

In the following we will assume that the analyst has good stopping power values. However, it must be pointed out that these are basic analytical data, which are not easy to obtain accurately. Therefore any critical work must take into account the uncertainties deriving from the stopping power database. Both Paul's work cited above and the SRIM database give these uncertainties in considerable detail: as a rough indication for the reader, a stopping power value is unlikely to be known much better than about 4%.

2.2. *Rutherford backscattering spectrometry*

Rutherford reasoned that the scattering of positively charged alpha particles must be due to the Coulomb repulsion of an atomic (positively charged) nucleus. He therefore derived the simple relation for the differential scattering cross-section $d\sigma/d\Omega$:

$$d\sigma/d\Omega = \{Z_1 Z_2 e^2 \mathrm{cosec}^2(\theta/2)/4E\}^2 \qquad (1)$$

where $d\Omega$ is the solid angle at the detector, θ is the angle of scattering, E is the particle energy at scattering, Z_i are the atomic numbers of the projectile and target nuclei and e is the charge on the electron. This formula was verified in detail by Geiger & Marsden in 1913 [34].

For simplicity, Eq. (1) is written in the *centre-of-mass frame of reference* and therefore has no mass dependence: in the *laboratory frame* it is rather more complicated, usefully given in a power series by Marion & Young [35]:

$$d\sigma/d\Omega \cong \{Z_1 Z_2 e^2/4E\}^2 \{\sin^{-4}(\theta/2) - 2r^{-2} + \cdots\}$$

$$r \equiv M_2/M_1$$

$$(2)$$

where M_1 and M_2 are the masses of the incident and target nuclei respectively.

The Rutherford formula is derived from the Coulomb repulsion of two like charges assuming that the two colliding nuclei are bare point charges. The electron screening that must shield the charges from each other until the nuclei are in very close proximity is usually rather a small effect which was determined in adequate detail by Andersen *et al.* in 1980 [36]. It is this screening correction that relieves the singularity at $\theta = 0$ where Eq. (1) makes the cross-section infinite: with no scattering the nuclei are so far apart that the nuclear charge is screened by the electron shells, and the cross-section vanishes.

The scattering event itself must conserve energy and momentum, and thus for an elastic scattering event the *kinematics* give the split of the initial energy E_0 between the scattered and the recoiled nuclei:

$$E \equiv kE_0 \tag{3}$$

$$k_s = \{(\cos\theta \pm (r^2 - \sin^2\theta)^{1/2})/(1+r)\}^2 \tag{4}$$

$$k_r = (4r\cos^2\phi)/(1+r)^2 \tag{5}$$

where k is known as the *"kinematical factor"* given (in the laboratory frame) by Eq. (4) for the scattered particle (with a scattering angle of θ), and by Eq. (5) for the recoiled particle (with a recoil angle of ϕ). The scattering and recoil angles are measured relative to the incident beam direction. Equation (4) is double-valued if $r < 1$; for $r > 1$ the positive sign is taken. Thus, for a head-on collision with $r > 1$ (e.g. He RBS), $\theta = 180°$ and $k_s = \{(M_2 - M_1)/(M_2 + M_1)\}^2$. In the case $r < 1$ (ERD), there can be no scattering into angles $\theta > \sin^{-1} r$. Note that the kinematical factor is *not* a function of beam energy: RBS spectra look qualitatively similar for all beam energies.

We can now give an example of RBS analysis. Figure 1 shows the RBS spectrum from an anti-reflection coating on a glass substrate, where the coating is about a micron thick: the detailed analysis fits 19 alternating layers of zirconia and silica, where these molecules and the substrate are treated as three

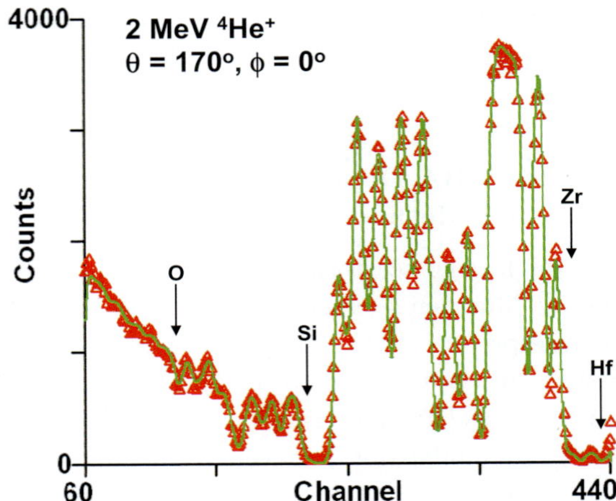

Fig. 1. Antireflection coating with alternate zirconia and silica layers on float glass. Normal incidence beam: the line through the points is the spectrum calculated for the fitted structure. The surface positions of Hf, Zr, Si, O are shown. Hf is a normal contaminant in Zr. (From Fig. 2 of Jeynes *et al.*, 2000 [37].)

logical elements for fitting purposes [37]. Notice that the fit (the line through the data points) is extremely good: this means that the model for the fitting is also very good, in fact, the layer thicknesses are determined with *sub-nm precision* even at the bottom of the coating (for details see the paper). We will discuss this exceptional precision below (Sec. 5.4).

2.3. Elastic (non-Rutherford) backscattering

As the beam energy is increased, Rutherford's approximation of point charges for the colliding nuclei fails, the Coulomb barrier is exceeded, and a proper quantum mechanical treatment of the interaction must be made. An estimate of the "actual Coulomb barrier" was made by Bozoian *et al.* using optical model calculations [38] (see Appendix 8 of both handbooks [3, 4]) but the optical model does not take into account specific features of nuclei (apart from the radius, A and Z) and the "actual Coulomb barrier" is not an identifiable potential useful for calculation [39]; this is emphasised by the calculations of the astrophysicists who calculate the probability of (p, γ) reactions at *stellar* temperatures ($\sim$10–30 keV!).

For example, the $^{12}C(p, \gamma)^{13}N$ reaction is critical to understanding stellar hydrogen burning in massive stars, initiating the CNO cycle. Figure 2

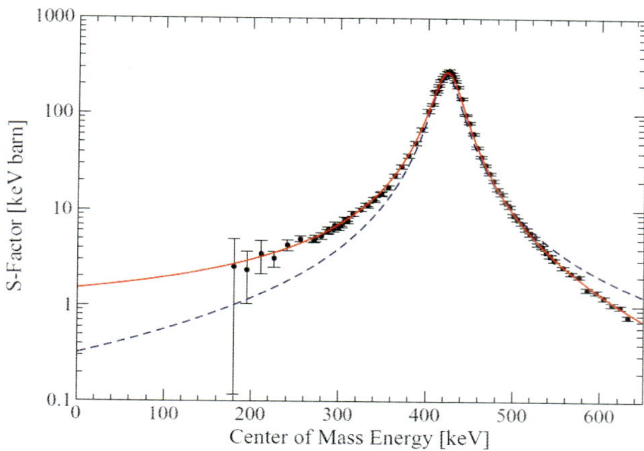

Fig. 2. "S-factor" calculated from the AZURE fit to the ^{12}C$(p, \gamma)^{13}$N data of J. Vogl (PhD Thesis, Caltech, 1963) (and using the ^{12}C$(p, p)^{12}$C data of Meyer *et al.*, 1976: see IBANDL at www-nds.iaea.org/ibandl). The red (solid) line indicates the best fit including external capture, which the blue (dashed) line neglects (reproduced from Fig. 3 of Azuma *et al.*, 2010 [40]). Note that the S-factor is significant right down to zero energy.

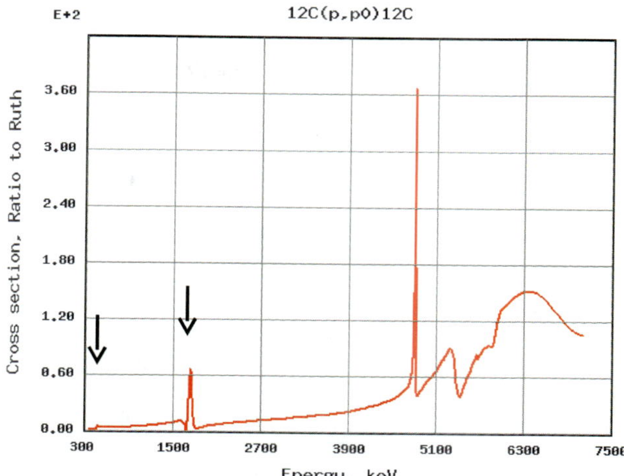

Fig. 3. Evaluated ^{12}C$(p, p)^{12}$C elastic scattering cross-section for $\theta = 180°$, relative to the Rutherford cross-section, by Gurbich and co-workers [41]. Note the resonances at 440 keV and 1734 keV (arrowed). Downloaded 21 July 2011 using the SigmaCalc calculator at www-nds.iaea.org/ibandl.

shows the "S-factor" for this reaction[e], near the resonance at 461 keV (centre of mass frame), which is also observable in the elastic scattering ^{12}C$(p, p)^{12}$C reaction channel at 440 keV (laboratory frame: see Fig. 3 [41]). This S-factor is calculated using the AZURE code of Azuma *et al.* [40]. The S-factor is the pre-factor in the expression for the cross-section which has an (approximately) exponential decrease with energy corresponding to the term for tunnelling through the Coulomb barrier. The point is that these cross-sections are dominated by the low energy tails of the resonances due to the nuclear structure.

These complicated elastic scattering cross-section functions can be calculated from nuclear models, using all available nuclear data (not only scattering cross-section data). We have already mentioned the AZURE code (see Fig. 2). The materials community, under the auspices of an IAEA CRP [42] has built the IBANDL website (www-nds.iaea.org/ibandl) which gathers together all the relevant cross-section measurements available [43]. IBANDL also gives access to the Sigma-Calc calculator (www-nds.iaea.org/sigmacalc) of Gurbich [44, 45] which is based on several codes in which various nuclear reaction models are implemented, the calculations being performed with individual sets

of parameters obtained through the evaluation procedure for each reaction considered. These nuclear models have been used to critically *evaluate* existing elastic scattering (and other) cross-section measurements, enabling nuclear parameters to be chosen such that the cross-section can be calculated for any scattering angle with a much smaller uncertainty than for any particular dataset.

As an example, Fig. 4 shows the strong angular dependence of the ^{14}N$(\alpha, \alpha)^{14}$N reaction [46]. It is clear that, were a nuclear model not available, the experimenter would have to rely on measured cross-sections only, and would be forced to either make measurements for the experimental geometry used, or set the scattering angle to match the existing measurements.

Figure 5 shows the evaluated elastic scattering cross-sections for protons on natural magnesium, relative to Rutherford, with a benchmark measurement shown in Fig. 6 [47]. Note the exceptionally strong and sharp resonance at 1483 keV. It is not trivial to calculate spectra which involve cross-sections with such sharp resonances, and special methods need to be used [48].

For both ^{12}C and ^{nat}Mg elastic scattering (p, p_0) cross-sections, optical model estimates of the

---

[e]The ordinate in Figs. 2, 4, 26 is in "barns" $\equiv 10^{-24}$ cm^2. The word "barn" (as in large farmyard building) is a joke of the nuclear physicists, and according to the Oxford English Dictionary first used by Holloway and Baker in 1942.

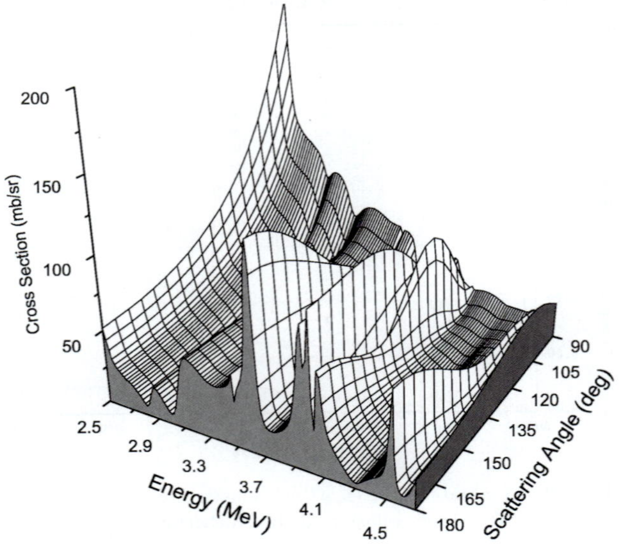

Fig. 4. Evaluated $^{14}N(\alpha, \alpha)^{14}N$ cross-sections as a function of scattering angle, reproduced from Fig. 4, Gurbich *et al.*, 2011 [46].

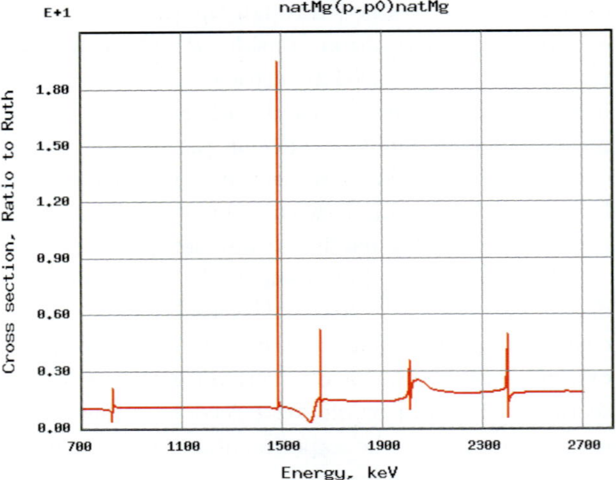

Fig. 5. SigmaCalc calculation of $^{nat}Mg(p, p)^{nat}Mg$ elastic scattering cross-section (relative to Rutherford) for $\theta = 180°$ [47]. Downloaded 21 July 2011 from www-nds.iaea.org/ibandl.

"actual Coulomb barrier" energy are wildly wrong, as expected [39], and as is clear from Table 1 which gathers available data together to estimate the minimum beam energy where the (p, p) cross-section differs significantly from Rutherford: the value of 4% deviation is chosen because it is not presently possible to specify the value at 1% (or even 2%) deviation with any confidence.

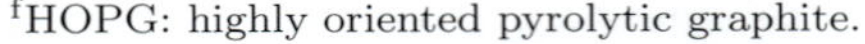

^{f}HOPG: highly oriented pyrolytic graphite.

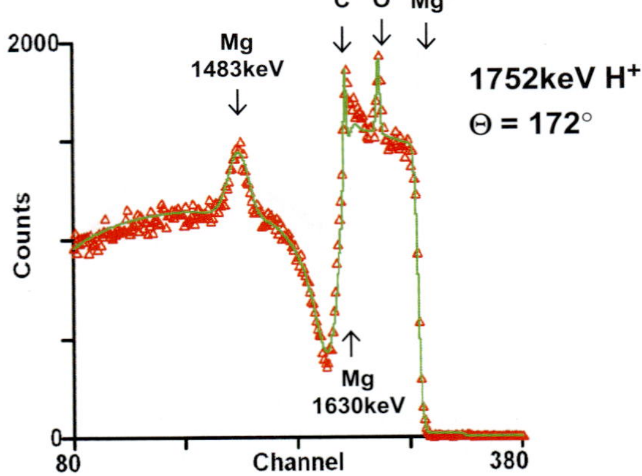

Fig. 6. Benchmark EBS measurement of bulk magnesium showing strong resonances at 1483 & 1630 keV (from Fig. 3 of Gurbich & Jeynes, 2007 [47]). An O peak from the surface oxide and a C peak from surface contamination are visible.

2.4. *Elastic recoil detection*

Equation (5) shows the kinematical factor for the recoil ion in the scattering event. In every elastic scattering event the kinematics requires that the target atom be recoiled with significant energy. If the target is thin, or the geometry is appropriate, such a recoil particle could escape the sample and be measured.

The earliest example of ERD, from the Montréal group in 1976, used a transmission target and a normally incident 35 MeV ^{35}Cl beam to detect Li recoiled from a LiF target [49], obtaining a depth resolution of $24\,\mu g/cm^2$ of copper. In 1998, Dollinger *et al.* [50] used a 60 MeV $^{127}I^{23+}$ beam on a thick HOPG^f sample tilted at $>85°$ to the beam and a recoil angle of $10°$, and energy analysed recoiled $^{12}C^{5+}$ ions with a magnetic spectrograph. They were able to distinguish the first four or five atomic layers, but noticed that the HOPG damaged rapidly under the beam with a sputter yield of 500/ion. This analysis gave a depth resolution of $73\,ng/cm^2$ of C. There are many subtleties of this complex measurement, which have been analysed in detail by Szilágyi [51]. One good reason for the growing interest in much lower energy beams for ERD is specifically that they are much less damaging, and the prospects for low energy ERD have been reviewed by Döbeli *et al.* (2005) [52].

Because ERD typically (with thick targets) uses glancing beam incidence and glancing recoil exit

Table 1. RBS/EBS boundary for protons: where the cross-section differs from Rutherford by 4%.

Z	A	Estimate 4% deviation		1st R keV	Minimum database energy (May 2011)			Comment
		OM keV	Data keV		σ calc keV	Data keV	σ (rtr)	
2	He	4	250				1600	90	(No data available)
3	Li	6	350	600	1400		500	1.01	Data at 156.7°
3	Li	7	343	407	440		373	1	Data at 81.3°
4	Be	9	444	219	257		143	1	Data at 170°
5	B	10	550	<500			500	1.24	Data at 170°
5	B	11	545	<500	525		500	1.35	Data at 170°
6	C	12	650	<360	440	360		1.04	SigmaCalc at 180°
7	N	14	750	400	1003	300		1.01	SigmaCalc at 180°
7	N	15	747		415				$p, \alpha\gamma$ resonance, no EBS data
8	O	16	850	<100	280	100		0.95	SigmaCalc at 180°
8	O	18	844	< 590(?)	620(?)		590	0.88	Data at 138.7°: strong resonance at 800 keV
9	F	19	947	552	623	552		0.95	SigmaCalc at 180°
10	Ne	20	1050	<1500	1160	1500		1.28	SigmaCalc at 180°: strong p, γ resonance at 1160 keV
10	Ne	22	1045	≪1574	1588		1574	0.1	Data at 150°: clear p, p_0 resonance at 1700 keV; strong p, p_1 resonance at 2130 keV
11	Na	23	1148	573	593	573		0.96	SigmaCalc at 180°
12	Mg	24	1250	700	821	700	400	1	SigmaCalc at 180°
13	Al	27	1348	936	937	900	500	1	SigmaCalc at 180°
14	Si	28	1450	1272	1608	1000		0.99	SigmaCalc at 180°
14	Si	29	1448	1219	1333		1219		Relative data at 140°
14	Si	30	1447	<1007	1018		1007	0.87	Data at 140°, strong resonance at 1290 keV
15	P	31	1548	1175	1251		1000	1	SigmaCalc at 180°
16	S	32	1650	1790	1889	1506		1	SigmaCalc at 180°
17	Cl	35	1749	1326	2100		1320	1.02	Data at 150°
17	Cl	37	1746	1338	1338		1130	1	Data at 172°
18	Ar	40	1845	1810	1858	1723	1020	1	SigmaCalc at 180°
19	K	39	1949	<1731	1770	1731		0.95	SigmaCalc at 180°
20	Ca	40	2050	2159	2383	1799		0.99	SigmaCalc at 180°
22	Ti	48	2246	1901	1901	1811		1	SigmaCalc at 180°
24	Cr	52	2446	<2600	2600	3200	2400	0.57	SigmaCalc at 180°; relative data at 141° show strong resonance at 2600 keV
27	Fe	56	2850	2050	2051	2011		0.993	SigmaCalc at 180°

The optical model ("OM") calculations of Bozoian [3, 4, 38] are compared with "Data" from either IBANDL (www-nds.iaea.org/ibandl) or SigmaCalc (www-nds.iaea.org/sigmacalc) calculations of Gurbich [44, 45] for a head-on collision. The position of the first resonance ("1st R") in the (p, p) elastic cross-section is shown, and the minimum energy for which data are available is also shown, distinguishing between SigmaCalc ("σ calc ") and measured data ("Data"). The value of the (p, p) cross-section relative to Rutherford at this minimum value ("σ (rtr)") is also shown. The "optical model" calculation is approximated by $Z_2 (M_1 + M_2)/(10M_2)$ [4].

angles, the resultant energy spectra have a significant component of plural scattering events. These will be considered in more detail in Sec. 5, but simple single scattering calculations that have dominated the interpretation of IBA data last century are not accurate enough to account for the detail of most ERD spectra. Figure 7 shows ERD from a study comparing the best depth resolution available from a ToF-ERD[g] system using a 6 MeV ^{35}Cl beam, with high resolution SIMS and high resolution RBS (using a magnetic spectrometer) [53]. With ToF-ERD a low energy beam is preferable since then the flight time is long compared to the time resolution of the system. This 2006 study highlights the need to interpret ERD spectra more closely, taking plural scattering properly into account. It shows that the apparent resolution of the N signal does not behave simply since it does not look significantly worse for a buried

---

[g]ToF: "time of flight": in this case the "spectrometer" has a flight tube (the "telescope") with a foil at the entrance that the particle must pass through, thus giving a "start" signal, and a detector at the end which gives a "stop" signal.

layer. However, a depth resolution $<1\,\mathrm{nm}$ is readily available near the surface.

Detectors for ERD pose interesting problems. The ubiquitous Si diode detectors work, but their energy resolution gets much worse for heavy recoil ions. Unfortunately, they are also susceptible to radiation damage, so for heavy ions they do not work for long! Si detectors have the great virtue of simplicity, but they have to be used with a so-called "range foil" to range out the forward scattered (relatively heavy) beam, and therefore simple range foil ERD has two major disadvantages: the loss of energy resolution due to the energy straggle and thickness

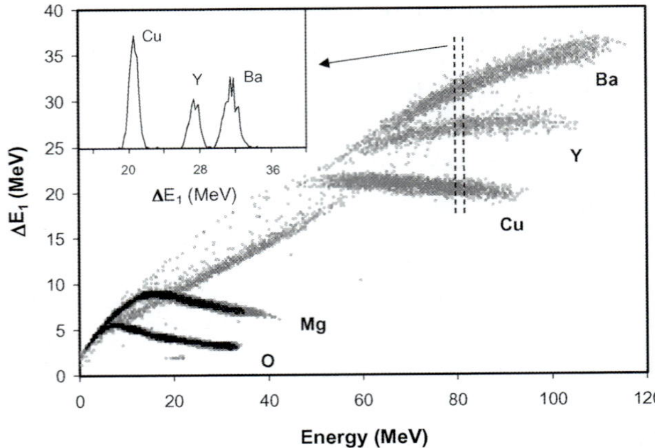

Fig. 8. $241\,\mathrm{MeV}\ ^{197}\mathrm{Au}$ ERD of YBCO/MgO using a gas ionisation (E-ΔE) detector. The scattered Au signal is kinematically forbidden. Reproduced from Fig. 3 of Timmers *et al.*, 2000 [61].

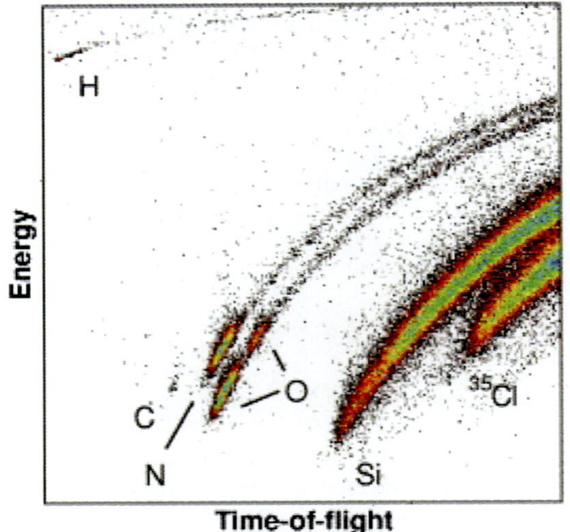

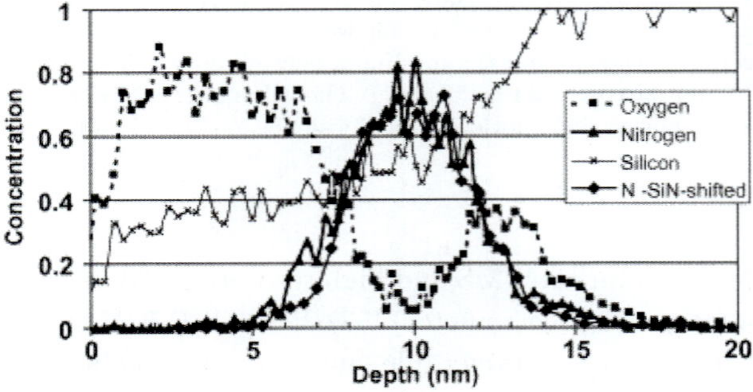

Fig. 7. $6\,\mathrm{MeV}\ ^{35}\mathrm{Cl}$ ToF-ERD analysis of a $15\,\mathrm{nm}$ 3-layer oxide/nitride stack ($\mathrm{SiO_2}$ $8\,\mathrm{nm}/\mathrm{Si_3N_4}$ $5\,\mathrm{nm}/\mathrm{SiO_2}$ $2\,\mathrm{nm}/\mathrm{Si}$ substrate). The analysis is repeated with the surface oxide etched off. *Above*: Time of flight spectra with $38.2°$ detector angle and $3.6°$ exit (beam take-off) angle. Both scattered $^{35}\mathrm{Cl}$ and recoiled Si, O, N, C and H can be seen. *Below*: Reconstruction of depth profile for unetched and etched samples. The N profile for the etched sample is marked "shifted" and aligns with original N profile. Reproduced from Figs. 1 and 2 of Brijs *et al.* (2006) [53].

inhomogeneity of the range foil [54], and the intrinsic indistinguishability of the signals for the various different recoiled target elements. Contrast this with both Figs. 7 and 8 where the detector distinguishes the different elements.

Nevertheless, for many purposes, especially H-isotope analysis (for which a He or Li beam is ideal: following a proposal in 2000 [55] there were two round robin reports in 2004 [56, 57]), range foil ERD with a Si detector remains very convenient and powerful. It is worth underlining the value of He-ERD for profiling H isotopes. Hydrogen is difficult to analyse with other methods, and it is a very important element in materials science. He-ERD was developed [58] and optimised [59] very early for H-profiling; it can very easily be used self-consistently with RBS to solve complex materials problems such as the formation of a-GaN by implantation of Ga into SiN_x:H [60].

ToF-ERD does not have the same problems, but (as with every telescope detector) the throughput is limited since there should only be one ion in the detector at a time; of course the longer the detector, the better the time resolution — equivalent to energy resolution. Gas ionisation detectors have a very long history, and have also been highly developed for ERD. They have the huge advantage that they are completely impervious to beam damage, since the gas can be changed continuously. Figure 8 shows the $E - \Delta E$ signal for such a detector using $241\,\mathrm{MeV}$ $^{197}\mathrm{Au}$ ERD of a superconducting $\mathrm{YBa_2Cu_3O_{7-\delta}}$ film on a MgO substrate [61]. The scattered Au signal is forbidden since for this detector angle the solution of Eq. (4) is imaginary.

The energy resolution of gas ionisation detectors has always been limited by the entrance window, which needs to be strong enough to withstand the (usually) atmospheric pressure of the gas. Recently, ultra-thin silicon nitride windows have been introduced, together with dramatic simplifications in the design which have been demonstrated to be nearly as good as more complex designs for low energy ERD [62]. These detectors can also be used for heavy-ion RBS, and have an energy resolution better than that of silicon for beams heavier than He. For He the resolution is comparable. Not only can these detectors support a very high count rate, they are also insensitive to light making them usable at high temperatures.

Finally, we should mention the ultimate sensitivity for hydrogen of $<5.10^{16}$ H/cc in polycrystalline diamond, obtained at the 15 MV superconducting tandem SNAKE[h] facility in München [63]. This was on diamond slices 55 μm thick using a 17 MeV H$^+$ microbeam and detecting H forward recoils and H forward scatters in coincidence. In a 3-D analysis they were able to demonstrate that H was located (in very low concentrations!) at grain boundaries. Note that the forward recoils are indistinguishable from the forward scatters, and this gives the Mott quantum interference term to the scattering cross-section function [12].

2.5. Nuclear reaction analysis

Ernest Rutherford first observed nuclear reactions, using the 4.87 MeV α particle from ^{226}Ra on nitrogen gas: the ^{14}N$(\alpha, p)^{17}$O reaction has a Q-value of -1.19 MeV, so fast protons were visible [64] [65]. Later, Cockcroft and Walton were the first to use an accelerator for nuclear reactions, demonstrating the ^{7}Li$(p, \alpha)^{4}$He reaction [66] (for which the Q-value is 17.347 MeV) which has a non-zero cross-section down to very low energies (0.27 mb/sr at 430 keV

[67]). The so-called "Q-values" of these reactions are determined by the mass differences, and can readily be calculated [68] using Einstein's $E = mc^2$ relation.[i]

Figure 9 shows an interesting example of the use of NRA methods to determine growth mechanisms. In this case the ^{18}O$(p, \alpha)^{15}$N reaction was used near the extremely sharp resonance at 151 keV [69]. Note that sharp resonances allow very high depth resolution near the surface. The beam energy is varied to collect an "excitation curve": this puts the

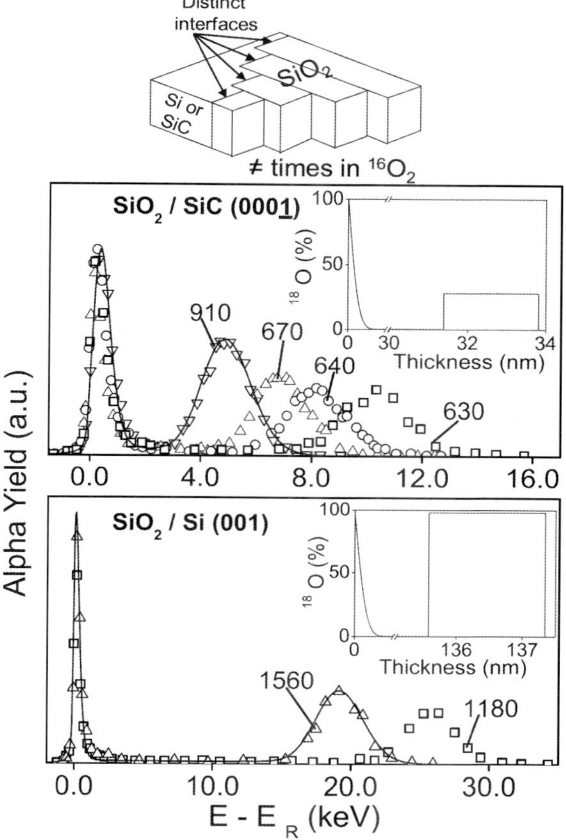

Fig. 9. Determination of oxide growth mechanism by NRA. Narrow resonance nuclear reaction profiling using the ^{18}O$(p, \alpha)^{15}$N resonance at 151 keV of SiO$_2$ films on SiC and Si. Oxidation was started in ^{16}O and then completed in ^{18}O. The ^{18}O migrates to the surface and the interface during oxidation. The numbers on the figure show the areas under the peaks. Reproduced from Vickridge *et al.*, Fig. 2 [69].

---

[h]SNAKE: Superconducting Nanoscope for Applied nuclear (KErn-) physics. This pp coincidence technique has to take account of the quantum mechanical indistinguishability of the two protons: the interference effects significantly change the differential scattering cross-sections (see Mott, 1930 [12])!

[i]The notation (for example) ^{18}O$(p, \alpha)^{15}$N means that a proton ("*p*") beam is used; it strikes target atoms ^{18}O; the reaction annihilates the proton which combines with the ^{18}O to form ^{19}F in an excited state; the excited ^{19}F relaxes by splitting into an alpha particle and ^{15}N, together with 3.98 MeV of energy (the "Q-value") which is split between the resultants using kinematics (conservation of energy and momentum). Thus, the alpha comes off very fast and can readily be recognised and counted.

resonance at various depths in the sample given by the energy loss, and the depth profile can then be reconstructed.

A similar analytical problem involved the tribology (wear resistance) of low-carbon steels, for which nitrogen ion implantation is used. To profile ^{15}N implanted into spherulitic cast iron (with heat treatment) the $^{15}N(p, \alpha\gamma)^{12}C$ reaction at the strong 898 keV resonance was used [70]. In this case the γ-rays were counted. This material is inhomogeneous both laterally and in depth, being full of graphite nodules. The analysis used a microbeam to profile nitrogen separately in the nodules and in the iron matrix (awkward, since because of the large chromatic aberration of the focussing lenses, every energy change required the beam to be refocussed). This showed that although N is mobile in Fe, it is relatively immobile in C.

There are many nuclear reactions that can be used. Both particle and photon reaction products are available, and cross-sections for these reactions can be calculated using optical model or R-matrix methods. Some evaluated (d, p) and (d, α) reactions can already be found in IBANDL, and some (p, γ) reactions have also been analysed. Of course, measured cross-sections can be used for analytical purposes, or reference standards can be used for relative measurements. The IAEA has established a new CRP to create a database for PIGE [71] as it has already done for EBS [42].

3. Other IBA Methods

Ion beam methods are not only used for thin film depth profiling, and depth profiling itself comes in various flavours. In this section we will (too briefly, alas) review a variety of other major techniques that should not be ignored.

Methods are available to investigate various aspects of crystals including: lattice site location and damage, and the true structure of the surface (using ion channelling, see Secs. 3.1 and 3.3); and electrically active defects (using ion beam induced charge measurements, see Sec. 3.5).

Low energy beams (requiring completely different detectors) can be used for sensitivity to the true surface (LEIS, see Sec. 3.2), and a sensitivity to the first few atomic layers (MEIS, see Sec. 3.3). These low energy RBS techniques can also be used for ultrahigh resolution thin film depth profiling.

Depth profiling itself is regularly done today with a low energy sputtering ion beam (SIMS, see Sec. 3.4). We briefly review this important IBA topic even though the main focus of the present work is MeV-IBA. SIMS is important, not only because it is a standard and widespread technique but because MeV-SIMS is a very interesting new technique giving molecular information that may prove as powerful as MALDI.

3.1. *Crystalline damage and impurities*

If an energetic ion beam is aligned with a major axis of a crystal it will "see" a significantly lower stopping power in the so-called channels between the strings of atoms. Moreover, a slightly misaligned beam will experience a *focussing effect* which tends to keep it in the crystal channel. It is this focussing effect that is called **channelling**. Stark and Wendt suggested the existence of this effect in 1912 [72], but Davies, Amsel and Mayer tell us [18] that it was first observed by Bredov and Okuneva in 1957 [73], but they did not recognise it, misinterpreting their data. It was in 1963 that the first channelling measurements were published by Davies and co-workers [74], who observed the exponential channelling tails of the implant profiles of 210 keV ^{222}Rn in polycrystalline Al by using the alphas emitted by the implanted radon.

Robinson and Oen's Monte Carlo calculations of the channelling effect also published in 1963 [76] showed that the exponential implant tails were expected. Figure 10 very clearly demonstrates the focussing effect of the channels in a crystal for low energy (1 keV) incident Cu atoms. In non-aligned directions, such an energetic beam would normally have a range of 1.2 nm, but the trajectories shown in this figure have a range of about 30 nm. Knowledge of the energy loss of ion beams channelling in single crystals is important when the depth distribution of impurities and defects is of interest, and this was explicitly considered by Kótai in 1996 [77].

Lindhard published his complete theoretical treatise on channelling in 1965 [78] but it was in 1962 that channelling experiments were first done using all three major IBA techniques: NRA, RBS and PIXE. Blocking was also demonstrated in 1962 to validate Lindhard's reversibility theorem between channelling and blocking. Channelling found an immediate use

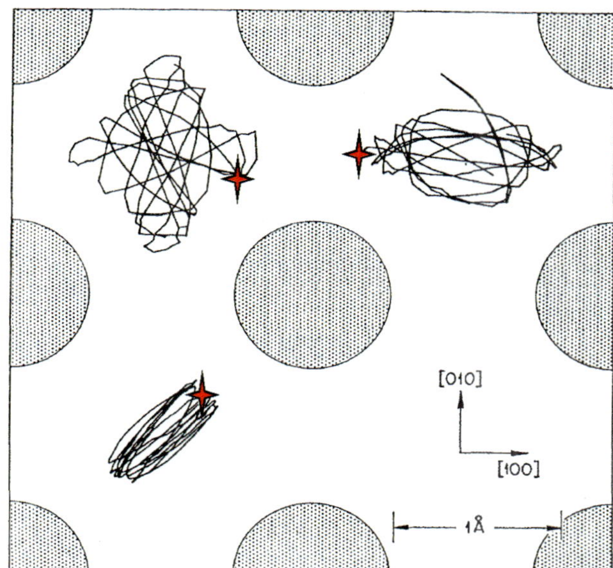

Fig. 10. Projection of three calculated trajectories of 1 keV Cu incident (at points marked with stars) along the $\langle 001 \rangle$ direction of a face centred cubic Cu crystal. The trajectories for 250 collisions are shown and the truncated Bohr potential is used. Reproduced from Fig. 2 of Robinson & Oen [76].

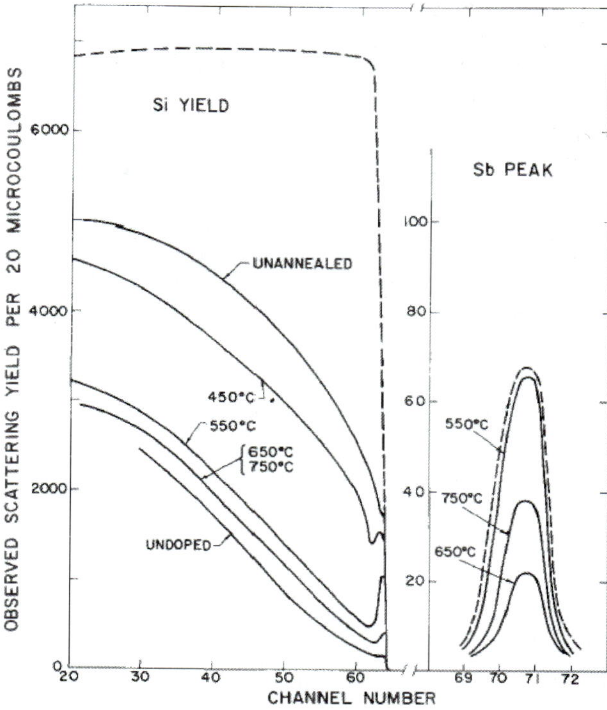

Fig. 11. 1 MeV proton RBS of $10^{15} \mathrm{Sb/cm^2}$ implanted at 40 keV into silicon and annealed at various temperatures. Channelling (in $\langle 111 \rangle$ direction) and non-aligned spectra are shown. Reproduced from Fig. 1 of Eriksson, Davies, Mayer *et al.* [75].

in the investigation of semiconductor doping by ion implantation and annealing. Figure 11 shows an early measurement of this annealing process.

There is now a very large literature on channelling (see Handbooks [2–4]). As an indication of the range of possibilities we mention: channelling to determine the defect profiles and defect type in a study of implanted CdTe [79] with quantitation by the code DICADA (see [80] for recent development of this code); also lattice site location of Ni (at low concentration <5 mg/kg) in diamond by PIXE channelling [81]. Monte Carlo codes for channelling include FLUX [82] and CASSIS [83]. RBX is an analytical general purpose IBA code that also able to calculate defect depth profiles and simulate channelled spectra [84].

Figure 12 shows an image of the spatial distribution of crystalline damage in silicon turned on an ultra-stiff lathe by a single-point diamond tool, where a scanning microbeam is used to form the image [85]. In this case channelling conditions were maintained over the scan area by rocking the beam about the principal axis of the Russian quadruplet focussing magnet using a dog-leg electrostatic scanner before the magnet. Different depths of cut

can be seen on the image: in this work amorphous layers up to 350 nm thick are observed, together with dislocation arrays of about $5.10^{10}/\mathrm{cm^2}$ under the amorphous region (quantified with the DICADA code). This is an example of *plastic* deformation in a *brittle* material.

3.2. *Catalysis and related studies (LEIS)*

RBS spectra can still be obtained for very low energy ion beams of a few keV; this is now called "low energy ion scattering". LEIS was first shown to be a technique selectively sensitive to the true surface (the outermost atomic layer) in 1971 [86]. In this sort of analysis one must use electrostatic detectors, which rely on ion focussing in electric fields, instead of semiconductor diode detectors sensitive to energy loss. Historically, rather simple detectors were used which had a low effective solid angle: thus counting times were relatively long and beam damage severely limited the usefulness of the technique. But in 1992, an efficient electrostatic detector called

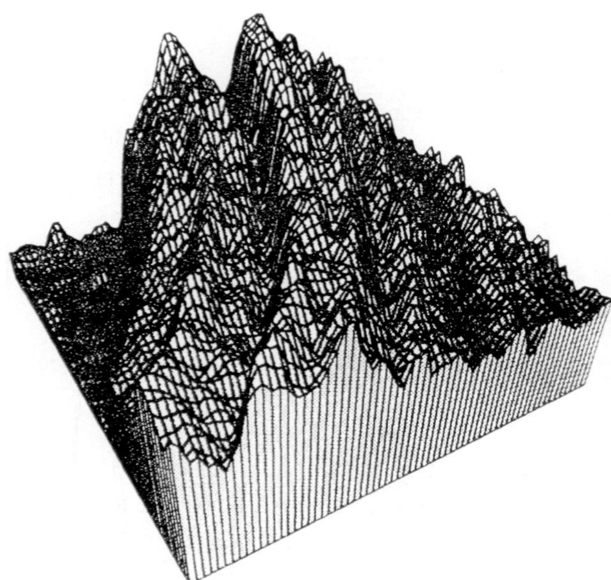

Fig. 12. Image (about 3 mm square) of circular damage tracks in a silicon sample turned on a lathe. 1.2 MeV ^{4}He$^+$ channelled normal to the sample and focussed to a 40 μm spot. Signal is the low-energy part of the RBS Si signal: high yield means high damage. 128 × 128 pixels, 0.15 nC/pixel, pixel area (23.4 μm)^2. Reproduced from Fig. 4 of Jeynes *et al.*, 1996 [85].

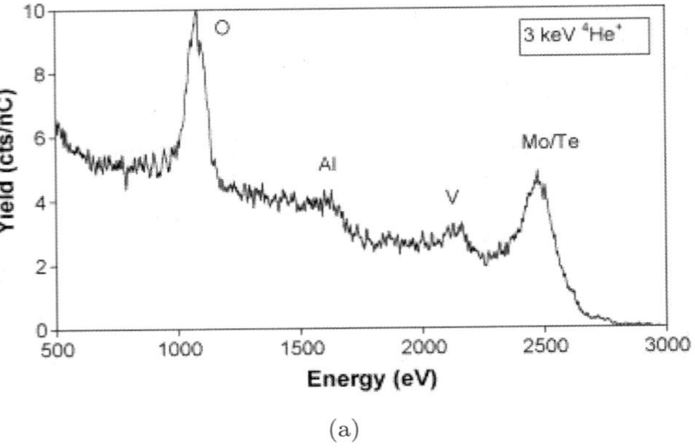

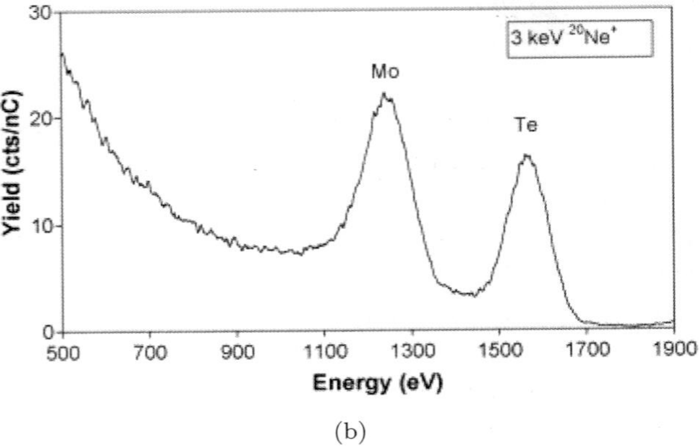

Fig. 13. LEIS of a mixed oxide catalyst using a beam of 3 keV He$^+$ (*above*) and 3 keV Ne$^+$ (*below*). The well-defined peaks are from single scattering events at the outer atomic layer. The heavier Ne$^+$ beam has mass resolution for Mo/Te but is not sensitive to the light O. (Reproduced from Fig. 2 of ter Veen *et al.* [89]).

EARISS[j] was reported with orders of magnitude better sensitivity [87, 88]. This has enabled the rapid development of a technique very powerful in applications which require knowledge of the true surface (the outer atomic layer).

Figure 13 shows typical LEIS spectra, using the high sensitivity detector, from a simple recent overview aimed at catalysis applications [89]. The well defined peaks have an area proportional to the number of scattering centres in the outermost atomic layers, and the collection times for these spectra was only a few minutes, giving minimal beam damage (sputtering only <1% of the outermost monolayer).

The low energy LEIS signal in Fig. 13 is due to more complex scattering events. LEIS necessarily uses an *electrostatic* detector, which means that the detected species must be an *ion*. Therefore the details of the neutralisation and re-ionisation of the scattered particles are essential to the technique. These have recently been comprehensively reviewed recently by Brongersma *et al.* [90]. It turns out that the observed fact that there usually appears to be little or no matrix effect on the binary collision peak area (for a comprehensive list of literature on this see Table 9.1 of [90]) is rather surprising and is still not well understood. Nevertheless, the behaviour of the LEIS signal *is* well-understood in a wide variety of applications. Noble gas primary ion beams are used because they are readily neutralised in sub-surface interactions, and hence give well-defined binary scattering peaks.

3.3. *Surface reconstruction studies* (*MEIS*)

The FOM group in Amsterdam developed medium energy ion scattering (MEIS) in the 1970s, as a development of the channelling science worked out in the 1960s [91]. They realised that the details of the scattering would give valuable information about surface relaxation effects in crystals. In particular, LEED

---

[j]EARISS: energy and angle resolved ISS.

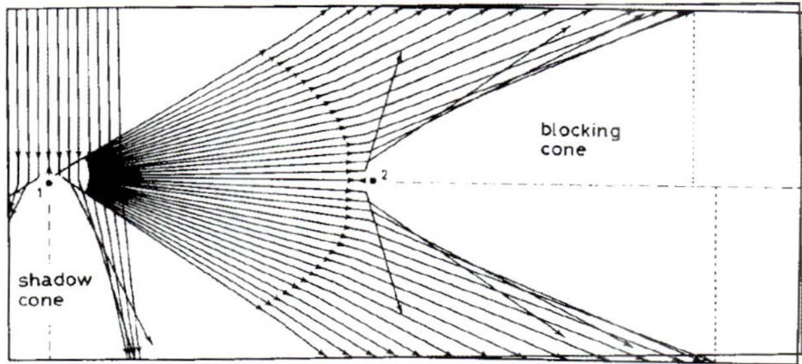

Fig. 14. Shadow and blocking cones typical of 200 keV proton MEIS. Reproduced from Fig. 19 of Turkenburg *et al.*, 1976 [91].

data for Cu, Ni and Al crystals were ambiguous and appeared to contradict theory, and another experimental probe was sought.

Figure 14 shows the shadow cone for an incident beam entering from the top (left) and passing close to a scattering centre (#1). The scattered beam will in its turn be blocked by a second scattering centre (middle; #2). The blocking and shadow cones are different phenomenologically — it is the blocked beam that is detected, and therefore the detected beam will have low yields in the blocked directions. There is also a real difference, since the shadow cone is for a parallel beam, where the blocking cone is for a beam scattered from a scattering centre that is close. But both effects rely for their detection on atomic coordination given by the crystal symmetry. The authors made it clear, for example, that there will be yield from the atoms near the surface which, relative to the beam, shadow the coordinated strings of atoms behind them, whereas no yield is expected from the shadowed strings. But the blocking effects give valuable information about correlations in the surface as well.

Figure 15 shows a recent example of a joint MEIS/LEIS study of the position of Sn atoms on the surface of Cu crystals, and the rearrangement of the Cu atoms that this implies [92]. Both MEIS and LEIS (CAICISS) are needed to resolve this complex structure; and AES with LEED are used in both instruments as monitors of, respectively, surface cleanliness and Sn deposition thickness.

3.4. *Sputtering methods (SIMS)*

Secondary ion mass spectrometry (SIMS) uses energetic ions to sputter the surface material from a sample. The ejected material is collected with the aid

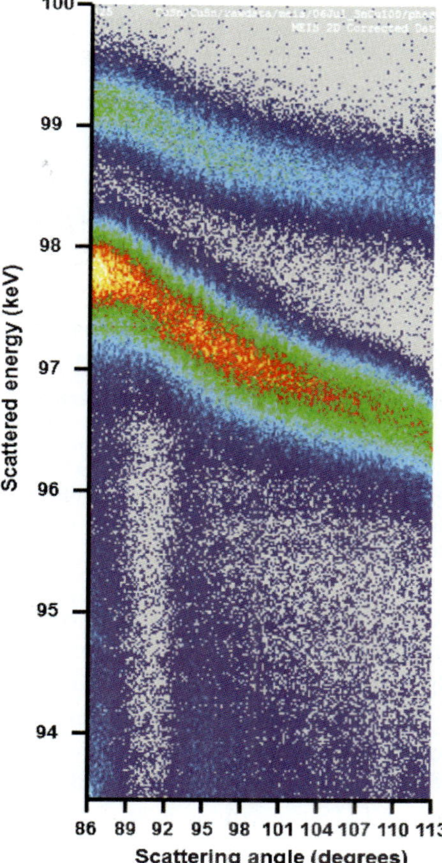

Fig. 15. MEIS of 100 keV He incident in the ⟨112⟩ direction on a (100) surface of a Cu crystal with sub-monolayer Sn coverage. The surface signals of the Cu and the Sn overlayer are marked, as is the blocking dip for 90° scattering. Reproduced from Fig. 3 of Walker *et al.* [92].

of a mass spectrometer to determine the constituent make-up of the sample.

Figure 16 sketches a SIMS tool: this technique has been well established for three decades [93] but is still under vigorous development. In particular, high throughput ToF instruments [94] are becoming available with a continuous primary beam and a pulsed spectrometer [95]. Other recent work includes complementary SIMS/LEIS to effectively follow the operation of electrochemical systems (e.g.: fuel cells!) by isotope exchange depth profiling [96], ultra-low energy SIMS for reliable depth profiling close to the surface (in this case for the tarnishing of silver coating layers on cultural heritage samples) [97], and high resolution EELS analysis of an adhesion problem to complement previous (inconclusive) ToF-SIMS work [98].

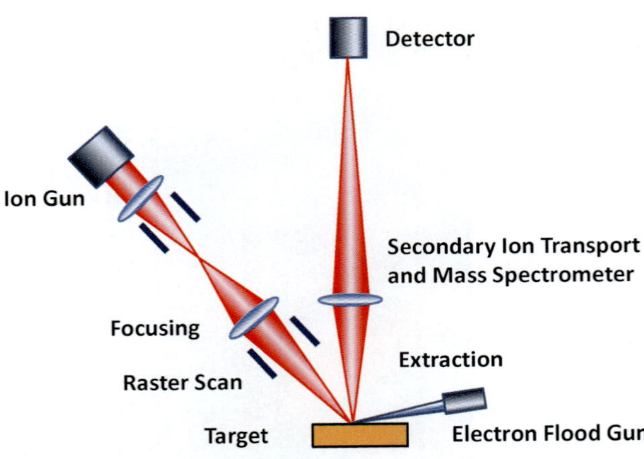

Fig. 16.　Simple schematic of a SIMS system. The scanning ion beam (typically 250 eV–40 keV O$^-$ or Cs$^+$) sputters the target. Sputtered ions are extracted into the MS (quadrupole or magnetic sector or ToF are used). The electron flood gun is used for charge compensation of insulating samples.

Static SIMS is a surface-sensitive technique where the primary ion fluence is kept below the *static limit*; the fluence at which less than 10% of the surface atoms have been displaced by the primary ion beam, typically less than about 10^{12} ions/cm^2, ensuring that the signal is from undamaged material. In ***Dynamic SIMS*** the sample is eroded by the sputtering beam and the mass spectra are recorded as a function of time, giving the elemental depth profile. Some molecular depth profiling is also available since sputtered ions show a molecular fragmentation pattern characteristic of the target. Some SIMS systems have been developed with both high flux (dynamic) and low flux (static) ion guns.

SIMS is not usually a fully quantitative analytical technique because the ionisation probability of the sputtered particles can vary by *orders of magnitude*, depending on the details of the surfaces. Moreover, not only are sputter rates for dynamic SIMS strongly composition-dependent, but also there are a variety of sputtering artefacts that can significantly modify the composition of the instantaneous surface. These problems are all particularly acute at interfaces. However, the SIMS community has succeeded in full quantification (with reference to certified standards) for many important materials systems.

A number of variations of the SIMS technique are in use to increase the fraction of secondary ions or to analyze all of the sputtered particles. Matrix Assisted SIMS uses a matrix (as MALDI — see below) to substantially increase the sputtered

ion fraction. Meta-SIMS uses metallic nanoparticles deposited on the surface to again enhance the yield of ionized particles emitted. Secondary neutral mass spectrometry (SNMS) uses a post-ionization chamber to ionize the neutrals sputtered from the surface. There are many more variants and combinations which are not really the focus of this article and the interested reader is directed to the various review articles and conference proceedings available: see for example the excellent old review of depth-profiling SIMS from 1982 [99], a recent review of depth-profiling SIMS using molecular beams [100], and a recent textbook [101].

Although SIMS is not a traditional MeV-IBA technique, MeV heavy ions were used in the 1970s to sputter molecules from insulating materials in a technique known as PDMS [102]. PDMS instruments were made using the fission fragments generated from a ^{252}Cf as a source of heavy ions, but it was found that a laser could effectively substitute for the radioactive source. The resulting technique, MALDI, has proved very successful especially for biomolecules [103]. Recently it has been demonstrated that SIMS with an MeV primary ion beam can be used to generate molecular concentration maps on surfaces [104], this is the first use of "MeV-SIMS". Employing relatively light ions such as ^{16}O at 10 MeV it has also been shown that MeV-SIMS, RBS, and PIXE spectra can all be recorded simultaneously with the same primary ion beam [105].

Recent work [106] has demonstrated that the damage cross-section of a sample of leucine (an essential amino acid) deposited on silicon caused by a 10 MeV ^{16}O^{4+} is comparable to that of conventional SIMS [107] using keV Ar$^+$ ions — 1.3×10^{-14} cm^2. This is shown in Fig. 17 where the logarithm of the leucine signal is plotted as a function of time. The signal decays as the molecular ion is fragmented by the primary ion beam.

The advantage of using low mass primary ions (e.g. 6 MeV O) is that it is possible also to obtain simultaneous RBS and PIXE spectra and images with the same beam and hence at the same time (see Fig. 5 of [105] for example). Experiments using higher mass primary ions (Cu) show substantially higher molecular yields and much lower fragmentation than has been achieved using low energy cluster ions. The relative yield of a fragment peak (at $m/z = 147$ Da) to the principal peak (at $m/z = 1047$ Da) for

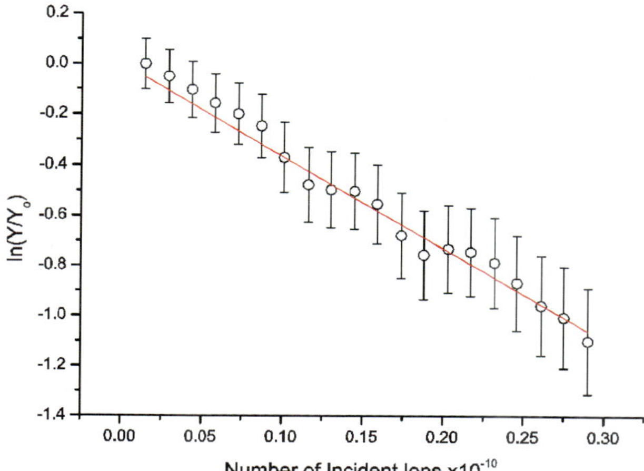

Fig. 17. The natural log of the normalized yield of the protonated leucine ions (132 amu) plotted as a function of the total number of incident ions ($10\,\mathrm{MeV}$ $^{16}\mathrm{O}^{4+}$). Reproduced from Fig. 1 of Jones *et al.*, 2011 [106].

an angiotensin II sample using a $6\,\mathrm{MeV}$ Cu^{4+} ion has been found to be up to 100 times that obtained using a $30\,\mathrm{keV}$ Bi_3^+ cluster (and 40,000 times greater than for $30\,\mathrm{keV}$ Bi^+) [108].

Simulation and modelling suggest that the process by which large molecular ions are ejected from the surface of molecular materials is similar for both MeV ion and keV cluster ion impacts (for the latter, see [109]). In Fig. 18, the volume of ejected and fragmented molecules is shown as predicted from molecular dynamics (MD) computer simulation for impacts on a frozen benzene (C_6H_6) target with keV cluster (C_{60}) and representative MeV ions. As can be seen, the ejection of material is from a conical region in both cases and the fragmentation occurs in a

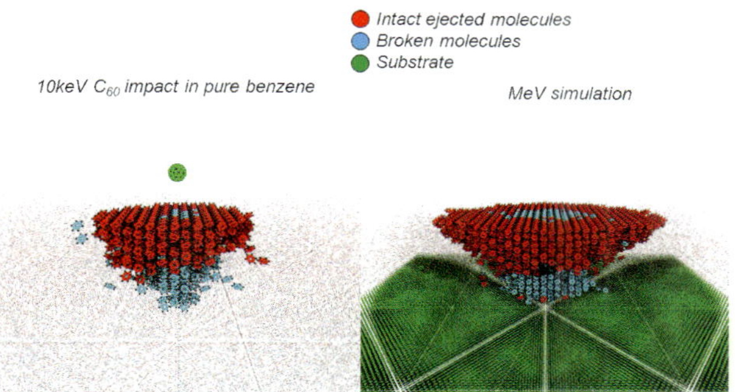

Fig. 18. Molecular dynamics simulations showing volume of ejected and fragmented molecules after $10\,\mathrm{keV}$ fullerene impacts in solid benzene, and after the energy loss expected from $6\,\mathrm{MeV}$ O impacts into 10 benzene layers on a silver substrate.

cylindrical region about the path of the impinging ion. The significant difference between the two results is that the length of the fragmentation cylinder is much longer for the MeV ions and the passage of the MeV ion causes substantial break-up of the molecular target over most of the path length of the ion. The path length of the keV cluster is relatively short, with the fragmentation cylinder also being short and barely extending below the ejection cone.

The implication of this is that depth profiling of molecular solids using keV cluster ions is feasible, with the possibility of maintaining reasonable signals from the principal sputtered molecular ion over large depths into the material. With MeV ions however, the molecular material will fragment at large depths and only lower mass fragments of the target should be observable as the sputtering proceeds. This has been borne out in experimental work using this technique in which molecular images of doped fingerprints on paper over and under ink were analyzed [110]. Figure 19 shows the fragment signals of the molecular material representative of ink and fingerprint as they were followed to obtain depth profile information.

One of the great potentials of MeV-SIMS is that it can in principle be used in air, as MALDI is already. This is by no means simple but measurements have already been achieved at base pressures up to $\sim 100\,\mathrm{Pa}$ ($\sim 1\,\mathrm{mbar}$) [111]. The main difficulty is that of maintaining a high spatial resolution through air of such high mass ions; this means that the sample must be placed very close to the exit nozzle of the MeV-ion beamline, and then there is very little space left for the spectrometer.

3.5. *Electrical characterisation (IBIC)*

Ion beams striking a semiconductor will deposit electron-hole pairs all down the ion track. If these charge carriers experience an electric field then their subsequent drift results in a current and hence a signal on the device output terminals, similar to the operation of conventional semiconductor particle or photon detectors. In this case single ion impacts are detectable, and only very low beam currents are needed. In any case, higher beam currents would introduce noticeable damage into the detector! An application of IBIC (ion-beam-induced charge) was first reported in 1992 [112].

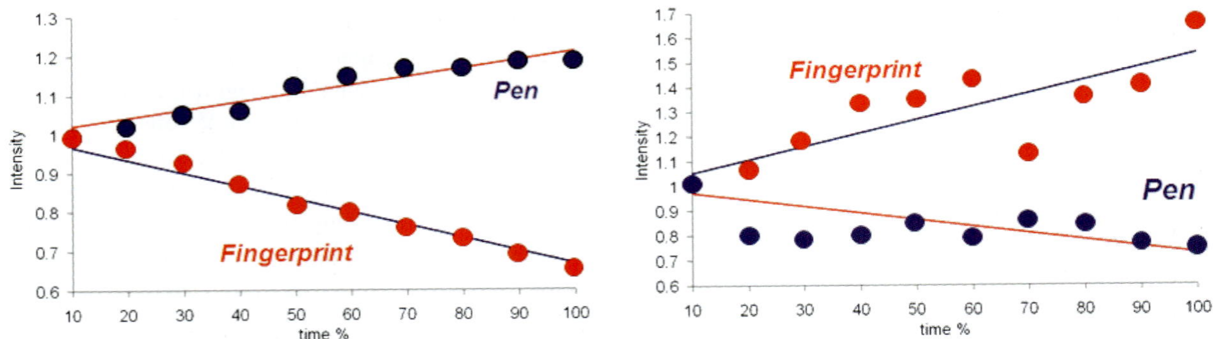

Fig. 19.   Depth profiles obtained from molecular fragment signals characteristic of fingerprint and ink. On the left the fingerprint is deposited over ink and on the right the ink is drawn over the print. That these cases can be distinguished has importance in forensics applications. Reproduced from Fig. 4 of Bailey *et al.*, 2010 [110].

The ion beam induced charge (IBIC) in the detector is immediately visible as a mirrored charge on the device output terminals (the Shockley Ramo Theorem [113]), and monitoring the size and time evolution of the IBIC signal gives valuable information about the electrical quality of the device. The technique is typically used in microbeam imaging mode to investigate the spatial variations of charge transport properties, like the charge carrier lifetime and velocity, and to correlate those variations with other imaging techniques. Detailed analysis as a function of applied bias often enables the researcher to infer information about the internal electric field profile, hence also the space charge and details on carrier drift or diffusion (see [114] for an example involving angle resolved IBIC analysis of 4H-SiC Schottky diodes). When analysing IBIC images, it is important to be aware that the volume that contributes to the formation of the signal is not limited to the range of the ion or its knock-on effects. On the contrary, the probed volume consists of all areas that the charge carriers drift through. Consequently the "spatial" resolution will vary with drift length and is often dominated by charge carrier diffusion rather than the ion beam diameter. Figure 20 shows the spatial variation of hole mobility in a high resistivity ($1.3 \times 10^{10}\,\Omega$.cm) CZT device, obtained by mapping the charge collection efficiency as a function of bias voltage and then obtaining the mobility pixel-by-pixel from a Hecht analysis [115].

The corresponding electron mobility map for this device is very uniform, showing the uniformity of the electric field strength of the applied bias. Therefore, the non-uniformity of Fig. 20 must be attributed to large scale variations of either drift mobility or

carrier lifetime in the bulk material. Both of these can also be mapped, giving much detailed and valuable information about the type of crystal defects in the device as a function of position.

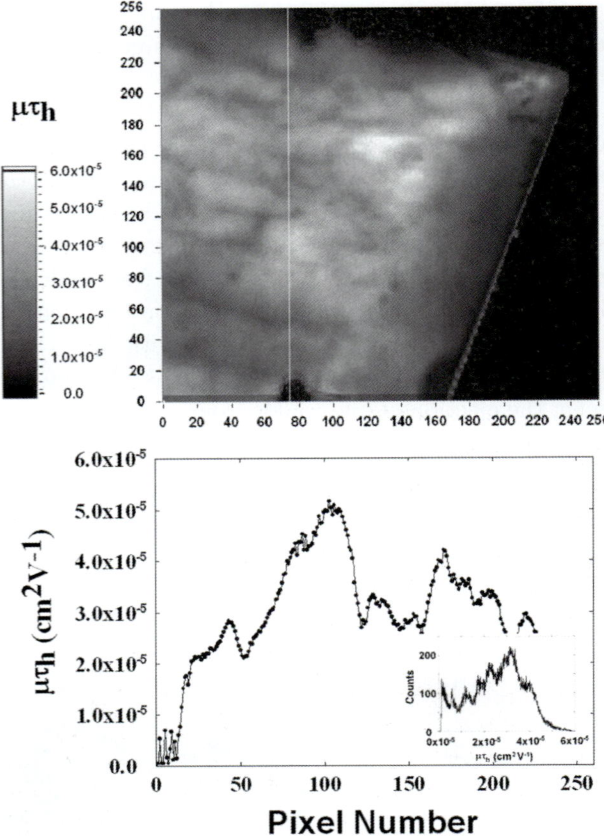

Fig. 20.   *Above*: Map of the hole mobility in $Cd_{0.9}Zn_{0.1}Te$ (with Au contacts) obtained from 2 MeV proton microbeam IBIC with spot size $2.5 \times 2.5\,\mu m$. *Below*: Variation of hole mobility along the vertical white line in the map. Reproduced from Veale *et al.*, 2008 [115].

Due to the low beam currents used, IBIC can often be considered as non-destructive. Radiation damage can however affect the IBIC signal as damage typically induces additional electrically active defects that can modify both the carrier lifetime and mobility, and often also the space charge in the sample ([116]: in this case divacancy traps are produced by the beam).

During an IBIC experiment the charge carriers are generated deep (tens of microns underneath the sample surface) compared with the equivalent electron beam technique (EBIC) which only probes near the surface [112], just under the contact electrode and in some cases not even penetrating the electrode. In contrast, the deep interaction of the ion with the sample has even enabled researchers to probe buried structures in semiconductor devices (power devices [117] and MOS devices [118]). Typical EBIC experiments operate in continuous current mode (compared to pulse-by-pulse or ion-by-ion acquisition of the IBIC signals) and thus cannot analyse the transient current response to each interaction. But similar information to IBIC can be gained from XBIC experiments that are typically carried out with highly focussed monoenergetic X-ray beams available at synchrotron facilities (for example described for diamond devices in [119, 120]). The synchrotron source typically provides bunches of photons whose time structure and intensity depends on the source itself. If the bunch structure is short enough, similar time resolved information (lifetime and velocity profiles) can be gained as in an IBIC experiment (for CZT [121]). The X-ray energy range typically employed is in the order of tens of keV, which means that charge carriers are often generated throughout more than a millimeter of the sample thickness. This often makes the experimental separation of electron and hole contributions to the signal impossible in thin samples, however X-ray induced charge carrier densities are lower compared to ion beam irradiation and therefore drift velocities extracted are less likely to be influenced by shielding of the effective fields due to the presence of the generated charge itself (subject to a sufficiently low X-ray flux). Both techniques require significant infrastructure and running costs and whether IBIC or XBIC is the method of choice will depend on the details of each individual study.

4. Particle-Induced X-Ray Emission

In a landmark pair of papers rapidly following Bohr's 1913 publication of his model of the atom, Henry Moseley investigated the characteristic X-rays produced when materials were bombarded with cathode rays (electrons). Since electrons are particles too, this is the first report of PIXE. In the first paper [123], he described the spectrometer (a crystal of potassium ferrocyanide), and pointed out that his "elemental" targets were contaminated with impurities saying, presciently: "The prevalence of [X-ray] lines due to impurities suggests that this may prove a powerful method of chemical analysis."

In his second paper [124], he systematically measured K- and L-series wavelengths (see Fig. 21). The first report of α-PIXE was by Chadwick in 1914 [125]. But Charles Barkla is responsible for the first recognition of characteristic X-ray lines of elements,

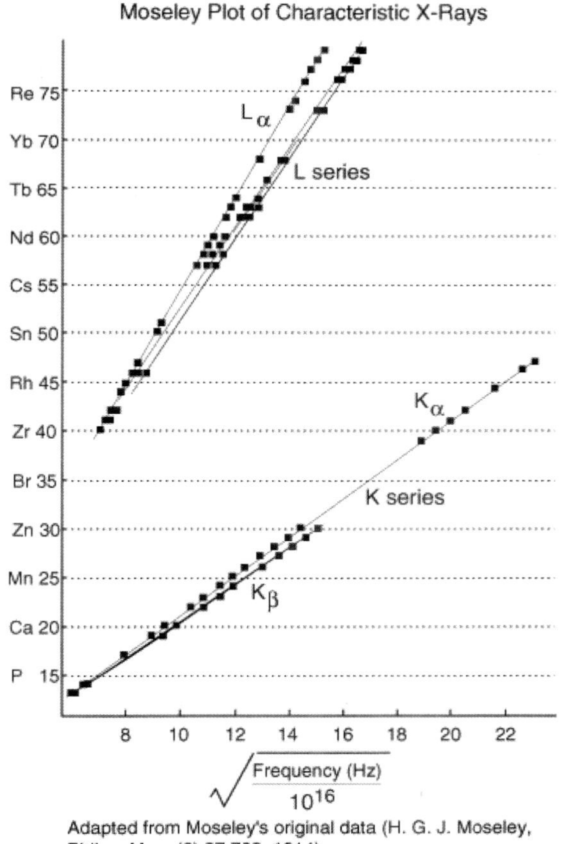

Adapted from Moseley's original data (H. G. J. Moseley, Philos. Mag. (6) 27:703, 1914)

Fig. 21. Henry Moseley's measurement of characteristic X-ray energies. Adapted by R. Nave [122] from Fig. 3 of Moseley [124].

for which he received the 1917 Nobel Prize: it was in his 1911 paper that he first named X-ray "fluorescence", and introduced the "K" and "L" notation: mid-alphabet letters being used since both longer and shorter wavelengths were expected [126]!

The first report of modern PIXE using cooled Si(Li) detectors [127] was by Johansson *et al.* in 1970 [128] who suggested that trace-element detection limits could be as low as ng/g, and analysed air pollution samples as an example. This rapidly led to a report of the variation of trace metal concentrations along single hairs [129]. Other highly cited examples using microbeam PIXE include measuring concentration gradients of pollutants in aqueous systems [130] and measuring the absence of Al in Alzheimer's disease samples [131] (see Fig. 22).

Moseley used wavelength dispersive spectrometry (WDX) which is a high resolution technique quite capable of picking up differences in the electronic structure of the atoms due to different bonding states. This valence information is regularly used in EPMA, the electron spectroscopies (XPS, AES), and the absorption spectroscopies (EELS, XAS). It can also be used in PIXE if a high resolution detector is used, which could be WDX [132, 133], or one of the new high resolution calorimetric EDS detectors [134]. Of course, high resolution also allows disentangling of overlapping peaks which often occurs, especially for the L lines, and is one main cause of the degradation of sensitivity [135].

Three physical effects have to be quantified to use PIXE for analysis: the ionisation cross-section, the fluorescence probability, and the mass absorption; these are all quite complex and need describing separately. To this needs to be added the energy loss of the incident particles in the sample, which is of course exactly the same as for the particle reaction techniques (see Sec. 2.1). We should note here that PIXE has its own ionisation physics, but shares the fluorescence (or, equivalently, Auger) probabilities with the other atomic excitation methods (EPMA, XRF, XPS, AES, EELS, XAS). The X-ray absorption coefficients are also needed by the X-ray methods (EPMA, XRF, XAS).

4.1. *Energy Loss (STIM)*

We have noted in Sec. 2.1 that the ion beam will lose energy inelastically as it passes through the sample. If the sample is thin enough to allow transmission

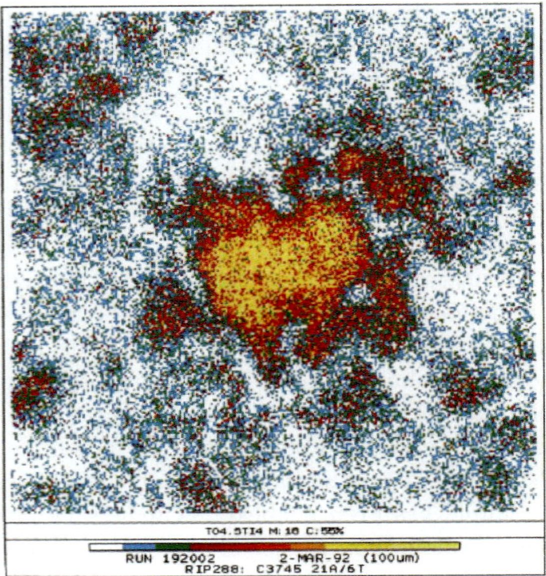

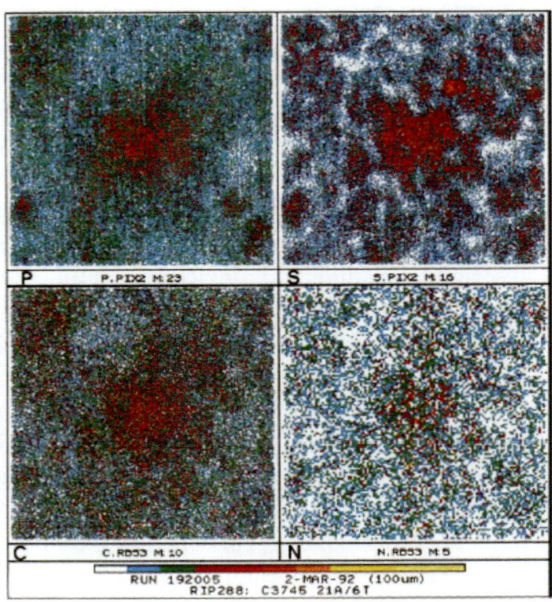

Fig. 22. Scanning 3 MeV proton microbeam STIM, PIXE and EBS images ($100\,\mu m \times 100\,\mu m$) from sections of unstained *post-mortem* tissue of a patient suffering Alzheimer's disease. The spot size was $1\,\mu m \times 1\,\mu m$. *Above*: STIM map of region containing neuritic plaque; *Below*: Maps of the same area for P & S (PIXE), and C & N (EBS). Reproduced from Fig. 2 of Landsberg *et al.* [131].

of the beam and a detector is placed behind it, then the average energy of the detected particles will be determined by the average sample thickness. If a microbeam is used, then an image of the sample density is projected onto the detector. This is the ion analogue of X-ray radiography.

Figure 22 shows STIM/PIXE/EBS maps of brain tissue in an important study which ruled out the presence of aluminium in brain tissue from Alzheimers patients at levels greater than 15 mg/kg. The difficulty with previous studies is that the plaques characteristic of the disease are almost impossible to see optically without staining. But using STIM they can be easily visualised. Notice that in this case the contrast with STIM is very much larger (with orders of magnitude smaller beam fluence) than for PIXE.

STIM requires a very low beam current ($\ll$pA) since every particle is detected. On the one hand this means that the technique is the least destructive of any IBA technique, but on the other it means that the PIXE maps need orders of magnitude increase in the beam intensity. Therefore STIM and PIXE must be done sequentially. But PIXE/BS maps are collected simultaneously.

STIM/PIXE can be collected simultaneously if the detector is put off-axis so that it does not see the direct beam. This was first reported by Orlić *et al.* in 1994 in a study of single aerosol particles [136].

4.2. *Ionisation cross-section*

Energetic particles implanted into a target will suffer inelastic energy loss as mentioned in Sec. 2.1. This energy is deposited mainly into the electrons of the sample. Thus, K-shell and L-shell (and outer shell) electrons can be ejected from their shells, leaving the atom in an excited state. Semi-empirical models for the calculation of the probability of K-shell ionisation have been established by Helmut Paul and co-workers [137] L-shell ionisation has been similarly determined by Miguel Reis and co-workers [138] with a polynomial representation for both K- and L-shells [139] and sub-shell ionisation cross-sections also extracted from the data [140]. Reliable data for M-shell ionisation is not currently available in the same form, but Fig. 23 shows working values obtained by interpolation and extrapolation from ECPSSR[k] calculations (Campbell *et al.* [141]).

4.3. *Fluorescence probability*

The relaxation mechanism of the excited atom is complicated. But atomic spectroscopy has a very

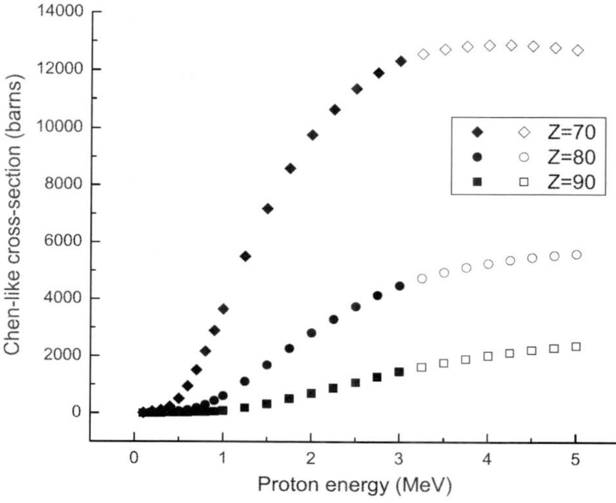

Fig. 23. M_1 sub-shell ionisation cross-sections for protons to 5 MeV. Reproduced from Fig. 2 of Campbell *et al.* [141].

long history. The Doppler effect was first proposed by Christian Doppler as a means of detecting the motion of binary stars in 1842 [142], and observed (for sound, not light) by John Russell in 1844 [143]. Stellar spectroscopy was responsible for the discovery of helium in 1868 independently by both Janssen and Lockyer [144]. Bohr's model of the atom [145] was a triumph in 1913 precisely because it solved the problem of the hydrogen Balmer and Rydberg lines.

There are a variety of competing mechanisms. First is the possibility of a non-radiative transition by the Auger process. The branching probability of fluorescence instead of Auger relaxation has been calculated by Chen and Crasemann using ECPSSR theory for the K-shell [146], L-subshells [147] and M-subshells [148]. There are also extensive experimental data for the L-shell transition probabilities which have been critically reviewed by Campbell [149, 150].

The reason for this experimental interest in the L-shell fluorescence probability is that it is greatly complicated by the existence of the so-called Coster-Kronig (CK) transitions. These are a special class of non-radiative transition that transfers the vacancy from the initial subshell to a higher subshell within the same shell; that is, a re-arrangement of the electronic structure of the excited atom. The energy balance is preserved by the loss of outer shell

---

[k]ECPSSR: Energy-loss Coulomb-repulsion perturbed-stationary-state relativistic theory.

electrons with appropriate energies; of course, quantum mechanical selection rules apply in all these electronic structure re-arrangements. There are very many CK transition probabilities to be determined, which can have a large effect on the relative X-ray intensities in the L, or M series; these intensities are therefore hard to determine, with large uncertainties remaining. Campbell and co-workers have given semi-empirical fitted data for the K [151] and L [152] series. Chen and Crasemann long ago calculated the relative line intensities for the M series [153]; this remains the best dataset available, since good experimental data for M-lines are hard to obtain. Therefore uncertainties for M-lines remain high.

In summary, the relaxation mechanism of the excited atom is far from simple. After the atom is ionised it can return to its ground state in a large number of different ways, many of which give X-rays at a variety of energies. This is a powerful probe of the atomic energy levels of the atom, but all of these effects must be understood reasonably well to use the fluorescence analytically. Recent work on L-lines using very high energy resolution EDS detectors has underlined the complexity of this process, and also the — potentially large — gaps in our understanding of it [154].

4.4. *Mass absorption coefficient*

After the X-rays are produced in the interaction volume they still have to escape to the detector. Since the range of energetic light ions is large (>0.1 mm for 3 MeV protons in most materials) there may be a large thickness of material to absorb the X-rays before they reach the detector. Clearly, any quantitative work must take this into account.

The mass absorption coefficient (μ) is defined from the fraction of an X-ray beam (intensity I_0) that is absorbed during transmission through thickness x of a material. The transmitted beam intensity I is given by:

$$\ln(I_0/I) = \mu x. \qquad (6)$$

The XCOM mass absorption coefficient database from NIST includes comprehensive values for μ and is kept up to date [155, 156].

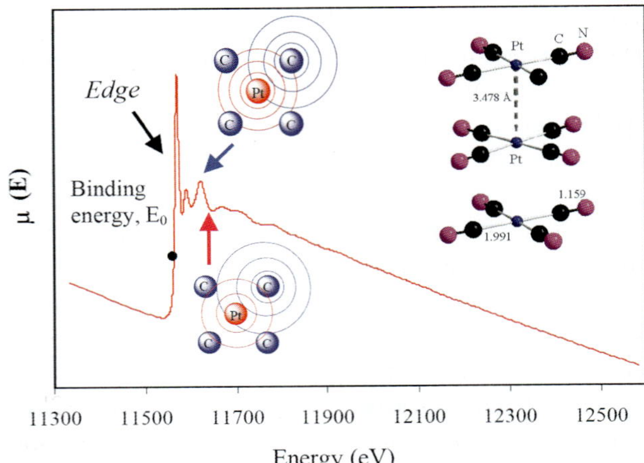

Fig. 24. X-ray absorption spectrum of potassium tetra-cyanoplatinate K$_2$[Pt(CN)$_4$] near the Pt L$_{III}$ edge showing constructive and destructive interferences in the NEXAFS signal (figure drawn by Farideh Jalilehvand, University of Calgary, and reproduced from [157]).

μ is a complicated function of X-ray energy. This is because absorption of the X-ray *suddenly* becomes possible as soon as the photon energy exceeds the binding energy of any particular atomic electron. Therefore there are absorption edges corresponding to all the atomic subshells. Actually, the situation is even more complicated than that since near the absorption edges there are quantum mechanical interference effects, as shown in Fig. 24.

Happily, these details are not usually important in PIXE, but the accuracy of the absorption database is of continuing concern in accurate work. Note that the absorption coefficient represents the sum of the cross-sections for three quantities: the photoelectric effect, and both of the incoherent and coherent scattering effects.

Work is continuing in the community to make further critical measurements of this (and other) important quantities. For example, the synchrotron group at the PTB, Berlin[1] recently used an absolutely calibrated instrument to make a determination of mass attenuation coefficients for Al [158] relative to previous values [159], finding internal inconsistencies in them of up to 10%. EXSA[m] has promoted the "International initiative on x-ray (*sic*[n]) fundamental parameters" which is coordinating efforts by all of

---

[1]PTB: Physikalisch-Technische Bundesanstalt, Germany's national metrology institute.
[m]EXSA: European X-ray Spectrometry Association.
[n]Properly, "X-ray" is capitalised, since the "X" is an abbreviation for the proper name Röntgen.

the PIXE, XRF and EPMA communities to improve the various databases [160].

4.5. *Quantitation and complementary methods*°

It is important for analysts to appreciate that there are a number of other analytical techniques involving exactly the same atomic relaxation and X-ray absorption physics. The excitation methods are comparable, but have to be treated separately to give the detail required for analytical purposes. It is worth pointing out here that atomic ionisation with ion beams has complexities not present for the simpler photoionisation processes since higher charge states are much more probable.

Except that the excitation is via photo-ionisation instead of particle impact, the physics of the X-ray fluorescence (XRF) technique and PIXE are similar, with similar spectra, and detection limits only somewhat worse due to the presence of background radiation originating directly from the exciting beam. Desktop XRF instruments are in wide use, but these do not give monochromatic beams and calibrating the tube spectra is difficult. With synchrotron XRF a tunable monochromatic source is available, making this an exceptionally powerful analytical tool. However, at present mapping is still rather slow, and XRF spectra give no direct access to depth profiling.

A PIXE code should also be able to interpret XRF spectra, provided the photoionisation can be properly handled. Because the information depth for the two techniques is complementary there is sometimes an advantage in an XRF/PIXE analysis, as shown in recent work on Roman silver coins [161] and ancient glass beads [162]. On the other hand, the instrumentation of the Mars Rover has generated mixed XRF/PIXE data, the analysis of which is a *tour de force* that has established the presence of hydrated minerals on Mars, an extraordinarily important result [163]. Note that PIXE will always excite secondary fluorescence by XRF.

Electron-probe microbeam analysis (EPMA) is a scanning electron microscope (SEM) method specialised for X-ray analysis. Typically, an EPMA instrument will have both EDS and WDX detectors.

Both EPMA and SEM-EDS (like XRF) have similar physics and observed spectra as PIXE: the only difference is that in this case the excitation is via electron impact rather than ion impact. But SEM methods have important analytical differences from PIXE. Electrons are 2000 times lighter than protons, and the primary Bremsstrahlung for protons is negligible. Therefore detection limits for SEM methods are orders of magnitude worse than for PIXE.

Also, because of the large lateral straggle for electron beams and for SEM energies (usually $< 30\,\mathrm{keV}$) and thick targets, the excitation volume for electrons is determined by the electron energy and not by the probe diameter. But proton beams penetrate to large depths from which few X-rays escape. So the excitation volume for protons is determined essentially by the probe size. If ion microbeams can be built with sub-micron spot sizes (see Sec. 6.4), then PIXE maps will have better spatial resolution than SEM-EDS X-ray maps.

SEM-EDS and PIXE are similar in that for both techniques there is a backscattered particle available: for PIXE it is the BS signal, and modern SEMs usually have a BSE detector for Z-contrast. But the SEM-BSE signal cannot be treated quantitatively without great difficulty (Monte-Carlo methods are needed), and therefore, as with XRF, any depth profile is not accessible directly with SEM methods.

EDS detectors are also often installed on TEM instruments. In this case the samples are always thin, but their absolute thickness is usually hard to obtain and so the X-ray spectra are rarely treated quantitatively. However, modern TEM instruments often also have an EELS attachment. This is the *inverse* process to AES: in EELS the effect of the atomic structure (including chemical effects) on the transmitted electron energy is observed.

The electron spectroscopies (XPS and AES) excite the atom with photons and electrons respectively, energy-analysing the electrons resulting from atomic relaxation. XPS is a one-electron process with the photoelectron observed directly. AES is a process involving at least three electrons, which occurs when the atom relaxes non-radiatively. Of course, Auger electrons are also observed in XPS spectra. Because the energy resolution available is high ($< 1\,\mathrm{eV}$) chemical effects are easy to observe.

°This section is something of an acronym soup, for which apologies. Please see Appendix B for a Glossary.

The X-ray absorption spectroscopies should also be mentioned (see Fig. 24 and an example on nickel oxy-compounds [164]). These include XANES, EXAFS and NEXAFS as synchrotron techniques, and are analogous to EELS in the same way that XRF (X-rays in, X-rays out) is analogous to AES (electrons in, electrons out).

These techniques with others are frequently used in conjunction. For example, a recent review of methods to visualise spatial distributions and assess the speciation of metals and metalloids in plants addressed: histochemical analysis, autoradiography, LA-ICP-MS,[p] SIMS, SEM-EDX, PIXE, XRF, XAS, and differential and fluorescence tomography [165].

4.6. *Depth profiling with PIXE*

PIXE spectra (like EPMA or XRF spectra) do not give information about possible depth profiles directly, although of course the line intensities are greatly modified by how much they are absorbed on their path to the detector; also, because the ionisation cross-sections are a strong function of energy, the number of X-rays produced is a strong function of depth. To access this information, the profiles must be inverted from the spectra, and this problem is in general mathematically ill-posed.

If the sample structure is known *a priori*, then the structure parameters (layer thicknesses etc.) can be modelled from the data. This is routine for XRF, for which software is commercially available. But if the sample structure is not known then single PIXE (or XRF or EPMA) spectra are almost always ambiguous.

XPS has exactly the same problem, and with the same solution. Angle-resolved XPS has received careful attention [166], and differential PIXE can vary either the beam energy or the angle of incidence: we mention recent work on multilayer targets [167] and a recent review of cultural heritage applications [168].

5. IBA Codes

The brief sketch we have given so far of the physics of the MeV-IBA methods used for depth profiling makes it very clear that computer codes are needed to make any systematic use of these methods for analytical purposes. In this section we will make general and technical remarks about the codes specifically for photon emission and particle reactions, and in Sec. 6, we will concentrate on integrating these methods.

We will consider stopping power and straggling effects (roughness, variable density, phase separation) in Sec. 5.3, since in principle these affect both the particle and the photon techniques.

We will not consider non-depth-profiling techniques (channelling, MEIS, IBIC); nor will we cover the low energy or sputtering techniques (LEIS, MEIS, SIMS). For further details see Sec. 3.

5.1. *Codes for nuclear reactions*

Codes for (mostly) depth profiling using particle scattering and non-resonant nuclear reactions were reviewed in IAEA-sponsored work by Rauhala *et al.* in 2006 [169]. A detailed Intercomparison also sponsored by the IAEA of selected codes was subsequently completed [170], with a brief summary also available [171]. The Intercomparison identified two "new generation" advanced codes (**SIMNRA** [172, 173, 174] and **DataFurnace**[q] [175, 176, 177]) that performed excellently and had all the facilities needed to treat the complex cases we will discuss below. SIMNRA and DataFurnace are analytical codes based on the single scattering approximation, with other effects introduced as perturbations. In most cases they give indistinguishable results, but DataFurnace is specifically aimed at handling multiple spectra self-consistently where SIMNRA is designed only for single spectra.

Both SIMNRA and DataFurnace allow fitting of RBS, EBS, ERD, and (non-resonant) NRA spectra to models using standard algorithms (Simplex for SIMNRA and grid-search for DataFurnace [179]). DataFurnace also allows model-free fitting with the simulated annealing algorithm [180], which facilitates the natural implementation of a Bayesian algorithm for the reliable estimation of the uncertainty of the resulting depth profile [181].

Both SIMNRA and DataFurnace use the accurate pulse pileup algorithm of Wielopolski & Gardner (1976) [182], and DataFurnace also implements both triple pileup [183] and Molodtsov & Gurbich's more accurate calculation [184]. DataFurnace also

---

[p]Laser ablation ICP-MS.
[q]DataFurnace is also known as NDF.

implements Lennard's [185] pulse height defect algorithm [186].

Both SIMNRA [187] and DataFurnace [188] can calculate the effect of roughness on backcattering spectra; SIMNRA is aimed at gross roughness such as is found on tokamak facing tiles, where DataFurnace is aimed at fine interface roughness effects such as is found in magnetic superlattice samples. But neither can correctly model really severe roughness such that the beam enters (or exits) the sample multiple times (through asperities). Molodtsov *et al.* [189] implement a general algorithm which will allow the effects of such severe roughness to be interpreted, but this algorithm is not yet implemented in any general purpose IBA code. SIMNRA allows calculation of the effect on spectra of roughness whose model parameters are known (perhaps from AFM measurements), using a spectral[r] summing method (see [190] for a recent example). DataFurnace allows fitting of given spectra, from the parameters of simple models for the roughness of interfaces or layer thickness inhomgeneity; sub-nm sensitivity was demonstrated for magnetic multilayers [191].

The algorithm of Stoquert & Szörenyi [192] allows the calculation of the effect of precipitates (density variation) on the spectra. DataFurnace [48] implements this algorithm and SIMNRA [193] implements a more accurate comparable algorithm. On the other hand, DataFurnace implements EBS resonances more correctly than SIMNRA (see Fig. 6 and [47], and see the Intercomparison [170] for more detail). The difference is noticeable for sharp resonances. DataFurnace also implements a good approximation for low energies [194] (see Fig. 25).

Finally, both SIMNRA [173] and DataFurnace [195] implement similar algorithms for double scattering. The symmetrical components of multiple scattering are handled with Szilágyi's algorithm [196, 197] by both codes. DataFurnace uses DEPTH [198] directly, and SIMNRA implements the algorithms (see the discussion in [199]). Kinematical broadening and related effects are also treated. These effects become large for glancing incidence geometries, such as are used for high resolution or ERD. See Fig. 25 for an example showing all these effects.

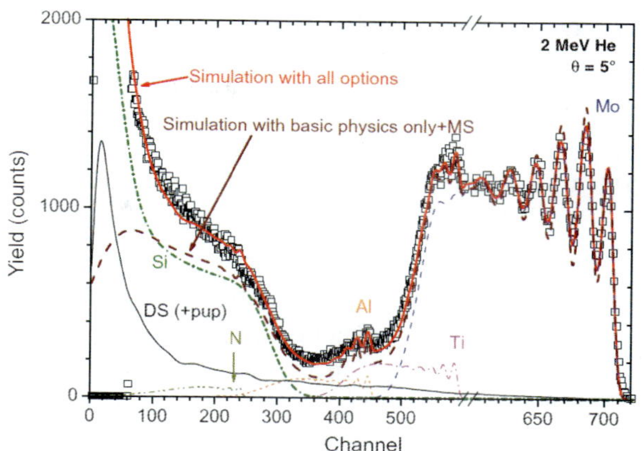

Fig. 25. High resolution (glancing incidence) analysis of a tribological coating: TiAlN/Mo multilayer on Si, modulation period 3.9 nm [178] (reproduced from Fig. 1 of Barradas & Jeynes, 2008 [176]). Simulations are shown: including only symmetrical effects due to multiple scattering; showing the double scattering and pileup background; the full simulation including the low energy yield (partial spectra for Ti, Al, Si, N also shown for this case).

For high depth resolution RBS, and for ERD especially HI-ERD, glancing incidence beams are typical. In these circumstances multiple scattering and related spectral distortion effects become important, and the uncertainty of the analytical codes' calculations increases. **CORTEO** [200, 201] is a Monte Carlo code designed for routine use, optimised for speed and entirely comparable to the code MCERD [202] described in the Intercomparison. It also enables the detailed simulation of geometrical effects, including the geometry of the detector (typically ToF for ERD).

For high resolution resonant NRA, sharp resonances (usually at low energies) are used to probe depth profiles. In these circumstances the approximations normally used for straggling fail close to the surface, where the fundamentally stochastic nature of the straggling cannot be approximated by a continuous (Gaussian) distribution. The **SPACES** code [203, 204] takes this into account. DataFurnace has been extended to treat resonant NRA, and by using a gamma function approximation to the straggling (as SIMNRA does for roughness) also treats high resolution NRA [205], giving results comparable to SPACES provided the Lewis effect [206] is not too large.

---

[r]The noun is *spectrum*; the adjective is *spectral*. From Newton's usage (p. 1) and the Latin, *specere*, "to see".

5.2. *Codes for atomic excitation*

The first full account in 1989 of the GUPIX[s] code [207] starts very helpfully for those wishing to appreciate what is involved in PIXE (or XRF, or EPMA) spectral analysis:

"The typical PIXE or micro-PIXE arrangement employs a Si(Li)[t] detector to record the spectrum induced in an analytical specimen by [typically] 2–4 MeV protons. Elements of low and medium atomic number each contribute about six distinct K X-ray lines and those of high Z each contribute about twenty L X-ray lines. The lines, approximately Gaussian in shape, are superimposed on a continuous background arising from various bremsstrahlung [*sic*[u]] processes [comprehensively described by Ishii & Morita, 1984 [208]] and sometimes also from nuclear reaction gamma rays. The resulting Si(Li) spectrum is complex and peak overlaps occur frequently."

GUPIX itself dates from 1984, and codes then available (including Clayton's PIXAN) were reviewed and compared in 1986 to evaluate PIXE as an analytical technique [209]. Campbell *et al.* go on to describe the way GUPIX handles the particle impact ionisation, the fluorescence probability and the absorption calculation databases. They consider the non-Gaussian Si(Li) lineshapes, the method for stripping the Bremsstrahlung background from the spectra, and the details of the complex fitting procedure used. Subsequent work implemented secondary fluorescence, and later versions of GUPIX (see citations in the 2010 paper [141]) have continued to refine this program.

Currently, GUPIX does not permit the calculation of PIXE spectra from general multilayer samples: multilayered samples are only supported provided that elements of interest appear in only one layer. This is an arbitrary restriction that will be lifted in future versions. The distributed version of GUPIX also does not support XRF spectra, although there is a version that does which was used for analysing the Mars Rover data [210, 211].

GeoPIXE [212] was first described in 1990 [213]. Unlike GUPIX, GeoPIXE is explicitly for microbeam applications, and includes true-elemental visualisation; in 1995, a version was described that was able to correct in a very elegant way for differential self-absorption in samples with different phases (very common in geological applications) [214]. GeoPIXE is still being actively developed, and is now supporting large detectors and high count rates [215], and STIM [216]. We should note that GeoPIXE also supports XRF spectra.

Axil[v] was originally a commercial XRF program from Canberra-Packard [217], it was rebuilt in 1994 [218], and explicitly applied to PIXE in 1996 [219]. It is now publicly available from the IAEA XRF Laboratory [220] and commercially (with a windows GUI) from Canberra [221]. Both versions are aimed at energy dispersive XRF applications.

OMDAQ[w] was first reported in 1995 [222]. This code was the first to make a systematic use of the particle BS spectrum to give a reliable charge normalisation and the sample thickness measurement essential to doing the X-ray absorption correction, and is specifically aimed at microbeam applications. For handling the PIXE spectrum OMDAQ relies on the GUPIX code.

Where OMDAQ is a microbeam PIXE code that makes (quite simple) use of the particle spectrum, **DataFurnace**, is originally a nuclear code (RBS/EBS/ERD/NRA) extended to accept PIXE data and designed to handle atomic and nuclear data self-consistently at the highest accuracy possible [223]. It relies on calling the open-source PIXE module **LibCPIXE** (2006) [224] which implements the 1996 **DATTPIXE**[x] code by Miguel Reis and co-authors [225]. As PIXE input, DataFurnace currently takes line areas obtained by preprocessing raw PIXE spectra using AXIL or GUPIX. At present,

---

[s] The "Guelph PIXE software package".

[t] Si(Li): lithium drifted silicon diode detector (pronounced "silly"). These are typically up to 5 mm thick, giving good absorption of X-rays below about 30 keV energy. It is hard to make such thick crystals without recombination sites in the electrically active region (destroying the linearity of the device) and the defects are passivated by a lithium drift process. There is a similar process for Ge(Li) ("jelly") detectors, but these have now been superceded by high purity (HPGe) crystals.

[u] As a German noun, "Bremsstrahlung" is properly capitalised.

[v] Axil: Analysis of X-ray spectra by Iterative Least squares fitting. Of course, GUPIX also used iterative least squares fitting, but was ignored by the XRF people since it was a PIXE code!

[w] OMDAQ: Oxford Microbeams data acquisition.

[x] TTPIXE: thick target PIXE, first published in 1992.

DataFurnace has not implemented the secondary fluorescence correction. Developments which will be applied in future to the DataFurnace PIXE module are continuing under the pressure of difficult high resolution spectra from an EDS detector based on a microcalorimeter. It is important to be able to handle raw spectra [226], since the recognition of the lines, and the determination of line ratios, cannot be done either by AXIL or GUPIX in a model-free way.

Finally, we should mention that **GEANT4** has now also implemented a PIXE module [227].

A fundamental limit to the accuracy of PIXE is down to the uncertainties in the various databases used, as highlighted by the Fundamental Parameters Initiative of EXSA [160]. The discussion here has shown that the calculations are extensive and rather complicated, and a "fundamental parameters" approach is not likely to yield uncertainties much smaller than 20%, except in particular cases where values have been specially determined (see [158] as an example).

Where better accuracy is required the usual procedure in the PIXE/XRF community is to calibrate the fundamental parameters used against standards: of course, relative accuracies <1% are easily obtainable in simple cases; even for complex samples, accuracies ∼5% can usually be demonstrated with appropriate standards.

The accuracy of the PIXE codes themselves has been confirmed in an IAEA-sponsored Intercomparison exercise [228, 229]. Subsequently, LibCPIXE and GUPIX (and SEM-EDS) calculations have been critically compared (for hydroxyapatite) [230].

5.3. *Artificial neural networks*

There is a new approach to IBA which may prove remarkably valuable. Artificial neural networks (ANNs) can be constructed that are capable of effectively analysing various classes of IBA data [231]. O'Connor & co-workers discuss this "inverse problem" with noisy data, and directly compare ANNs and simulated annealing [232]. Figure 26 shows real-time data obtained to determine the detailed annealing kinetics of the nickel silicide system [233]. This is a simple example: there are a number of other examples not so simple. Much intervening work has shown that ANNs can be trained to handle multiple spectra, or multiple techniques; and it is clear that any sort of IBA can be implemented in an ANN for which

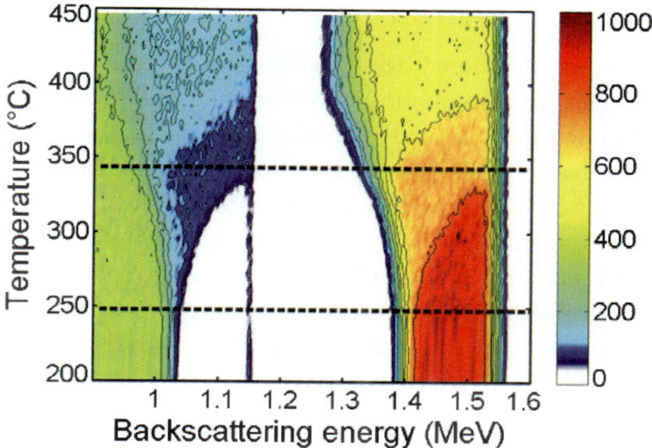

Fig. 26. Composite of 250 RBS spectra taken during a Ni silicidation reaction of an 80 nm Ni film on silicon, capped with 7 nm of Si and annealed at 2°C/min. The colour code on the right shows spectral intensity in arbitrary units. The dashed lines indicate the initiation of the mono- and di-silicides. (Reproduced from Fig. 2 of Demeulemeester *et al.*, 2010 [233].)

a valid training set can be defined (although an ANN system for PIXE has not yet been reported).

The point here is that once the ANN has been trained a solution of a spectrum is obtained completely automatically and effectively instantaneously. This solution is itself remarkably accurate and, provided that the ANN training is adequate, will be qualitatively correct. Such a solution can be given back to the analytical codes for post-processing to automatically obtain the best possible solution, together with robust estimates of the uncertainty. Thus, the qualitative recognition of a set of spectra by an ANN can be used as the basis of an automatic and fully quantitative machine analysis of the dataset. To check whether individual spectra are actually validly analysed by the ANN (based on the sampling space of its training set), prior to passing them to it, an additional ANN can be trained that classifies spectra as "acceptable" or "not acceptable" (see [231]). Thus, the ANN cannot only recognise a given spectrum, but can also recognise whether or not it has been trained to recognise it.

The aim is a push-button (or turnkey) system suitable for non-experts (*"IBA without Humans"*! [234]), similar to the systems already available for SEM-EDS, or EPMA or XRF, or even AMS. Such IBA systems would be essential for a more general industrial acceptance of the technique, and seem to be feasible.

5.4. *Accurate IBA*

It is extraordinary that the idea of an uncertainty budget [235] to quantify experimental and traceability uncertainties for IBA was only recently published by a National Metrology Institute (NMI) [236]. However, no NMI now has IBA capability despite their previous interest in the use of RBS in particular for metrology (NPL for a Ta_2O_5 standard material [237] and a metrology exercise on the native oxide of Si [238]; IRMM and BAM for the certified reference materials used for fluence standards in IBA: the old "Harwell" Bi-implanted standard [239] and the new Sb-implanted certified reference material ERM-EG001 from IRMM/BAM [240]).[y] The situation incidentally is entirely different for XRF, for which the PTB has been active in obtaining ISO 17025 certification and where the "Fundamental Parameters Initiative" is supported by no less than three National Standards Institutes (PTB, LNE, NIST [160]). Of course, this FP initiative has as much relevance to PIXE as it has to XRF.

The qualification of implantation fluence is valuable for many different purposes, and is the simplest of RBS problems. We have already demonstrated an absolute accuracy of about 4% (95% confidence) in its determination [241–243], where the cited uncertainty was dominated by the uncertainty in the stopping powers used. However, we have demonstrated that for 1.5 MeV He in Si, the SRIM2003 stopping powers are accurate at 0.8% (1σ, see Fig. 1 in [171], but note that there are extra unaccounted uncertainties here [244]), and therefore the absolute traceable accuracy of RBS should approach 1% (1σ) if this beam is used with a-Si substrates to determine the actual charge·solid-angle product for a given spectrum. RBS with 1% accuracy is rather hard to achieve: 1% accuracy has often been claimed (the report by Turkovich *et al.* on the analysis of moon rocks by Surveyor V in 1967 is an important early example [21]), but there has been only one such previous *critical* report [245], and that was the rather special case of the RBS determination of the In:Ga ratio in InGaAs samples. There is no other thin film technique that can match this level of absolute accuracy for the determination of the quantity of material. To achieve this accuracy it is necessary to correct properly for pulse pile-up, and to correctly determine the electronic gain of the detectors, including the appropriate pulse height defect correction (see Sec. 5.1).

For any sort of accurate IBA, an essential prerequisite is the availability of statistically robust estimates of the uncertainty of the result. Such an estimate is entirely absent in the classical approach to solving RBS spectra where manual simulations are made until a plausible fit is obtained. Such an approach (discussed at length in [175]) is not even adequate to explore the intrinsic ambiguity of the data! A fully Bayesian (maximum entropy) approach to this has been made [246] but has not proved to be generally usable. Although astonishingly good results can be achieved, the calculations are very time consuming and the information needed about the system is prohibitively detailed. However, a cruder approach using the DataFurnace code, still Bayesian but not using maximum entropy, has proved to be of general use in IBA depth profiling.

This approach was used for the example shown in Fig. 1: hence the extraordinary precision for the layer thicknesses obtained in this multilayer sample. Note that one of the virtues of this approach is that it allows the systematic evaluation of explicitly stated prior assumptions imposed on the data. In this case sub-nm precision in the layer thickness determination, even at the bottom of the stack, is obtained *assuming* that the sample has alternating layers of zirconia and silica. Then the precision represents the precision with which the signal areas can be obtained, *given the structure*. Analysts using informal fitting methods invariably have assumptions they impose on the data: unfortunately, too often these assumptions are hidden, and the analyst is unaware of them.

It is worth emphasising that this approach to uncertainty in the DataFurnace code is quite general. It is applied to all the data submitted for self-consistent analysis, currently including PIXE, RBS, EBS, ERD, PIGE, NRA and NDP (neutron depth profiling, see [247] for an example of this technique).

---

[y]NPL, IRMM, BAM, PTB, LNE, NIST: *National Physical Laboratory* in London, *Institute for Reference Materials & Measurement* in Geel, *Bundesanstalt für Materialforschung und –prüfung* and the *Physikalisch-Technische Bundesanstalt* in Berlin, *Laboratoire national de métrologie et d'essais* in Paris, *National Institute for Science & Technology* in Gaithersburg.

We should point out that for a valid estimate of the total combined uncertainty, an estimate of the database uncertainty is also required. The stopping powers used are usually from a semi-empirical compilation whose uncertainty in any particular case can generally be estimated quite well. But in cases where the stopping power is not known, or not known well enough, a Bayesian method of extracting the stopping power from thick sample spectra [248] is particularly valuable, since it also gives the uncertainty. EBS cross-sections on the other hand should properly be obtained using thin film samples, although some evaluations also make use of thick film data inverted to obtain the cross-section (see [249, 250] as examples). But evaluations of EBS cross-sections (see [44]) do not as yet include any reliable estimate of the uncertainty since these are very complicated to calculate correctly. Even the RBS/EBS boundary is not known with precision (see Table 1). Clearly this is a major problem for accurate (traceable) analysis that needs evaluated EBS cross-sections. We believe that one useful approach to determining such uncertainties could include uncertainties obtained from benchmark measurements which use this Bayesian method [251]. But this approach is still to be properly developed.

6. "Total IBA"

We now put together all the ideas presented so far, and consider the benefits of combining IBA techniques to increase their power.

Our thesis here is that the old approach of IBA labs, where RBS was mainly on offer in one lab and PIXE mainly in another, is not sustainable in the second decade of the 21st century. RBS is good for heavy elements in a light matrix and typically the mass resolution is not very good, so that only fairly simple things can be said about fairly simple samples. On the other hand, PIXE cannot compete on price against the almost equivalent XRF (there is even an explicit comparison of PIXE with XRF showing their near-equivalence [252]), and for half the price of microbeam PIXE equipment would you not be better off investing in a micro-XRF? Why bother with IBA at all?

But we believe that if an integrated approach is taken, where multiple detectors are used with every analysis beam so that some combination of PIXE/RBS/EBS/ERD/PIGE/NRA is always systematically done, then not only does the range of samples for which IBA is appropriate increase dramatically but also the quality of information about each sample also increases. We will mention several cases where neither the backscattered particle signal nor the emitted photon signal by itself could solve the sample; but where the solution is straightforward when multiple signals are treated self-consistently.

Why have the laboratories using PIXE and backscattering (BS) historically been so separate? There are good and bad reasons for this. Undeniably, for samples where the trace element content is important, BS often adds little useful information; similarly for PIXE where a layer thickness is required. Also, PIXE quantification is rather troublesome, with quite a long traceability chain; so that if accuracy is required for the major elements seen by BS, then PIXE adds no information.

Moreover, because the cross-section for PIXE is large relative to RBS, microbeam maps obtained from the particle detector have a very low number of counts compared to those obtained from the X-ray detector. Do very noisy spectra have negligible information? This is emphatically denied by a Bayesian analysis [181] of the complicated three-layer mixed Co:Fe silicide shown in Fig. 27! The qualitative structure could easily be resolved from the RBS spectrum with a very small charge solid-angle product of only $0.3\,\mu\text{C.msr}$ (readily obtained in μbeam-PIXE). It is very clear that it is completely *false* to assume that the noisy BS spectra obtained in regular μbeam-PIXE mapping are effectively information-free: on the contrary, these spectra frequently contain crucial information. For example, in such mixed silicides the absorption of the metal K lines even in relatively thin layers will be significantly different for different layer structures. This conclusion is underlined by recent work that shows that robust information is available even in the presence of 10% Poisson noise [232].

For PIXE, a very common beam to use is 3 MeV H^+. This is because the production cross-section goes up with beam energy, but beyond this energy nuclear reaction products tend to also decrease the signal to noise ratio. Thus this beam usually gives about the best available sensitivity. However, for this beam the particle scattering is non-Rutherford up to at least Fe (see Table 1). In the last ten years the (non-Rutherford) differential scattering cross-sections for

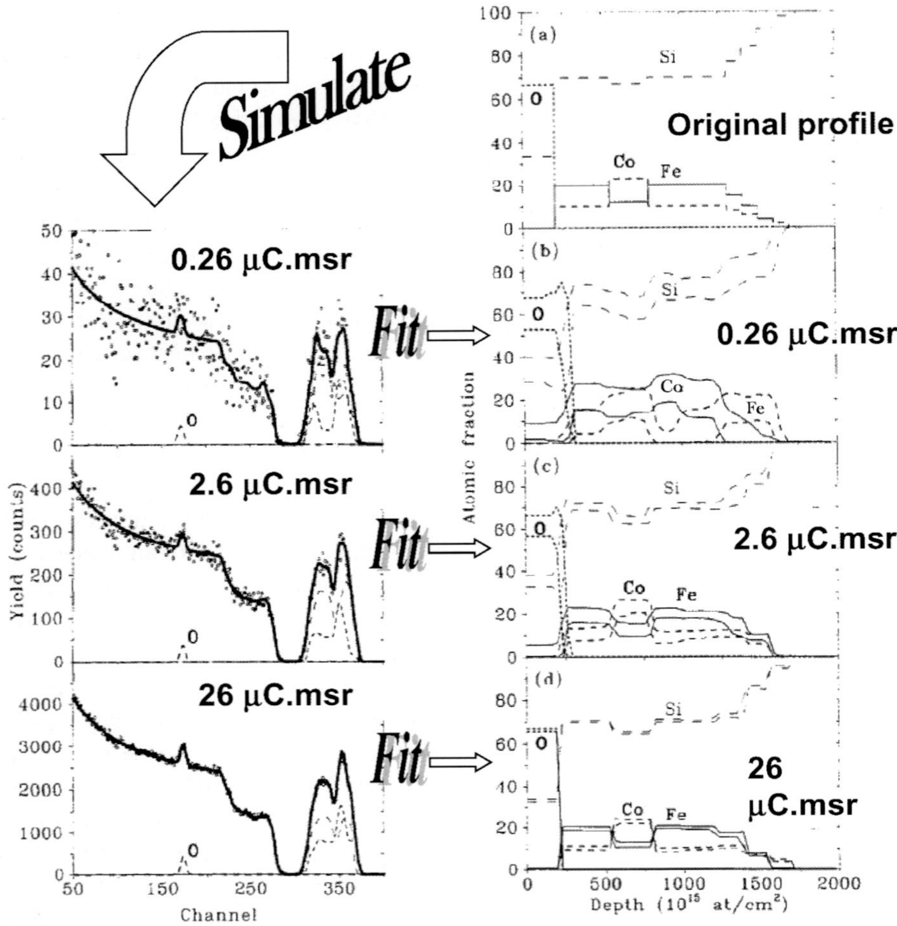

Fig. 27. Noisy spectra can contain crucial information! 1.5 MeV He RBS spectra of a mixed Fe:Co silicide on a Si substrate are simulated from an initial structure (*top right*), given various charge.solid-angle products (*left hand column*). These simulated spectra are treated as data, fitted (lines), and hence inverted back to depth profiles, with uncertainties given by a Bayesian analysis (*right hand column*). Modified from Figs. 1 and 2 of Barradas *et al.*, 2000 [181].

most of these reactions have been measured and evaluated, so that today we can usually do analyses with an accuracy previously unthinkable, and even at sharp resonances. So although in the quite recent past it was reasonable to discount the particle spectra on the grounds that they were uninterpretable, today this would be a grave mistake.

Therefore, although even in the quite recent past it was understandable that self-consistent PIXE/BS was often judged to be too problematical to be worth the effort, this is not the case today. We should point out that self-consistent PIXE/BS has been available for a long time time (OMDAQ [253] and "IBAlab" [254]), and accurate analysis using "Total IBA" has been applied recently to applications in geochemistry (PIXE/BS/ERD [255]), biomedical surfaces (PIXE/BS [256]), Alzheimer's disease (mapping STIM/PIXE/BS [257]), amino-functionalised gate oxides (RBS/ERD [258]), fish otolith composition (microbeam BS/ERD/PIXE [259), cultural heritage studies in the Louvre Museum (BS/PIXE [260]), and multilayer materials (RBS/EBS/PIXE/HR-PIXE [261]): in the latter case high energy resolution PIXE was also essential.

6.1. *Nuclear and atomic methods compared*

We believe that great synergistic advantage results when photon and particle spectra are analysed self-consistently: while BS spectra typically have rather poor mass resolution, PIXE spectra usually have excellent mass discrimination. Conversely, while depth information is implicit in PIXE spectra, it is explicit in BS spectra.

There are two important ways where the various IBA methods do not match well. Roughly speaking, PIXE X-ray production cross-sections are typically of the order of thousands of barns, RBS scattering cross-sections of the order of barns and NRA reaction cross-sections of the order of milli-barns. Clearly trace element analysis using NRA with microbeams is not usually going to work very well. And handling the noisy BS spectra that are common in μPIXE work has its own pitfalls. Nevertheless, we here show many cases that powerfully demonstrate that important results are available despite this mismatch in signal size.

Just to emphasise that this mismatch between PIXE and PIGE is not limiting, there are *hundreds* of papers using both techniques, from the Namur group on ancient gold jewelry in 1982 [262] to the Florentine group on atmospheric aerosols in 2010 [263], and of course the reason for this is precisely that PIXE runs out of sensitivity for the light atoms (F, Na, Mg, Al) for which there are good PIGE reactions and which are of crucial importance in many materials.

The second mismatch is in information depth. Roughly speaking, the information depth for PIXE is around $20\,\mu$m (more or less depending on the X-ray line), for RBS it is $10\,\mu$m or so (for a proton beam, some five times less for a He beam), this is because the scattered particle may not have the energy to reach the detector from deep in the sample. For NRA it depends on the reaction. But PIGE is very sensitive to the light elements, and γ-rays are excited (for non-resonant reactions) along most of the length of the ion track. γ-rays are not absorbed readily and in most cases it is a reasonable approximation to assume no absorption on the path length to the detector. So, although for BS/PIXE the analyst must beware that the PIXE spectrum is giving information for a much thicker layer than for the BS spectrum, for PIGE/PIXE this mismatch can be a positive advantage. For the important element Na, for example, the PIXE spectrum information depth is very small where that for PIGE is very large: the PIGE/PIXE combination then gives differential information about surface Na loss which is rather hard to obtain otherwise. This is one of the essential points made by both the PIXE/PIGE papers cited.

6.2. *Nuclear and atomic methods combined*

The examples already mentioned of multiple IBA techniques used together mostly handle the individual spectra individually. We will here give a number of examples of *"total IBA"* showing the extra analytical power available where multiple spectra can be handled together self-consistently.

Protein crystallography is an important example where IBA is used to interpret another technique: in this case the bound metal ions or co-factors cannot be uniquely determined by X-ray or NMR methods. PIXE has the sensitivity to detect the metals, which are usually in trace concentrations, and uncertainty is avoided by the use of the intrinsic protein S content as an internal standard for normalisation. The critical feature of this method, on which its accuracy depends, is the use of the BS spectrum by the OMDAQ code for an internally consistent X-ray absorption correction [264] (and see [265] for a recent example on a variety of *E-coli*).

The DataFurnace code was used to analyse Niepce's heliograph of 1827 [266], a 19th century reproduction of Frans Hals' *La Bohémienne* [267], oxidation of carbon nanotubes [268], and photovoltaic and ferroelectric materials [269–271]. The *La Bohémienne* analysis followed a PIXE/BS analysis which was not self-consistent [272], but was itself flawed by an incorrect treatment of the sample roughness. Using a correct analysis of this same data, Molodtsov *et al.* [189] have shown that gross surface roughness can be treated correctly; moreover, IBA can be used in principle with good sensitivity to determine the average roughness parameters of a sample without any prior model and also without any surface contact!

A number of different forensics applications are being developed at present (see a recent collection of papers [274]): in principle the analysis of samples for forensic purposes should be fully quantitative and non-destructive. The previous discussion makes it clear that IBA can be accurate even for complex samples requiring the use of multiple IBA techniques. Gunshot residue (GSR) analysis by IBA looks very promising: interestingly, current police practice uses exclusively qualitative SEM-EDS as a characterisation technique. Different GSR particles which can be recovered from the crime scene and from suspects can be shown to distinguish the primers for the explosive charge used by different gun manufacturers. Electron-induced PIXE has no explicit depth information, and poor sensitivity because of the high primary electron Bremsstrahlung background. SEM-EDS therefore cannot discriminate many modern primers, limiting its usefulness for forensic evidence. We have shown that PIXE/BS can give very sensitive *quantitative* information on many types of primer [275, 276] thereby giving IBA substantially increased discrimination power over SEM-EDS.

6.3. *3D PIXE/BS, the Darwin glass example*

Another interesting example is of the so-called Darwin glasses [277] which are impact glasses resulting from a meteor strike 800,000 years ago near Mt. Darwin in Tasmania. The geologist who collected these glass samples subsequently used one of them as an amorphous standard for setting up his XRD kit, and was astonished to see the diffraction spots of quartz. These crystals — unexpected in a glass! — turned out to be inside *inclusions* in the glass.

But the nature of these inclusions was then entirely unknown to the geologists. IBA analysis unequivocally demonstrated them to be carbonaceous, a result that initially baffled the geologists, for whom such a sample was unprecedented. Moreover, not only did the microbeam PIXE/BS determine the main constituents and demonstrate the great heterogeneity of the samples (both laterally and in depth: see as examples Fig. 28, and Fig. 1 of [277], for example), but the IBA data could be completely quantitatively analysed without any presupposed model despite the heterogeneity.

This sort of mapping microbeam data is effectively a 3D (three-dimensional) data cube, with 128×128 pixels and a PIXE and BS spectrum pair for each pixel. The data cube, intractable as it stands, can be analysed into its principal components (one of which is shown in Fig. 29) by using a multivariate image analysis program such as AXSIA ("Automated eXpert Spectral Image Analysis [278]). These principal components are determined in the spatial domain [279] each giving a pair of PIXE/BS spectra characteristic for a given area of the map, and which can be directly interpreted by DataFurnace. The results of such an analysis are shown in Fig. 30. In principle, the depth profile at each pixel can be

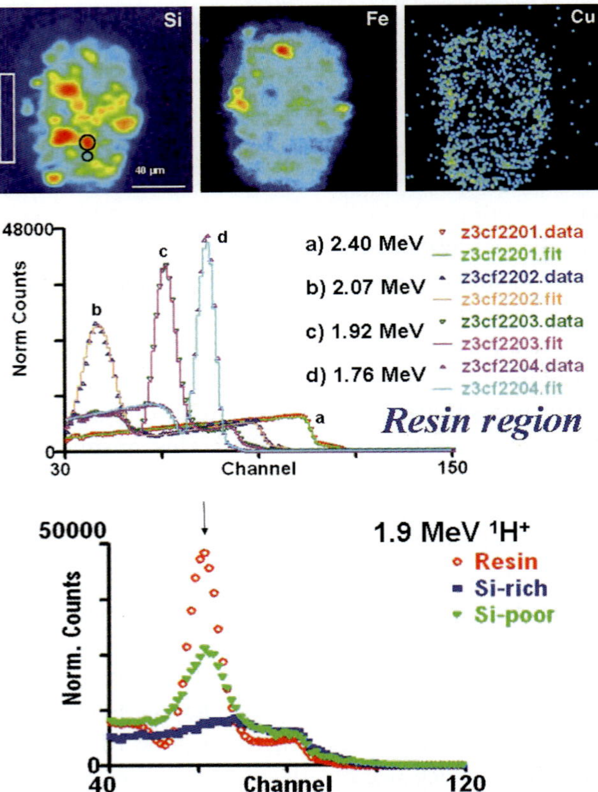

Fig. 28. "Total IBA" of an inclusion in a Darwin glass (see text). *Above*: selected PIXE maps showing distribution of Si, Fe, Cu; *Centre*: BS spectra at varying energies of the resin region showing the $^{12}C(p, p_0)^{12}C$ resonance at 1734 keV; *Below*: BS spectra at 1.9 MeV for three areas marked on the Si PIXE map (above, left). (See Bailey *et al.*, *Nucl. Instrum. Methods B* **267**, 2009, 2219 [277]). Reproduced (and corrected) from Fig. 1 of [273].

reconstructed from a linear combination of the principal components, thereby completely solving the 3D structure of the sample.

6.4. *Mapping and tomography*

Much of the PIXE literature already cited is for microbeam applications. This is accentuated for modern applications since PIXE on its own is analytically very similar to XRF, and cheap desktop XRF tools are now available. The great advantage of PIXE is the ease with which a microbeam can be formed; for XRF, polycapillary lenses are needed: these are not cheap and they have rather poor transmission, and so the probe size available is still not much below 20 µm. For PIXE, 1 µm microbeams have been routine for nearly 20 years, and there are

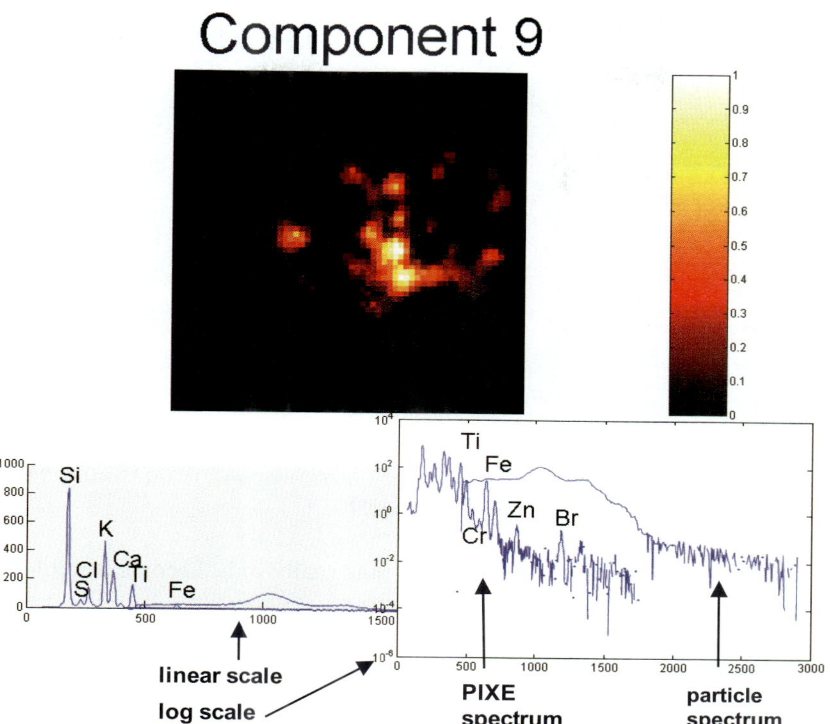

Fig. 29. One component from the principal component decomposition of the data cube of Fig. 28 using AXSIA (see Doyle *et al.*, *Nucl. Instrum. Meth. B* **249**, 828 (2006) [278]). This component is one of the several Si-rich components. Reproduced from Fig. 2 of [273].

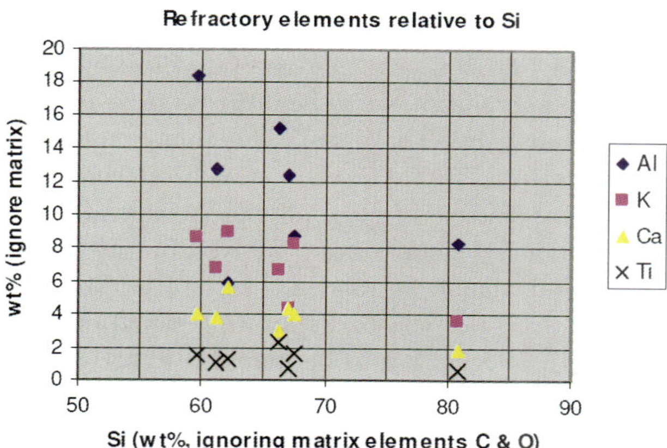

Fig. 30. Composition plot of the abundance of minor elements relative to silicon in seven principal components of a mapped carbonaceous inclusion of a Darwin glass sample (see Fig. 28) analysed by DataFurnace from a decomposition of the data cube by AXSIA; see Fig. 29 and text. Reproduced from Fig. 3 of [273].

persistent promises of much smaller beams. But a 2001 review points out that it has been much harder than expected to get smaller beams: with only a few reports at 500 nm [280]. Nevertheless, several groups are working on beamlines designed to give deep submicron spot sizes [281]. The Singapore group have shown that low current beams down to about 20 nm are feasible (see Fig. 31): PIXE microscopy should be feasible at 100 nm resolution.

The Darwin glass example in the last section points towards tomography. X-ray tomography (XR-T) is already established (see a recent description of a computed tomography system [282], a recent application on sea-urchin teeth [283], and a direct comparison of XR-T with IBA [284]), and STIM-T is an almost equivalent (and solved) problem [285]. Great strides have also been made towards a PIXE-T (see the summary in a recent review [286]), which is qualitatively much more complex than STIM-T (or, equivalently, XR-T). We have shown that in principle IBA-T (that is, using the BS signals as well as the PIXE signals) is already achievable in principle, and should be significantly more efficient (and therefore much faster!) than pure PIXE-T since a single slice already has (nearly) complete 3D information. This is important as tomography is rather slow, and its importance is increased since

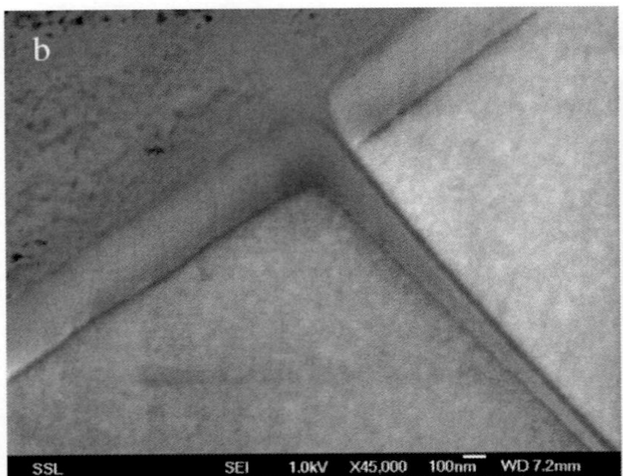

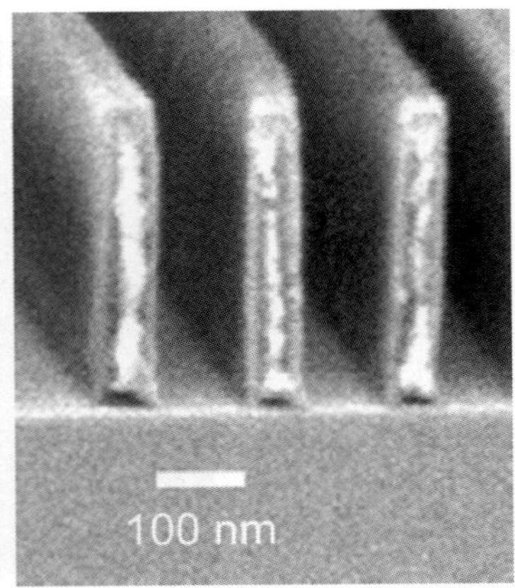

Fig. 31.   *Left*: 22 nm line written in 850 nm thick HSQ[y] with a 1 MeV proton beam and an aspect ratio of about 37. Reproduced from Fig. 3 of van Kan *et al.*, 2006 [288]; *Right*: Parallel lines written in 350 nm thick PMMA with a 2 MeV proton beam and an aspect ratio of about 7. Reproduced from Fig. 4 of van Kan *et al.* [289].

it seems that beam damage severely limits the use of a pure PIXE-T for important classes of samples [287]. In principle, using the depth information available explicitly in IBA-T (from the particle signals) must be quicker than unfolding the depth information available only implicitly (and at much lower resolution) in the PIXE signals.

7. Summary

Thin film depth profiling using MeV-IBA is a major application field for small accelerators. We have here briefly reviewed it, giving a flavour for its development and history, and providing summary details of the critical parameters (theory and databases) needed to obtain reliable analyses. (A further fine review from a different point of view has also recently appeared [290].)

Unusually, we have covered both the photon and the particle techniques in a unified approach, for the reason that we wish to emphasise the enormous benefit available from treating spectra from multiple techniques self-consistently.

There have been really dramatic advances over the last decade in the detailed implementation and in the reliability of the IBA codes used, also in the new availability of reliable EBS cross-sections, and therefore we believe that the approach we have sketched has really only become feasible in the last five years or so.

A modern treatment of IBA spectra would expect to extract depth profiles which give calculated spectra effectively identical to the data: thus we are now usually able to extract from the spectra almost all the information they contain. Where photon or particle spectra separately are ambiguous, jointly analysed (perhaps with multiple energies or beams or geometries) they become unambiguous.

We even expect that systematic procedures could be found to do this analysis pixel-by-pixel with the microbeam, with the ultimate prize of IBA tomography (IBA-T). Where X-ray tomography (or, equivalently, STIM-T) are now standard techniques, PIXE-T is feasible in principle but so slow in practice that most samples are destroyed by the beam. But IBA-T should be (at least) an order of magnitude faster, since the backscattering signal enables the depth information to be unfolded directly from the PIXE data.

We hope that the new power of these old techniques will inspire a new generation of analysts.

Acknowledgments

Chris Jeynes wishes to thank B. L. Doyle, A. F. Gurbich, N. P. Barradas, M. A. Reis, and E. Szilágyi

---

[y] PMMA: poly-methyl methacrylate; HSQ: hydrogen sil-sesquioxane. Note that electron-beam lithography feature size and aspect ratio are both limited by the proximity effect, and deep UV lithography is limited by diffraction effects. These structures cannot be formed by e-beam or UV lithography. Note also that although the e-beam spot size in SEM can be nanometres, the X-ray SEM-EDS maps have a resolution given by the e-beam interaction volume, typically microns. For PIXE the interaction volume is defined by the beam spot size.

for helpful comments, and D. Jeseršek for pointing him to Newton's use of the word "spectrum". Surrey Ion Beam Centre is supported by the European Community as a TransNational Access Infrastructure of the Integrating Activity "Support of Public and Industrial Research Using Ion Beam Technology (SPIRIT)" under EC contract no. 227012. We also acknowledge the support of EPSRC through contract EP/D032210/1 and various EPSRC contracts in the UK academic community.

Appendix A: Glossary of IBA Techniques

ANN   Artificial neural networks. Black boxes containing image recognition software (and no physics!) that instantly recognises (without calculation) features of spectra (layer thicknesses etc.). The performance of ANNs depends critically on their training.

BS   (Elastic) backscattering. Can be either RBS or EBS.

Blocking: Inverse of channelling. Crystal channels have high transparency to collimated ion beams, major axis strings of atoms block the ion beam.

CAICISS: Coaxial impact collision ISS. A variety of LEIS.

Channelling: Damage in single crystals is frequently quantified with ion channelling, where the well collimated ion beam is aligned with major crystallographic axes. Can readily be used with STIM, PIXE, BS, NRA. Channelling and blocking patterns are essential to MEIS.

EBS   Elastic (non-Rutherford) backscattering. The scattering cross-section is given by the elastic scattering channel of the reaction, and depends on the nuclear structure of the two nuclei. Cross-section can be calculated by R-matrix or other methods which have nuclear data (energy levels etc.) as input, but the calculations must be informed by direct cross-section measurements. Measurements and evaluations are on the IBANDL website.

EDS   Energy dispersive (X-ray) spectrometry.

ERD   Elastic recoil detection. Follows the recoiled rather than the scattered ion in the elastic collision. He-ERD is valuable for analysing H isotopes. HI-ERD (heavy ion ERD) typically uses primary beams of ~ 1 MeV/amu, and ToF (time of flight) or gas detectors for the heavy recoils.

ERDA   Elastic recoil detection analysis. Synonym for ERD and preferred in Finland, sounding better in Finnish.

FIB   Focussed ion beam. See SIMS.

FRS   Forward recoil spectrometry. Synonym for ERD.

HI-ERD   Heavy ion ERD. Gas ionisation or ToF detectors, and energies comparable to 1 MeV/ nucleon are usually used.

HI-RBS   Heavy ion RBS (gas ionisation detectors are ideal). A Li beam is often used for better mass resolution.

IBA   MeV Ion Beam Analysis, including STIM, IBIL, PIXE, BS (RBS or EBS), ERD, NRA, PIGE, MeVSIMS. MEIS and LEIS are lower energy versions of RBS. Commercial SIMS instruments use keV energy primary beams.

IBIC   Ion beam induced current. Low fluence technique for non-destructively characterising semiconductor device quality: carrier lifetime, carrier mobility, charge trapping defects etc. Ion analogue of EBIC and XBIC.

IBIL   Ion beam induced luminescence. The ion analogue of cathodoluminescence (electron-induced) and photoluminescence.

ISS   Ion scattering spectrometry. Synonym for one of LEIS, MEIS, RBS, depending on the energy regime.

LEIS   Low energy ion scattering. RBS using keV ion beams. New high sensitivity detectors make this a rapid technique which looks at the outermost layer of the sample. Thus complementary (with higher depth resolution) to XPS.

MEIS   Medium energy ion scattering. RBS using ~ 100 keV ion beams. Gives information on the crystallography and composition of the near-surface region (~ 100 nm). Microbeam Ion beams can be readily focussed with quadrupole triplets (or multiplets) to a focus of $\sim 1\,\mu$m. It is thought that 100 nm is feasible for PIXE and 30 nm is feasible for STIM.

NRA   (Inelastic) nuclear reaction analysis. NRA cross-sections can also be calculated (as well as measured of course) and a few evaluations are on the IBANDL website together with many measured cross-sections.

PDMS   Plasma desorption mass spectrometry. Forerunner of MALDI.

PESA Proton elastic scattering analysis. Synonym for proton EBS, except that PESA can also be at forward angles.

PIGE Particle induced gamma ray emission. A special case of NRA where a gamma ray results.

PIXE Particle induced X-ray emission. The ion analogue of XRF or, since today PIXE is usually used with a scanning microbeam, SEM-EDS or EPMA. Note that EPMA is also PIXE, since electrons are also particles!

RBS Rutherford backscattering spectrometry. Scattering cross-section is analytical, and given by the Coulomb potential (with screening). Ion analogue of the BSE signal in SEM. Switches to EBS when the Coulomb barrier is increased. Called MEIS for beams near the stopping power maximum ($\sim$100 keV), and LEIS for keV beams.

SIM Scanning ion microscopy. Ion analogue of SEM.

SIMS Secondary ion mass spectrometry. Another form of IBA using (for example) a 30 keV ion source for sputtering. The secondary (sputtered) ions are mass analysed. One important variant is FIB (focussed ion beam machining) which uses a high intensity (and very bright) nano-focussed liquid metal ion source (usually Ga): another is MeV-SIMS, where the sputtering results from electronic energy loss, not the nuclear collision cascade.

STIM Scanning transmission ion microscopy. Typically looks at the energy loss of primary beam particles transmitted through thin samples, so that it is similar to EELS in the TEM (but with much lower energy resolution).

ToF Time of flight. ToF-SIMS and ToF-ERD are standard techniques.

WDX Wavelength dispersive X-ray spectrometry (high resolution).

Appendix B: Glossary of Related Techniques

AES Auger electron spectrometry. Also SAM: scanning Auger microscopy. Electrons in, electrons out. Same electron spectrometer as XPS, same EMFP, thus also a true surface technique. AES is really SEM in UHV (ultra-high vacuum), but looking at the (high energy) Auger electrons rather than the number of (low energy) secondary electrons. Chemical shifts are also present in AES, but are more complex than in XPS.

AFM Atomic force microscopy. One of a number of scanning probe microscopies, including the original STM (scanning tunnelling microscopy).

AMS Accelerator mass spectrometry. A form of IBA (the accelerator is the same) where the sample goes in the source. Used routinely for ^{14}C and similar isotopic analyses.

BSE Backscattered electron signal available on SEMs. This carries qualitative Z-contrast.

DESI Desorption electrospray ionisation. As MALDI.

DART Direct analysis in real time. As MALDI.

EBIC Electron beam induced current. Analogue of IBIC.

EDS Energy dispersive X-ray spectrometry ("EDX" is a tradename). Usually used as "SEM-EDS".

EELS Electron energy loss spectrometry. A TEM technique. See XAS.

EPMA Electron probe microanalysis: just an SEM specialised for X-ray analysis, generally with one or more WDXs (wavelength dispersive X-ray spectrometers). See also SEM-EDS.

ESCA Electron spectroscopy for chemical analysis. The photoelectron process can show pronounced chemical shifts for difference valence states. Synonym for XPS.

EXAFS Extended X-ray absorption fine structure. Incident X-ray energy $\sim$50 eV–$\sim$1000 eV above the absorption edge, giving high energy photo-electrons for which *single* scattering dominates. See NEXAFS and XAS.

FTIR Fourier transform infra-red spectrometry. One of a large class of emission and absorption spectrometries sensitive, like Raman spectroscopy, to atomic and molecular vibration modes.

ICP-MS Inductively coupled plasma mass spectrometry. One of a large class of mass spectrometries sensitive to ng/g (and better) where PIXE is only sensitive to mg/kg (at best). But ICP-MS analyses trace elements in bulk samples whose gross composition is known.

LEED Low energy electron diffraction. Often used as a surface monitor in UHV (ultra-high vacuum) deposition systems.

MALDI   Matrix-assisted laser desorption ionization. An in-air spectrometry sensitive to molecules of high molecular weight. SIMS (with keV ions) must be done in vacuum; also gives molecular ions, but is a much more energetic sputtering technique and fragments the sputtered ions more for larger molecules.

NEXAFS   Near-edge extended X-ray absorption fine structure. Incident X-ray energy $\sim$10 eV–$\sim$50 eV above the absorption edge, giving low energy photoelectrons for which *multiple* scattering dominates. See EXAFS and XAS.

NMR   Nuclear magnetic resonance or MRI (magnetic resonance imaging) in imaging mode.

SAM   Scanning AES.

SEM   Scanning electron microscopy for imaging surface topography, primarily looking at the secondary electron signal. Often comes with EDS (or EDX: energy dispersive X-ray spectrometry) and often has a BSE (backscattered electron) signal too. The X-ray detector is the same as usually used for PIXE. The scanning ion microbeam (SIM) is thus an analogue of SEM, EDS and BSE being analogues of PIXE and EBS. And often a secondary electron detector is included in an SIM chamber to see the topography directly.

TEM   Transmission electron microscopy for imaging in both real and reciprocal space: always includes SAD (selected area electron diffraction). Also XTEM for cross-sectional TEM, and HR-TEM for high (atomic) resolution TEM. Often has EDS and EELS.

WDX   Wavelength dispersive X-ray spectrometry. Very high energy resolution is available. See EDS and EPMA.

XANES   X-ray absorption near-edge structure. In the XANES region the energy of the incident beam is $\sim$10 eV from the absorption edge, and transitions of core electrons to non-bound levels with close energy occur with high probability. See XAS.

XAS   X-ray absorption spectrometry. See EELS for the low resolution TEM technique. See XANES, NEXAFS and EXAFS for the high resolution synchrotron techniques which need high intensity monochromatic beams.

XBIC   X-ray beam induced charge (synchrotron technique). Analogue of IBIC.

XPS   X-ray photoelectron spectrometry. X-rays in, photoelectrons out. Because the EMFP (electron mean free path) is only a few nm this is a true surface technique, but sputtering is frequently used to give depth profiles. Synonym for ESCA.

XRD   X-ray diffraction for observing crystalline structure. A very wide variety of methods are in use including thin-film variants.

XRF   X-ray fluorescence. Like PIXE and EPMA but excited by X-rays. Same physics as XPS and AES but looks at the X-ray resultant, not the Auger or photo-electron one, and is therefore a "bulk", not a surface technique.

References

[1] A Letter of Mr. Isaac Newton, Mathematick Professor in the University of Cambridge; containing his New Theory about Light and Colors [*sic*]: Where Light is declared to be not Similar or Homogeneal, but consisting of difform rays, some of which are more refrangible than others: And Colors are affirm'd to be not Qualifications of Light, derived from Refractions of natural Bodies (as 'tis generally believed;) but Original and Connate properties, which in divers rays are divers; Where several Observations and Experiments are alledged to prove the said Theory, *Philos. Trans.* **6** (February 19.1671/72) 3075–3087.

[2] J. W. Mayer and E. Rimini (eds.), *Ion Beam Handbook for Materials Analysis* (Academic Press, New York, 1977). Product of the previous IBA conference at Catania.

[3] J. R. Tesmer and M. Nastasi (eds.), *Handbook of Modern Ion Beam Analysis* (Mat. Res. Soc., Pittsburgh, 1995).

[4] Y. Q. Wang and M. Nastasi (eds.), *2010 Handbook of Modern Ion Beam Analysis*, 2nd edn. (MRS, Pittsburgh).

[5] J. W. Mayer and J. F. Ziegler (eds.), *Proc. Int. Conf. Ion Beam Surface Layer Analysis* (Yorktown Heights, New York, 18–20 June 1973), *Thin Solid Films* **19** (December 1973).

[6] R. Webb, M. Bailey, C. Jeynes and G. Grime (eds.), 12th International Conference on Ion Beam Analysis, University of Cambridge (7–11 September 2009), *Nucl. Instrum. Methods B* **268**, 1703 (2010).

[7] G. W. Grime (ed.), 19th International Conference on Particle-induced X-ray Emission and Its Analytical Applications, University of Surrey (27 June–2 July 2010), *X-Ray Spectrom.* **40**, 119 (2011).

[8] L. Giuntini, *Anal. Bioanal. Chem.* **401**, 785 (2011).

[9] H. Geiger and E. Marsden, *Proc. Roy. Soc. Ser. A* **82**, 495 (1909).

[10] E. Rutherford, *Philos. Mag. Ser. 6* **21**, 669 (1911).

[11] J. Chadwick and E. S. Bieler, *Philos. Mag. Ser. 6* **42**, 923 (1921) .

[12] N. F. Mott, *Proc. Roy. Soc. A* **126**, 259 (1930).

[13] C. Gerthsen, *Annalen der Physik* **401**, 769 (1931).

[14] M. A. Tuve, N. P. Heydenburg and L. R. Hafstad, *Phys. Rev.* **50**, 806 (1936).

[15] G. Breit, E. U. Condon and R. D. Present, *Phys. Rev.* **50**, 825 (1936).

[16] R. A. Laubenstein, M. J. Laubenstein, L. J. Koester and R. C. Mobley, *Phys. Rev.* **84**, 12 (1951).

[17] S. Rubin, T. O. Passell and E. Bailey, *Anal. Chem.* **29**, 736 (1957).

[18] J. A. Davies, G. Amsel and J. W. Mayer, *Nucl. Instrum. Methods B* **64**, 12 (1992).

[19] G. Amsel, P. Baruch and O. Smulkowsi, *Nucl. Instrum. Methods* **8**, 92 (1960).

[20] A. Turkevich, *Science* **134**, 672 (1961).

[21] A. L. Turkevich, E. J. Franzgrote and J. H. Patterson, *Science* **158**, 635 (1967).

[22] J. Gyulai, O. Meyer and J. W. Mayer, *Appl. Phys. Lett.* **16**, 232 (1970).

[23] W. H. Bragg and R. Kleeman, *Philos. Mag.* **10**, 318 (1905).

[24] SRIM — The Stopping and Ranges of Ions in Matter: website http://www.srim.org (downloaded 21 July 2011).

[25] J. F. Ziegler, *Nucl. Instrum. Methods B* **219**, 1027 (2004).

[26] J. F. Ziegler, J. P. Biersack and M. D. Ziegler, SRIM — The Stopping and Range of Ions in Matter (2008) (published by print-to-order) http://www.lulu.com/content/1524197

[27] H. Paul, Stopping Power for Light Ions (downloaded 21 July 2011) http://www.exphys.uni-linz.ac.at/stopping/

[28] H. Paul, *Nucl. Instrum. Methods B* **268**, 3421 (2010).

[29] P. D. Bourland and D. Powers, *Phys. Rev. B* **3**, 3635 (1971).

[30] M. Msimanga, C. M. Comrie, C. A. Pineda-Vargas and S. Murray, *Nucl. Instrum. Methods B* **268**, 1772 (2010).

[31] Z. Siketić, I. Bogdanović Radović, E. Alves and N. P. Barradas, *Nucl. Instrum. Methods B* **268**, 1768 (2010).

[32] S. Damache, D. Moussa and S. Ouichaoui, *Nucl. Instrum. Methods B*, **268**, 1759 (2010).

[33] J. F. Ziegler, M. D. Ziegler and J. P. Biersack, *Nucl. Instrum. Methods B* **268**, 1818 (2010).

[34] H. Geiger and E. Marsden, *Philos. Mag. Ser. 6* **25**, 604 (1913).

[35] J. B. Marion and F. C. Yough, *Nuclear Reaction Analysis, Graphs and Tables* (Wiley, New York, 1968).

[36] H. H. Andersen, F. Besenbacher, P. Loftager and W. Möller, *Phys. Rev. A* **21**, 1891 (1980).

[37] C. Jeynes, N. P. Barradas, H. Rafla-Yuan, B. P. Hichwa and R. Close, *Surf. Interface Anal.* **30**, 237 (2000).

[38] M. Bozoian, K. M. Hubbard and M. Nastasi, *Nucl. Instrum. Methods B* **51**, 311 (1990).

[39] A. F. Gurbich, *Nucl. Instrum. Methods B* **217**, 183 (2004).

[40] R. E. Azuma, E. Uberseder, E. C. Simpson, C. R. Brune, H. Costantini, R. J. de Boer, J. Görres, M. Heil, P. J. LeBlanc, C. Ugalde and M. Wiescher, *Phys. Rev. C* **81** (2010) 045805.

[41] D. Abriola, A. F. Gurbich, M. Kokkoris, A. Lagoyannis and V. Paneta, *Nucl. Instrum. Methods B* **269**, 2011 (2011).

[42] http://www-nds.iaea.org/iba/(downloaded 21 July 2011) — Development of a Reference Database for IBA, *An IAEA Nuclear Data Section Co-Ordinated Research Project*, 2005–2009 (F41023).

[43] D. Abriola, N. P. Barradas, I. Bogdanović Radović, M. Chiari, A. F. Gurbich, C. Jeynes, M. Kokkoris, M. Mayer, A. R. Ramos, L. Shi and I. Vickridge, *Nucl. Instrum. Methods B* **269**, 2972–2978 (2011).

[44] A. F. Gurbich, *Nucl. Instrum. Methods B* **268**, 1703 (2010).

[45] A. F. Gurbich, *The Interaction of Charged Particles with Nuclei*, Chap. 3 in *2010 Handbook of Modern IBA*, eds. Y. Q. Wang and M. Nastasi, 2nd edn., Chap. 3 (MRS, Pittsburgh).

[46] A. F. Gurbich, I. Bogdanović Radović, Z. Siketić and M. Jakšić, *Nucl. Instrum. Methods B* **269**, 40 (2011).

[47] A. F. Gurbich and C. Jeynes, *Nucl. Instrum. Methods B* **265**, 447 (2007).

[48] N. P. Barradas, E. Alves, C. Jeynes and M. Tosaki, *Nucl. Instrum. Methods B* **247**, 381 (2006).

[49] J. L'Ecuyer, C. Brassard, C. Cardinal, J. Chabbal, L. Deschênes, J. P. Labrie, B. Terreault, J. G. Martel and R. St.-Jacques, *J. Appl. Phys.* **47**, 381 (1976).

[50] G. Dollinger, C. M. Frey, A. Bergmaier and T. Faestermann, *Nucl. Instrum. Methods B* **136–138**, 603 (1998).

[51] E. Szilágyi, *Nucl. Instrum. Methods B* **183**, 25 (2001).

[52] M. Döbeli, C. Kottler, F. Glaus and M. Suter, *Nucl. Instrum. Methods B* **241**, 428 (2005).

[53] B. Brijs, T. Sajavaara, S. Giangrandia, T. Janssens, T. Conard, K. Arstila, K. Nakajima, K. Kimura, A. Bergmaier, G. Dollinger, A. Vantomme and W. Vandervorst, *Nucl. Instrum. Methods B* **249**, 847 (2006).

[54] E. Szilágyi, F. Pászti, V. Quillet and F. Abel, *Nucl. Instrum. Methods B* **85**, 63 (1994).

[55] J. F. Browning, R. A. Langley, B. L. Doyle, J. C. Banks and W. R. Wampler, *Nucl. Instrum. Methods B* **161**, 211 (2000).

[56] J. C. Banks, J. F. Browning, W. R. Wampler, B. L. Doyle, C. A. LaDuca, J. R. Tesmer, C. J. Wetteland and Y. Q. Wang, *Nucl. Instrum. Methods B* **219–220**, 444 (2004).

[57] G. Boudreault, R. G. Elliman, R. Grötzschel, S. C. Gujrathi, C. Jeynes, W. N. Lennard, E. Rauhala, T. Sajavaara, H. Timmers, Y. Q. Wang and T. D. M. Weijers, *Nucl. Instrum. Methods B* **222**, 547 (2004).

[58] B. L. Doyle and P. S. Peercy, *Appl. Phys. Lett.* **34** 811 (1979).

[59] F. Pászti, E. Kótai, G. Mezey, A. Manuaba, L. Pócs, D. Hildebrandt and H. Strusny, *Nucl. Instrum. Methods B* **15**, 486 (1986).

[60] N. P. Barradas, S. A. Almeida, C. Jeynes, A. P. Knights, S. R. P. Silva and B. J. Sealy, *Nucl. Instrum. Methods B* **148**, 463 (1999).

[61] H. Timmers, R. G. Elliman and T. R. Ophel, *Nucl. Instrum. Methods A* **447**, 536 (2000).

[62] M. Mallepell, M. Döbeli and M. Suter, *Nucl. Instrum. Methods B* **267**, 1193 (2009).

[63] P. Reichart, G. Datzmann, A. Hauptner, R. Hertenberger, C. Wild and G. Dollinger, *Science* **306**, 1537 (2004).

[64] E. Rutherford, *Philos. Mag.* **37**, 581 (1919).

[65] E. Rutherford and J. Chadwick, *Philos. Mag.* **44**, 417 (1922).

[66] J. D. Cockcroft and E. T. S. Walton, *Nature* **129**, 649 (1932).

[67] W. E. Sweeney Jr. and J. B. Marion, *Phys. Rev.* **182**, 1007 (1969).

[68] See the Q-value calculators at (for example) http://nucleardata.nuclear.lu.se/database/masses/ or http://www.nndc.bnl.gov/qcalc/

[69] I. C. Vickridge, I. Trimaille, J.-J. Ganem, S. Rigo, C. Radtke, I. J. R. Baumvol and F. C. Stedile, *Phys. Rev. Lett.* **89**, 256102 (2002) [4 pages].

[70] A. P. Matthews, C. Jeynes, K. J. Reeson, J. Thornton and N. M. Spyrou, *Nucl. Instrum. Methods B* **64**, 452 (1992).

[71] http://www-nds.iaea.org/pige/publicPIGE.html (downloaded 21 July 2011): IAEA Coordinated Research Project on *"Reference Database for Particle-Induced Gamma Ray Emission (PIGE) Spectroscopy"*.

[72] J. Stark and G. Wendt, *Annalen der Physik* **343**, 921 (1912).

[73] M. M. Bredov and I. M. Okuneva, *Sov. Phys. Dok.* **113**, 795 (1957).

[74] B. Domeij, I. Bergstrom, J. A. Davies and J. Uhler, *Arkiv Fysik* **24**, 399 (1963).

[75] L. Eriksson, J. A. Davies, J. Denhartog, J. W. Mayer, O. J. Marsh and R. Markarious, *Appl. Phys. Lett.* **10**, 323 (1967).

[76] M. T. Robinson and O. S. Oen, *Phys. Rev.* **132**, 2385 (1963).

[77] E. Kótai, *Nucl. Instrum. Methods B* **118**, 43 (1996).

[78] J. Lindhard, *K. Dan. Vidensk. Selsk. Mat. Fys. Medd.* **34**(14), 1 (1965).

[79] C. W. Rischau, C. S. Schnohr, E. Wendler and W. Wesch, *J. Appl. Phys.* **109**, 113531 (2011).

[80] K. Gärtner, *Nucl. Instrum. Methods B* **227**, 522 (2005).

[81] V. S. Drumm, A. D. C. Alves, J. O. Orwa, D. N. Jamieson, J. C. McCallum, S. Prawer and C. C. G. Ryan, *Phys. Stat. Sol. A* **208**, 42 (2011).

[82] L. Rebouta, P. J. M. Smulders, D. O. Boerma, F. Agulló-Lopez, M. F. da Silva and J. C. Soares, *Phys. Rev. B* **48**, 3600 (1993).

[83] A. Kling, *Nucl. Instrum. Methods B* **102**, 141 (1995).

[84] E. Kótai, RBX, Computer methods for analysis and simulation of RBS and ERDA spectra, *Proc. 14th Int. Conf. Application of Accelerators in Research and Industry*, 6–9 November, 1996, Denton, Texas USA, **CP392**, eds. J. L. Duggan and I. L. Morgan (AIP Press, New York, 1997), pp. 631–634.

[85] C. Jeynes, K. E. Puttick, L. C. Whitmore, K. Gärtner, A. E. Gee, D. K. Millen, R. P. Webb, R. M. A. Peel and B. J. Sealy, *Nucl. Instrum. Methods B* **118**, 431 (1996).

[86] D. P. Smith, *Surf. Sci.* **25**, 171 (1971).

[87] H. H. Brongersma, R. H. Bergmans, L. G. C. Buijs, J.-P. Jacobs, A. C. Kruseman, C. A. Severijns and R. G. van Welzenis, *Nucl. Instrum. Methods B* **68**, 207 (1992).

[88] See the Calipso website for up-to-date information on this technique: http://www.calipso.nl/

[89] H. R. J. ter Veen, T. Kim, I. E. Wachs and H. H. Brongersma, *Catal. Today* **140**, 197 (2009).

[90] H. H. Brongersma, M. Draxler, M. de Riddera and P. Bauer, *Surf. Sci. Rep.* **62**, 63 (2007).

[91] W. C. Turkenburg, W. Soszka, F. W. Saris, H. H. Kersten and B. G. Colenbrander, *Nucl. Instrum. Methods* **132**, 587 (1976).

[92] M. Walker, M. G. Brown, M. Draxler, M. G. Dowsett, C. F. McConville, T. C. Q. Noakes and P. Bailey, *Phys. Rev. B* **83** (2011) 085424 (8pp).

[93] J. C. Vickerman, A. Brown and N. M. Reed, *Secondary Ion Mass Spectrometry* (Oxford Science Publications, 1989).

[94] *ToF SIMS: Surface Analysis by Mass Spectrometry*, eds. J. C. Vickerman and D. Briggs (IM Publications, 2001), ISBN 1-901019-03-9.

[95] R. Hill, P. Blenkinsop, S. Thompson, J. C. Vickerman and J. S. Fletcher, *Surf. Int. Anal.* **43**, 506 (2011).

[96] J. A. Kilner, S. J. Skinner and H. H. Brongersma, *J. Solid State Electrochem.* **15**, 861 (2011).

[97] M. G. Dowsett, A. Adriaens, M. Soares, H. Wouters, V. V. N. Palitsin, R. Gibbons and R. J. H. Morris, *Nucl. Instrum. Methods B* **239**, 51 (2005).

[98] J. Bertho, V. Stolojan, M.-L. Abel and J. F. Watts, *Micron* **41**, 130 (2010).

[99] C. W. Magee and R. E. Honig, *Surf. Interface Anal.* **4**, 35 (1982).

[100] J. S. Fletcher, N. P. Lockyer and J. C. Vickerman, *Mass Spectrom. Rev.* **30**, 142 (2011).

[101] H. F. Arlinghaus, *Surface and Thin Film Analysis. Principles, Instrumentation, Applications*, eds. H. Burbert and H. Jennete (Wiley-VCH, Weinheim, 2002), pp. 132–141.

[102] D. F. Torgerson, R. P. Skowronski and R. D. Macfarlane, *Biochem. Biophys. Res. Commun.* **60**, 616 (1974).

[103] R. D. Macfarlane, *Brazilian J. Phys.* **29**, 415 (1999).

[104] Y. Nakata, Y. Honda, S. Ninomiya, T. Seki, T. Aoki and J. Matsuo, *J. Mass Spectrom.* **44**, 128 (2009).

[105] B. N. Jones, V. Palitsin and R. P. Webb, *Nucl. Instrum. Methods B* **268**, 1714 (2010).

[106] B. N. Jones, J. Matsuo, Y. Nakata, H. Yamada, J. Watts, S. Hinder and V. Palitsin, *Surf. Interface Anal.* **43**, 249 (2011).

[107] W. Sichtermann and A. Benninghoven, *Int. J. Mass Spectrom. Ion Processes.* **40**, 177 (1981).

[108] Y. Wakamatsu, H. Yamada, S. Ninomiya, B. N. Jones, T. Seki, T. Aoki, R. P. Webb and J. Matsuo, *Ion Implantation Technology*, eds. J. Matsuo, M. Kase, T. Aoki and T. Seki (AIP, 2010), **CP1321**, pp. 223–236.

[109] M. F. Russo Jr. and B. J. Garrisson, *Anal. Chem.* **78**, 7206 (2006).

[110] M. J. Bailey, B. N. Jones, S. Hinder, J. Watts, S. Bleay and R. P. Webb, *Nucl. Instrum. Methods B* **268**, 1929 (2010).

[111] H. Yamada, Y. Nakata, S. Ninomiya, T. Seki, T. Aoki and J. Tamura, *Surf. Interface Anal.* **43**, 363 (2011).

[112] M. B. H. Breese, P. J. C. King, G. W. Grime and F. Watt, *J. Appl. Phys.* **72**, 2097 (1992).

[113] Z. He, *Nucl. Instrum. Methods A* **463**, 250 (2001).

[114] A. Lo Guidice, Y. Garino, C. Manfredotti, V. Rigato and E. Vittone, *Nucl. Instrum. Methods B* **249**, 213 (2006).

[115] M. C. Veale, P. J. Sellin, J. Parkin, A. Lohstroh, A. W. Davies and P. Seller, *IEEE Trans. Nucl. Sci.* **55**, 3741 (2008).

[116] Ž. Pastuović, E. Vittone, I. Capan and M. Jakšić, *Appl. Phys. Lett.* **98**, 092101 (2011).

[117] M. Zmeck, L. J. Balk, T. Osiepwicz, F. Watt, J. C. H. Phang, A. M. Khambadkone, F.-J. Niedernostheide and H.-J. Schulze, *J. Phys. Condens. Matter.* **16**, S576 (2004).

[118] G. Vizkelethy, D. K. Brice and B. L. Doyle, *J. Appl. Phys.* **101**, 074506 (2007).

[119] J. Morse, M. Salomé, E. Berdermann, M. Pomorski, W. Cunningham and J. Grant, *Diamond Relat. Mater.* **16**, 1049 (2007).

[120] J. Bohon, E. Muller and J. Smedley, *J. Synchrotron Radiat.* **17**, 711 (2010).

[121] S. Camarda, A. E. Bolotnikov, Y. Cui, A. Hossain, S. Awadalla, J. MacKenzie, H. Chen and R. B. James, *IEEE Trans. Nucl. Sci.* **55**(6), 3725 (2008).

[122] Adapted by R. Nave (downloaded 21 July 2011 from Georgia State University website), hyperphysics.phy-astr.gsu.edu/hbase/quantum/moseley.html

[123] H. G. J. Moseley, *Philos. Mag. Ser. 6* **26**, 1024 (1913).

[124] H. G. J. Moseley, *Philos. Mag. Ser. 6* **27**, 703 (1914).

[125] J. Chadwick, *Philos. Mag. Ser. 6*, **25**, 193, (1913).

[126] C. G. Barkla, *Philos. Mag. Ser. 6*, **22**, 396 (1911).

[127] H. R. Bowman, E. K. Hyde, S. G. Thompson and R. C. Jared, *Science* **151**, 562 (1966).

[128] T. B. Johansson, R. Akselsson and S. A. E. Johansson, *Nucl. Instrum. Methods* **84**, 141 (1970).

[129] V. Valkovic, D. Miljanic, R. M. Wheeler, R. B. Liebert, T. Zabel and G. C. Phillips, *Nature* **243**, 543 (1973).

[130] W. Davidson, G. W. Grime, J. A. W. Morgan and K. Clarke, *Nature* **352**, 323 (1991).

[131] J. P. Landsberg, B. McDonald and F. Watt, *Nature* **360**, 65 (1992).

[132] J. Hasegawa, T. Tada, Y. Oguri, M. Hayashi, T. Toriyama, T. Kawabata and K. Masai, *Rev. Sci. Instrum.* **78**(7), 073105 (2007).

[133] M. Kavcić, A. G. Karydas and C. Zarkadas, *X-Ray Spectrom.* **34**, 310 (2005).

[134] M. A. Reis, L. C. Alves, N. P. Barradas, P. C. Chaves, B. Nunes, A. Taborda, K. P. Surendran, A. Wu, P. M. Vilarinho and E. Alves, *Nucl. Instrum. Methods B* **268**, 1980 (2010).

[135] M. Kavcić, M. Zitnik, K. Bucar and J. Szlachetko, *X-Ray Spectrom.* **40**(1), 2 (2011).

[136] I. Orlic, F. Watt, K. K. Loh and S. M. Tang, *Nucl. Instrum. Methods B* **85**, 840 (1994).

[137] H. Paul and J. Sacher, *A. Data Nucl. Data Tables* **42**, 105 (1989).

[138] M. A. Reis and A. P. Jesus, *At. Data Nucl. Data Tables* **63**, 1 (1996).

[139] A. Taborda, P. C. Chaves and M. A. Reis, *X-Ray Spectrom.* **40**, 127 (2011).

[140] G. Lapicki and J. Miranda, *X-Ray Spectrom.* **40**, 122 (2011).

[141] J. L. Campbell, N. I. Boyd, N. Grassi, P. Bonnick and J. A. Maxwell, *Nucl. Instrum. Methods B* **268**, 3356 (2010).

[142] C. Doppler, Über das farbige Licht der Doppelsterne und einiger anderer Gestirne des Himmels, *Proceedings of the Bohemian Society of Sciences* (1843). Doppler published this privately in 1842.

[143] J. S. Russell (1848), *Report of the 18th Meeting of the British Association for the Advancement of Science* (John Murray, London) **18**, 37 (1849).

[144] See W. Thomson, *Rep. Brit. Assoc.* **xcix** (1872).

[145] N. Bohr, *Philos. Mag.* **26**, 1 (1913).

[146] M. H. Chen, B. Crasemann and H. Mark, *Phys. Rev. A* **21**, 436 (1980).

[147] M. H. Chen and B. Crasemann, *Phys. Rev. A* **24**, 177 (1981).

[148] M. H. Chen and B. Crasemann, *Phys. Rev. A* **27**, 2989 (1983).

[149] J. L. Campbell, *At. Data Nucl. Data Tables* **85**, 291 (2003).

[150] J. L. Campbell, *At. Data Nucl. Data Tables* **95**, 115 (2009).

[151] J. L. Campbell, P. L. McGhee, J. A. Maxwell, R. W. Ollerhead and B. Whittaker, *Phys. Rev. A* **33**, 986 (1986).

[152] J. L. Campbell and J.-X. Wang, *At. Data Nucl. Data Tables* **43**, 281 (1989).

[153] M. H. Chen and B. Crasemann, *Phys. Rev. A* **30**, 170 (1984).

[154] M. A. Reis, P. C. Chaves and A. Taborda, *X-Ray Spectrom.* **40**, 141 (2011).

[155] M. J. Berger and J. H. Hubbell, *NIST X-ray and Gamma-ray Attenuation Coefficients and Cross Sections Database, NIST Standard Reference Database 8, Version 2.0*, National Institute of Standards and Technology, Gaithersburg, MD (1990).

[156] M. J. Berger, J. H. Hubbell, S. M. Seltzer, J. Chang, J. S. Coursey, R. Sukumar, D. S. Zucker and K. Olsen, XCOM: Photon Cross Section Database (v. 1.5). http://physics.nist.gov/xcom (downloaded 23 July 2011). National Institute of Standards and Technology, Gaithersburg, MD, 2010.

[157] F. Jalilehvand, University of Calgary: www.chem.ucalgary.ca/research/groups/faridehj/xas.pdf, downloaded 23 July 2011.

[158] B. Beckhoff, *J. Anal. At. Spectrom.* **23**, 845 (2008).

[159] B. L. Henke, E. M. Gullikson and J. C. Davis, *At. Data Nucl. Data Tables* **54**, 181 (1993).

[160] International Fundamental Parameters Initiative http://exsa.pytalhost.net/FP-XRF by EXSA http://www.exsa.hu/ (downloaded 27 July 2011).

[161] F. Rizzo, G. P. Cirrone, G. Cuttone, A. Esposito, S. Garraffo, G. Pappalardo, L. Pappalardo, F. P. Romano and S. Russo, *Microchem. J.* **97**, 286 (2011).

[162] D. Sokaras, A. G. Karydas, A. Oikonomou, N. Zacharias, K. Beltsios and V. Kantarelou, *Anal. Bioanal. Chem.* **395**, 2199 (2009).

[163] J. R. Michalski and P. B. Niles, *Nat. Geosci.* **3**(11), 751 (2010).

[164] A. N. Mansour and C. A. Melendres, *J. Phys. Chem. A* **102**, 65 (1998).

[165] E. Lombi, K. G. Scheckel and I. M. Kempson, *Environ. Exp. Botany* **72**, 3 (2011).

[166] M. P. Seah and S. J. Spencer, *Surf. Interface Anal.* **37**, 731 (2005).

[167] M. A. Reis, N. P. Barradas, P. C. Chaves and A. Taborda, *X-Ray Spectrom.* **40**, 153 (2011).

[168] P. A. Mando, M. E. Fedi and N. Grassi, *Eur. Phys. J. Plus* **126**(4), 41 (2011).

[169] E. Rauhala, N. P. Barradas, S. Fazinić, M. Mayer, E. Szilágyi and M. Thompson, *Nucl. Instrum. Methods B* **244**, 436 (2006).

[170] N. P. Barradas, K. Arstila, G. Battistig, M. Bianconi, N. Dytlewski, C. Jeynes, E. Kótai, G. Lulli, M. Mayer, E. Rauhala, E. Szilágyi and M. Thompson, *Nucl. Instrum. Methods B* **262**, 281 (2007).

[171] N. P. Barradas, K. Arstila, G. Battistig, M. Bianconi, N. Dytlewski, C. Jeynes, E. Kótai, G. Lulli, M. Mayer, E. Rauhala, E. Szilágyi and M. Thompson, *Nucl. Instrum. Methods B* **266**, 1338 (2008).

[172] M. Mayer, *Technical Report IPP9/113*, Max-Planck-Institut für Plasmaphysik, Garching, Germany (1997).

[173] M. Mayer, *AIP Conf. Proced.* **475**, 541 (1999).

[174] M. Mayer, SIMNRA Version 5.0 User's Guide, max-Planck-Institut für Plasmaphysik, 1997–2002 (156pp) www2.if.usp.br/~lamfi/guia-simnra.pdf (downloaded 25 July 2011).

[175] C. Jeynes, N. P. Barradas, P. K. Marriott, G. Boudreault, M. Jenkin, E. Wendler and R. P. Webb, *J. Phys. D: Appl. Phys.* **36**, R97 (2003).

[176] N. P. Barradas and C. Jeynes, *Nucl. Instrum. Methods B* **266**, 1875 (2008).

[177] www.surreyibc.ac.uk/ndf (downloaded 25 July 2011).

[178] C. J. Tavares, L. Rebouta, E. Alves, N. P. Barradas, J. Pacaud and J. P. Rivière, *Nucl. Instrum. Methods B* **188**, 90 (2002).

[179] W. H. Press, B. P. Flannery, S. A. Teukolsky and W. T. Vetterling, *Numerical Recipes* (Cambridge University Press, Cambridge, New York, 1988).

[180] N. P. Barradas, C. Jeynes and R. P. Webb, *Appl. Phys. Lett.* **71**, 291 (1997).

[181] N. P. Barradas, C. Jeynes, M. Jenkin and P. K. Marriott, *Thin Solid Films* **343–344**, 31 (1999).

[182] L. Wielopolski and R. P. Gardner, *Nucl. Instrum. Methods* **133**, 303 (1976).

[183] N. P. Barradas and M. A. Reis, *X-Ray Spectrom.* **35**(4), 232 (2006).

[184] S. L. Molodtsov and A. F. Gurbich, *Nucl. Instrum. Methods B* **267**, 3484 (2009).

[185] W. N. Lennard and G. R. Massoumi, *Nucl. Instrum. Methods B* **48**, 47 (1990).

[186] C. Pascual-Izarra and N. P. Barradas, *Nucl. Instrum. Methods B* **266**, 1866 (2008).

[187] M. Mayer, *Nucl. Instrum. Methods B* **194**, 177 (2002).

[188] N. P. Barradas, *J. Phys. D: Appl. Phys.* **34**, 2109 (2001).

[189] S. L. Molodtsov, A. F. Gurbich and C. Jeynes, *J. Phys. D: Appl. Phys.* **41**, 205303 (2008) (7pp).

[190] Y. Sunitha, G. L. N. Reddy, S. Kumar and V. S. Raju, *Appl. Surf. Sci.* **256**, 1553 (2009).

[191] N. P. Barradas, *Nucl. Instrum. Methods B* **190**, 247 (2002).

[192] J. P. Stoquert and T. Szörenyi, *Phys. Rev. B* **66**, 144108 (2002).

[193] M. Mayer, U. von Toussaint, J. Dewalque, O. Dubreuil, C. Henrist, R. Cloots and F. Mathis, *Nucl. Instrum. Methods B*, doi:10.1016/j.nimb. 2011.07.045. Available online 27 July 2011.

[194] N. P. Barradas, *Nucl. Instrum. Methods B* **261**, 418 (2007).

[195] N. P. Barradas, *Nucl. Instrum. Methods B* **225**, 318 (2004).

[196] E. Szilágyi, F. Pászti and G. Amsel, *Nucl. Instrum. Methods B* **100**, 103 (1995).

[197] E. Szilágyi, *Nucl. Instrum. Methods B* **161–163**, 37 (2000).

[198] E. Szilágyi, The DEPTH code homepage. http://www.kfki.hu/~ionhp/doc/prog/mdepth.htm (downloaded 25 July 2011).

[199] M. Mayer, *Nucl. Instrum. Methods B* **266**, 1852 (2008).

[200] F. Schiettekatte, *Nucl. Instrum. Methods B* **266**, 1880 (2008).

[201] F. Schiettekatte, The CORTEO code homepage. www.lps.umontreal.ca/~schiette/index.php?n= Recherche.Corteo (downloaded 26 July 2011).

[202] K. Arstila, T. Sajavaara and J. Keinonen, *Nucl. Instrum. Methods B* **174**, 163 (2001).

[203] I. Vickridge and G. Amsel, *Nucl. Instrum. Methods B* **45**, 6 (1990).

[204] Link on SPIRIT website (downloaded 26 July 2011) http://www.spirit-ion.eu/Techniques/IBA-software.html

[205] N. P. Barradas, R. Mateus, M. Fonseca, M. A. Reis, K. Lorenz and I. Vickridge, *Nucl. Instrum. Methods B* **268**, 1829 (2010).

[206] H. W. Lewis, *Phys. Rev.* **125**, 937 (1962).

[207] J. A. Maxwell, J. L. Campbell and W. J. Teesdale, *Nucl. Instrum. Methods B* **43**, 218 (1989).

[208] K. Ishii and S. Morita, *Phys. Rev. A* **30**, 2278 (1984).

[209] J. L. Campbell, W. Maenhaut, E. Bombelka, E. Clayton, K. Malmqvist, J. A. Maxwell, J. Pallon and J. Vandenhaute, *Nucl. Instrum. Methods B* **14**, 204 (1986).

[210] J. L. Campbell, J. A. Maxwell, S. M. Andrushenko, S. M. Taylor, B. N. Jones and W. Brown-Bury, *Nucl. Instrum. Methods B* **269**(1), 57 (2011).

[211] J. L. Campbell, A. M. McDonald, G. M. Perrett and S. M. Taylor, *Nucl. Instrum. Methods B* **269**(1), 69 (2011).

[212] GeoPIXE home page, downloaded 26 July 2011: http://nmp.csiro.au/GeoPIXE.html

[213] C. G. Ryan, D. R. Cousens, S. H. Sie, W. L. Griffin, G. F. Suter and E. Clayton, *Nucl. Instrum. Methods B* **47**, 55 (2000).

[214] C. G. Ryan, D. N. Jamieson, C. L. Churms and J. V. Pilcher, *Nucl. Instrum. Methods B* **104**, 157 (2005).

[215] C. G. Ryan, R. Kirkham, R. M. Hough, G. Moorhead, D. P. Siddons, M. D. de Jonge, D. J. Paterson, G. De Geronimo, D. L. Howard and J. S. Cleverley, *Nucl. Instrum. Methods A* **619**, 37 (2010).

[216] J. Pallon, C. G. Ryan, N. A. Marrero, M. Elfman, P. Kristiansson, E. J. C. Nilsson and C. Nilsson, *Nucl. Instrum. Methods B* **267**, 2080 (2009).

[217] J. Boman and J. Isakson, *X-Ray Spectrom.* **20**, 305 (1991).

[218] B. Vekemans, K. Janssens, L. Vincze, F. Adams and P. Van Espen, *X-Ray Spectrom.* **23**, 278 (1991).

[219] K. Janssens, B. Vekemans, F. Adams, P. van Espen and P. Mutsaers, *Nucl. Instrum. Methods B* **109**, 179 (1996).

[220] IAEA X-Ray Fluorescence Laboratory: Quantitative X-Ray Analysis System (QXAS) and AXIL tools available www.iaea.org/OurWork/ST/NA/ NAAL/pci/ins/xrf/pciXRFdown.php (downloaded 26 July 2011).

[221] WinAxil X-Ray Analysis Software (S-5005) http://www.canberra.com/products/438012.asp (downloaded 26 July 2011).

[222] G. W. Grime and M. Dawson, *Nucl. Instrum. Methods B* **104** 107 (1995).

[223] C. Pascual-Izarra, M. A. Reis and N. P. Barradas, *Nucl. Instrum. Methods B* **249**, 780 (2006).

[224] C. Pascual-Izarra, N. P. Barradas and M. A. Reis, *Nucl. Instrum. Methods B* **249**, 820 (2006).

[225] M. A. Reis, L. C. Alves and A. P. Jesus, *Nucl. Instrum. Methods B* **109–110**, 134 (1996).

[226] M. A. Reis, P. C. Chaves, L. C. Alves and N. P. Barradas, *X-Ray Spectrom.* **37**, 100 (2008).

[227] A. Mantero, H. B. Abdelouahed, C. Champion, Z. El Bitar, Z. Francis, P. Guèye, S. Incerti, V. Ivanchenko and M. Maire, *X-Ray Spectrom.* **40**(1), 2 (2011).

[228] M. Blaauw, J. L. Campbell, S. Fazinić, M. Jakšić, I. Orlić and P. Van Espen, *Nucl. Instrum. Methods B* **189**, 113 (2002).

[229] IAEA-TECDOC-1342: *Intercomparison of PIXE Spectrometry Software Packages* (IAEA, Vienna, 2003).

[230] M. J. Bailey, S. Coe, D. M. Grant, G. W. Grime and C. Jeynes, *X-Ray Spectrom.* **38**(4), 343 (2009).

[231] N. P. Barradas, A. Vieira and R. Patricio, *Phys. Rev. E* **65**(6), 066703 (2002).

[232] M. M. Li, W. Guo, B. Verma, K. Tickle and J. O'Connor, *Neural Comput. Appl.* **18**(5), 423 (2009).

[233] J. Demeulemeester, D. Smeets, N. P. Barradas, A. Vieira, C. M. Comrie, K. Temst and A. Vantomme, *Nucl. Instrum. Methods B* **268**, 1676 (2010).

[234] N. P. Barradas, A. Vieira and R. Patricio, *Nucl. Instrum. Methods B* **190**, 231 (2002).

[235] *Guide to the Expression of Uncertainty in Measurement*, International Organization for Standardization, Geneva, Switzerland (1995), ISBN 92-67-10188-9.

[236] K. A. Sjöland, F. Munnik and U. Wätjen, *Nucl. Instrum. Methods B* **161**, 275 (2000).

[237] M. P. Seah, D. David, J. A. Davies, C. Jeynes, C. Ortega, C. Sofield and G. Weber, *Nucl. Instrum. Methods B* **30**, 140 (1988).

[238] M. P. Seah *et al.*, *Surf. Interface Anal.* **36**, 1269 (2004).

[239] U. Wätjen and H. Bax, *Nucl. Instrum. Methods B* **85**, 627 (1994).

[240] K. H. Ecker, U. Wätjen, A. Berger, L. Persson, W. Pritzcow, M. Radtke and H. Riesemeier, *Nucl. Instrum. Methods B* **188**, 120 (2002).

[241] G. Boudreault, C. Jeynes, E. Wendler, A. Nejim, R. P. Webb and U. Wätjen, *Surf. Interface Anal.* **33**, 478 (2002).

[242] C. Jeynes, N. Peng, N. P. Barradas and R. M. G. William, *Nucl. Instrum. Methods B* **249**, 482 (2006).

[243] C. Jeynes and N. P. Barradas, Pitfalls in Ion Beam Analysis, Chapter 15 in *2010 Handbook of Modern IBA*, eds. Y. Q. Wang and M. Nastasi, 2nd edn. (MRS, Pittsburgh).

[244] H. Paul, private communication (1st July 2011).

[245] C. Jeynes, Z. H. Jafri, R. P. Webb, A. C. Kimber and M. J. Ashwin, *Surf. Interface Anal.* **25**, 254 (1997).

[246] R. Fischer, M. Mayer, W. von der Linden and V. Dose, *Phys. Rev. E* **55**, 6667 (1997).

[247] M. Kopecek, L. Bacakova, J. Vacik, F. Fendrych, V. Vorlicek, I. Kratochvilova, V. Lisa, E. Van Hove, C. Mer, P. Bergonzo and M. Nesladek, *Physica status solidi (a)* **205**, 2146 (2008).

[248] Z. Siketić, I. B. Radović, E. Alves and N. P. Barradas, *Nucl. Instrum. Methods B* **268**(11–12), 1768 (2010).

[249] E. Rauhala, *Nucl. Instrum. Methods B* **12**(4), 447 (1985).

[250] R. Salomonović, *Nucl. Instrum. Methods B* **82**(1), 1 (1993).

[251] N. P. Barradas, A. R. Ramos and E. Alves, *Nucl. Instrum. Methods B* **266**(8), 1180 (2008).

[252] G. Calzolai, M. Chiari, F. Lucarelli, F. Mazzei, S. Nava, P. Prati, G. Valli and R. Vecchi, *Nucl. Instrum. Methods B* **266**, 2401 (2008).

[253] G. W. Grime, *Nucl. Instrum. Methods B* **109–110**, 170 (1996).

[254] I. Orlić, S. J. Zhou, J. L. Sanchez, F. Watt and S. M. Tang, *Nucl. Instrum. Methods B* **150**, 83 (1999).

[255] H. Bureau, C. Raepsaet, H. Khodja, A. Carraro and C. Aubaud, *Geochimica et Cosmochimica Acta* **73**(11), 3311 (2009).

[256] J. Lao, J. M. Nedelec and E. Jallot, *J. Phys. Chem. C* **112**(25), 9418 (2008).

[257] R. Rajendran, M. Q. Ren, M. D. Ynsa, G. Casadesus, M. A. Smith, G. Perry, B. Halliwell and F. Watt, *Biochem. Biophys. Res. Commun.* **382**(1), 91 (2000).

[258] M. Arroyo-Hernandez, M. Manso-Silvan, E. Lopez-Elvira, A. Munoz, A. Climent and J. M. Martinez Duart, *Biosensors Bioelectron.* **22**(12), 2786 (2007).

[259] R. Huszank, A. Simon, E. Szilágyi, K. Keresztessy and I. Kovács, *Nucl. Instrum. Meth. B* **267**, 2132 (2009).

[260] L. de Viguerie, L. Beck, J. Salomon, L. Pichon and Ph. Walter, *Anal. Chem.* **81**, 7960 (2009).

[261] M. A. Reis, L. C. Alves, N. P. Barradas, P. C. Chaves, B. Nunes, A. Taborda, K. P. Surendran, A. Wu, P. M. Vilarinho and E. Alves, *Nucl. Instrum. Methods B* **268**(11–12), 1980 (2010).

[262] G. Demortier and F. Bodart, *J. Radioanal. Chem.* **69**, 239 (1982).

[263] F. Lucarelli, S. Nava, G. Calzolai, M. Chiari, R. Udisti and F. Marino, *X-Ray Spectrom.* **40**(1), 162 (2011).

[264] E. F. Garman and G. W. Grime, *Progr. Biophys. Mol. Biol.* **89**, 173 (2005).

[265] S. C. Willies, M. N. Isupov, E. F. Garman and J. A. Littlechild, *J. Biol. Inorg. Chem.* **14**(2), 201 (2009).

[266] C. Pascual-Izarra, N. P. Barradas, M. A. Reis, C. Jeynes, M. Menu, B. Lavedrine, J. J. Ezrati and S. Röhrs, *Nucl. Instrum. Methods B* **261**, 426 (2007).

[267] L. Beck, C. Jeynes and N. P. Barradas, *Nucl. Instrum. Methods B* **266**, 1871 (2008).

[268] J. C. G. Jeynes, C. Jeynes, K. J. Kirkby, M. Rümmeli and S. R. P. Silva, *Nucl. Instrum. Methods B* **266**, 1569 (2008).

[269] V. Corregidor, P. C. Chaves, M. A. Reis, C. Pascual-Izarra, E. Alves and N. P. Barradas, *Mat. Sci. Forum* **514**, 1603 (2006).

[270] C. Jeynes, G. Zoppi, I. Forbes, M. J. Bailey and N. Peng, Characterisation of thin film chalcogenide PV materials using MeV IBA, *IEEE Proc. Supergen Conf.*, Nanjing (April 2009), DOI:10.1109/ SUPERGEN.2009.5348162.

[271] M. A. Reis, N. P. Barradas, C. Pascual-Izarra, P. C. Chaves, A. R. Ramos, E. Alves, G. Gonzalez-Aguilar, M. E. V. Costa and I. M. M. Salvado, *Nucl. Instrum. Methods B* **261**, 439 (2007).

[272] L. Beck, P. C. Gutiérrez, J. Salomon, Ph. Walter and M. Menu, in *Proceedings of the XI International Conference on PIXE and Its Analytical Applications*, Puebla, Mexico (May 25–29, 2007).

[273] C. Jeynes, M. J. Bailey, N. J. Bright, M. E. Christopher, G. W. Grime, B. N. Jones, V. V. Palitsin and R. P. Webb, *Nucl. Instrum. Methods B* **271**, 107–118 (2012), doi: 10.1016/j.nimb.2011.09.020.

[274] M. J. Bailey (ed.), *Surf. Interface Anal.* **42**(5), 339 (2010).

[275] M. J. Bailey, K. J. Kirkby and C. Jeynes, *X-Ray Spectrom.* **38**(3), 190 (2009).

[276] M. J. Bailey and C. Jeynes, *Nucl. Instrum. Methods B* **267**, 2265 (2009).

[277] M. J. Bailey, K. T. Howard, K. J. Kirkby and C. Jeynes, *Nucl. Instrum. Methods B* **267**, 2219 (2009).

[278] B. L. Doyle, P. P. Provencio, P. G. Kotula, A. J. Antolak, C. G. Ryan, J. L. Campbell and K. Barrett, *Nucl. Instrum. Methods B* **249**, 828 (2006).

[279] M. R. Keenan, *Surt. Interface Anal.* **41**, 79 (2009).

[280] N. David and N. Jamieson, *Nucl. Instrum. Methods B* **181**, 11 (2001).

[281] M. J. Merchant, G. W. Grime, K. J. Kirkby and R. Webb, *Nucl. Instrum. Methods B* **260**, 8 (2007).

[282] O. Hagiwara, M. Watanabe, E. Sato, H. Matsukiyo, A. Osawa, T. Enomoto, J. Nagao, S. Sato, A. Ogawa and J. Onagawa, *Nucl. Instrum. Methods B* **638**(1), 165 (2011).

[283] C. E. Killian, R. A. Metzler, Y. T. Gong, T. H. Churchill, I. C. Olson, Trubetskoy, M. B. Christensen, J. H. Fournelle, F. De Carlo, S. Cohen, J. Mahamid, A. Scholl, A. Young, A. Doran, F. H. Wilt, S. N. Coppersmith and P. U. P. A. Gilbert, *Adv. Funct. Mater.* **21**(4), 682 (2011).

[284] I. Gomez-Morilla, T. Pinheiro, S. Odenbach and M. D. Y. Alcala, *Nucl. Instrum. Methods B* **267**, 2103 (2009).

[285] T. Satoh, M. Oikawa and T. Kamiya, *Nucl. Instrum. Methods B* **267**, 2125 (2009).

[286] C. G. Ryan, *Nucl. Instrum. Methods B* **269**, 2151–2162 (2011), doi: 10.1016/j.nimb.2011.02.046 (available online 27 February 2011).

[287] T. Andrea, M. Rothermel, R. Werner, T. Butz and T. Reinert, *Nucl. Instrum. Methods B* **268**, 1884 (2010).

[288] J. A. van Kan, A. A. Bettiol and F. Watt, *Nano Lett.* **6**, 579 (2006).

[289] J. A. van Kan, A. A. Bettiol, K. Ansari, E. J. Teo, T. C. Sum and F. Watt, *Int. J. Nanotechnol.* **1**, 464 (2004).

[290] M. Mayer, W. Eckstein, H. Langhuth, F. Schiettekatte and U. von Toussaint, *Nucl. Instrum. Methods B* **269**, 2972–2978 (2011), doi:10.1016/j.nimb.2011.04.066.

Chris Jeynes studied Physics at Bristol University, taking a PhD there in 1981 ("A proposed diamond polishing process", *Phil. Mag. A*, **48**, 1983, 169–197). He joined the University of Surrey in 1981, and the Ion Beam Centre in 1982, immediately developing an interest in accurate ion beam analysis. In 1997, with Nuno Barradas (now in Lisbon) and Roger Webb, he published the basic DataFurnace paper (*Appl. Phys. Lett.* **71**, 1997, 291–293). This code has become (with Matej Mayer's SIMNRA) one of the two "New Generation" codes and is the only one to support self-consistent fitting of PIXE and particle scattering.

Roger Webb took his PhD in 1980 from the University of Salford in mathematical modelling of atomic collisions in solids, an interest he pursued for the next three years in the Naval Postgraduate School in Monterey (California), and still the main area of his research activities. He joined the Ion Beam Centre in 1983, becoming Professor in 2002. He is now the Ion Beam Centre Director, and enthusiastically developing the exciting new field of MeV-SIMS.

Annika Lohstroh took her Diploma in Physics at Göttingen University. She developed an interest in solid state physics during a 6-month exchange programme with the Catholic University Louvain. She subsequently obtained her PhD at the University of Surrey, researching charge transport properties in wide band gap semiconductors for radiation detection applications with a particular emphasis on CVD diamond. She extended these studies as a Research Fellow, and was appointed as Lecturer in the Physics Department in 2008. Her scientific aim is to understand the limitations of current radiation detector performance and to find ways to overcome these limits.

Reviews of Accelerator Science and Technology
Vol. 4 (2011) 83–101
© World Scientific Publishing Company
DOI: 10.1142/S1793626811000495

Neutrons and Photons in Nondestructive Detection

J. F. Harmon,* D. P. Wells† and A. W. Hunt‡

Idaho Accelerator Center, Idaho State University,
921 So. 8th Avenue, Stop 8263

Physical Sciences, Idaho State University,
921 So. 8th Avenue, Stop 8106,
Pocatello, ID 83209, USA
**harmon@iac.isu.edu*
†wells@iac.isu.edu
‡alwhunt@iac.isu.edu

Active, nondestructive interrogation with neutrons and photons has seen a renaissance in recent years, owing to a broad spectrum of important applications in security, nuclear nonproliferation, contraband detection and materials analysis. Active methods are of high interest for such applications because they provide at least an order of magnitude greater sensitivity than passive methods. Accelerator-based neutron and photon active methods exploit two important factors to attain greater sensitivity: these are (i) the control of interrogating beam properties such as directionality, energy, intensity, polarization and the temporal distribution of radiation; (ii) well-founded, low energy nuclear physics that yields distinct "signatures" for elemental and isotopic content. This review addresses accelerator-based neutron and photon nondestructive testing methods and issues when applied to modern and emerging wide-ranging challenges in nondestructive detection.

Keywords: Active interrogation; contraband detection; materials accountability; fissile material accountability.

1. Introduction

This review article is concerned with active interrogation methods using accelerator-produced radiation, neutrons produced by either plasmas or positive beams on targets, or photons produced by electron beams on converter targets. These neutrons or photons stimulate nuclear reactions in objects. The detection of radiation from interrogation-induced reactions constitutes the evidence used to determine the composition or contents of the object. Active techniques which make use of neutrons and photons from radioisotopic sources are not discussed, although the physical setup may be similar to those used in accelerator methods. In active interrogation schemes the intent is that no damage is done to the sampled object and, aside from induced short-lived radioactivity, the process is nondestructive. Active interrogation does expose the object and its surroundings with ionizing radiation. This consequence raises health and safety concerns in certain applications. Consideration of this aspect of the technology must be part of any application planning.

In most cases active detection schemes provide at least an order of magnitude greater sensitivity than passive detection techniques [1], particularly for detection of fissionable material in challenging and shielded environments. Active processes are a subclass of nondestructive evaluation (NDE) methods and are often referred to as active inspection, nondestructive evaluation and, if quantitative, nondestructive assay (NDA). Past reviews and detailed discussion of specific devices for nuclear fuel assay, fissile material detection etc. can be found in reports [2] and books [3, 4] which carry information up to 1981. The Gozani work [4] has a "biblical" reputation in the field. As new situations appear requiring nondestructive methods, there is often in published reports a "rediscovery" of previously known techniques set in the new context. Frequently in these reports, proper citation is not given to the original work. Because of this, unless the novelty of the method of application is significantly different from the original, the work is not cited in this review. Further, in the past decade there has been a significant increase in the commercial aspects of the development of many of these technologies, and this too has led to inadequate reporting in the scientific literature. As the physics that has been exploited is rather limited, what has brought the major innovation in the past 30–40 years has been the occasional introduction of new detector types and new accelerator types and, most recently, vastly increased computer power for control and data handling. A principal interest here is in manifestations of NDE techniques in "out of laboratory" or "in field" type situations. Not considered are topics such as low energy x-ray radiography, elemental detection based on characteristic x-rays and the like.

1.1. *Early development*

Perhaps the earliest use of an accelerator-based NDE system in a field setting was the University of Illinois Betatron, transplanted to Los Alamos for the Manhattan Project, and used in the early 1940s for high speed radiography in implosion experiments. After WWII the onset of large scale nuclear power development, in the period 1950–1970, was the impetus for the growth of both passive and active inspection. Inspection techniques using radioactive sources and, later, accelerators, were employed for nondestructive assay of nuclear materials in safeguards, nonproliferation, waste assay, nuclear fuel burnup measurements, etc. Initial work on active methods was done at LANL in the 1950s, with further development at commercial laboratories; Monsanto, General Atomics, Varian Associates and others. A major need was to account for nuclear material in, for example, the examination of fuel elements for fissile content and general safeguard concerns [2–4]. Neutron and photon-based techniques for the nondestructive elemental characterization of materials in processing applications have been in commercial operation dating from the same period, radioactive waste remediation and online analysis in the minerals industry being examples [5]. Increased concerns about illicit drug trafficking and terrorism in recent decades have prompted much interest in exploiting these techniques for the nonintrusive detection of explosives, bioagents, drugs, and special nuclear materials hidden at various transport venues, as in Fig. 1 [6, 7]. Reconsideration of used nuclear fuel recycling by the US has again brought to the fore the need for nondestructive, real time accountancy

Fig. 1. An experimental apparatus to demonstrate large scale interrogation technologies, located at the Idaho Accelerator Center. The container can be moved at various speeds, the shed houses accelerator hardware and the frame can be used to mount detector arrays.

and control of nuclear materials in reprocessing scenarios [8].

The principal technological advance that made possible active forms of NDE/NDA using neutrons and electrons/photons was the development, and subsequent commercialization, of practical accelerator sources of neutrons and electron/photons for inducing useful signals from targeted substances. The DT neutron generator came from work done at Los Alamos National Laboratory in the early 1950s [9], and later in that decade from the same laboratory came the coupled cavity electron linear accelerator (linac) [10]. The latter device was adapted rapidly by several companies for medical application in radiation oncology. The oil well services industry adapted the DT neutron generator as a down-hole tool for well logging. As a consequence of these commercial interests and others, the technology of high output neutron generators and compact linacs became available from various vendors in the 1960s.

1.2. *General considerations*

Neutron- and photon-based probes detect the presence of materials by stimulating a variety of nuclear reactions. These processes can be categorized by the type of probing source particle, by the type of signature particles, and by the energy and behavior in time of these quantities. Unlike chemical methods such as optical spectroscopy, nuclear methods can, at best, determine only elemental and isotopic information. Chemical compositions of drugs and explosives may be inferred from elemental composition to first order, i.e. detection of chlorine as a sign of narcotics, or ratios of elemental constituents, such as in explosive detection using carbon, nitrogen and oxygen ratios. Only if the chemical compositions of the materials are sufficiently different from the surrounding matrix can an accurate identification be made with reasonable probability. As in all nuclear measurements the desired signatures must compete with background signals so that the sensitivity and specificity of a particular examination system hinge on its ability to distinguish the signature of the material of interest from the background.

Generally speaking, the nuclear physics that can be practically applied to the problems confronted in NDE is limited to nuclear phenomena that take place at relatively low stimulation neutron or photon energies — a few to a few tens of MeV. These phenomena are reasonably well understood, even though some needed basic data remain undetermined. The following two sections review the production of stimulating radiations, neutrons and photons, and the important nuclear reactions they induce. In the first section will be found brief discussions on neutron sources and reactions which neutrons undergo, along with examples of applications. Discussion of electron sources and their use in the production of photons and applications of photonuclear reactions follows.

2. Neutron Sources and Induced Nuclear Reactions

The typical compact DT neutron generator takes advantage of the fusion reactions, ^{3}H(d, n)^{4}He (DT) or the ^{2}H(d, n)^{3}He (D, D) reaction, produced by deuterons impinging on a tritium- or deuterium-loaded target [11]. The former produces 14.1 MeV neutrons, the latter 2.5 MeV neutrons. The cross sections of these reactions allow considerable neutron production even with low accelerating potentials of a few hundred kilovolts, the DT reaction about 100 times more prolific than the DD reaction at 300 keV deuteron energy. The neutrons produced from both reactions are termed "fast neutrons."

Sealed tube versions of these small "accelerators" incorporating ion source and target are powered by pulsed power supplies and can provide fluencies up to $\sim 10^9$ n/s for DT. Often these devices are configured in a cylindrical form a few inches in diameter and two or three feet long, with the pulsed

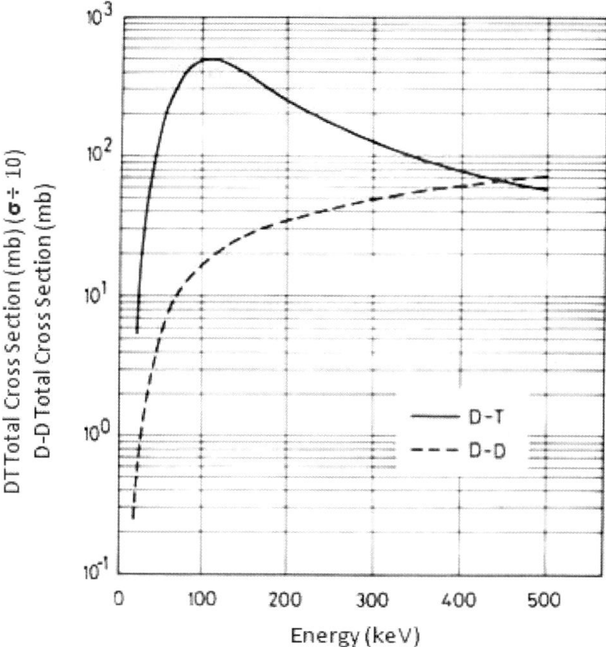

Fig. 2. Total cross sections of the DT and DD reactions. Note that for DT the values on the vertical axis must be multiplied by 10. (Figure adapted from Ref. 11.)

power supply built into the device. The pulses of neutrons produced are typically variable, from a few microseconds to tens of microseconds, repeated with variable rates up to a few kilohertz. For the DT generator the energy spectrum is dominated by 14 MeV neutrons with a small background of lower energy ^{2}H(d, n)^{3}He neutrons produced by the buildup of deuterium in the tritium-loaded target. Modern versions of this type of neutron source are commercially available. They may also be equipped with an internal particle detector to take advantage of associated particles produced along with neutrons, the signals from which form the basis for an elemental imaging NDE technique discussed in Subsec. 4.1. The DT systems so equipped are run in a nonpulsed continuous mode to allow for coincidence detection counting techniques. Plasmas formed in deuteron and deuteron–tritium mixtures by plasma focus and other high energy density schemes can produce neutron pulses. While outputs from this type of device can be highly reproducible shot to shot, heat production and limited lifetime become issues.

Neutrons can be produced by other positive-ion-induced reactions at higher energies; protons and deuterons above a few MeV on lithium and beryllium

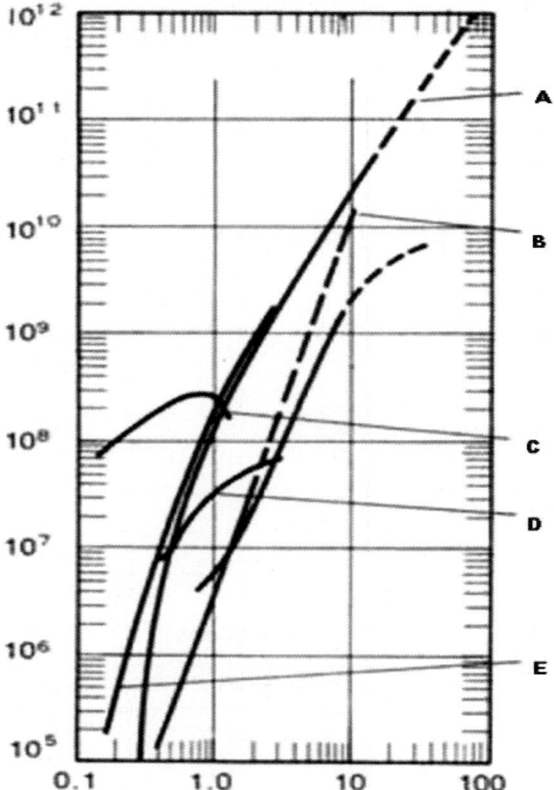

Fig. 3. Neutron yields from thick natural targets. Units of the vertical axis are neutrons/microcoulomb and those of the horizontal axis are MeV. A: Be(d, n). B: C(d, n). C: D, T (titanium tritide target). D: D, D (titanium deuteride target). E: Li(d, n). Unmarked curve: Be(alpha, n). (Figure adapted from Ref. 11.)

can produce a neutron yield of order 10^{11} $\mathrm{n^{-1}kW^{-1}}$ of beam power. Neutron yields from various light element targets are depicted in Fig. 3.

The nature of the accelerators required increases the cost per neutron significantly over the DT source, and for some applications the size of the required apparatus can be limiting. This is not to say that these factors are prohibitive; Gozani and his colleagues have developed and deployed contraband detection systems based upon tandem electrostatic accelerators in the 4 MeV class which can produce 8 MeV deuteron beams. The physics of this technology will be discussed below. An important feature of most of the low energy neutron sources is that the neutrons are emitted into 4π steradians and thus large losses accrue when attempts are made to form directed beams. This statement is modified for neutrons produced by DD reactions produced by deuteron beams of several MeV (as used by Gozani

et al. and discussed below); in this case there are appreciable forward-directed kinematic effects on the emitted neutrons.

2.1. *Neutron sources and induced nuclear reactions*

A large number of neutron-based NDE techniques have been proposed and some have reached deployment in various NDE applications. Neutrons introduced into matter will with very high probability be involved in nuclear reactions. The reactions will depend on the incident energy of the neutron and on the nuclear characteristics of the target nuclide. Most nuclides exhibit distinctive structure in their reaction probabilities (cross sections) which enhances the probability of certain reactions occurring at specific incident neutron energies and scattering angles. For example, major constituents of illicit drugs and explosive substances (H, C, N and O) differ in properties sufficiently to allow them to be discriminated from one another. The exploited reactions between a neutron and the nucleus are:

(1) *Elastic scattering* (n, n). The neutron is involved in a scattering process that does not alter the internal energy of the nucleus involved. These processes tend to equilibrate the neutron kinetic energy with that of the scattering matter, i.e. "thermalize" the neutron energy.

(2) *Inelastic scattering* $(n, n\gamma)$. The neutron scatters and leaves the nucleus in an excited state. In subsequent de-excitation a gamma ray is emitted. These reactions exhibit energy thresholds and generally require fast neutrons.

(3) *Particle emission* $(n,$ *charged particle*$)$. After the interaction the de-excitation of the target nucleus involves the emission of a charged particle, usually a proton. The reaction transmutes the target nucleus. The transformed nucleus may undergo further radioactive decays.

(4) *Radiative capture* (n, γ). The most common neutron-induced nuclear reaction. The compound nucleus emits a gamma photon. The probability of radiative capture increases as the neutron energy decreases, although there can be strong resonances in the capture probability as a function of neutron energy that are unique to the nucleus.

Table 1.   Essentials of selected neutron-based interrogation methods.

Technique	Neutron energy	Reaction	Source
Thermal neutron analysis (TNA)	14 MeV	(n, γ) activation	(D, T)
Fast neutron analysis (FNA)	14 MeV	$(n, n'\gamma)$	(D, T)
Pulsed fast neutron analysis (PFNA)	8–10 MeV	$(n, n'\gamma)$	(D, D)
Associated particle imaging (API)	14 MeV	$(n, n'\gamma)$	(D, T)

(5) *Fission* (n, f). This is a unique property of fissionable nuclei that includes those materials that can fuel nuclear weapons. After neutron excitation the signatures of prompt and delayed neutrons and gamma rays are clear indications of the presence of fissionable material in the target matter.

The ways in which the possible neutron-induced nuclear reactions are used in various active interrogation techniques are briefly discussed below and some of the characteristics of each are summarized in Table 1, where the abbreviations refer to the names of the various techniques.

2.2. *Thermal neutron method*

In thermal neutron analysis (TNA), neutrons generated by any process are thermalized by elastic scattering, sometimes using moderating hydrogenous materials, and sometimes by the matrix of the interrogated object itself. This process takes advantage of the increase in radiative capture probability for thermal neutrons. After undergoing capture, gamma rays characteristic of the target nuclides are detected with a high resolution detector(s). For example, the presence of explosives may be inferred based on the identification of nitrogen and hydrogen using detection of the 2.22 and 10.83 MeV capture gamma rays from hydrogen and nitrogen, for narcotics one looks for chlorine [12]. Good energy resolution in the detector is clearly necessary in order to pick out the capture gamma rays of the target material from those of the matrix. Arrays of detectors can allow some spatial discrimination within the interrogated object. Unlike neutron activation using a reactor, short exposure to

neutrons means only short half-life activities are produced. Applications of TNA are generally restricted to small objects of luggage size because of the limited volume over which a thermal neutron field can be established using a practical accelerator-driven neutron source.

2.3. *Fast neutron methods*

The fast neutron methods discussed here are fast neutron analysis (FNA), pulsed fast neutron analysis (PFNA) and associated particle imaging (API).

2.3.1. *Fast neutron analysis*

In FNA a pulse of 14 MeV neutrons from a (D, T) neutron source microseconds in length is directed into the object [13]. The (D, T) production reaction gives an isotropic distribution of neutrons which is collimated into an appropriate beam. Clearly, considerable neutron intensity must be sacrificed in beam formation, increasing as the beam size is reduced to improve the resolution of image formation. The neutron beam interacts in the object, beginning the sequence of nuclear reactions as the neutrons, on average, lose energy by elastic scattering in the object. The reactions occurring during the neutron pulse are mostly inelastic scattering when the neutrons have energy above inelastic thresholds. As the neutrons thermalize in the object after the pulse, neutron capture reactions increasingly take place. Distinctions can be made among these events based on time and energy using detector arrays and high speed electronics to extract information sufficient for constructing two-dimensional images. In Fig. 4 is shown the inelastic-scattering-produced energy spectral data of a narcotic analog. The data in time, as the neutron beam moves past the contraband, would show changes in the peaks. Depending on the matrix material, the chlorine peaks might disappear, the carbon peaks change in amplitude relative to the oxygen peaks, etc. [13]. Since fast neutrons have a greater range in objects than do thermal neutrons, larger objects can be examined than in the case of TNA.

2.3.2. *Pulsed fast neutron analysis*

Nanosecond pulses of nearly monoenergetic fast neutrons are used in PFNA, which provides sharp timing

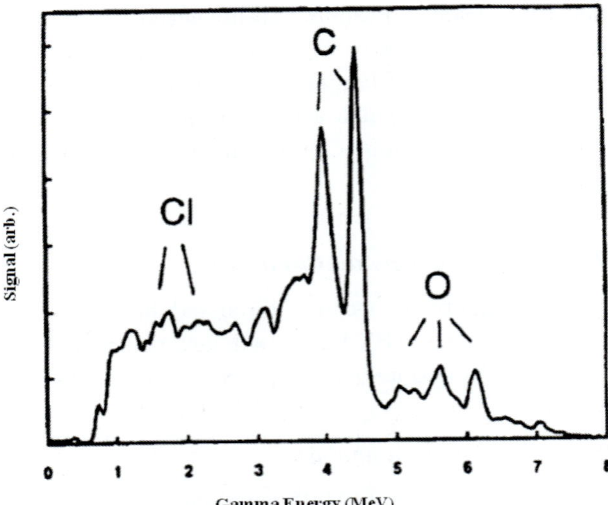

Fig. 4. NaI (Tl) gamma ray spectra data from a narcotic analog sample in a PFN experiment.

signals, allowing neutron time of flight to be coupled with the penetrating and multi-interaction capability of fast neutrons. A tandem electrostatic accelerator is employed to produce nanosecond bunches of 8.5 MeV neutrons using deuterons on a deuterium gas target. The kinematics of this neutron production reaction give forward-directed neutrons which can be collimated to a defined beam with reasonably high neutron efficiency. All of the interactions that occur in FNA can take place but there is now a time correlation between the initiating neutron pulse and the fast-neutron-induced inelastic scattering gamma rays. This can give location information in three dimensions of the nuclei producing the fast-neutron-induced gamma rays, because the time of flight of neutrons and inelastic gammas are known, thus facilitating three-dimensional elemental imaging. High energy neutrons efficiently used, along with large efficient detector arrays, allow cargo-container-sized objects to be examined (Figs. 5 and 6) [14–16].

2.3.3. *Associated particle imaging*

API is based on the observation that a 3.5 MeV alpha particle is emitted simultaneously with the emission of the 14 MeV neutron in the DT reaction. Kinematics demand that the two particles be emitted on paths 180° apart. The associated particle interacts with a position-sensitive detector within the vacuum

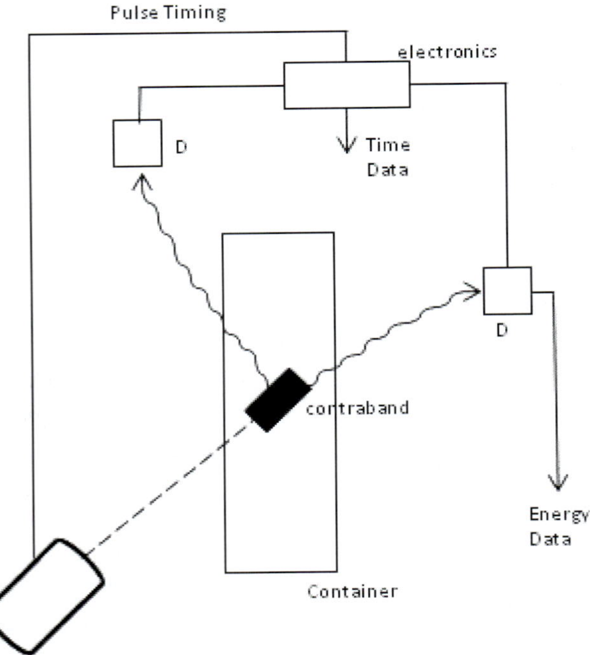

Fig. 5. Schematic of the PFNA method. Coupling inelastic gamma detection with time of flight of photons and neutrons relative to initiating accelerator neutron pulses can provide three-dimensional information about the elemental composition of the inspected item.

Fig. 6. A 4 MV terminal tandem electrostatic accelerator of the type used for PNFA. The tank, about 25 ft long and 8 ft in diameter, is the accelerator. The equipment in the foreground is the ion source; the neutron production target would be found on the far end of the tank.

space near the tritium target in the accelerator, as sketched in Fig. 7. As noted above, neutron DT generators with internal alpha detectors are commercially available. The solid angle intercepted by

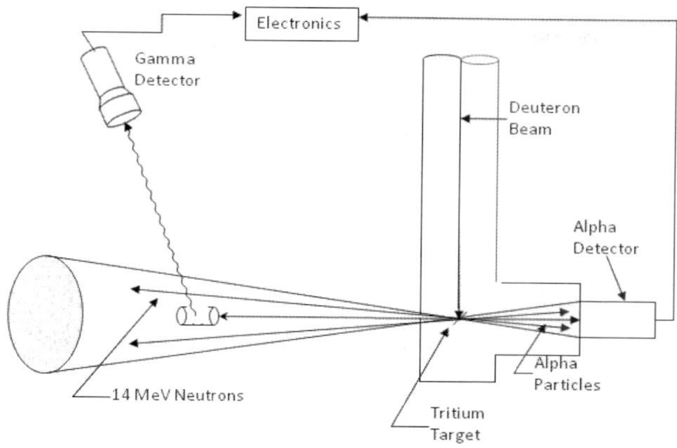

Fig. 7. Schematic of the associated particle method. Note that coincidence detection requires a continuous beam, limiting the lifetime of the sealed tube DT generator.

the alpha detector defines the number and direction of the tagged neutrons. Should one of these tagged neutrons undergo an $(n, n'\gamma)$ reaction in the interrogated object and the emitted gamma ray being detected, its origin can be correlated with the alpha signal by direction and time of flight giving the location at which the $(n, n'\gamma)$ scattering took place.

There is considerable elegance to the API method; inelastic scattering events are detected in a similar manner to PFNA but very fast pulses are not needed to set time on events. Now the accelerator is a simpler DT generator that has a continuous beam. However, the price of simplicity is a serious limitation on count rates. Neutron fluencies are necessarily kept low ($\sim 10^6\,\mathrm{n^{-1}s^{-1}}$) to limit accidental coincidence events, which means that long measurement times are needed [17].

Several other neutron-based interrogation techniques, which are to a large extent variations and combinations of the above, have been advanced in recent years, largely as a response to the needs of cargo screening for contraband and border security [13].

2.4. *Fissile material detection using neutrons*

An NDE application of neutron-induced fission, (n, f), is found in differential die-away analysis, a sensitive method for detecting fissile nuclei in a mass of other materials [22]. A measurement is initiated in a material by a pulse of neutrons from a neutron source, often a DT generator. Due to absorption and diffusion, the population of fast neutrons exponentially decays as the neutrons scatter down in energy after the pulse, as shown by the neutron detector response in Fig. 8(a) (the so-called neutron die-away time). The — now thermalized — neutrons from the source can cause fissions in fissile isotopes like ^{235}U or ^{239}Pu. A properly shielded neutron detector that is preferentially sensitive epithermal prompt fission neutrons will respond to the induced fissions [Fig. 8(b)]. Repeating accelerator pulses allows the data to be accumulated and measurement statistics to improve. A drum scanner that uses the differential die-away technique is shown in Fig. 7. This is an example of a type of assay system that has been used in the DOE complex for several decades to determine the mass of fissile isotopes ^{235}U, ^{239}Pu, etc. in waste drums [3, 4, 18]. High precision and sensitivity can be achieved, and $\sim 10\,\mathrm{mg}$ of U or Pu can be detected in many cases.

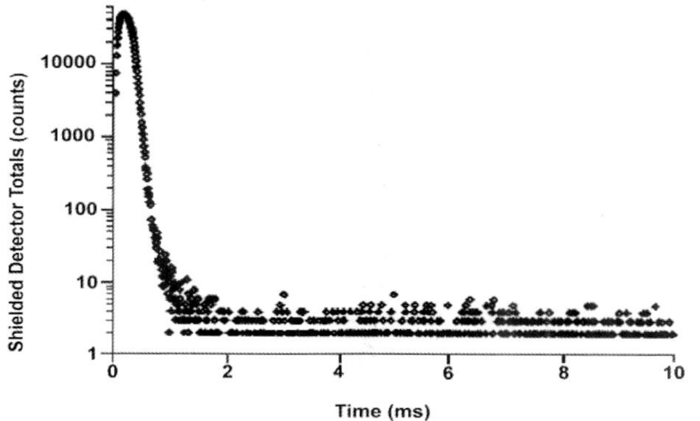

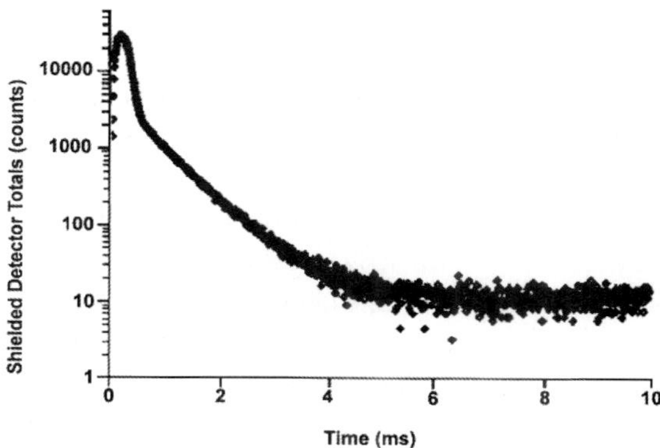

Fig. 8(a). Differential die-away in a blank on the left and on the right with 11 g of plutonium oxide [18].

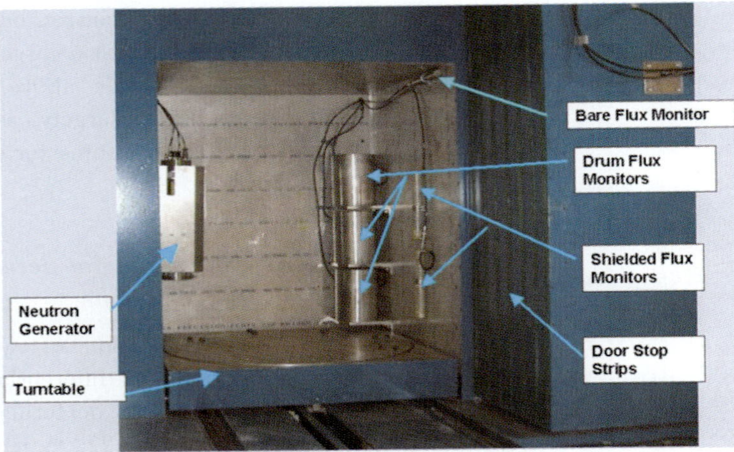

Fig. 8(b).   The interior of Los Alamos National Laboratory's differential die-away waste drum assay device. The turntable supports a standard 50-gallon drum.

3. Electron Accelerators, Photon Production and Nuclear Reactions

Restricting the discussion to available compact accelerators that are useful for interrogation, which are those with output beam energies above 5 MeV, reduces choices considerably. Also eliminated from this discussion due to their bulk are direct voltage machines such as Cockcroft–Walton or Van de Graaff. This leaves resonance devices; betatrons, rodatrons, microtrons and linacs. A compact betatron is shown in Fig. 9; this version is based on an air core magnet and is lightweight, but beam power is quite low at a few 10s of watts. Rodatrons are available up to 10–15 MeV but are large at these energies; however, beam duty factors are high, so that beam power may be tens of kilowatts. Microwave linacs can produce a wide range of energies and are used for the bulk of interrogation applications.

3.1. *Electron linacs*

Linacs typically produce electron beams in pulses of duration of a few microseconds or less. These are repeated with a period ranging from seconds to milliseconds (thus the frequency of beam pulses can be up to several kHz). The resulting duty factors, the fraction of time the beam is on, are typically less than 1%. The pulsed nature of these devices is necessitated by the high microwave powers required in the accelerating structures during acceleration. Typically, the microwave energy is supplied by klystrons or magnetrons at the 2–5 MW peak

Fig. 9.   A commercial 2.5 MeV betatron type electron accelerator. Higher energy models with up to 7 MeV are available. Photon beams are emitted from the area defined by the square marking on the lower left at the case. Power supplies are not shown.

level at frequencies of ~ 3 GHz. High peak powers are achievable for reasonable average line power consumption using energy compression techniques. In spite of the pulsed nature of the output, and in some cases because of it, linacs have a number of attractive features as active inspection interrogation sources.

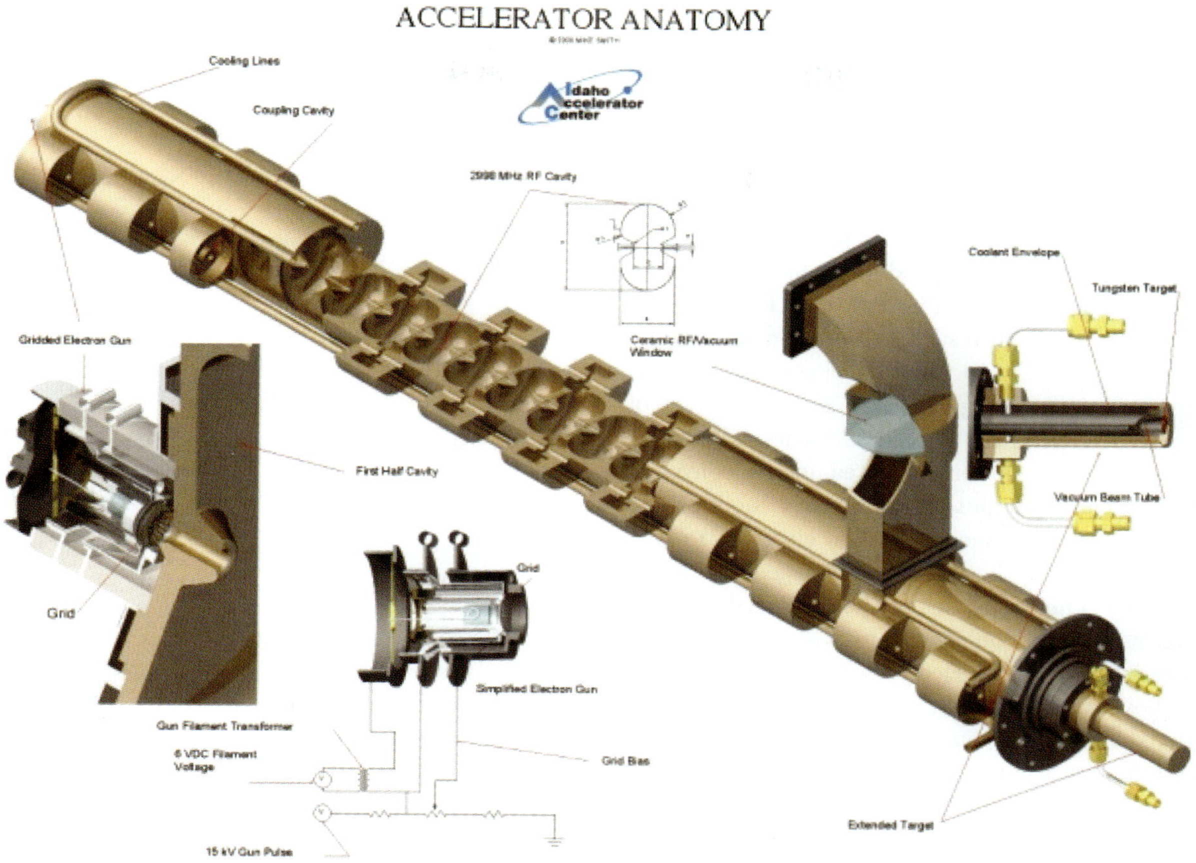

ACCELERATOR ANATOMY

Fig. 10. Drawing of a microwave-powered linac; the number of microwave cavities is appropriate for $\sim 10\,\mathrm{MeV}$. Input microwave power of $\sim 2\,\mathrm{MW}$ (peak) would be supplied by a klystron or magnetron.

Most importantly, there are several useful signature signals prompted by the interrogating signals available. In addition:

(1) The accelerators with electron energies up to $\sim 25\,\mathrm{MeV}$ or more are reliable, compact, mobile and are available essentially off the shelf. Tens of thousands are used for radiation oncology and as irradiators for food and other products throughout the world.

(2) If fitted with appropriate converter targets, photon radiation can be produced that exhibits a high degree of directivity and good penetration into all types of materials.

(3) Converter targets can be designed so that neutrons can be produced, $\sim 10^{12}\,\mathrm{ns^{-1}kW^{-1}}$ at $20\,\mathrm{MeV}$. These neutrons have a fission like energy spectrum into 4π steradians.

(4) Measurements of the stimulated signals can be made during the accelerator on time ("flash"),

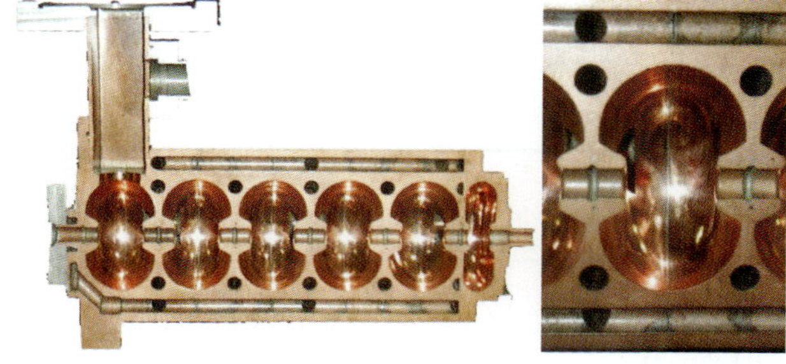

Fig. 11. On the left is a cross-sectional cut through a $4\,\mathrm{MeV}$ side-coupled cavity electron linac; the electrons enter from the right. The right hand picture is a closeup of a single cavity. A peak voltage of $\sim 800\,\mathrm{kV}$ is developed across each cavity for the typical $\sim 2\,\mathrm{MW}$ peak input power.

between flashes, or after a specified period of irradiation.

(5) Electron beam energy, output current, pulse width and other parameters can be controlled from pulse to pulse if needed and pulse widths,

amplitudes and rates can be varied over fairly wide dynamic ranges.

3.2. *Photon production*

When employed as an interrogation source electron, accelerators are generally used to produce photon beams. An electron beam falling on an appropriate converter target produces photons via the bremsstrahlung process. The converter is typically a high-Z refractory metal such as tantalum or tungsten a few millimeters thick. The thickness is adjusted to optimize the energy-spectral characteristics of the photons, and the thickness varies with electron energy. The power deposited in the converter is usually high enough to require water cooling. The optimum target thickness will allow electrons to pass through; so, often, the converter is backed by a low-Z beam stop. The properties of the produced photons change as the electron energy increases. Above a few hundred kilovolts the electrons become relativistic and above a MeV the electrons are highly so. Due to the relativistic nature of the individual electron interactions in the converter, the photons produced are highly forward-directed, and with proper converter design this feature can be maintained macroscopically. The half-angle of the forward "spike" is about $\approx E(\text{MeV})/100$ in degrees. The percentage of electron energy converted to photon energy in the high-Z target (e.g. tungsten) increases with electron kinetic energy, being ~ 10–20% at $5\,\text{MeV}$ and up to 60% or so at $50\,\text{MeV}$. Figure 12 shows the

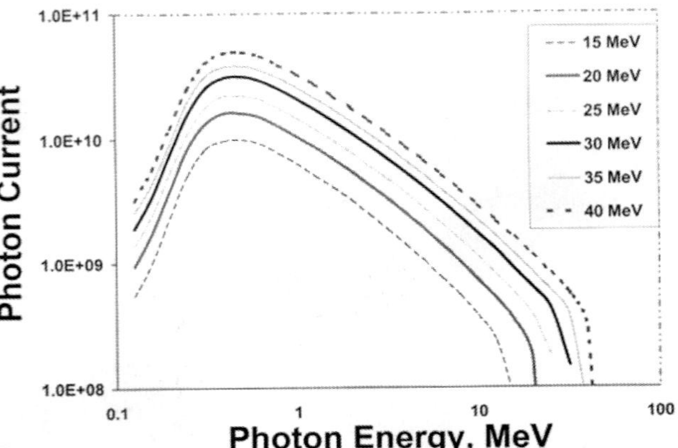

Fig. 12. Sketch of bremsstrahlung spectra from a converter target as a function of electron beam energy [photon current $(\text{s}^{-1}\text{mA}^{-1}\text{MeV}^{-1})$ vs. electron beam energy]. The family of curves represents successively higher electron energies. The photon energy extends up to the end point energy of the incident electrons.

form of the energy spectrum for bremsstrahlung; the photon energy extends up to the end point energy of the incident electrons. A useful compilation of bremsstrahlung properties can be found in Ref. 19.

3.3. *Photon interactions with nuclei*

A sketch of the photon cross section of a mid-to-high-Z nucleus is shown in Fig. 13, illustrating the individual narrow nuclear resonance florescence lines at low energy below the onset of the photonuclear giant dipole resonance (GDR). Then the GDR is followed by higher energy reactions of smaller cross sections.

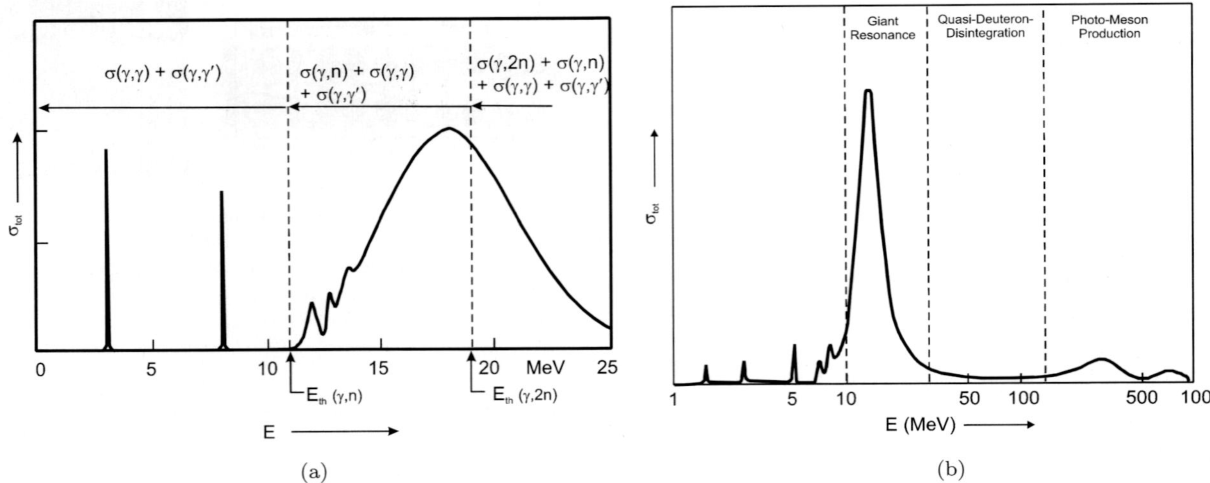

Fig. 13. Sketch of the photon nucleus cross section as a function of energy, showing (a) the narrow nuclear resonance florescence lines at low energy below the onset of the GDR, then the GDR, and (b) the GDR followed by weaker higher energy reactions.

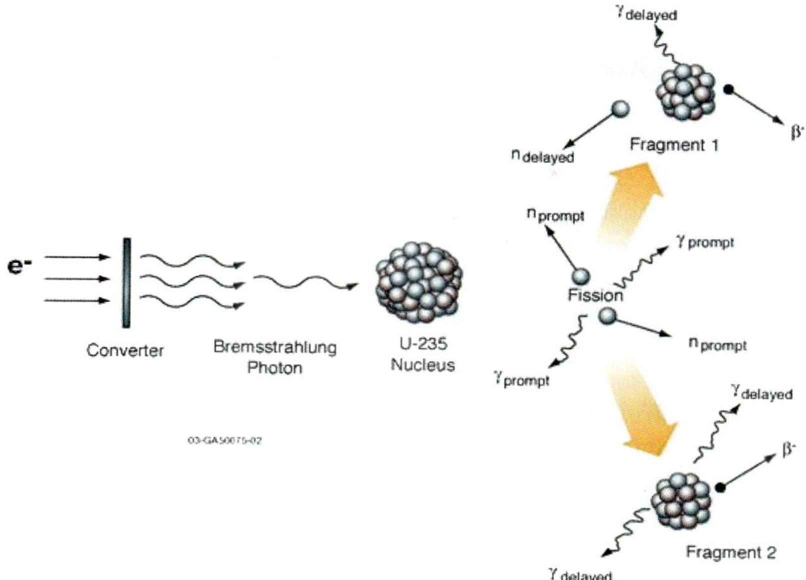

Fig. 14. Sketch of the photofission process showing the various prompt and delayed emitted particles. Very large energies are released, $\sim 200\,\mathrm{MeV/fission}$.

For fissionable nuclei the GDR region also includes a photofission resonance whose magnitude is about half that of the GDR. One notes that for electron beam energies $> 5\,\mathrm{MeV}$ the bremsstrahlung photon spectrum begins to overlap the energies of the various nuclear excitations, giving rise to photon-induced reactions:

(1) (γ, γ): nuclear resonance florescence (NRF) and (γ, γ') inelastic scattering. The stimulated signals will appear during accelerator on time (termed "flash" for a pulsed accelerator).

(2) (γ, f); includes photofission processes via the GDR producing prompt and delayed neutrons, as well as prompt and delayed gammas. Figure 14 depicts the various photofission particle yields. The prompt signals appear during accelerator on time; the delayed signals at various times after or between pulses.

(3) (γ, xn), (γ, xp), (γ, np), ...; excluding fission but including processes involved in the GDR energy region of nuclei. The products neutrons and or radioactive product nuclei can be detected after irradiation or between pulses of radiation. The produced radionuclides are the basis for photon activation analysis (PAA).

These reactions lead to four categories of penetrating outgoing radiations: prompt gamma emission, prompt neutron emission, delayed gamma emission and delayed neutron emission. In this context "prompt" and "delayed" are defined as occurring on a timescale less than (greater than) the pulse width of the accelerator. Emitted particle energies in the above reactions can be very high, such as photofission fragments that are emitted with a total kinetic energy $\sim 200\,\mathrm{MeV}$, and quite low, such as gammas from low-lying nuclear de-excitations ($\sim 100\,\mathrm{keV}$). In addition, it should be pointed out that secondary processes such as neutron capture also produce potentially useful signals that fit into the four categories of prompt and delayed neutron and gamma emissions.

3.3.1. *Nuclear de-excitation*

The basic unique nuclear de-excitation emissions determine the manner of detection:

(1) Prompt neutron emission (PNE) includes (γ, n), $(\gamma, 2n)$, etc., as well as $(\gamma, \text{fission})$ processes. Detection and energy spectroscopy of these neutrons can be done by neutron time of flight (NToF), or via proton recoil pulse height analysis, or by exploiting energy excitation thresholds of target or detector material, etc. These measurements may then be used as a measure of neutron energy distributions and thus as an

identification technique. Among the many technical challenges associated with this technique is "flash recovery" in a pulsed bremsstrahlung environment, as noted below for prompt gamma detection.

(2) Prompt gamma emission (PGE) includes gamma-producing processes that include photons from (γ, γ'), photofission or the (γ, x) event itself, as well as prompt photons emitted from highly excited fission fragments. These emissions can, in principle, be used as a high resolution, species-specific quantification method for many nuclei if the spectroscopy instrument can recover from the bright bremsstrahlung "flash" associated with pulsed beams.

(3) Delayed neutron emission (DNE) includes post-fission neutrons from highly excited fission fragments and any beta-delayed or spin-delayed neutron emissions that follow. With few exceptions, only fission fragments emit such neutrons [one of a few exceptions is ^{17}N, $t_{1/2} = 4\,\mathrm{s}$, from the ^{18}O(γ, p) reaction]. Measurements of these delayed neutrons will include energy spectroscopy and "temporal spectroscopy," meaning measurements of yield versus time, and the extraction of "effective lifetimes" (photoinduced group parameters, which are different than neutron-induced group parameters). This delayed neutron lifetime (DNL) analysis has been shown to have the potential to be used for parent fissile nucleus detection and identification, as discussed below.

(4) Delayed gamma emission (DGE) processes are either γ-delayed or isomer-delayed gamma rays emitted by fission products, from (γ, xn) and related processes, and from neutron capture gammas. In any of these cases, temporal and energy spectroscopy, as with delayed neutrons, shows promise for discriminating photofissionable material from other materials and may be able to discriminate fissile material (^{233}U, ^{235}U, ^{239}Pu) from other photofissionable material.

4. Examples of Photon-Based Techniques

In this section several examples of photon interrogation schemes using the reactions discussed above are outlined. Some have been deployed in various applications and some are still in development.

4.1. *Nondestructive analysis using PAA*

Photon activation analysis (PAA) is a useful and sensitive method of elemental analysis for many elements [20] based on the [photon in particle(s) out] reactions discussed in Subsec. 3.31. It allows multiple elements to be determined simultaneously, to the part per million to part per tens of billons level nondestructively. Generally PAA does not require chemical separation procedures. When high energy photons are absorbed by a nucleus, $(\gamma, \text{particle})$ reactions occur (a neutron is the most probable particle) and the nucleus is transmuted. Often the newly formed nucleus is radioactive and in decaying subsequently releases gamma rays that are associated with the specific isotope produced in the transmutation reaction. Thus, unknown nuclides in a sample are transferred to an excited radioactive state. As these radionuclides decay, measurement of the unique gamma ray spectrum emitted is used as a signature of the elemental composition of the sample. Note that exposure times to the photons must be a reasonable fraction of the half-life of the targeted constituents to allow sufficient production. The gamma rays are detected off-line and analyzed in energy by a high resolution high purity germanium detector system. Quantification can be made using standards. PAA can complement, or compete with, the many methods in use today, such as laser ablation, x-ray fluorescence and neutron activation analysis (NAA). While these latter methods provide accurate results, the analyzed samples are often damaged or destroyed in the preanalysis chemical processes needed. Additionally, the results sometimes represent only the surface concentrations, and not the object as a whole. In PAA, photons easily pass completely through smaller objects, and can penetrate deeply into larger objects. Thus, the size or shape of an object that can be analyzed is not necessarily a limitation.

An accelerator that produces 30–40 MeV electrons is optimum for PAA, as these energies are sufficiently high to activate all nuclides via various $(\gamma, \text{particle})$ reactions. Higher beam energies can cause multiple particle emission reactions which complicate the interpretation of the emitted gamma spectrum but, in some cases, may provide additional identification information.

There are obvious parallels between NAA and PAA besides the basic difference in these two methods activate the sample: (n, γ) for NAA and

(γ, particle) for PAA. The gamma spectrometry analysis hardware required for the two methods is identical; only the gamma spectrum library in the analysis software differs. The clear advantage of PAA, however, is the use of a much less expensive and more benign activating source (an accelerator rather than a nuclear reactor), and matrix effects which can adversely affect NAA by flux depression are not present. Compared to NAA, PAA can be used to determine elemental concentrations of lighter elements, such as F, O and C. In addition, the number of heavy elements that can be analyzed is greater for PAA than for NAA. Another advantage of PAA is that portable electron accelerators allow PAA, in principle, to be performed on site. Lastly, the half-lives of post-PAA radionuclides are generally much shorter than for NAA, reducing analysis time and postanalysis radiation issues. PAA has particular importance for biological and geological environmental samples as well as historical and cultural artifacts, owing to the particular elements where it is more sensitive.

4.2. *Fissile material detection using photons*

Detection and identification of fissionable material is a strong suit of the use of electron accelerator interrogation. Several examples are given, beginning with an experimental cargo scanning system using delay fission neutron detection to identify fissionable contraband to discussions on more sophisticated methods under development. As described in

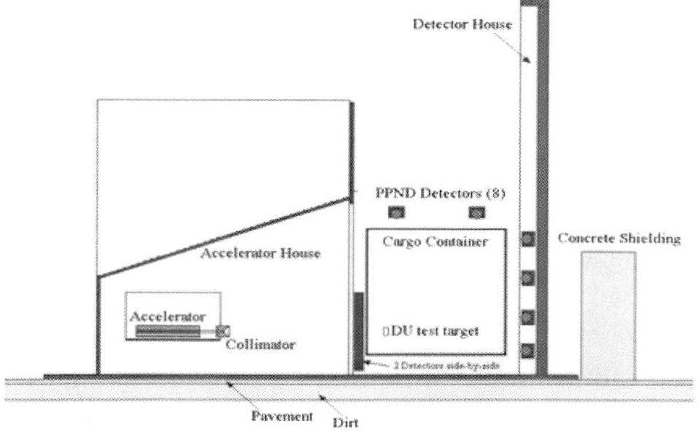

Fig. 15.   Sketch of a cross section of the IAC cargo scanner apparatus in Fig. 1 set up for delayed neutron detection from fissionable material.

Subsec. 3.3 and sketched in Fig. 15, high energy photons transverse materials containing fissionable isotopes can cause photofission reactions releasing on average 200 MeV of energy per fission reaction. The majority of this energy is in the kinetic energy of the very short range fission fragments but as these highly excited fission fragments decay they emit the following secondary radiation:

(1) Prompt neutrons. Yield: 2–3 per fission. Timescale: $\sim 10^{-14}$ s.
(2) Prompt γ-rays. Yield: ~ 7 per fission. Timescale: $\sim 10^{-14}$ s.
(3) Delayed neutrons. Yield: 0.4%–5% of fissions: Timescale: ~ 100 ms to 55 s.
(4) Delayed γ-rays. Yield: ~ 7 per fission. Timescale: ~ 100 ms to years.

These four emissions can be detected and form the basis for most of the techniques that have been studied and utilized for detecting fissionable materials.

4.2.1. *Cargo container scanner*

An application of the DNE phenomena is the detection of fissionable material. As pointed out, postinterrogation delayed neutrons come from highly excited neutron-rich fission fragments and any beta-delayed or spin-delayed neutron emissions that follow in the decay chains. A cargo container scanning device that utilizes DNE for detection of fissionable contraband is described. Figure 1 is an image of the IAC cargo scanning facility used in this demonstration experiment. Figure 15 depicts the arrangement of the accelerator and the detectors. The detector arrays consisted of neutron-sensitive ^{3}He proportional counters shielded to be sensitive to only epithermal neutrons [21]. The accelerator used in this experiment was a 10 MeV linac that produced 10 MeV end point bremsstrahlung photons for photoexcitation. The photons were also used to perform simultaneous radiographic imaging. The container was moved through the collimated photon beam at a constant rate so that variations in signals with time gave two-dimensional information on variations in space of the container contents, both in radiographic density and photofission source content.

Figure 16 shows the radiograph and the neutron detector signal as a function of position in the

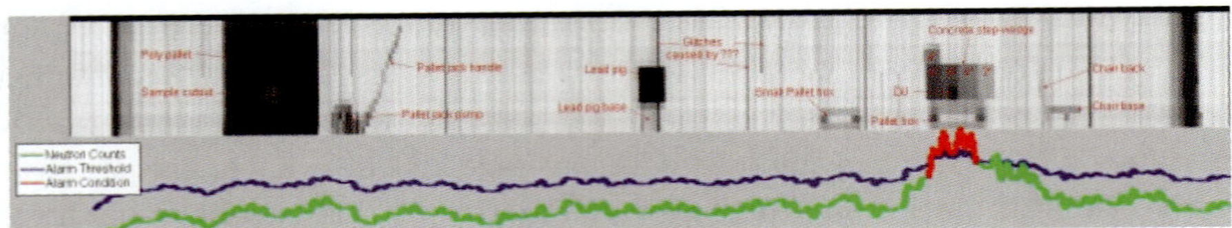

Fig. 16. The top of this figure shows a radiographic image of the bottom half of a moving cargo container in the IAC scanner (Figs. 1 and 14) with various items inside, pallet jack, lead pig, chair and 4 kg of shielded uranium 238. The lines along the bottom of the figure are neutron detector signals. When the photon beam crosses the position of the uranium, the signal goes to the alarm state. This scan took 80 s.

container. The detection limit with the accelerator operating at a 25 Hz rep. rate ($\sim$8 microamp average beam current) was 2 kg of bare depleted uranium (DU) at a cargo container rate of 2 ft per second, indicating a 80 s total scan time for a 40 ft container.

4.2.2. Fissionable time constants of delayed neutrons

Delayed neutrons are emitted during the decay of neutron-rich fission fragments that are left in a particle unstable state. The delayed neutron emission rate is determined by the half-lives and yield of the fission fragments that precede neutron emission; the timescale for emission ranges from hundreds of milliseconds to $\sim$55 s [23–27]. Since over 200 delayed neutron precursors have been identified, they have traditionally been categorized into six groups based on their decay constant, λ_i, with each group having a relative yield of α_i [4, 28–30]. Assuming that the precursors are produced instantaneously, the delayed neutron emission rate is then given by a sum of simple exponential decays,

$$R_{\mathrm{dn}} = N_{\mathrm{dn}} \sum_{i=1}^{6} \alpha_i \lambda_i e^{-\lambda_i \cdot t},$$

where N_{dn} is the total number of precursors produced. Of these groups, the shortest half-life [i.e. $t_{\mathrm{half}} = \lambda^{-1} \ln(2)$] is typically 150 ms and the longest half-life is 55.6 s being attributed to the ^{87}Br fission fragment. These group decay constants and relative yields have been measured by many researchers for a variety of fissionable isotopes, and are available in the literature and tabulated nuclear data files. Due to the importance of delayed neutrons for reactor operations, the majority of these parameters were determined from neutron-induced fission data, and specific photofission data are not as abundant [4, 24, 31].

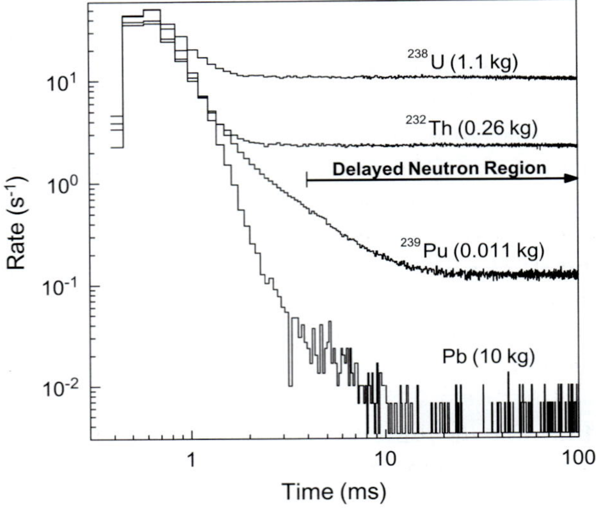

Fig. 17. Neutron detection rates versus time following a 15 MeV bremsstrahlung pulse for the ^{238}U, ^{239}Pu and Pb targets. The data are presented as a histogram, with each channel being 130 μs wide. The neutrons were counted between bremsstrahlung pulses over a 300 s period for a total of 2.25×10^3 pulses.

Since the fission reaction is essentially the only available reaction that emits neutrons on these timescales, the detection of delayed neutrons provides a unique signature of fissionable materials, despite their relatively low yield. As an example of this unique signature, Fig. 17 compares the neutron detection rates from 0.26 kg ^{232}Th, 1.1 kg ^{238}U, 11.0 g ^{239}Pu and 10 kg Pb targets, measured between intense $\sim$15 MeV bremsstrahlung pulses produced by a linac operated at 7.5 Hz. The neutrons were counted between bremsstrahlung pulses over a 300 s period for a total of 2.25×10^3 pulses.

The data in Fig. 17 were collected by shielded ^{3}He detectors, which are highly insensitive to γ-rays and low energy neutrons [21]. In the first $\sim$600 μs the detection rate rises as the efficiency of the detectors increases during their recovery from the intense bremsstrahlung flash. The rate then decreases rapidly as neutrons created during the flash scatter from objects in the environment and return to the detectors. For the Pb target, this rate decreases over three orders of magnitude between its peak and $\sim 10\,\mathrm{ms}$ after the bremsstrahlung pulse. For the fissionable materials, the neutron detection rate stays elevated beyond $\sim 10\,\mathrm{ms}$ due to the emission of delayed neutrons following the induced photofission reactions.

4.2.3. *Fissionable material identification using time constants of delayed γ-rays*

As with delayed neutrons, delayed γ-rays are emitted following the γ-decay of fission fragments left in excited states that then decay by photon emission. Since the excitation energy required for photon emission is considerably less than for neutron emission, the delayed γ-ray yield can be up to three orders of magnitude larger than the delayed neutron yield, with approximately seven delayed γ-rays emitted per fission [32]. The average energy of these photons is $\sim 1\,\mathrm{MeV}$ and the time scale for emission extends to years (i.e. long-lived fission products). If these delayed γ-rays with their larger yield can be utilized for fissionable material detection, identification and quantification, several advantages would be possible, including an increase in sensitivity, an increase in accuracy or an increase in standoff distance from the contraband. Unlike delayed neutrons, a unique delayed γ-ray signature must discriminate

between delayed fission γ-rays and those γ-rays emitted from other photonuclear reactions that occur in nonfissionable materials. Fission results in a significant γ-ray yield above $\sim 2.5\,\mathrm{MeV}$, as seen in the data in Fig. 18, which compares the γ-ray spectrum from ^{238}U, Pb and ^{9}Be.

These data were collected by a well-shielded high purity germanium (HPGe) detector between $14.5\,\mathrm{MeV}$ bremsstrahlung pulses [34]. While the yield of γ-rays above $2.5\,\mathrm{MeV}$ is significantly larger for ^{238}U, both Pb and ^{9}Be produce γ-rays in excess of $2.5\,\mathrm{MeV}$. Other nonfissionable materials produce similar spectra, and hence large quantities of nonfissionable materials could potentially confound a fissionable material detection and quantification system. However, as shown in Fig. 19, for the nonfissionable materials (Pb and Be) the γ-ray yield above $2.5\,\mathrm{MeV}$ drops by two orders of magnitude but the ^{238}U yield is over three orders of magnitude larger so that the fissionable material identification is the γ-ray signature of $> 2.5\,\mathrm{MeV}$ γ-rays at long times after the interrogating pulse.

In a manner similar to that of delayed neutrons, the decay of delayed γ-ray emission is inherently exponential in nature but is more complicated due to the large number of fission fragments that contribute. The emission rate appears constant in Fig. 18(a) due to the short interval over which these measurement were performed and the longer half-lives of the fission fragments associated with high energy γ-ray emission. To see the full exponential decay, longer timescales must be examined. Figure 18(b) shows the

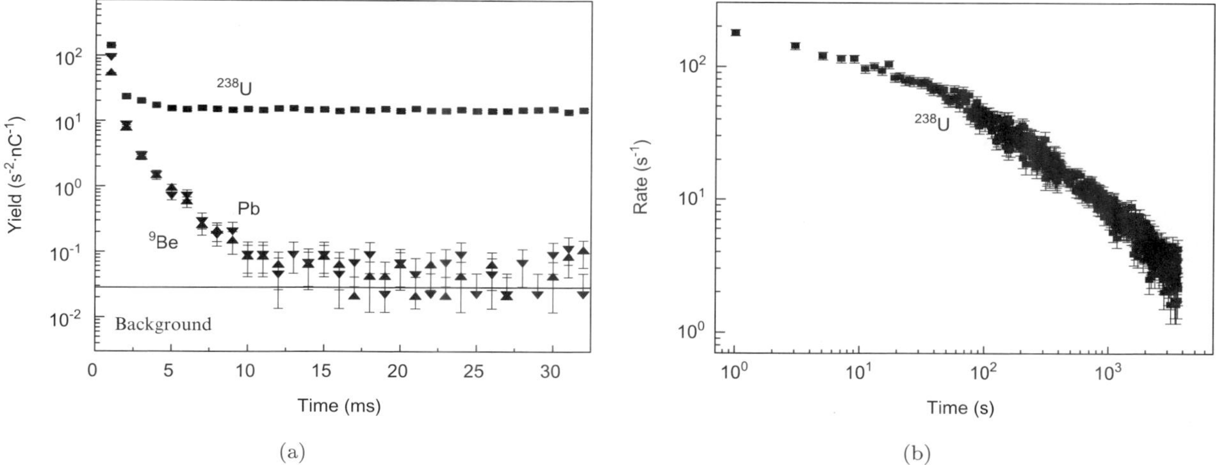

(a)

(b)

Fig. 18.   Photon energy spectra for all detected x-rays (a) and those detected $13\,\mathrm{ms}$ after the bremsstrahlung pulse for ^{238}U, Pb and ^{9}Be. These spectra were collected between $14.5\,\mathrm{MeV}$ bremsstrahlung pulses.

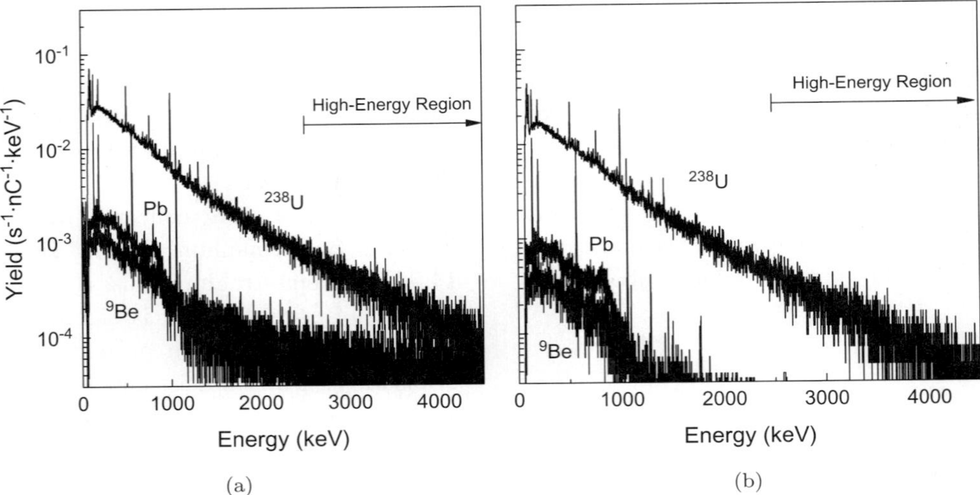

(a)   (b)

Fig. 19.   High energy ($>2.5\,\mathrm{MeV}$) γ-ray differential emission rate between bremsstrahlung pulse (a) and after a 57 min continuous irradiation (b).

high energy γ-ray detection rate for ^{238}U from 1 s out to ~ 1 h, following a 57 min 14.5 MeV bremsstrahlung irradiation. The use of a simplified group model has not been as widely carried out for delayed γ-rays as for delayed neutrons. Nevertheless, at long times (i.e. $t > 20$ min), a single exponential decay does seem to appear with an effective half-life of ~ 27 min. Unlike delayed neutrons, this long half-life cannot be attributed to a single fission fragment and clearly, even at these long times, the emission decay is still a combination of a large number of different fission fragments.

These experiments were performed with an ~ 3 m radiator to the target distance, 10 min assay time and approximately 18 times less bremsstrahlung intensity than was used in the delayed neutron experiment. The detector was a high purity germanium (HPGe) detector. But since high energy resolution is not needed, it can be replaced with an array of more efficient γ-ray detectors such as NaI or $\mathrm{Bi_4Ge_3O_{12}}$. Large increases in efficiency are thus possible, allowing larger standoff distances, decreased inspection times or increased fissionable material sensitivity. Furthermore, the delayed γ-ray signature would complement a delayed neutron signature, which could be measured simultaneously with appropriate detectors.

4.2.4. *Fissionable material identification using prompt neutrons*

Unlike delayed neutrons or gamma rays, prompt neutrons are emitted less than 10^{-14} s after a fission

reaction, which for almost all practical purposes is instantaneous emission [33]. Hence, any temporal spread in prompt neutron detection comes from differing flight times from the target to the detector. Figure 20(a) compares neutron time-of-flight spectra from Pb and ^{238}U targets collected using a ~ 2 ns bremsstrahlung pulse and a neutron time-of-flight spectrometer with an ~ 1.1 m flight path.

The end point energy of the 2 ns bremsstrahlung beam pulses was 9 MeV and had a repetition rate of 120 Hz [34]. In these time spectra, zero was set to the time when the bremsstrahlung pulse strikes the target and the narrow first peak at 3.7 ns is from photons scattering from the target and traveling to the detector at the speed of light. The broad second peak is from neutrons created in the target traveling to the detector at nonrelativistic speeds. A direct correlation exists between the neutron flight time and the neutron kinetic energy, so the excess neutrons detected from the ^{238}U target between ~ 30 ns and 70 ns compared to the Pb target indicate that more high energy neutrons are emitted from the fissionable material. In contrast to using the delayed neutron yield to detect ^{239}Pu and ^{235}U, the fission process emits two or three prompt neutrons which are over two orders of magnitude larger [32]. If these prompt neutrons with their larger yields can be utilized for fissionable material detection and quantification, several advantages would be possible, including an increase in sensitivity, an increase in accuracy and an increase in standoff distance.

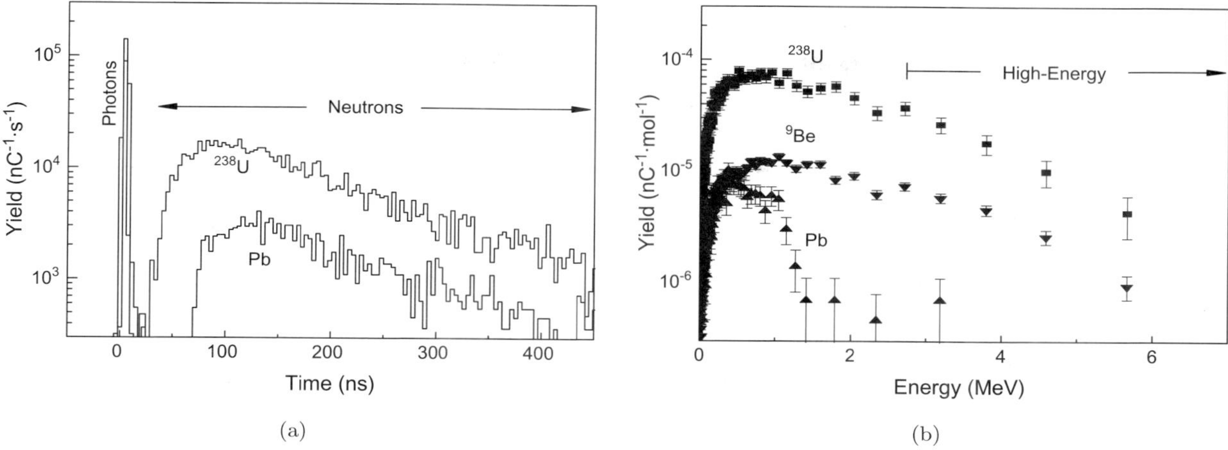

Fig. 20. (a) Neutron time-of-flight spectra from ^{238}U and Pb targets. The yields are normalized to the electron charge on the radiator. (b) Neutron energy spectra from ^{238}U (■), Pb (▲) and ^{9}Be (▼) targets. The high energy neutron region ($E_\mathrm{n} > 2.3\,\mathrm{MeV}$) is indicated. These yields are normalized to the total electron charge on the radiator and moles of the target.

As with delayed γ-rays, a fission signature based on prompt neutron detection needs to discriminate between prompt fission neutrons and those neutrons resulting from (γ, n), $(\gamma, 2n)$ and other reactions that may occur in nonfissionable materials. Prompt fission neutrons can have energies in excess of $\sim 10\,\mathrm{MeV}$, even when fission is induced by low energy incident neutrons or γ-rays [32]. In contrast, neutrons produced through other reactions have distinct maximum energy limits that depend upon binding and incident particle energies. For example, for $9\,\mathrm{MeV}$ photons the most energetic neutron that can be emitted from natural Pb is $2.3\,\mathrm{MeV}$. This is seen in Fig. 20(b), which converts the neutron time-of-flight spectra of to neutron energy. The region above $2.3\,\mathrm{MeV}$, as indicated in Fig. 20(b), is dominated by prompt fission neutrons for the ^{238}U spectrum. In this high energy neutron region, the yield from ^{238}U is 40 times larger than the background. However, some nonfissionable materials, like beryllium, can also emit high energy neutrons, as seen in Fig. 20(b), due to their extremely small neutron-binding energies of $1.7\,\mathrm{MeV}$.

4.3. *More speculative schemes for fission NDE*

Investigating nuclear photofission at a fundamental level may reveal phenomena that will provide new techniques for identifying and quantifying actinides. As an example, there is the possibility that effects like asymmetries in the fission process induced by active interrogation of various forms will result in useful signatures.

4.3.1. *Spatial distribution and asymmetries of prompt neutron*

There is a potential use of spatial asymmetries of prompt neutrons in fissionable isotope identification based on theoretical predictions of spatial/angular asymmetries in prompt photofission neutrons. It has long been known that the nuclear fragments resulting from photon-induced fission of heavy nuclei are emitted anisotropically when measured with respect to the incident photon direction. The first such measurement in 1956 [35], performed with unpolarized photons, found symmetrical angular distributions that were dependent on the energy of the incident photon, the target, and the type of fission fragment observed. The introduction of linear photon polarization breaks the symmetry, as observed in 1982 [36] with the measurement of fission fragments from the polarized photofission of thorium.

For any object thicker than a few mg/cm^2, the fission fragments are not easily detectable, due to their heavy ionization loss and the resulting short range when traversing the target. If, however, the angular asymmetry in the fission fragments is manifest in the angular and energy distributions of the prompt neutrons emitted in the fission event, a possible unique signature for the presence of photofission could be had. Such a technique exploits the kinematics of the fission process in conjunction

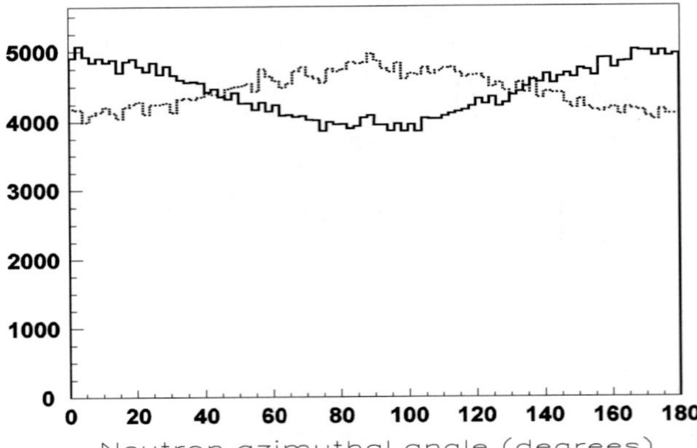

Fig. 21. Monte Carlo calculation of $\square$ distribution of fission neutrons with 30% polarization for $K = 0$ and $K = 1$.

with the high penetrability of the emitted fission neutrons.

The calculated angular distributions of the fission fragments are shown in Fig. 21 for fragments emitted parallel and perpendicular to the polarization vector for a photon polarization of 30% and for both $K = 0$ and $K = 1$ (the quantum number K describes the projection of the angle momentum on the symmetry axis of the fragment). It can be seen that the asymmetries are large, even for gamma polarizations of only 30%.

These asymmetries are merely representative of what one *might* observe in fission of actinides induced by a bremsstrahlung beam of continuous energy. In practice, fission may proceed via multiple channels, and one cannot directly calculate the expected neutron asymmetries.

5. Conclusion

Modern and emerging wide-ranging challenges in nondestructive detection require active, nondestructive interrogation with neutrons and photons, because such methods provide at least an order of magnitude greater sensitivity than do passive methods. Accelerator-based neutron and photon active methods exploit two important factors to attain greater sensitivity for important applications in security, nuclear nonproliferation, contraband detection and materials analysis. The first of these is the control of interrogating beam properties such as directionality, energy, intensity, polarization and the temporal distribution of radiation. The second is the well-established, low energy nuclear physics that

yields distinct "signatures" for elemental and isotopic content. The current state of knowledge in these areas is such that numerous important applications are practical today. However, both of these factors, and important new applications, will continue to benefit from advances in the state of knowledge of accelerator and nuclear physics.

Acknowledgments

This work is partially supported by DoE Contract No. DE-FC07-06ID14780 and DoD Contract No. FA8650-04-2-6541. Thanks are also due to our many colleagues at other institutions and the national laboratories, and to Prof. Daniel Dale, Idaho State University Department of Physics, for advance notice of his interesting and important work.

References

[1] D. Reilly, N. Ensslin, H. Smith and S. Kreiner, Passive nondestructive assay of nuclear materials, Tech. Rep. NUREG/CR-5550 (United States Regulatory Commission, Washington D.C., 1991).

[2] G. R. Keepin, LA-3741, TID-4500 (1967).

[3] S. Untermyer and R. Sher, American Nuclear Society #300017 (1980).

[4] T. Gozani, NUREG/CR-0602 (United States Regulatory Commission, Washington, D.C., 1981).

[5] B. Sowerby, *Nucl. Geophys.* **5**, 491 (1991).

[6] T. Gozani, *Nucl. Instrum. Methods B* **213**, 460 (2003).

[7] L. Grodzins, *Nucl. Instrum. Methods B* **57**, 829 (1991).

[8] L. E. Smith *et al.*, AFCI Safeguards Enhancement Study: Technology Development Roadmap (PNNL-18099, Dec. 2008).

[9] J. Gow and J. Foster, *Rev. Sci. Instrum.* **24**, 606 (1953).

[10] E. Knapp and B. Knapp, *Rev. Sci. Instrum.* **39**, 979 (1968).

[11] J. Csikai, *CRC Handbook of Fast Neutron Generators*, Vols. I and II (CRC, Boca Raton, Florida, 1987). See also: W. Bygrave, P. Treado and J. Lambert, Accelerator Nuclear Physics, High Voltage Engineering Corp., Burlington, MA.

[12] W. Lee, D. B. Mahood, P. Ryge, P. Shea and T. Gozani, *Nucl. Instrum. Methods B* **99**, 739 (1995).

[13] A. Buffler, *Rad. Phys. Chem.* **71**, 853 (2004).

[14] Z. Sawa, T. Gozani and P. Ryge, US Patent No. 5,076,993 (1991).

[15] D. Brown, T. Gozani, R. Loveman, J. Bendahan, P. Ryge, J. Stevenson, F. Liu and M. Sivakumar, *Nucl. Instrum. Methods A* **353**, 684 (1994).

[16] D. Brown and T. Gozani, *Nucl. Instrum. Methods B* **99**, 753 (1995).

[17] E. Rhodes, C. E. Dickerman, A. DeVolpi and C. W. Peters, APSTNG. *IEEE Trans. Nucl. Sci.* **39**, 1041 (1992).

[18] See also *Nucl. Instrum. Methods B* **261**, 365ff (2007).

[19] Radiation Protection Design Guidelines for Particle Accelerator Facilities, NCRP Rep. #51 (2003).

[20] C. Segebade, H. P. Weise and G. J. Lutz, *Photon Activation Analysis* (Walter De Gruyter, New York, 1988).

[21] J. L. Jones, W. Y. Yoon, K. J. Haskell, D. R. Norman, J. M. Hoggan, C. E. Moss, C. A. Goulding, C. L. Hollas, W. L. Myers and E. Franco, Tech. Rep. INEEL/EXT-02-01406 (2002).

[22] M. T. Kinlaw and A. W. Hunt, *Appl. Phys. Lett.* **86**, 254104 (2005).

[23] G. R. Keepin, T. F. Wimett and R. K. Zeigler, *Phys. Rev.* **107**, 1044 (1957).

[24] L. A. Kull, R. L. Bramblett, T. Gozani and D. E. Rundquist, *Nucl. Sci. Eng.* **39**, 163 (1970).

[25] L. Tomlinson, *Nucl. Technol.* **13**, 42 (1972).

[26] R. J. Tuttle, *Nucl. Sci. Eng.* **56**, 37 (1975).

[27] R. W. Waldo, R. A. Karam and R. A. Meyer, *Phys. Rev. C* **23**, 1113 (1981).

[28] T. A. Parish, W. S. Charlton, N. Shinohara, M. Andoh, M. C. Brady and S. Raman, *Nucl. Sci. Eng.* **131**, 208 (1999).

[29] D. J. Loaiza and F. E. Haskin, *Nucl. Sci. Eng.* **134**, 22 (2000).

[30] M. B. Chadwick, P. Oblozinsky, M. Herman *et al.*, *Nucl. Data Sheets* **107**, 2931 (2006).

[31] M. C. Brady and T. R. England, *Nucl. Sci. Eng.* **103**, 129 (1989).

[32] M. F. James, *J. Nucl. Energy* **23**, 517 (1969).

[33] T. M. Snyder and R. W. Williams, *Phys. Rev.* **81**, 171 (1951).

[34] E. T. Reedy, S. J. Thompson and A. W. Hunt, *Nucl. Instrum. Methods A* **606**, 811 (2009).

[35] Winhold and Halpern, *Phys. Rev.* **103**, 990 (1956).

[36] R. Ratzek, W. Wilke, J. Drexler, R. Fischer, R. Heil, K. Huber, U. Kneissl, H. Ries, H. Stroher, R. Stock and K. Z. Wienhard, *Z. Phys. A* **308**, 63 (1982).

J. Frank Harmon is the founding Director of the Idaho Accelerator Center (IAC) at Idaho State University and now Senior Scientist at IAC. He is Professor Emeritus in the Physics Department, which he joined in 1969 after a postdoctoral year at the University of Utah. His Ph.D. in Physics is from the University of Wyoming. He has extensive expertise in accelerator technology, instrumentation and nuclear measurements applied to industrial and security needs. He has served as a consultant for Idaho National Laboratory, Los Alamos National Laboratory, and various private companies.

Doug P. Wells is the Director of the Idaho Accelerator Center, which performs R&D advancing applications of accelerators and nuclear science in diverse areas. His Ph.D. in Physics is from the University of Illinois. He was a postdoctoral research associate at the University of Washington and worked for the Washington State Department of Health as an environmental health physicist. He came to Idaho State University in 1996. His work has focused on photon-based active accelerator methods for material interrogation. He is a Professor and the former chair of the ISU Department of Physics.

Alan W. Hunt is Deputy Director of the Idaho Accelerator Center and an Associate Professor in the Department of Physics. He received his Ph.D. in Physics from Harvard University. He held positions at Lawrence Livermore National Laboratory and Washington State University before joining the staff at the Idaho Accelerator Center in 2001. His focus is on nuclear methods for national security applications and advanced detector technology.

Reviews of Accelerator Science and Technology
Vol. 4 (2011) 103–116
© World Scientific Publishing Company
DOI: 10.1142/S1793626811000574

Review of Cyclotrons for the Production of Radioactive Isotopes for Medical and Industrial Applications

Paul Schmor

Advanced Applied Physics Solutions, Inc.,
4004 Westbrook Mall, Vancouver, BC, V6T 2A3, Canada
schmor@triumf.ca

Radioactive isotopes are used in a wide range of medical, biological, environmental and industrial applications. Cyclotrons are the primary tool for producing the shorter-lived, proton-rich radioisotopes currently used in a variety of medical applications. Although the primary use of the cyclotron-produced short-lived radioisotopes is in PET/CT (positron emission tomography/computed tomography) and SPECT (single photon emission computed tomography) diagnostic medical procedures, cyclotrons are also producing longer-lived isotopes for therapeutic procedures as well as for other industrial and applied science applications. Commercial suppliers of cyclotrons are responding by providing a range of cyclotrons in the energy range of 3–70 MeV for the differing needs of the various applications. These cyclotrons generally have multiple beams servicing multiple targets. This review article presents some of the applications of the radioisotopes and provides a comparison of some of the capabilities of the various current cyclotrons. The use of nuclear medicine and the number of cyclotrons supplying the needed isotopes are increasing. It is expected that there will soon be a new generation of small "tabletop" cyclotrons providing patient doses on demand.

Keywords: Low energy cyclotrons; commercial cyclotrons; radioisotope production; commercial radioisotopes; radionuclides; PET isotopes; SPECT isotopes; radio therapeutics; nuclear medicine; isotope tracers; nuclear diagnostics.

1. Introduction

1.1. *Radioisotopes*

The application of nuclear technology to benefit society was one of the more significant developments of the 20th century. In 1911, G. de Hevesy, one of the early pioneers in applying radioisotopes as a diagnostic tool, used naturally occurring radioactive isotopes as a tracer to confirm his suspicions that his boarding house was using leftovers in his meals [1]. The advent of particle accelerators in the first half of the 20th century, and their subsequent use in nuclear transmutation, made available a wide variety of new short-lived isotopes. Researchers have been examining this landscape, looking for radioisotopes that have the best characteristics for their particular application. In particular, they are interested in decay properties of targeted atomic elements.

Radioisotopes are unstable atomic nuclei that undergo decay with emission of gamma rays or subatomic particles. They can be created when accelerated energetic particles impinge on stable isotopes.

The unique properties of radionuclides have led to numerous applications in pure and applied scientific research, medicine, and industry. Nuclear medicine relies on the decay of radioisotopes for the diagnosis and treatment of diseases. In nuclear medicine procedures, radionuclides are combined in compounds to form radiopharmaceuticals that can localize in specific organs or cells. In industry and in other scientific research, radioisotopes are primarily used as tracers to monitor processes and as probes to measure the magnetic and electric fields in condensed matter.

1.2. *Radioisotope production*

Radionuclides can be produced in nuclear reactors and with particle accelerators. In general, accelerator-produced radionuclides can provide isotopes with more suitable decay properties than reactor-produced nuclides as a result of the accelerator proximity to the user. A key consideration in many applications is the specific activity, which is a measure of the relative amount of radioactive isotopes in a sample. For many of the radioisotopes of interest, particularly in PET procedures, the specific activity (SA) of accelerator-produced PET isotopes is high. Access to nuclear reactors is limited and the rapid delivery of short-lived radioisotopes to hospitals and clinics can be problematic. Hospital-based or region-based accelerators can produce many of the desired short-lived radionuclides on demand.

1.3. *Cyclotrons*

Cyclotrons are particle accelerators that use a fixed magnetic field to constrain charged particles to nearly circular orbits that are synchronous with a radio frequency (rf) accelerating field. The charged particles repeatedly traverse the electric field produced by the rf electrodes, increasing in energy at each traversal. That is to say, the charged particle period of revolution in the fixed magnetic field of the cyclotron is independent of the particle's momentum. The cyclotron was conceived by E. Lawrence in 1930 [2] and started operating in 1932 [3].

Cyclotrons have become the tool of choice for producing the short-lived proton-rich radioactive isotopes used in biomedical applications [4, 5]. Industry has responded with a variety of cyclotrons to address the particular needs of different user groups. Most of these machines have been installed in hospitals, institutes for academic research, and commercial facilities specializing in producing and selling radioactive isotopes. Cyclotrons for biomedical radionuclide production are generally compact, accelerate light ions (proton, deuteron or helium), and are primarily used to produce short-lived, proton-rich radionuclides. The main use of these unstable isotopes produced by industrial cyclotrons is for tracers, diagnostics and therapy in biomedicine. Other fields using radioactive nuclides as tracers include agriculture (biokinetics in plants and soil), biology (biochemical and toxicological studies), ecology (pollution, environmental impact, and ecology studies), geology (migration of elements in soils and waters) and pharmacology (metabolic studies). The use and need of radioactive isotopes for biomedical applications continues to increase worldwide. However, the list of radioactive nuclides being produced and the applications have not changed significantly over the past 20 years [6, 7].

Five years after demonstrating the first cyclotron in 1932, E. Lawrence's brother, G. Lawrence, was using a cyclotron to produce phosphous-32, which by 1939 [8] had been used for injection into a patient with chronic leukemia. Other isotopes generated by the cyclotron also had important applications in medicine. However, E. Lawrence's vision of developing a radiopharmaceutical industry at the Radiation Laboratory to help sustain accelerator physics in the late 1930s was an idea ahead of its time.

As early as 1938, J. Chadwick published an article [9] in *Nature* entitled "The Cyclotron and Its Applications," in which he points out that the induced nuclear transmutation from accelerated particles has application in the biological sciences as a source of secondary particles (neutrons in particular), as a labeling agent (tracer) and as a therapeutic in treating disease in targeted areas of the body. In 1941, the first cyclotron dedicated to the production of radioisotopes was installed at Washington University, St. Louis, with funds provided by the Rockefeller Foundation [10], and was used to produce isotopes of phosphorus, iron, arsenic and sulfur [11, 12]. In the mid-1950s, a group at Hammersmith Hospital in London put into operation the first hospital-based cyclotron wholly dedicated to radionuclide production [13]. Scanditronix was founded in 1965 as a private company to commercialize cyclotrons for use in the medical field [14].

Radioactive isotopes have both diagnostic and therapeutic applications in nuclear medicine. PET (positron emission tomography), combined PET/CT (PET combined with computed tomography) and SPECT (single photon emission computed tomography) are the main diagnostic procedures in nuclear medicine. Radioactive isotopes for biomedical applications are produced in nuclear reactors and with accelerators. Of particular importance for PET are the short-lived positron emitters, ^{11}C, ^{13}N, ^{15}O and ^{18}F. Carbon, nitrogen, and oxygen represent basic constituents of organic matter, and this characteristic permits the diagnostic labeling of a great variety of radiopharmaceuticals that are important to diagnostic and therapeutic medicine. SPECT uses medium-lived radionuclides that are predominately single photon emitters. This is a technique in which a gamma camera moves (and rotates) around the patient taking "pictures," which a computer transforms into a cross-sectional tomographic image.

Therapeutic applications prefer to use radionuclides that have a high linear energy transfer associated with their decay products and that can be chemically attached to a biologically active molecule which preferentially attaches to a tumor site.

Accelerator-produced radioisotopes exhibit a high specific activity (SA) that can be obtained as a consequence of using a nuclear reaction that optimizes the production of an isotope of interest to yield a radioisotope that generally is chemically different from the target element. The desired radioisotope can then be chemically separated efficiently from other unwanted elements in the irradiated target, thereby yielding a product with high SA. By comparison, nuclear fission of uranium in a reactor produces many different isotopes of a desired element, each having similar chemical properties to the desired element. (SA is a measure of the number of radioactive decays from an isotope of interest per unit weight of an irradiated sample.) Another significant difference is the smaller amount of radioactive waste generated in particle reactions using accelerators compared to reactor fission-produced radioactive isotopes. Most of the reactions used for radioisotope production with accelerators are of the form (p, n), (p, 2n), (p, α), (p, xn), where protons (p) are the incident charged particles and alphas (α) or neutrons (n) are in the exit channel. To a lesser extent, reactions involving D, ^{3}He and ^{4}He as the projectile are used as well. Measured cross sections for many of these reactions along with references for the measurements can be found in an IAEA report titled "Cyclotron-Produced Radio-Nuclides: Physical Characteristics and Production Methods" [15].

For proton/deuteron acceleration, the negative ion is preferred in order to reduce activation/beam loss around the cyclotron beam extraction region, whereas for helium acceleration the positive ions must be accelerated. The accelerated negative ions can be extracted from the cyclotron with nearly 100% efficiency by using a thin foil to "strip" the electrons from the accelerated particle and thereby alter the particle's trajectory to exit the cyclotron. For positively charged ions, it is necessary to either place the target in the cyclotron or extract the beam from the cyclotron by altering the particle's trajectory with electrostatic and magnetic devices. Both techniques for using positively

charged ions lead to cyclotron activation. In 2006, an IAEA report estimated that, worldwide, there were about 350 cyclotrons that were primarily used for the production of radionuclides [16]. Nearly 50% of these were in the 10–20 MeV energy range and about 75% of the cyclotrons were being used to produce ^{18}F for FDG. (Others estimate that this number is increasing by about 50 cyclotrons per year, in response to the growth of nuclear medicine.)

Cyclotrons and linear accelerators are also used to produce radioisotopes by fission, spallation and fragmentation of solid, liquid and gaseous targets. ISOL (isotope separation on-line) and fragmentation facilities have provided short-lived nuclides for research in fundamental and applied applications. The cost of using a dedicated facility to produce reasonable yields of radioisotopes by fission, spallation or fragmentation for industrial applications is prohibitively high and limits the industrial applications at these facilities to "parasitic" use.

2. Accelerator-Produced Isotopes

2.1. *Radioisotopes for scientific research*

Accelerators in rare isotope research facilities produce isotopes either by the ISOL technique or by projectile fragmentation. The isotopes of interest to nuclear physicists at these facilities are too short-lived to collect for off-site applications and must be quickly delivered to experimental stations. The primary use of these rare isotope beams is to study nuclear structure, nuclear astrophysics and fundamental symmetries.

There is considerable interest in using similar accelerators to produce intense energetic ($\sim$1 GeV) ions for use in advanced reactor fuel cycles to transmute waste fuel and thereby better use the reactor fuel and also reduce the amount of long-term waste [17, 18].

The short-lived radioisotopes are also used in numerous condensed matter science techniques, such as photoluminescence, perturbed angular correlation, Mössbauer spectroscopy and beta-nuclear magnetic resonance (β-NMR). The science of rare isotope facilities can be found in a recent publication by the National Research Council of the National Academies [19].

2.2. *Radioisotopes for pharmaceuticals, medical sciences, applied sciences and industry*

The radioisotopes in this category have half-lives that are long enough for them to be isolated, chemically attached to appropriate molecules and observed *in situ* for a length of time commensurate with the process of interest. They are primarily used as tracers and as a diagnostic tool [20].

There is an increasing expectation for more use of the radioisotopes for medical therapy. Therapeutic radiopharmaceuticals make use of radioactive isotopes that can be localized near weakened malfunctioning cells to destroy them with the decay radiation.

A "tracer" is a radioactive isotope that is used to monitor the function of an organ, a living plant, a physical or chemical process. The radionuclide can be injected, ingested, embedded or induced by activation with energetic charged particles or neutrons.

In biological applications, it is assumed that the radioactive isotopes are chemically identical with stable isotopes of the same element and that the tracer does not alter the chemical and physical properties of the molecule it is attached to.

Radioisotopes can be and are produced within irradiated samples and components for activation analysis. Activation analysis is a sensitive technique for determining the atomic composition of a sample [21–23].

Various accelerator-produced radionuclides in this category and some of their primary areas of application are presented below. The nuclear reactions and production methods are listed in the IAEA reports [15, 24]. The nuclides have been arranged alphabetically.

(1) Actinium-225 (10 d) and its daughter bismuth-213 are used for cancer therapy

(2) Arsenic-73 (80.3 d) is used for labeling and PET imaging

(3) Arsenic-74 (18 d) is used for PET imaging of cancer cells

(4) Astatine-211 (7.2 h) is an alpha emitter that is used for therapy

(5) Beryllium-7 (53 d) is used in single photon emission studies in the earth sciences

(6) Bismuth-213 (46 m) is used in alpha immunotherapy

(7) Bromine-75 (97 m)

(8) Bromine-76 (16 h)

(9) Bromine-75 (97 m)

(10) Bromine-77 (16 h)

(11) Cadmium-109 (461 d) is used in fluoroscopy to determine the metallic composition of samples

(12) Cobalt-57 (272 d) is used as a marker to estimate organ size

(13) Carbon-11 (20 m) is used in PET (positron emission tomography), for imaging of tumors and therapy

(14) Chlorine-34 m (32 m) positron emitter

(15) Cobalt-55 (17.6 h) positron emitter

(16) Cobalt-57 (270 d) is used as a marker to estimate organ size and for measuring vitamin B12 uptake

(17) Copper-61 (3.4 h) positron emitter

(18) Copper-64 (13 h) is used to study copper metabolism, for PET imaging of tumors and therapy

(19) Copper-67 (2.6 d) is used in therapy

(20) Flourine-18 (110 m) is used in FDG (flourodeoxyglucose) for detecting cancers and monitoring the progress of treatment

(21) Gallium-67 (78 h) is used for lung tumor imaging and localization of lesions

(22) Gallium-68 (68 m) is a used in PET and PET/CT for biomedical imaging

(23) Germanium-68 (271 d) is used in a generator to produce gallium-68

(24) Indium-110 (ground state 4.9 h, metastable state 69 m)

(25) Indium-111 (2.8 d) is used as a diagnostic in brain and colon studies

(26) Indium-114 m (50 d)

(27) Iodine-120 g (1.35 h) positron emitter

(28) Iodine-121 (2.1 h), primarily a gamma emitter

(29) Iodine-123 (13 h) is used for diagnosis of thyroid function

(30) Iodine-124 (4.2 d) is used as a tracer

(31) Iron-52 (8.3 h) positron emitter

(32) Iron-55 is used in metabolism studies, to analyze electroplating solution and to detect airborne sulfur

(33) Krypton-81 m (13 s) is used to study the function of pulmonary ventilation and for the early diagnosis of lung diseases, and is produced as a daughter of rubidium-81 (4.6 h)

(34) Lead-201 (9.3 h)

(35) Lead-203 (52 h)

(36) Mercury-195 m (41 h)

(37) Nitrogen-13 (10 m) is used with PET as a blood flow tracer

(38) Oxygen-15 (2 m) is used to label gases for inhalation to measure blood flow and oxygen consumption

(39) Palladium-103 (17 d) is used in brachytherapy seeds in the treatment of rapidly proliferating tumors

(40) Rubidium-81 (4.6 h) generator for krypton-81m

(41) Rubidium-82 (1.3 m) is used in PET for myocardial perfusion imaging

(42) Sodium-22 (2.6 y) is used as a positron source in calibrating ion chambers and PET cameras

(43) Strontium-82 (25 d) is used as a generator of rubidium-82

(44) Technetium-94 m (53 m) can be used as a substitute for technetium-99 m

(45) Technetium-99 m (6 h) is used for imaging the brain, liver, bone marrow, kidney and heart

(46) Thallium-201 (73 h) is used for diagnosis of coronary artery disease

(47) Tungsten-178 (21.6 d)

(48) Vanadium-48 (16 d)

(49) Xenon-122 (20 h) used to produce I-122

(50) Xenon-127 (34.4 d) is used as a single photon emitter

(51) Yttrium-86 (14.7 h) is used as a tracer in bone pain palliation

(52) Yttrium-88 (106 d)

(53) Zinc-62 (9.2 d)

(54) Zinc-63 (38 m) positron emitter

(55) Ziconium-89 (3.3 d) is used with PET for labeling proteins

3. Production Techniques

The production of radionuclides requires projectiles of sufficient energy and intensity in order to produce viable quantities of the desired nuclide. The projectiles are generally light ions such as protons, deuterons and alpha particles. The energy of the accelerated particles must exceed a threshold energy for a given nuclear reaction by an amount that depends on the energy dependence of the particular cross-section of the reaction in order to optimize the product yield. For example, for fluorine-18, the $^{18}O(p, n)^{18}F$ reaction has a threshold reaction of

about 2.5 MeV and is optimally produced by proton beams in the 8–15 MeV range. For astatine-211, the $^{209}Bi(\alpha, 2n)^{211}At$ reaction has a threshold around 20 MeV and is optimally produced by alphas (^4He) in the 30–40 MeV range.

Targets to produce radioisotopes are placed into the path of the accelerated beam. The target can be liquid, solid or gas and is located within a container that is generally separated from the accelerator vacuum by a thin window. It frequently uses isotopically enriched elements to improve the yield and reduce the production of contaminants. The cost of the enriched isotope in the target is often an important parameter in choosing a practical nuclear reaction for producing the radioactive isotopes. The window design is critical to effective isotope production. Not only must it withstand a pressure difference but it must also take into account the projectile energy loss-related issues, such as thermal conductivity, tensile strength, chemical reactivity, activation and melting temperature. The targets are designed to be quickly removed and moved to hot cells after irradiation. They are frequently tilted with respect to the beam in order to reduce the power density deposited on the vacuum window and the target. Other considerations in target design include the particle range, the stopping power, charged particle interactions, small angle scattering and the chemistry taking place in the target.

The beam profile and beam uniformity at the target is an important factor in determining yields. Hot spots lead to target failure. Collimators are often needed to achieve a uniform density over the target. Raster scanning and beam wobbling have also been used to obtain greater uniformity.

4. Accelerators for Radioisotope Production

4.1. *Accelerator options*

There are many types of accelerators that are used for the production of radionuclides [16]. Each has its own advantages and disadvantages.

- A cyclotron accelerating positive ions is versatile, reliable, compact and energy-efficient. It can be designed to accelerate and extract the variety of light ions needed for the various nuclear reactions used in radioisotope production. These cyclotrons can have the targets placed either within the cyclotron (internal targets) or in beam lines external to the cyclotron (external targets). Both schemes lead to component activation. Extracting the beam into external targets generally has significant beam loss, along the beam trajectory out of the cyclotron, which can lead to component failure and high activation, effectively limiting the maximum accelerated current. The use of internal targets avoids the loss at extraction and the current limit of the beam extraction system but does result in high residual activation of the cyclotron.

- A cyclotron accelerating negative ions can achieve very high extraction efficiency and is capable of generating much higher extracted beam intensities. By allowing the beam to traverse a thin "stripping" foil, the electrons are stripped from the ion and the now-positive ion trajectory is altered to exit the cyclotron. The extracted beam has excellent beam qualities and generally must be enlarged to meet the energy density limits of the target. In order to avoid activation, the cyclotron must have very good vacuum conditions. The most cost-effective cyclotrons have internal ion sources but this solution limits the maximum achievable current and also introduces a gas source that leads to some gas neutralization of the accelerated particles, which in turn causes machine activation. Cyclotrons using external negative ion sources overcome these problems, albeit with greater accelerator complexity, and typically accelerate larger currents with less activation of the cyclotron.

- A superconducting cyclotron is very compact and uses less wall power compared to a warm cyclotron magnet, but is more expensive. Most users prefer to avoid the complications of dealing with liquid helium issues and the higher capital cost of the equipment even with the lower operating cost.

- Linacs are used but have not seriously penetrated the commercial market because they generally have higher operating costs for cw operation at a fixed energy and adversely impact the civil construction cost because they require a larger footprint than cyclotrons. Linacs usually operate in a pulsed mode and the instantaneous power deposited on the target forces complicated engineering solutions in the target.

- Electrostatic accelerators are low-cost and power-efficient, but do not provide adequate current at the energies of interest and therefore have limited applicability.

4.2. *Cyclotron design characteristics*

Cyclotrons are the most extensively used type of particle accelerator for radioisotope production. Manufacturers have a long history of adapting the cyclotron technology to meet the needs of their customers. The cyclotrons have become reliable, requiring minimal maintenance, with a well-distributed service network.

For the customer, the primary cyclotron specification is provided by the target yield for a given radioisotope in the form of a particular molecule within a specified irradiation time. For the cyclotron designer, this user-provided specification leads to decisions on the ideal nuclear reaction (projectile) to be used, the target material, the target chemistry, the projectile energy and the projectile current (flux) on target. The accelerated particles must have adequate energy for the chosen nuclear reaction and adequate current to provide the required yields at that energy. Higher energies (at the same current) generally result in a greater product yield, but often lead to increased impurities.

Hence, there is a tradeoff between purity, energy and current. Each product has its own ideal energy and current. Accelerators that provide beams of variable energy are able to optimize for different radioisotopes, albeit with greater complexity and added cost. Secondary specifications include, the number of targets, the target dimensions, the number of extracted beams (ports), shielding (facility) requirements, activation, and operating costs. Invariably, compromises are needed in order to arrive at a manageable facility budget.

Table 1 lists some of the important characteristics of the cyclotrons offered by several of the current industrial suppliers. These suppliers and their products are discussed further in Subsec. 4.4. The most common choice for the magnet is a conventional room temperature coil. Most manufacturers have the option of self-shielding for the cyclotrons when the maximum energy is less than 15 MeV. For the lower energy cyclotrons, the customer must choose between vault (concrete) shielding and the generally more

Table 1.   A list of the current manufacturers of cyclotrons used in radioisotope production for medical applications, their cyclotron models and some selected cyclotron specifications. (V is for acceleration in a vertical plane, and H is for acceleration in a horizontal plane.)

Company	Cyclotron model	Particles	Energy (MeV)	Beam current (μA)	Ion source type	Peak field (T)	Hill–valley ratio	Radio freq. (MHz)	Plane of acceleration	Cyc. weight (t)	Shield weight (t)	Power (kW)
ACSI	TR14	H$^-$	14	>100	Cusp	2.1		74	V	22	40	60
ACSI	TR19/(9)	H$^-$/(D$^-$)	19/(9)	>300/100	Cusp	2.1		74/37	V	22		65
ACSI	TR24	H$^-$	24	>300	Cusp	2.1	4	83.5	H or V	84		80
ACSI	TR30/(15)	H$^-$/(D$^-$)	30/(15)	1600/400	Cusp	1.9			H	56		150
ABT	Tabletop	H$^+$	7.5	5	PIG	1.2		72	H	3.2	7.6	10
Best	BSCI 14p	H$^-$	14	100 (400)	PIG (Cusp)			73	H	14		60
Best	BCSI 35p	H$^-$	15–35	1500	Cusp			70	H	55		280
Best	BSCI 70p	H$^-$	70	800	Cusp	1.6		58	H	195		400
CIAE	CYCCIAE14	H$^-$	14	400	Cusp							
CIAE	CYCCIAE70	H$^-$	70	750	Cusp							
NIIEFA	CC-18/9	H$^-$/D$^-$	18/9	100/50	Cusp			38.2	V	20		60
EUROMEV	Isotrace	H$^-$	12	100	Cusp	2.36	(SC)	108	V	3.8		40
GE	MINItrace	H$^-$	9.6	>50	PIG	2.2	~2	101	V	9	40	35
GE	PETtrace	H$^-$/D$^-$	16.5/8.6	>100/65	PIG	1.9	~1	27.2	V	22	47	70
IBA	Cyclone 3	D$^+$	3.8	60	PIG	1.8	1	14	H	5		14
IBA	Cyclone 10/5	H$^-$/D$^-$	10/5	>100/35	PIG	1.9	~5	42	H	12	40	35
IBA	Cyclone 11	H$^+$	11	120	PIG	1.9	~5	42	H	13	52	35
IBA	Cyclone 18/9	H$^-$/D$^-$	18/9	150/40	PIG	1.9	~5	42	H	25		50
IBA	Cyclone 30	H$^-$/D$^-$	30/15	1500/?	Cusp	1.7	14		H	50		180
IBA	Cyclone 30XP	H$^-$/D$^-$/^{4}He^{++}	15–30/7.5–15/30	350/50/50								
IBA	Cyclone 70 (Arronax)	H$^-$/D$^-$/H$_2$+/He^{++}	30–70/15–35 /17.5/70	2 × 350/50/50/35		1.7	14	66/30	H	125		350
KIRAMS	KIRAMS-30	H$^-$	15–30	500	Cusp	1.9	8	64	H			
KIRAMS	Kotron-13	H$^+$	40372	100	PIG	2	2	77.3	H	20	80	18?
Siemens	Eclipse RD	H$^-$	11	2 × 40	PIG	1.9	~13		H	11	39	35
Siemens	Eclipse HP/ST	H$^-$	11	2 × 60	PIG	1.9	~13	72				35
Sumitomo	HM-7	H$^-$/D$^-$	7.5/3.8						V		30	
Sumitomo	HM-10	H$^-$/D$^-$	9.6/4.8						V		52	
Sumitomo	HM-12/S	H$^-$/D$^-$	12/6	>60/30	PIG	2	~5	45	V	11	56	45
Sumitomo	HM-18	H$^-$/D$^-$	18/10	>90/50	PIG	2	~4	45	H	24	86	55

expensive but compact arrangement of close-packed steel around the cyclotron and targets.

Earlier generations of cyclotrons accelerated mostly positive ions. Today, most of the cyclotrons accelerate the negative H^- and/or D^- ions because of the ease of extracting multiple beams, each with variable current, into targets that are usually separately shielded by steel or concrete. The choice of either internal or external targets impacts the footprint of the system and also the radiation exposure for personnel servicing the cyclotron. In addition, for H^- currents less than $200\,\mu A$, an internal PIG (Penning ionization gauge) ion source [25] is the preferred cost-effective choice. If H^- currents greater than about $200\,\mu A$ are required, then an external (Cusp) ion source is required [25]. With internal targets and H^+ acceleration, high currents are readily achievable with internal PIG type ion sources. Extraction of intense H^+ beams invariably leads to severe component activation in the cyclotron.

4.3. *Cyclotron categories*

It is convenient to categorize the cyclotrons into three broad (proton) energy ranges based on their primary function [16]. (For reasons based on efficiency and cost considerations, some facilities/manufacturers have chosen cyclotrons that also accelerate other light ions. Some preferred nuclear reactions use deuterons and helium.) These proton energy ranges are:

(I) Cyclotrons with proton energy less than 20 MeV are primarily used for producing positron-emitting radionuclides. These PET isotopes tend to have short half-lives and the cyclotrons are located in regional centers/hospitals determined by the yield loss due to the delivery time from cyclotron to patient. Many of the cyclotrons have the capability of being shielded with close-packed steel and thereby reduce the need for the user to provide a heavily shielded bunker. The delivery time of the radioisotope, the patient dose requirement and the number of doses needed per day lead to a cyclotron providing at least a $50\,\mu A$ per target (for fluorine-18). Many of the current cyclotrons have the capability of using multiple targets on each of two or more extracted beams.

(II) Cyclotrons with proton energies between 20 and 35 MeV are primarily used to produce many of the gamma-emitting radioisotopes (commonly used as imaging radioisotopes for SPECT), as well as several other PET isotopes. The SPECT isotopes have medium half-lives and production generally takes place in dedicated facilities. The longer half-lives permit isotope delivery to more distant users and this leads to dedicated production facilities with high power targets and larger throughput.

(III) Cyclotrons providing protons with energies greater than 35 MeV are employed in the production of a number of the isotopes used for radiotherapy. The primary need is for high current cyclotrons with currents in the 1 mA range.

4.4. *Cyclotron suppliers*

Commercial firms have responded to the varied user specifications with a number of basic cyclotrons with optional add-ons in an attempt to satisfy each particular need and budget. What follows is a list of several companies, arranged alphabetically, and their key product specifications. A short comparative summary can be found in Table 1.

4.4.1. *Advanced Biomarkers Technology*

Advanced Biomarkers Technology (ABT) [26] has recently introduced a new low power, small "tabletop" cyclotron with internal targets, shown in Fig. 1. The company website notes that the accelerator has an on–off switch, a target selector switch and a beam current switch. The cyclotron accelerates H^+ from an internal ion source. It has a fixed energy of 7.5 MeV, a target volume of $15\,\mu L$, and is designed to provide single patient doses (mCi) of ^{18}F and ^{11}C. At $1\,\mu A$ and 7.5 MeV, 1 mCi/min of ^{18}F is produced. Currents are in the range of 1–$5\,\mu A$. With shielding the total weight of the accelerator is 10.8 tons and it occupies only about $2\,m^2$. The cyclotron is self-shielded and weighs about 7.6 tons. The wall plug power is only 10 kW.

4.4.2. *Advanced Cyclotron Systems Inc.*

Advanced Cyclotron Systems Inc. (ACSI) [16, 27] fabricates and sells 14 MeV, 19 MeV (fixed and

Fig. 1.   A compact cyclotron designed for the production of single-patient doses of ^{18}F and ^{11}C. *Photograph provided by Advanced Biomarkers Technology.*

Fig. 2.   The TR30 during installation at Nordion. *Photograph provided by Nordion.*

variable energy), 24 MeV and 30 MeV cyclotrons. Simultaneous dual extraction is available in each of the models. The 14 MeV cyclotrons are available in self-shielded and unshielded configurations. A new product line is the TR24, which provides high current, variable energy extracted proton beams in the energy range 15–24 MeV to partly bridge the energy gap between the firm's 19 and 30 MeV models. The TR30 offers variable energy extraction (15–30 MeV) with currents exceeding 1600 μA and simultaneous dual beams. The TR30 (TR30/15) can also be configured to provide dual-particle acceleration, albeit not simultaneously. Figure 2 shows a typical TR30, installed at the Nordion site in Vancouver.

4.4.3. *Best Cyclotron Systems Inc.*

Best Cyclotron Systems Inc. (BSCI) [28, 29] is bringing three new cyclotron models to market. It is fabricating a fixed energy 14 MeV H$^-$ cyclotron with an internal ion source, having a total extracted current of up to 100 μA (or, optionally, with a Cusp ion source up to 400 μA), into four external beams. Its 35 MeV cyclotron has an external Cusp type ion source, two simultaneous extracted beams with energy variable from 15 to 35 MeV and an advertised maximum extracted current greater than 1.5 mA. A 70 MeV model accelerates H$^-$ from an external ion source, and has simultaneous variable energy extraction from 35 to 70 MeV into two external beam lines.

4.4.4. *China Institute of Atomic Energy*

China Institute of Atomic Energy (CIAE) has plans for 14 MeV and 70 MeV cyclotrons for the production of medical isotopes [30–32]. Both have external ion H$^-$/D$^-$ ion sources. The CYCIAE-70 will provide 700 μA of extracted H$^+$ at 35–70 MeV and 40 μA of D$^+$ at 18–33 MeV. The CYCIAE-14 is designed to provide two dual extraction ports servicing four different targets. It is a fixed energy cyclotron that will provide up to 400 μA at 14 MeV.

4.4.5. *Efremov (NIIEFA)*

The Efremov Institute supplies a cyclotron that accelerates negative hydrogen and deuterium ions, in a vertical plane, to energies of 18 and 9 MeV, respectively. The cyclotron uses a Cusp type external ion source and provides extracted currents of H/D at 100/50 μA [33].

4.4.6. *EuroMeV*

EuroMeV offered the ISOTRACE superconducting cyclotron (based on the OSCAR-12 [34], initially developed by Oxford Instruments). This cyclotron provided an extracted beam current of up to 100 μA at a fixed energy of 12 MeV. It weighed only 3.8 tonnes and had a total operating power consumption of 40 kW. Recently EuroMeV has terminated production.

4.4.7. *GE Healthcare*

GE Healthcare [16, 35] has two cyclotron products for PET, namely MINItrace and PETtrace. MINItrace accelerates H^- in a vertically oriented cyclotron that provides $50\,\mu A$ at a fixed energy of 9.6 MeV. PETtrace also accelerates in a vertical plane either H^-/D^- up to a fixed energy of 16.5/8.6 MeV, with extracted currents of $100/65\,\mu A$. The MINItrace cyclotron features integrated shielding and fully automated operation during startup, tuning and operation.

4.4.8. *IBA*

IBA [16, 36–40] markets cyclotrons at 3 (D), 10/5 (H/D), 11 (H), 18/9 (H/D), 30 (H) and 70/35 (H/D) MeV. The Cyclone 3D was originally marketed to address the need for ^{15}O in the early 1990s. The original four-pole geometry has been replaced by three poles to provide additional vertical focussing. The Cyclone 10/5 is a fixed energy cyclotron with four extraction ports. The Cyclone 11 is a new product line that features a fixed energy, self-shielded cyclotron that accelerates H^- up to 11 MeV [39]. The design is based on the Cyclone 10/5 cyclotron. The Cyclone 18/9 is a fixed energy cyclotron that accelerates H^- up to 18 MeV and D^- up to 9 MeV. The Cyclone 18 Twin is a fixed energy cyclotron that accelerates H^- up to 18 MeV and that improves uptime and reliability by using two independent ion sources. The Cyclone 30 is a fixed energy cyclotron that accelerates H^- up to 30 MeV and can extract two independent beams. The Cyclone 30 XP is another new product line; it accelerates proton and alpha beams up to 30 MeV and deuterons up to 15 MeV. Protons and deuterons are accelerated as negative ions and extracted by stripping and at variable energy. The alpha beam is accelerated as He^{++} and extracted with an electrostatic deflector [40]. The Cyclone 70 is a multiparticle, fixed energy cyclotron that accelerates H^- up to 70 MeV as well accelerating alpha beams. The specifications for the Cyclone 70 installed in Arronax had the H^- at variable energy from 35 to 70 MeV, with a maximum current of $750\,\mu A$, D^- from 17 to 25 MeV at $50\,\mu A$, He^{++} at 70 MeV and 35 μA, and H_2^+ at a fixed 35 MeV and a maximum current of $50\,\mu A$.

4.4.9. *KIRAMS*

The Cyclotron Application Laboratory (KIRAMS) has developed two cyclotrons for radioisotope

Fig. 3.   An IBA Cyclone 30-HC model (with D-Pace ion source), installed at the Turkish Atomic Energy Commission in Ankara. *Photograph provided by IBA.*

Fig. 4.   The IBA Cyclone 70 XP. The intended use is for nuclear medicine, radiochemistry and production of Sr/Rb, Ge/Ga generators for PET. *Photograph provided by IBA.*

production. The KIRAMS-30 accelerates H$^-$ from an external ion source up to 30 MeV. The cyclotron extracts protons of up to 500 μA into through two ports at energies from 15 to 30 MeV [41]. The KIRAMS-13 accelerates H$^-$ from an internal PIG ion source up to 13 MeV and extracts currents of up to 80 μA through two ports.

4.4.10. *Siemens*

Siemens [16] markets the Eclipse brand cyclotrons, which were initially developed by CTI and sold under the RDS label. The Eclipse cyclotrons accelerate H$^-$ up to a fixed energy of 11 MeV. The Eclipse HP (shown in Fig. 5) provides 60 μA into each of two beam lines and onto a carousel that holds up to two targets, whereas the RD provides 40 μA on a carousel that holds up to eight targets in each of two beam lines. The Eclipse ST is a self-shielded HP.

4.4.11. *Sumitomo*

Sumitomo [16] has built a series of cyclotrons under the names HM-7, HM-10, HM-12S, HM-12 and HM-18. Each accelerates both H$^-$ and D$^-$. The HM-7 is a fixed energy cyclotron that has a self-shielding option and provides H/D at an energy of 7/3.5 MeV. The HM-10 is also a fixed energy cyclotron; it accelerates H$^-$/D$^-$ to 9.6/4.8 MeV, can be equipped with five targets and has a self-shielding option. The HM-12 accelerates H$^-$/D$^-$ up to 12/6 MeV, and has two extraction ports that can each be equipped with up to four targets. The HM-12S is

Fig. 5. The Siemens Eclipse HP cyclotron. *Photograph provided by Siemens.*

the shelf-shielded version of the HM-12. The HM-18 is also a fixed energy cyclotron, accelerating H$^-$/D$^-$ up to 18/10 MeV, and has two extraction ports, each accommodating four targets. All of the Sumitomo cyclotrons use internal PIG ion sources.

4.4.12. *Others*

There are numerous facilities still using cyclotrons that were manufactured by companies that no longer offer cyclotrons for sale. These cyclotrons include products originally manufactured and sold by Scanditronix (MC series cyclotrons), The Cyclotron Corporation (TCC series cyclotrons), Japan Steel Works (JSW/BC series cyclotrons) and Computer Technology and Imaging Inc. (CTI with their CTI-RDS, Radio Isotope Delivery System cyclotrons).

4.4.13. *Research cyclotrons*

Cyclotrons at TRIUMF [42, 43], RIKEN [44, 45], VECC [46, 47], LNS [48], SPIRAL [49] and NSCL [50] are being used to produce rare isotopes for fundamental research science and other applications.

5. Industrial Applications

5.1. *Agriculture*

Radiation has been used to genetically alter grain crops to produce higher yields with superior quality [51].

5.2. *Plant physiology*

The radioisotopes C-11 and N-13 have been used to study and optimize the uptake of fertilizer in plants [52].

5.3. *Insect control*

Radiation from isotopes has been used to sterilize insects in the lab and, when released, reduce the offspring from mating [53].

5.4. *Engine wear and friction*

Cyclotrons have been used to irradiate piston rings of engines in order to quantify the wear. A 9 MeV deuteron beam can be used to produce cobalt-57 in steel. A measurement accuracy of 0.1 microns to

1 mm per year is achieved. The wear of hardened machine tools has also been measured. Radioisotopes are used to measure the wear of metals, tire rubber and engine oil [22, 54].

5.5. *Oilfields*

Radioisotope tracers have been used to monitor the movement of oil and gas between adjacent wells. The flow rates of liquids and gasses in pipelines can be accurately measured [22, 55].

5.6. *Civil construction structural integrity*

Radioactive sulfur and phosphorus have been used to find fissures in rock. Photographic plates can then be used to identify fissures and possible weak points in the structures [22, 55].

5.7. *Soil and air pollution*

Radioactive tracers have been used to monitor the dispersion of pollutants in water, soil and air. Radiotracers are used to locate sources of pollution [23, 55, 56].

5.8. *Industrial process monitoring*

Tracers are used in industrial blenders to ensure complete mixing of compounds. The extent of termite infestation can be measured by tracers [57].

5.9. *Gauging*

The thickness of materials is measured by the attenuation of decaying radioisotopes. This technique is used to measure the thickness of plastic film during the manufacturing procedure. It was also used to measure the height of coal in a hopper. (A light beam cannot perform this function in the dusty environment.) Backscattering is used to measure the thickness of coatings [58].

5.10. *Gamma radiography*

Radioisotopes are used to test material for flaws. Isotopes can be easily transported to remote locations for radiography tests [57].

5.11. *Consumer products*

Smoke detectors rely on a radioactive isotope to detect the smoke. Radioisotopes are used to shrink plastic (shrink-wrap) instead of heat. They have been also used to measure and control how much soda there is in soft drink bottles. (Tracers are chosen so that their activity lasts for the length of measurement and there is no measurable residual activity in the tested material when the measurement is finished.) Photocopiers often use radioactive isotopes to reduce the problems associated with static charge building up on the paper.

6. Prospects

The total number of cyclotrons producing radionuclides is gradually growing, primarily to meet the expanding needs of nuclear medicine. In some regions, physicians now require a PET/CT scan before setting up a treatment protocol for certain diseases. This growth in nuclear medicine that has been observed in the past decade is expected to continue. New cyclotron models adapted to current needs are being designed.

The recent shortage of reactor-produced 99-molybdenum used in technetium-99 m generators has revived interest in the possibility of direct accelerator production. Mo-99 is currently produced in older research reactors using highly enriched uranium (HEU). Both the security issue surrounding the safe storage of the HEU and the expected lifespan of these older reactors have given urgency to pursuing alternate production methods. Approximately 80% of nuclear medicine procedures currently use ^{99m}Tc. The molybdenum isotope of mass 100 (Mo-100) bombarded with $\sim$14–19 MeV protons could be used to supply regional amounts of ^{99m}Tc directly through the (p, 2n) reaction [59, 60]. High current cyclotrons are needed to meet the demand and to avoid the buildup of ground state contamination, but the high power target technology will be challenging.

The primary PET isotopes are C-11 and F-18. In most cases the cyclotrons producing these isotopes run in batch mode, producing 10 s of curies per run. Patient doses, on the other hand, tend to be a factor of 1000 smaller; i.e. 10 s of millicuries. Hospital clinical users have expressed interest in small machines that provide single patient doses with the touch of a button. Visionaries are suggesting that we will soon have small "tabletop" accelerators with targets using microfluidics to carry out the chemistry on a "chip," providing the radiopharmaceuticals in patient doses prior to each procedure. One supplier (Advanced Biomarkers Technology, USA) has already developed a tabletop cyclotron which promises to provide patient doses on demand [26].

Radionuclides in nuclear medicine have been predominately employed as diagnostics. The use of targeted radiotherapy is increasing. Of special interest is α-particle radiotherapy, particularly with ^{211}At [61]. Astatine-211 can be produced with accelerators through the ^{209}Bi$(\alpha, \text{n})^{211}$At reaction with α energies in the range of 30–40 MeV. Until recently, studies with ^{211}At have been inhibited by the non-availability of adequate amounts of the radionuclide with high specific activity. This shortage should be addressed with the advent of cyclotrons such as the IBA Cyclone 30XP [38].

References

[1] L. S. Graham *et al.*, *RadioGraphics* **9**, 1189 (1989).

[2] E. O. Lawrence and N. E. Edlefson, *Science* **72**, 376 (1930).

[3] E. O. Lawrence and M. S. Livingston, *Phys. Rev.* **40**, 19 (1932).

[4] D. L. Friesel and T. A. Antaya, *Reviews of Accelerator Science and Technology*, Vol. 2 (2009), pp. 133–156.

[5] B. F. Milton, Commercial compact cyclotrons in the 90s, in *Proc. 14th Int. Conf. Cyclotrons* (World Scientific, 1996).

[6] T. J. Ruth *et al.*, *Nucl. Med. Biol.* **16**, 323 (1989).

[7] C. Birattari *et al.*, *J. Med. Eng. Technol.* **11**, 166 (1987).

[8] J. H. Lawrence *et al.*, *Radiology* **35**, 51 (1940).

[9] J. Chadwick, *Nature* **142**, 630 (1938).

[10] The Rockefeller Foundation Annual Report (1939), pp. 42, 235–237.

[11] M. C. Kamen, *Ann. Rev. Biochem.* **55**, 1 (1986).

[12] R. G. Evans, *AJR* **160**, 1343 (1993).

[13] http://www.imperial.nhs.uk/nhs60/past60years/timeline/index.htm#top.

[14] B. Anderberg, Technology transfer — Experience at Scanditronix, in *Proc. EPAC 94*, p. 350.

[15] Cyclotron Produced Radionuclides: Physical Characteristics and Production Methods, IAEA Tech. Rep. No. 468 (Vienna, 2009).

[16] Directory of Cyclotrons Used for Radionuclide Production in Member States, 2006 Update. IAEA Tech. Rep. IAEA-DCRP/2006 (Vienna, 2006).

[17] A Roadmap for Developing Accelerator Transmutation of Waste (ATW) Technology. DOE/RW-0519 (1999).

[18] M. Skalberg *et al.*, *Partitioning and Transmutation (P&T): A Review of the Current State of the Art*, SKB study (1995).

[19] *Scientific Opportunities with a Rare-Isotope Facility in the United States* (The National Academies Press, 2007).

[20] T. J. Ruth, *Reviews of Accelerator Science and Technology*, Vol. 2 (2009), pp. 17–33.

[21] D. D. Cohen *et al.*, Ion beams for materials analysis, *Encycl. Phys. Sci. Technol.* (Academic, 2001).

[22] T. Delvigne *et al.*, Modern Radiotracing Tools for the Development of Engines, Lubricants, and After Treatment Systems, KSAE09-B0066 (2009), pp. 414–420.

[23] J. Guizerix *et al.*, *IAEA Bull.* 20–24 (1987).

[24] Cyclotron Produced Radionuclides: Principles and Practice. IAEA Tech. Rep. No. 465 (Vienna, 2008).

[25] *The Physics and Technology of Ion Sources*, 2nd, revised and extended edn., ed. I. G. Brown (Wiley-VCH Verlag, KGaA, 2004).

[26] ABT Molecular Imaging, Inc., Louisville, Tennessee, USA, http://advancedbiomarker.com/

[27] Advanced Cyclotron Systems, Richmond, BC, Canada, http://www.advancedcyclotron.com/

[28] Best Cyclotrons Systems, http://www.bestcyclotron.com/index.html

[29] R. R. Johnson *et al.*, New high intensity compact negative hydrogen ion cyclotrons, in *Cyclotrons 2010 — Proceedings* (Lanzhou, China, 2010).

[30] J. Zhong *et al.*, The physics design of magnet in CYCIAE-14, in *Cyclotrons 2010 — Proceedings* (Lanzhou, China, 2010).

[31] S. An *et al.*, Physics design and calculation of the CYCIAE-70 extraction system, in *Cyclotrons 2010 — Proceedings* (Lanzhou, China, 2010).

[32] M. Li *et al.*, The injection line and central region design of CYCIAE-70, in *Cyclotrons 2010 — Proceedings* (Lanzhou, China, 2010).

[33] P. V. Bogdanov *et al.*, New compact cyclotron CC-18/9 designed and manufactured in NIIEFA, in *Proc. RuPACC 2006* (Novosibirsk, Russia), pp. 70–72.

[34] R. Griffiths, *Nucl. Instrum. Methods B* **40–41**(part 2), 881 (1989).

[35] GE Healthcare, http://www.gehealthcare.com/euen/fun_img/products/radiopharmacy/products/cyclotrons_index.html

[36] IBA Group, http://www.iba-worldwide.com/iba-solutions/cyclotron-solutions

[37] S. Zaremba *et al.*, Upgrade of the IBA Cyclone 3D Cyclotron, in *Cyclotrons 2010 — Proceedings* (Lanzhou, China, 2010).

[38] W. Kleevan *et al.*, Recent development and progress of IBA cyclotrons, *Nucl. Instrum. Methods B*, in press.

[39] V. Nuttens *et al.*, Design of IBA cyclone 11 cyclotron magnet, in *Cyclotrons 2010 — Proceedings* (Lanzhou, China, 2010).

[40] E. Forton *et al.*, Design of IBA Cyclone 30XP cyclotron magnet, in *Cyclotrons 2010 — Proceedings* (Lanzhou, China, 2010).

[41] K. Kang *et al.*, Status of KIRAMS-30 commissioning, in *Cyclotrons 2010 — Proceedings* (Lanzhou, China, 2010).

[42] http://www.triumf.ca/research/research-topics/rare-isotope-beams

[43] P. Bricault, Progress towards new RI and higher RIB intensities at TRIUMF, in *Cyclotrons 2010 — Proceedings* (Lanzhou, China 2010).

[44] Y. Yano *et al.*, Status of the RIKEN RIB factory, in *IEEE Proc. PAC07* (2007), pp. 700–701.

[45] O. Kamigaito *et al.*, Status of RIBF accelerators at RIKEN, in *Cyclotrons 2010 — Proceedings* (Lanzhou, China, 2010).

[46] A. Bandyopadhyay *et al.*, Rare ion beam facility at Kolkata, in *Proc. PAC09* (2009), pp. 5053–5055.

[47] http://www.vecc.gov.in/rib.php

[48] http://www.lns.infn.it/FRIBS/fribshome.html

[49] http://www.ganil-spiral2.eu/spiral2-us

[50] http://frib.msu.edu/content/scientific-opportunities

[51] R. Tanaka *et al.*, Recent technical progress in application of ion accelerator beams to biological studies in TIARA, in *Proc. Asian Particle Accelerator Conference*, http://accelconf.web.chern.ch/AccelConf/a98/APAC98/7C003.PDF (1998).

[52] A. D. M. Glass *et al.*, *J. Exp. Botany* **53**, 885 (2002).

[53] L. J. Bruce-Chwatt, *Bull. WHO* **15**, 491 (1956).

[54] How Isotopes Benefit Industry, http://www.iaea.org/Publications/Magazines/Bulletin/Bull091/09105001319.pdf

[55] *Proceedings of Radioisotope Tracers in Industry and Geophysics* (IAEA, Prague, 1966).

[56] V. T. Dubinchuk *et al.*, Nuclear techniques for investigating migration of pollutants in ground water, *IAEA Bull.* 4/1990, 16 (1990).

[57] Industry: Use of Tracers, http://www.aboutnuclear.org/view.cgi?fC=Industry,Use_of_Tracers

[58] G. A. Johansen and P. Jackson, *Radioisotope Gauges for Industrial Process Measurements* (Wiley, 2004).

[59] B. Scholten *et al.*, *Appl. Radiat. Isot.* **51**, 60 (1999).

[60] A. Mushtaq, *Non-Proliferation Rev.* **16**, 285 (2009).

[61] M. R. Zalusky *et al.*, *Nucl. Med. Biol.* **32**, 779 (2007).

Paul Schmor is currently the Chief Scientific Officer (CSO) for Advanced Applied Physics Solutions, Inc. (AAPS). AAPS is a federally funded Canadian Centre of Excellence for Commercialization and Research, located at TRIUMF in Vancouver. He was chair of the LINAC' 08 conference in Victoria, BC and the PAC' 09 conference in Vancouver, BC. Previously, he worked for TRIUMF as an accelerator physicist. He was project leader for ISAC and also head of the TRIUMF Accelerator Division. Dr. Schmor has experience in a wide range of accelerator physics topics, with particular expertise in high intensity H$^-$ ion sources, including H$^-$-polarized ion sources.

Reviews of Accelerator Science and Technology
Vol. 4 (2011) 117–145
© World Scientific Publishing Company
DOI: 10.1142/S1793626811000586

Development of Accelerator Mass Spectrometry and Its Applications

Jiaer Chen,* Zhiyu Guo[†] and Kexin Liu[‡]

*State Key Laboratory of Nuclear Physics and Technology,
Peking University, Beijing 100871, China*
*chenje@pku.edu.cn
[†]zhyguo@pku.edu.cn
[‡]kxliu@pku.edu.cn

Liping Zhou

*Laboratory for Earth Surface Processes,
Department of Geography, Peking University,
Beijing 100871, China*
lpzhou@pku.edu.cn

With over 30 years' development, accelerator mass spectrometry (AMS) technology has been applied in almost all domains of science and technology. In this article, we first provide a brief description of the basic principles of AMS and then a review of recent technical developments. Selected examples of applications are given to show the impact and potential of AMS in earth and environmental sciences, archeology, art authentication, biomedical sciences, nuclear astrophysics and nuclear environmental safeguards. Finally, the trends and challenges of AMS development are briefly discussed.

Keywords: Accelerator mass spectrometry (AMS); cosmogenic radionuclides; anthropogenic radionuclides; applications.

1. Introduction

Accelerator mass spectrometry (AMS) is a powerful means of measuring the isotopic abundance of long-lived radionuclides.

The long-lived radionuclides can be divided into different categories, depending on their origin [1]. The primordial radionuclides, such as ^{40}K, ^{238}U, ^{235}U and ^{232}Th, were formed during the process of element formation in the early universe, with half-lives on the order of 10^9 a (a is the unit of year). They have not fully decayed since the formation of the Earth, so their abundance in nature is relatively high. The existent radionuclides with half-lives of 10^8 a and less, such as ^{10}Be, ^{14}C, ^{26}Al, ^{36}Cl, ^{41}Ca and ^{129}I, are mostly cosmogenic or anthropogenic. Cosmogenic nuclides are formed by nuclear reactions induced by cosmic rays mainly in the Earth's atmosphere, and then distributed into the hydrosphere, lithosphere and biosphere. During the transportation and deposition, the radionuclides are mixed with their stable isotopes, and their isotopic ratios are

usually very low. The decay rate of the radionuclide is not influenced by any external variables, while the concentration R decreases exponentially:

$$R = R_0 \exp(-\lambda t), \tag{1}$$

where R_0 is the concentration of the radionuclide at $t = 0$, and λ is decay constant. Such a nuclide has the potential to serve as a dating tool, and the age A can be calculated according to

$$A = -\tau \ln\left(\frac{R}{R_0}\right), \tag{2}$$

where τ is the mean lifetime of the radionuclide, $\tau = 1/\lambda = t_{1/2}/\ln 2$ and $t_{1/2}$ is the half-life of the radionuclide.

Apart from their use in dating, the cosmogenic nuclides can be employed as a powerful tracer to track changes in the terrestrial or marine environmental systems.

Usually the short-lived radionuclides can be measured by the decay-counting method easily, due to their high decay rate. The primordial radionuclides can be measured by conventional mass spectrometry (MS), due to their high abundance. By contrast, the measurements of long-lived cosmogenic radionuclides are more difficult for both decay counting and MS, due to their low decay rate and low abundance. Unlike in MS measurements, the ions in AMS are accelerated to several MeV or even several hundred MeV, which makes it possible to detect the radioactive nuclides with the isotopic ratios as low as 10^{-12}–10^{-16}. This is achieved by the application of various techniques at higher energy to eliminate and suppress the molecular and isobar interferences effectively. Therefore, AMS measurement has ultra-high sensitivity.

For the measurement of ^{14}C, whose half-life is 5730 a, the decay-counting method can also be used. In fact, liquid scintillation counting (LSC) and the gas proportional counter (GPC) have been widely used for ^{14}C measurements with satisfactory precision for the past several decades. However, AMS has shown distinct advantages over the decay-counting method. The ^{14}C detection sensitivity of AMS is much higher than decay counting, which makes it possible to perform ^{14}C measurements with a milligram carbon material instead of grams for decay counting and half-an-hour counting times instead of several days [2]. For the radionuclides with half-lives longer than ^{14}C, only AMS can be used to detect their natural concentrations.

Although the first AMS measurement can be traced back to 1939, when Alvarez measured the ^{3}He/^{4}He ratios in helium gas using a cyclotron [3], the extensive use of AMS started almost 40 years later. In 1977 two groups, at McMaster University and Rochester University, reported successful detection of ^{14}C from natural samples, respectively [4, 5]. The tandem-accelerator-based AMS was used in these experiments. The benefit of such an approach is that ^{14}N does not form stable negative ions and the interfering molecules such as ^{12}CH$_2^-$ and ^{13}CH$^-$ can be easily broken up by the stripping process in the tandem terminal. Following these initial successful experiments, many nuclear physics laboratories re-equipped their existing tandem accelerators with a terminal voltage of 6–14 MV to meet the requirements of AMS. Meanwhile, the first dedicated AMS facility specifically for ^{14}C measurements was developed by Purser in 1980 [6]. Since then, efforts to develop a dedicated ^{14}C AMS machine have been made continuously. The AMS measurement techniques for other radionuclides, such as ^{36}Cl [7], ^{129}I [8], ^{10}Be [9], ^{26}Al [10] and ^{41}Ca [11], were developed subsequently. The common radionuclides measured by AMS are listed in Table 1. At present, ^{14}C remains the most important nuclide, taking up more than 90% of AMS measurements.

The development of AMS and its applications have been reviewed in many papers during the past 30 years [1, 2, 18–23]. This review is mainly for those readers who might not be very familiar with AMS but would like to know how it works and what kind of problems can be solved with it. The basic principles of AMS, the technical progress of modern AMS instruments and the applications of AMS are discussed in the following sections. However, AMS has developed so rapidly and its application fields are immensely wide. It is impossible to give complete coverage in this article. We had to make a selection, especially in the area concerning applications. Any omission, therefore, does not imply less importance.

2. Basic Principles of AMS

AMS is essentially an atom-counting technology using a particle detector. In order to realize the ultrahigh sensitivity, the molecular, isobar and the abundant stable isotope interferences should be eliminated before the detector or suppressed with

Table 1.   Common radionuclides measured by AMS.

Nuclide	Half-life (a)	Abundant isotope	Isobar	Sample	Extracted ion	Background	Precision (%)
^{10}Be	1.4×10^6	^{9}Be	^{10}B	BeO	BeO$^-$	5×10^{-16} [12]	1
^{14}C	5730	^{12}C, ^{13}C	^{14}N	Graphite/CO$_2$	C$^-$	8×10^{-16} [13]	0.3
^{26}Al	7.1×10^5	^{27}Al	^{26}Mg	Al$_2$O$_3$	Al$^-$/AlO$^-$	1×10^{-15} [14]	3
^{36}Cl	3.0×10^5	^{35}Cl, ^{37}Cl	^{36}S	AgCl	Cl$^-$	1×10^{-16} [15]	3
^{41}Ca	1.04×10^5	^{40}Ca	^{41}K	CaH$_2$/CaF$_2$	CaH$_3^-$/CaF$_3^-$	6×10^{-16} [16]	3
^{129}I	1.7×10^7	^{127}I	^{129}Xe	AgI	I$^-$	1×10^{-15} [15]	3
^{236}U	2.34×10^7	^{235}U, ^{238}U	—	U$_3$O$_8$	UO$^-$	1×10^{-13} [17]	4

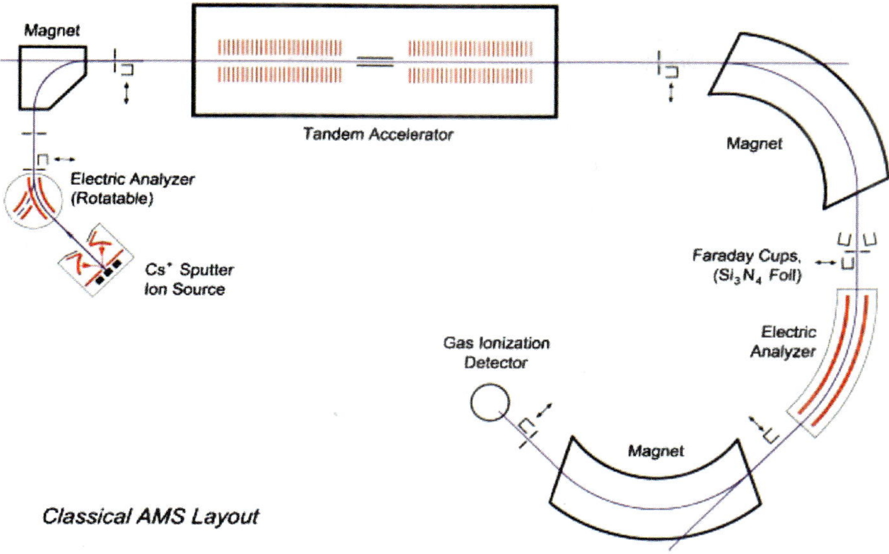

Fig. 1. Schematic diagram of a classical AMS layout based on a tandem accelerator [23].

a detector which provides particle identification. Another important issue concerns fractionation, which makes the isotope ratio shift from its real value. These issues are fundamental and must be well considered when designing an AMS facility.

2.1. *Typical tandem AMS system*

Figure 1 shows a schematic diagram of a classical AMS layout based on a tandem accelerator [23], which consists of a sputter ion source, a low energy mass spectrometer, a tandem accelerator, a high energy mass spectrometer and a detector.

2.1.1. *Ion source*

The ion source of AMS should have high beam current, high negative ion yield, small beam emittance, good stability, multitargets and low memory effect. The high intensity Cs sputter sources are most often used for tandem AMS, due to their versatility and flexibility, low background and low cross contamination, as well as their reasonable yields. A target in the source is bombarded with positive Cs ions, and atoms of the target material are sputtered out. Part of the emitted atoms and molecules becomes negatively charged and is extracted by an electric field [24]. The sample materials and extracted ions for different nuclides are given in Table 1.

For ^{14}C, ^{26}Al and ^{129}I, isobar interferences are eliminated in the ion source, as ^{14}N, ^{26}Mg and ^{129}Xe do not form stable negative ions. Since beryllium does not form stable negative atomic ion, its samples are prepared in the form of BeO, and the molecular ion BeO$^-$ is used for analysis. For ^{41}Ca measurement, although the Ca$^-$ ion is stable, its binding energy is too low to generate enough current in the sputter ion source. In this case, the molecular ions are used: either CaH$_3^-$ ions extracted from CaH$_2$ samples [16] or CaF$_3^-$ ions extracted from CaF$_2$ samples [25]. Both molecular ions can lead to very efficient suppression of the interfering isobar ^{41}K. The CaF$_2$ samples can be prepared much more easily than the CaH$_2$ samples, and CaH$_2$ decomposes very quickly when exposed to air, so CaF$_2$ is used for most applications.

The gas-fed sputter ion source has been studied since the 1980s, especially for ^{14}C measurements [26–29]. The advantages of introducing CO$_2$ directly into the sputter source are that graphitization is omitted and the sample size can be reduced down to 20 μg [29]. The source can also be coupled with other equipment for analysis, such as an elemental analyzer and gas chromatographs. However, the C$^-$ current from a gas source is lower than that from a graphite target, and the cross-contamination is more pronounced. These problems can be solved with improved gas-handling systems and appropriate measuring procedures including presputtering of Ti holders.

2.1.2. *Injection system*

The injection system of tandem AMS, which ideally consists of a magnetic analyzer and an electrostatic analyzer, as shown in Fig. 1, acts as a low energy mass spectrometer. The ion beam extracted from the sputter ion source is not monoenergetic. The abundant isotope peaks have long tails toward both the lower energy and high energy sides, which can cause interferences to neighboring rare isotopes. Therefore, a magnetic analyzer with sufficient mass resolution is essential, and the combination of magnetic and electrostatic spectrometers will select only the ions which have correct energy, and then the tails can be significantly suppressed [30].

In order to reduce the effects on the isotopic ratios caused by transmission variations, both the rare and the abundant isotopes should be injected into the accelerator on the same beam pass and analyzed on the high energy side. Ion injection can be performed in two ways, i.e. sequentially or simultaneously. Dedicated AMS facilities usually use the fast-cycling sequential injection with a cycle period around 100 ms. During this period, most of the time is used for the measurement of the rare isotope and only a very short fraction is for the stable isotope measurements. This procedure leads to significant reduction in the beam load on the accelerator. To realize the sequential injection, the vacuum box of the injection magnet is electrically isolated and two gap lenses are inserted at the entrance and exit of the isolated box. A bias voltage is then applied periodically to change the ion energy so that the different isotopes can go through the identical trajectories sequentially [31].

Another approach is to inject the rare and abundant isotopes simultaneously. In such a system, a sequence of dipole magnets and lenses are used, in which arrangement the different isotopes extracted from the ion source go through different trajectories and then are recombined at the entrance of the accelerator [32].

2.1.3. *Tandem accelerator*

In the tandem accelerator, the injected negative ions are accelerated to the positive high voltage terminal, where they are converted into positive ions in a stripping gas canal and further accelerated to ground potential. One advantage of the tandem accelerator is that both ion source and detector are located on ground potential, which makes operation and maintenance easier. Clearly, with this type of accelerators much higher beam energies can be produced compared to single stage electrostatic accelerators. Hence, tandem accelerators are mostly used for AMS.

Another advantage is that the stripping process of the tandem accelerator can eliminate the molecular interferences, such as $^{12}CH_2^-$ and $^{13}CH^-$ in the ^{14}C measurement, because multicharged molecules ($q \geq 3+$) are not stable. But the fragments of the molecules may become new isotope interferences through the subsequent multiple charge exchange and make a contribution to the continuum in the high energy analysis system. The relatively narrow gas stripping canal and the low vacuum condition in the accelerating tube will lead to limited beam transmission through the tandem. Bonani *et al.* [33] showed that remarkable improvement in the beam transmission can be achieved by the use of turbo-molecular pumps to recirculate the stripping gas and an enlarged stripping canal.

The tandem accelerator used for AMS can be grouped into two categories. One comprises those operational accelerators for nuclear physics with a terminal voltage of 6–14 MV. Modifications are necessary for a machine of this kind, especially for the injection system. The higher energy available from these larger machines has the advantages for measuring ^{36}Cl and ^{41}Ca with a lower background, and is suitable for developing new nuclide AMS measurement methodology, like the 14UD at ANU, Canberra [15] and the MP at Rochester University [34].

Another category is made up of the dedicated AMS facilities. In the last decade, two main commercial manufacturers, High Voltage Engineering Europa (HVEE) and the National Electrostatics Corporation (NEC), have been involved. At the beginning, the dedicated machines were mainly for ^{14}C measurements. Efforts have been made to reduce the size and the cost of the dedicated ^{14}C AMS. The terminal voltage has been reduced successfully from 2.5–3 MV down to 200 kV during the past 30 years.

The dedicated AMS for other nuclides has also been developed. One option is to use a midsized tandem with a terminal voltage of 3–6 MV, such as VERA in Vienna [35] and SUERC in Glasgow [14], which can measure a variety of nuclides. Recently, successful use of small tandems to measure ^{10}Be, ^{26}Al and ^{129}I has also been achieved [36].

2.1.4. *High energy mass spectrometry*

After acceleration, the rare nuclide with a selected charge state is analyzed by a magnet in the high energy beam line, and the stable isotopes with the same charge state are measured by off-axis Faraday cups after the magnet. The width of the magnet poles and the vacuum box must be large enough.

There are two types of background ions in the high energy analysis system. One is from the molecular fragments having similar E/Q and M/Q values to the rare ions of interest, and the other is from the tails generated by the scattering of nearby major beams with similar E/Q and M/Q. A magnetic analyzer alone is not sufficient for performing the final mass separation. To suppress the continuum of abundant isotopes, at least two different types of analyzers should be used in cascade. The second analyzer is usually an electrostatic deflector. Occasionally a Wien filter is also used. The double charge exchanging process may also occur in the magnetic or electrostatic analyzer, which could cause unwanted ions getting through together with the ions of interest [23]. Therefore, a third analyzer may be required, as shown in Fig. 1.

2.1.5. *Detector*

As the ions are accelerated to higher energy, the $\Delta E - E$ gas counter, which is commonly used in nuclear physics experiments, has been successfully employed in AMS. Such a detector acquires not only the signals of the total energy of the ions, but also the rate of energy loss. The energetic ions ionize the gas along its track, and electrons produced by ionization drift toward the anode in the transverse electric field. A fine-meshed grid with high transmission (called a Frisch grid) is placed in front of the anode to separate the fast electron signal from the slow ion signal [1]. By dividing the anode into sections, the ΔE signals can be collected. The rate of energy loss depends on the atomic number Z of the ion, so the element can be identified and the isobars, such as ^{36}Cl and ^{36}S, with identical energies can be separated in an energy spectrum [37].

Another kind of gas counter is the Bragg curve detector, in which the electric field is parallel to the ion track and the energy loss information is obtained from the time development of the signal at the anode.

If the isobar has been effectively eliminated, for example in the ^{14}C measurements, the semiconductor detector can be used and it is especially beneficial at low energies.

For heavy ions, such as ^{236}U, there are intense sources of the interfering background from neighboring masses, which cannot be fully separated by the high energy spectrometer. Therefore, a high resolution velocity measurement by time of flight (TOF) should be used to obtain mass information. A TOF consists of a start detector and a stop detector which are installed with a distance of 1–3 m. Thus, the velocity of ions can be determined [38]. The disadvantage of TOF is that the foil in the start detector may cause beam loss, making its efficiency only 50–80%.

2.2. *Background*

The lowest isotopic ratios measured with AMS will be limited by the background. The AMS background can be divided into two categories: the intrinsic background and the interference background. The intrinsic background is due to the radionuclide atoms which do not come from the original sample material. Contamination may be introduced during sample collection in the field, sample preparation in the laboratory, or the cross-talk between samples in the ion source. Once the intrinsic background is introduced in the sample, it cannot be removed by analysis.

The interference background is the atoms of different nuclides, which reach the final detector but cannot be identified. The interference ions include molecular ions, abundant isotopes and isobar ions, which should have been suppressed in the ion source, stripping process, low energy and high energy mass spectrometry, and the detector. Because the rare nuclide concentration is very low, it is possible that some residual interference ions enter the area set for rare nuclides in the two-dimensional spectrum of the detector.

During AMS measurements, the background levels are determined by measuring blank samples. The blank samples are made of "dead" material, i.e. the radionuclides of interest have fully decayed so that their isotopic ratio is much lower than for the samples to be measured. The blank samples are prepared and measured through the same procedure as the unknown samples. The possible lowest backgrounds of common radionuclides measured by AMS are listed in Table 1. The actual background level depends on the specific AMS system.

2.3. *Measurement uncertainty*

Measurement uncertainty includes two different concepts: precision and accuracy. Precision reflects the statistical uncertainty of the measurement, which is determined by the square root of ion counts N. In order to obtain higher precision, high beam current (for regular samples) or high sample utilization efficiency (for small samples) is essential. The possible highest precision of common radionuclides is listed in Table 1.

Accuracy depends on the systematic errors in measurements, which may come from the machine instability, contamination and fractionation. It is particularly important for ^{14}C measurements. Fractionation is a mass selective effect, which means that the isotopic ratio changes during a physical or chemical process. The fractionation effect may have the following causes [1]:

(a) *Natural fractionation.* This is related to the mechanism as to how the isotopes are incorporated into the sample. For ^{14}C measurements the natural fractionation of ^{14}C/^{12}C can be corrected by means of δ^{13}C.
(b) *Fractionation in sample preparation.* This depends on the particular chemical reaction and its controlling factors.
(c) *Fractionation in the ion source.* The generation of negative ions in the sputter process is velocity-related, which means being mass-dependent at the same energy. Also, a crater may be created at the target surface during the sputter process, which may change the beam emittance, and so the crater effect can introduce time-dependent fractionation.
(d) *Fractionation in the stripping process.* The stripping yields are also velocity-related.
(e) *Fractionation in beam transportation.* The effect of the magnetic field on ions is momentum-related, which means being mass-dependent at the same energy. It is better not to use magnetic quadrupoles and steers on the common beam line, in which all the isotopes pass through. The

Earth's magnet field and stray magnetic field in accelerator tubes may also introduce fractionation of the beam and tend to cause the transverse shift of the beam orbits of different isotopes. To avoid such an effect, all the slits and apertures should be widely opened to realize the so-called "flat top transmission," which means that the curve of beam transmission versus adjusted parameters (magnet field, quadrupoles, steers, terminal voltage, etc.) has a flat top so that the beam transmission is not sensitive to those parameters within a certain range. The transportation fractionation can also occur when the beam transmissions after the first magnet analyzer in high energy mass spectrometry are not the same for different isotopes. In some AMS machines, current-dependent fractionation has been found, which is generated by the space charge effect of the high current of abundant isotopes [39].

It is impossible to eliminate completely all the fractionation effects mentioned above. Therefore, standard samples, which have known isotope ratios and natural fractionation values prepared through the same process with the unknown samples and measured under the same operation conditions, must be used for correction. The measurement results of the unknown samples should be corrected with standard samples, blank samples and the natural fractionation factors. For example, the ^{14}C age of a sample, which is defined as an age before the present (BP), and here "the present" is taken as 1950, can be calculated as [40]

$$A = -8033 \ln\left\{\frac{1}{0.95}\left(\frac{R_x - R_b}{R_s - R_b}\right)\frac{1 - \frac{2(25 + \delta^{13}C_x)}{1000}}{1 - \frac{2(19 + \delta^{13}C_s)}{1000}}\right\},$$

$$(3)$$

where R_x, R_s and R_b are the ^{14}C/^{12}C ratios of the unknown, standard and blank samples respectively, $\delta^{13}C_x$ and $\delta^{13}C_s$ are the natural fractionation values of unknown and standard samples respectively, and 8033 is the mean lifetime of ^{14}C which corresponds to the conventional Libby half-life, 5568 a. δ^{13}C represents the deviation of the sample ^{13}C/^{12}C ratio relative to an international standard, whose value is given in per mil. The natural fractionation is a linear effect, so δ^{14}C $= 2\delta^{13}$C. The common standard material is OX-I (made by NIST), whose δ^{13}C value should be $-19‰$. The coefficient 0.95 in Eq. (3) is a correction factor of the $^{14}C/^{12}C$ ratio in atmosphere in 1950 over the $^{14}C/^{12}C$ ratio of OX-I in 1950. The ^{14}C age calculated by Eq. (3) is not the calendar age, but is an age derived under the hypothesis that the ^{14}C production rate is constant and the initial concentration $^{14}C/^{12}C$ in the sample is a constant too. However, this hypothesis is not robust and a calibration curve should be used to convert the ^{14}C age to the calendar age; see Subsec. 4.1. Because the calibration curve based on a tree ring is normalized to a $\delta^{13}C$ value $-25‰$, the $^{14}C/^{12}C$ ratio of the unknown sample should be corrected to a $\delta^{13}C$ value of $-25‰$ in Eq. (3).

2.4. *Other accelerators*

In addition to tandem accelerators, other types of accelerators have also been investigated as the main accelerator for AMS.

The cyclotron has good mass selection capacity, which can suppress the molecular interference effectively, but its mass resolution is not sufficiently high to separate isobars. In the early stage of AMS developments, the cyclotron had been used to measure ^3H, ^{10}Be, ^{14}C and ^{26}Al [41]. Subsequently, the tandem accelerator became dominated for AMS. However a minicyclotron AMS was developed in Shanghai in the 1990s [42]; it is able to measure ^{14}C at the natural level by injecting negative ions.

The radio frequency quadrupole (RFQ) linac can also be used for AMS. An RFQ-based ^3H AMS system was developed at LLNL in the late 1990s [43], and it can accelerate ^3H ions to 1.5 MeV with 84% transmission and ^1H to 0.5 MeV with 22% transmission simultaneously. The feasibility of using RFQ to construct the ^{14}C AMS system has also been investigated [44].

However, as compact ^{14}C AMS systems have been developed rapidly, both minicyclotron and RFQ ^{14}C AMS systems seem to have lost their competitiveness.

Because noble gases do not form negative ions, the tandem AMS cannot be used for noble gas measurements. The K1200 cyclotron at MSU with full stripping has been used for ^{81}Kr measurement, while the RF linear accelerator ATLAS at ANL with a gas-filled magnet has been used for measurement of ^{39}Ar [45].

In recent years, a single stage electrostatic accelerator AMS has been developed by NEC. It is described in Subsec. 3.1.

3. Technical Progress of Modern AMS Instruments

3.1. *Compact AMS system*

A significant technical development of AMS in the recent decade is the trend toward a smaller AMS facility. It has reduced substantially the size, complexity and cost of the facility, while maintaining high output, high sensitivity and low background.

3.1.1. *AMS with a sub–1 MV tandem accelerator*

Historically, over the first 20 years of developments, AMS had a golden rule of using charge states 3^+ or higher to eliminate molecular ions, since they are unstable and decay instantly. On the contrary, molecules in charge state 1+ and 2+ are stable and not easily destroyed. Systematic studies on cross sections for dissociation of mass 14 molecules with an increasing density of the stripper showing their interference with radiocarbon can be reduced over 11 orders of magnitude at a moderate density of the stripper gas [46]. This has made it possible to use charge state 1^+ and therefore a smaller tandem accelerator. The first AMS system using charge state 1^+ became operational in 1998 [47]. It was shown that 1^+ technology can have 42% overall beam transmissions and background levels of $(2–8) \times 10^{-15}$, which means that it is capable of competing with traditional AMS systems using charge state 3+ or higher [48]. Following the first construction of low energy AMS with an ~ 0.5 MV tandem accelerator at ETH in Zurich, NEC has commercialized this technology and built quite a number of compact AMS systems which are distributed worldwide, with satisfactory precision and background for ^{14}C measurement. Figure 2 shows the layout of NEC's compact AMS at Peking University [49]. It has an overall footprint of about 4.7 m × 5.5 m.

Efforts have also been made to measure other isotopes with small accelerators. ^{10}Be measurement at a terminal voltage below 1 MV has become possible by using a gas ionization detector with a very thin and homogeneous SiN window and a cooled FET preamplifier for the detection of low energy

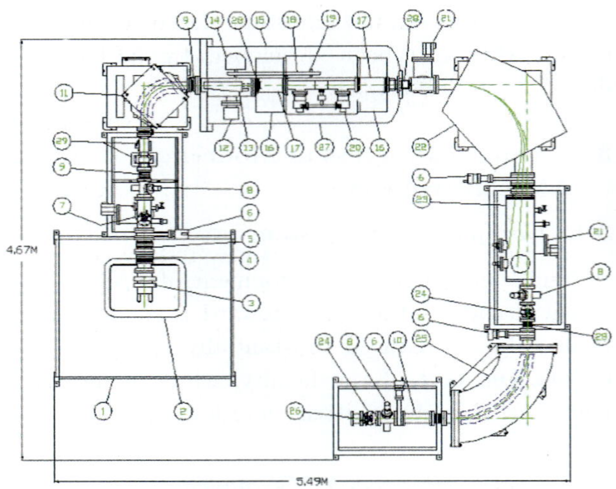

Fig. 2. Schematic layout of the compact AMS at Peking University.

Fig. 3. The BioMICADAS [56].

ions [50]. ^{10}Be measurement with the upgraded compact AMS system TANDY at ETH, Zürich at a terminal voltage of only 0.53 MV is fully competitive with some larger AMS facilities concerning the background and yield [51]. The main modifications of TANDY include a thin SiN degrader foil placed in front of the ESA in order to reduce the ^{10}B intensity and an additional 130° magnet mounted after the ESA to remove the ^{9}Be background. The transmission efficiency and the background in routine measurements are 10–11% and $\sim 1 \times 10^{-15}$, respectively. The measurement method for ^{129}I has also been established at the 0.5 MV TANDY AMS system and the achieved performance is satisfactory for measurement of most ^{129}I samples [35].

In 2006, Klein *et al.* [52] reported on a 1 MV compact multielement AMS system, built by HVEE in Amersfoort, the Netherlands, with a footprint of 3.8 m × 6.3 m. The first of such HVEE machines is in operation in Seville, Spain [53]. It is claimed that this system is capable of measuring ^{10}Be, ^{14}C and ^{129}I with high precision and low background. The system has also demonstrated impressive capabilities of much better transmission efficiency at low energies in measuring actinide nuclides [54].

3.1.2. *Mini-AMS at very low energy*

To explore the lowest voltage limit, the process of molecular interference was studied systematically and tests on very low voltage were conducted. For

a 200 kV tandem accelerator, a background ^{14}C/^{12}C ratio of $(0.4$–$1.3) \times 10^{-14}$ can be achieved and the stripping yields for the 1^+ charge state are close to 50% and an overall transmission for ^{12}C is about 41–43% [55]. Since then, small accelerators running at voltages as low as 200 kV for ^{14}C dating and biomedical applications have been commercially available. One of the typical examples is the BioMICADAS at the Laboratory of Vitalea Science in Davis, USA [56]. It has an overall footprint of only 2.5 m × 3 m, with a 200 kV high voltage platform for tandem-based ion acceleration, as shown in Fig. 3.

The accelerator is placed in the middle, the high energy magnet on the left side and the low energy magnet with the shielding for the beam switching system on the right side. In this mini-AMS system, both the high pressure tank and the accelerator tube are omitted. Instead, it consists of two gap lenses and a charge exchange channel that provides sufficient density of the stripper gas to eliminate molecular interferences. The potential is generated by a commercial high voltage power supply.

3.1.3. *Single stage AMS*

A single acceleration stage AMS (SSAMS) was developed by NEC [57] in parallel with MICADAS to replace the two-step tandem accelerator, and the first commercial one was installed at Lund University [58], as shown in Fig. 4. It operates at a maximum voltage of 250 kV. The stripper is maintained at a high voltage, and high energy mass spectrometry is performed

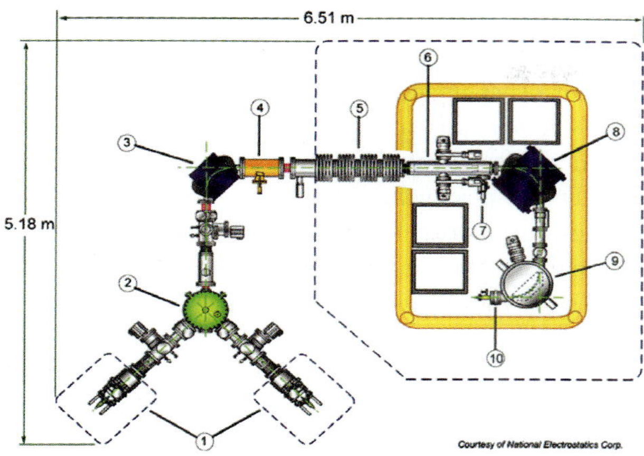

Fig. 4.   Schematic layout of the SSAMS at Lund [58].

following the stripper without any further acceleration. The performance of the SSAMS is similar to that of the 0.5 MV machine mentioned previously and can thus be a very useful tool for biomedical applications, as well as for radiocarbon dating. Compared to the compact 0.5 MV system, the SSAMS requires approximately 40% less floor space and $\sim 30\%$ lower investment cost, which has resulted in quite a number of new AMS installations [59, 60].

3.2. *Isobar separation*

Isobar separation is a key technique of AMS, and in many cases insufficient separation of the isobar is the main limitation on AMS sensitivity. There are quite a few techniques developed to suppress isobar interference. In addition to the two approaches of negative ions and the rate of energy loss described in the previous sections, the following methods have also been developed.

(a) *Chemical separation in sample preparation.* This is necessary, especially if the isobar is an abundant isotope. An attenuation factor of 10^4–10^6 can be expected.

(b) *Passive absorber.* This technique is suitable when the isobar has a higher Z than the radionuclide of interest. For example, ^{10}B is a main source of background in ^{10}Be measurements. With suitable density, it can be stopped in a gas absorber, while ^{10}Be ions can penetrate with a certain energy loss [9]. This has become a standard method for ^{10}Be measurements when the energy of ^{10}Be ions is sufficiently high. For ^{10}Be measurements with lower energy, it is necessary

to insert a thin foil in front of the final analyzer in order to achieve a sufficient separation of ^{10}B and ^{10}Be [61].

(c) *Complete stripping.* This technique is suitable when the isobar has a lower Z than the radionuclide of interest. For example, an isobar of ^{36}S can be easily suppressed when the fully stripped ions of ^{36}Cl^{17+} are selected for measurements, as sulfur ions can reach only charge state 16+. Obviously, very high energy (155 MeV) is required in this case. This technique can be used in the measurement of light nuclides such as ^{7}Be with lower energy [62].

(d) *Gas-filled magnet.* When ions pass through a gas-filled region in a magnetic field, their trajectories depend on their average charge states, which are related to Z. So the isobar and the measured nuclide will be separated at the exit of the magnetic field [63]. This technique is suitable for medium mass radionuclides, such as ^{32}Si, ^{36}Cl, ^{41}Ca, ^{53}Mn, ^{59}Ni and ^{60}Fe, as long as sufficient ion energy is available.

(e) *Projectile X-ray.* When a multiple MeV ion passes through a foil, it may be excited and then emit the characteristic X-ray, so its Z can be identified [64]. The drawback of this technique is that the efficiency is rather low and there is no discrimination between different isotopes of the same element.

(f) *RF field.* The RF accelerators (linac and cyclotron) have the effect of a mass filter. The high energy RF linac can be used for isobar separation [65].

3.3. *Sample preparation*

AMS measurements are made with targets that consist of a purified substance or compound of the nuclides in the study. One basic requirement for the target material is its capacity to generate negative ions that are intense enough for extraction under the bombardment of the sputtered beams. Chemical methods are employed to extract or concentrate specific compounds from natural samples, which vary with the types of nuclides. Some of the samples require mechanical or chemical pretreatments to remove obvious contaminations that are incorporated into the studied samples during the processes of formation, transportation, deposition or storage. These procedures are essential for the production of uniform and high quality targets. They must be applied stringently, as small sizes of samples are involved in the AMS measurements.

For ^{14}C, the standard chemical treatment of organic matter (e.g. charcoal, textiles, animal or plant remains) is the use of the acid–alkali–acid (AAA) method, which removes contamination by carbonates and humic acids, and then the carbonates that precipitated from modern atmospheric CO_2 [66, 67]. A modified method, acid base oxidation (ABOX), which involves a final oxidizing step in the treatment, was developed and found particularly useful for older materials [68]. The chemical treatment for inorganic carbon samples (e.g. shells of mollusks, foraminifera, carbonates) is simpler and involves mainly the mechanical removal of attached particles and the chemical removal of minute organic fractions and secondary carbonates precipitated. Samples used for carbon cycle studies require treatments that depend on the specific fractions to be utilized in the respective studies.

The next step in sample preparation for ^{14}C analysis is the production of pure graphite which is used in AMS measurements. The chemically treated samples are combusted (for organic matter) or acidified (for inorganic samples) to release CO_2. After the purification, the CO_2 gas is reduced, in the presence of a catalyst, to carbon in the form of graphite [66, 69]. The reduction is most commonly done in the presence of hydrogen. Recently, Xu *et al.* [70] modified the zinc reduction method of Vogel [71], which is simpler and quicker than the hydrogen reduction method. As they showed convincingly, the zinc reduction method is compatible with the hydrogen method in both background and precision for the production of high quality graphite [70, 72].

Some studies require analysis of specific materials which may need pretreatment at the molecular level where specific compounds are isolated [73, 74]. Such analyses as increasingly required in environmental studies would need an AMS facility that is able to handle samples as small as a few micrograms of carbon. The recent development of gas ion sources described above will meet such a challenge and allow routine measurements on CO_2 samples at μg levels. Another type of samples which require special treatments is the seawater or lake water in which dissolved organic carbon needs to be oxidized using ultraviolet

radiation and the released CO_2 is then converted to graphite for ^{14}C analysis [75, 76].

For other nuclides, such as ^{10}Be, ^{26}Al, ^{36}Cl and ^{129}I, the sample preparations involve a series of chemical treatments including ion exchange that in essence are designed to extract as pure as possible nuclides of interest by removing or suppressing contaminations that may distort or interfere with the signals of the studied nuclides [77]. In some cases (e.g. ^{10}Be), carriers of a stable or well-characterized isotope (e.g. ^{9}Be) need to be added in order to obtain sufficient current during the AMS measurements. In other cases, purity can be best achieved by hand-picking of individual mineral grains, such as in the case of AMS ^{26}Al–^{10}Be measurements in burial dating [78]. The purified substance of the specific nuclides is often mixed with powders of a metal, e.g. Ag, Nb or Co, to increase the heat conductivity of the targets.

Along with the preparation of samples to be studied, blanks are prepared in parallel under the same experimental conditions and are measured several times in order to assess the level of contamination introduced during the sample preparation. Similarly, standards with known isotope ratios and/or concentrations are prepared and measured, so that the unknown samples can be analyzed relative to specific standard values.

4. Applications

4.1. *Distribution of radionuclides in nature*

According to their production sites, cosmogenic nuclides fall into three categories: atmospheric origin, *in situ* origin and extraterrestrial origin.

Atmospheric origin ^{14}C is produced by the ^{14}N $(n, p)^{14}$C reaction on atmospheric nitrogen, and then it is quickly oxidized into $^{14}CO_2$ and equilibrates with volatile carbon species. The ^{14}C/C ratio in the atmosphere is quite homogeneous, and a natural value derived from the measurement of 1860s wood is about 1.2×10^{-12}. However, its production rate is not constant but depends on the intensity of the geomagnetic field and solar modulation. This variation is recorded in the tree rings, as shown in Fig. 5. The atmospheric ^{14}C is transferred into the hydrosphere through atmosphere–water boundary gas exchange and into the biosphere

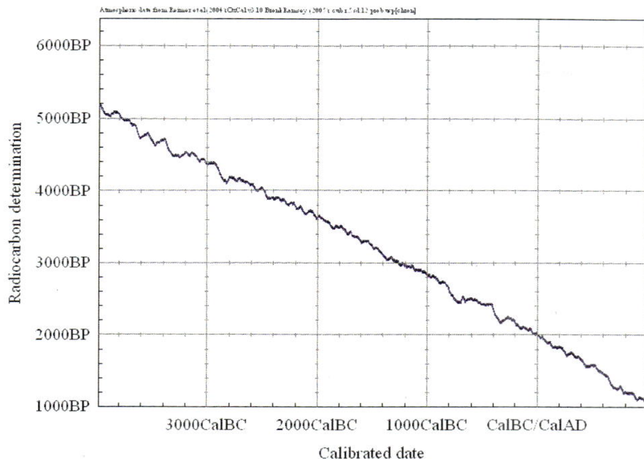

Fig. 5. A section of the radiocarbon calibration curve Int-Cal04 from 4000 cal BC to 1000 cal AD from the OxCal program [79, 80]. Here BP is the unit of ^{14}C age, and cal BC and cal AD are the units of calibrated dates.

through photosynthesis. Subsequently, the ^{14}C is further transferred into soils and sediments. All these transfers form an important part of the global carbon cycle.

Other atmospheric origin cosmogenic nuclides are produced mainly through spallation reaction in the atmosphere induced by high energy cosmic rays. For example, ^{10}Be is produced by spallation of nitrogen and oxygen, and the spallation of ^{40}Ar can produce ^{26}Al, ^{32}Si, ^{35}Cl and ^{39}Ar. ^{10}Be has a short atmospheric residence time and can be easily bonded to aerosols and then reach the Earth's surface by wet or dry precipitation. In contrast to ^{14}C, ^{10}Be does not mix homogeneously with ^{9}Be in the atmosphere. Nevertheless, the situation is different in the ocean, where ^{10}Be and ^{9}Be are well mixed, and their ratios can be used under some conditions for dating sediments or manganese crusts [81].

The cosmogenic nuclides can also be produced in rocks at or near the Earth's surface (*in situ*) through the nuclear reaction induced by secondary cosmic rays. For example, the spallation of oxygen and silicon can produce ^{10}Be and ^{26}Al. The *in situ* production rates are quite low and depend on the latitude and altitude. If the *in situ* production rate is known, the concentration of nuclides in rocks can be used to study the exposure age and erosion rate of the rocks. It is important that the sample should be taken from a mineral with a well-defined chemical composition and this mineral does not incorporate ^{10}Be which is introduced by water or other processes. Quartz

(SiO_2) is the most suitable mineral for such a study. After a certain time interval, a saturation value is reached in the ^{10}Be concentration, when the production and removal (decay and erosion) are equal. Therefore, the dating limit is 5–10 Ma for ^{10}Be.

The production rates of extraterrestrial origin nuclides are usually very high, because the extraterrestrial objects are directly exposed under cosmic rays without attenuation by the atmosphere. Those objects include meteorites, interplanetary dust and lunar rocks which were brought to the Earth by *Apollo*.

Anthropogenic nuclides are formed as the result of human activities, such as nuclear weapons testing, reactor emission or leakage from nuclear waste storage. Between 1950 and 1963, the intense atmospheric nuclear weapon tests produced the so-called "bomb peak" of radioisotopes in the atmosphere. By 1963 the ^{14}C content in the atmosphere had increased by a factor of 2 above the natural concentration, as shown in Fig. 6. Among the radioisotopes produced, ^{14}C turned out to serve as a particularly useful tracer for studying the global carbon cycle dynamics [82]. In addition, this ^{14}C bomb peak has been employed in detection of the art forgeries manufactured after 1955 (see Subsec. 4.4) and in the studies of forensic medicine [83]. Similarly, the bomb peak ^{36}Cl can be used as a hydrological tracer. Another important anthropogenic nuclide is ^{129}I, which was released into the ocean from nuclear fuel reprocessing plants at Sellafield in England and La Hague in France since 1954 and 1970 respectively, and is being utilized as an oceanographic tracer (see Subsec. 4.2).

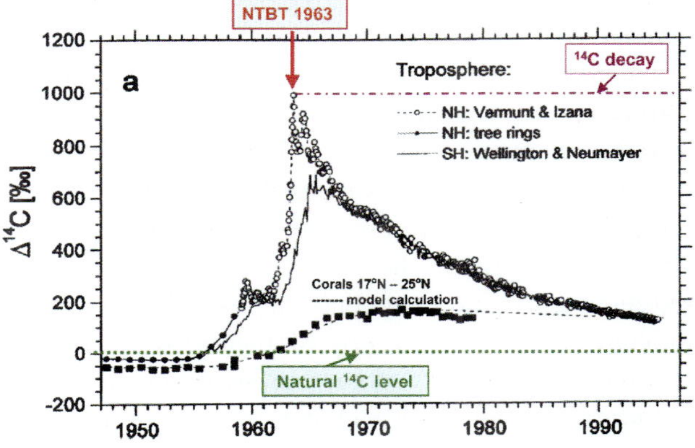

Fig. 6. The ^{14}C bomb peak in atmospheric CO_2 of the Northern Hemisphere and Southern Hemisphere [82].

Our physical world can be divided into seven large domains: atmosphere, biosphere, hydrosphere, cryosphere, lithosphere, cosmosphere and technosphere [2], where we can find radionuclides in virtually every terrestrial or extraterrestrial material. The advancement of AMS technology described above has led to substantial expansion in the applications of the radionuclides, such as ^{14}C, ^{10}Be, ^{26}Al and ^{36}Cl, into a wide variety of fields related to these domains [67, 77, 84–89]. Some examples of AMS applications are presented in the following subsections.

4.2. *Geological and environmental sciences*

The applications of AMS measurements of radionuclides in geosciences and environmental sciences can be divided into three categories: dating, tracing, and production rate and variations studies [90].

4.2.1. ^{14}C

The dating of cultural remains from the archeological context has dominated the applications since the invention of AMS and will be discussed in Subsec. 4.3. Due to its wide presence, ^{14}C is also measured to characterize samples from sedimentary environments, including deposits from archeological sites. The main scientific questions regarding these applications are concerned with the determination of the ages of these samples. An accurate chronology for geological samples from sedimentary contexts is crucially important and often forms the most challenging part of quaternary science investigations. One of the most exciting examples of such applications is the dating of fossil microfauna — mostly foraminifera — retrieved from marine sediments. Oceans act not only as basins to receive sediments deposited through seawater, but also as natural archives to record environmental change through geological time. The ages provided by the measurements of ^{14}C in calcium carbonate shells of the foraminifera are used to provide time constraints for the younger part of the marine sedimentary sequences which themselves were accumulated under different environmental conditions in the past. With the decrease in sampling intervals, higher resolution climate change signals are revealed from marine and ice core records. This has enabled paleoclimatologists to show that the Earth's climate

system underwent regular flips between warm and cold modes on the millennial timescale. Dating of such high frequency climate change records and correlation on regional and global scales have been made with the measurements of ^{14}C [91, 92], which allow geoscientists to advance our understanding of the mechanisms of climate change. Wang *et al.* [93] applied the AMS method to date 40 foraminifera shells from a sediment core in the South China Sea. The chronology based on the ^{14}C measurements allowed the authors to reconstruct the detailed monsoon climate change history over the last 10,000 years and to exploit the relationship between monsoon change and the development of Chinese history.

Many ^{14}C dating studies on samples from lakes and oceans have utilized bulk organic matter, which is highly complex in composition. With the development of AMS instrumentation as described above, measurements of samples containing micrograms of carbon are now becoming routine in many laboratories. This will encourage exploitation of ^{14}C dating of individual organic compounds [74]. ^{14}C dating of lake sediments has been a major application in geosciences. However, due to the "hard water effect" (i.e. the water carrying dead carbon), core top sediment samples from many lakes yielded apparent ^{14}C ages as old as over 2 ka (see e.g. Wu *et al.* [94]) and thus only macrofossil remains of terrestrial plants retrieved from the sediments have yielded meaningful ^{14}C ages [66]. In the absence of such remains, attempts have been made to use specific organic compounds originating from terrestrial plants for dating [74, 95]. The possibility of this kind of ^{14}C analysis at the molecular level has posed challenges for both sample preparation using chromatographic separation and AMS measurements, as only a tiny amount of pure compounds may be obtained, as demonstrated in marine sediments by Curie *et al.* [73]. This is currently a very active field of research and is likely to expand with further development of AMS technology for smaller samples.

The ^{14}C measurements have also contributed to the investigations of important scientific questions concerning the global carbon cycle on different timescales [67, 87]. The ocean is the largest carbon reservoir and ^{14}C dating provides a means of tracing the pathways of water masses. For example, the ^{14}C age difference between the planktonic (surface-living) and the benthic (bottom-dwelling) foraminifera from deep-sea sediment cores was used for investigating the history of changes in the meridional overturning circulation, which transports warm and highly saline surface water northward to cooler regions, where it sinks and forms "deep water" that eventually cycles back to the surface [96, 97]. As the ocean circulation is linked to the global climate, such applications will prove not only extremely valuable for understanding past climate change but also useful for predicting the future.

Although much has been learned from numerous innovative applications of the ^{14}C measurements, many challenging scientific questions remain. One example is the ongoing quest to explain the steep 15% drop in the ^{14}C/C ratio that occurred during the "mystery interval" between 17.5 and 14.5 ka ago [87]. Given the importance of this time interval for ^{14}C dating and for climate change studies, it is conceivable that many future AMS ^{14}C measurements will be associated with the endeavor for a better understanding of the mysterious ^{14}C decline. Another example is related to the dissolved organic carbon (DOC) in the ocean, which is comparable in magnitude to the total inventory of atmospheric CO_2 [98]. While it is widely held that the biogeochemical behavior of the DOC is critically important for understanding the global carbon cycle and climate change, little is known about the residence time of such a vast amount of organic carbon in the ocean. With improvement in the methodology of DOC sample preparation for ^{14}C measurements [76] and development of more compact AMS instrumentation, it is highly likely that the next few years will witness wider applications of AMS ^{14}C in microbial oceanography.

An application of ^{14}C measurement relevant to the present-day and the future carbon cycle is its use in monitoring the isotopic composition of atmospheric CO_2 on different spatial scales [99, 100]. AMS ^{14}C is also used as a unique source apportionment tool to estimate the relative contributions of fossil and contemporary carbon sources for atmospheric carbonaceous particles [101, 102]. This approach has a certain advantage over the traditional methods (such as by using a flask for gas collection) for investigating fossil-fuel-derived CO_2 in the atmosphere if one wants to collect CO_2 samples from different locations within a large area at the same time. Carbonaceous aerosol is directly emitted into the atmosphere

from various emission sources, such as fossil fuel combustion, biomass burning, or secondary organic aerosol from precursor gases [103, 104]. Analysis of ^{14}C in carbonaceous aerosol samples provides a convenient means of identifying the source of CO_2 [105].

Measurements of ^{14}C in annual plants provide an alternative and more economic means of monitoring fossil-fuel-derived CO_2 in the atmosphere. The time integration feature of the annual plants makes it possible to analyze the average seasonal and annual level of Δ^{14}C [106, 107]. Xi *et al.* [108] made a first test of this method in China by analyzing Δ^{14}C values of corn leaves from the Beijing area. Their results show a decreasing trend of Δ^{14}C from the outer suburb to the inner suburb and then to the urban center.

AMS ^{14}C measurements have played an important role in the study of the storage and turnover of natural organic matter in soils [109, 110]. As soil organic matter forms the largest carbon reservoir in rapid exchange with atmospheric CO_2, understanding its potential role as a source and sink of carbon to the atmosphere against the background of global warming is of immense relevance to future climate change. Such applications in soil carbon dynamics are expected to expand in the years to come [111].

4.2.2. ^{10}Be and ^{26}Al

The applications of ^{10}Be in Earth science can be divided into two main categories: one is the use of ^{10}Be produced in the Earth's atmosphere (atmospheric ^{10}Be) and the other is the use of *in-situ*-produced ^{10}Be (*in situ* ^{10}Be). The concentration of the two ^{10}Be signals in a sample may vary by several orders of magnitude and hence their measurements require different sensitivity for the AMS.

Atmospheric origin ^{10}Be has been used to study the changes in its production rate, which is modulated by the changes in the heliomagnetic field and geomagnetic field [112–114]. Such applications, sometimes combined with ^{14}C measurements, in natural archives (e.g. polar ice cores, marine and lake sediments, or loess), have played an important role in gaining a better understanding of the changes of solar activity in the past and their relationship with climate change [115, 116]. This has in turn contributed to the study of the role of ocean circulation in redistributing radiogenic nuclides as well as in long-term climate change [117, 118].

The applications of *in situ* cosmogenic nuclides including mainly ^{10}Be, ^{26}Al and ^{36}Cl have revolutionized Earth surface research over the last two decades [119]. The applications of *in situ* ^{10}Be cover the entire realm of geomorphology and beyond, allowing Earth scientists to address more challenging scientific questions of a highly interdisciplinary nature. Exposure dating, which measures the amount of ^{10}Be produced as a result of continuous exposure to cosmic rays, has now been used to date a variety of geomorphic surfaces or sedimentary units created by different processes ranging in age from a few hundred years to a few million years, e.g. bedrock landforms and surfaces, terraces, fault faces, debris flows or alluvial fans, and landslide bedrock surfaces [77, 120, 121].

Traditionally, ^{10}Be exposure dating has been considered a tool useful merely for geomorphology. Here we show two examples of recent applications of ^{10}Be exposure surface dating in climate change science.

The Antarctic ice cores displayed a brief return to colder conditions from about 14,540 to 12,760 years ago during the last deglaciation [122]. To understand the geographic extent of such a prominent cold event (called the Antarctic cold reversal) throughout the Southern Hemisphere, Putnam *et al.* [123] applied ^{10}Be dating to two glacial moraine sets from the Southern Alps, New Zealand. Their 27 high precision ^{10}Be surface exposure dates revealed widespread glacier resurgence around 13 ka ago, at the peak of Antarctic cooling. Therefore, the climate deterioration associated with the Antarctic cold reversal seemed to have extended into the southern midlatitudes of the southwestern Pacific Ocean. Putnam *et al.* [123] attributed the extensive cooling to the northward migration of the southern subtropical front, and concomitant northward expansion of cold Southern Ocean waters.

The timings of interhemispheric climate changes during the Holocene (the past 11.5 ka) have been a hot topic in climate science [124]. Recently, a high resolution ^{10}Be chronology of glacier fluctuations in New Zealand's Southern Alps over the past 7 ka obtained by Schaefer *et al.* [125] showed that the extents of glacier advances decreased from the middle to the late Holocene, in contrast with the Northern Hemisphere pattern and several glacier advances in New Zealand over the last 4 ka that occurred even during classic northern warm periods. The same authors also obtained ^{10}Be exposure ages as young as a couple of hundred years with remarkable consistency for individual moraines, demonstrating the potential of the method for studying high frequency Holocene glacier fluctuations in the future.

Burial dating, which uses the decay of previously *in-situ*-produced cosmogenic nuclides, has emerged as an extremely powerful tool for quantifying surface processes such as erosion, accumulation and river incision [126, 127], which are in turn related to geomorphology and archeology. Many types of deposits, such as cave fillings, alluvial fans, river terraces, delta deposits and dunes, have been measured [126, 128]. The ^{10}Be/^{26}Al nuclide pair is more widely used than other nuclides, due to their established production pathways [127]. In principle, the use of the ^{10}Be/^{26}Al nuclide pair allows dating of quartz-bearing materials from ~ 100 ka to 5 Ma, opening up a new possibility of dating wider and older materials. Here, we show an example in the study of long term climate change.

Although variations of global ice volume have been reconstructed from marine isotope records [129, 130], little is known about the extent of individual ice sheets on land, such as at the time of certain climate transitions. Balco and Rovey [131] applied a ^{26}Al–^{10}Be burial isochron method to the glacial sedimentary sequences in central Missouri, USA. Their burial ages suggested that the Laurentide ice sheet reached 39°N, i.e. near the extreme southern limit of North American glaciations, around 2.4 Ma and 1.3 Ma. These ages coincided with the timings of the intensification of the Northern Hemisphere glaciations and the mid-Pleistocene transition respectively, as inferred from the marine records, hence providing a fine example for a more direct comparison of the discontinuous terrestrial and continuous marine records of glaciation. These findings also offered a new clue for understanding the increase in the global ice volume at the mid-Pleistocene transition, which may be at least in part the result of an expanded Laurentide ice sheet [131].

4.2.3. ^{36}Cl and ^{129}I

^{36}Cl is another cosmogenic nuclide that has been widely used in surface dating. It is particularly suitable for carbonate rocks and basalts, being mainly produced by cosmic ray spallation reactions from potassium and calcium [112]. Fault scarps created by vertical tectonic movements in limestone have been common targets for ^{36}Cl dating [119, 132]. When the pre-exposure component can be precisely assessed, useful information about earthquake history may be obtained [133, 134]. Another major application of ^{36}Cl has been found in the study of groundwater. The ^{36}Cl technique provides a useful tool for determining the ages of groundwater as a decrease in ^{36}Cl/Cl along a hydraulic gradient may be interpreted as representing ^{36}Cl decay. However, processes such as mixing between different Cl sources can make the interpretation of the face values of the measured ^{36}Cl/Cl ratio highly complicated. In many of the applications, additional chemical and isotopic analyses are employed and information about the history and flow velocity of underground aquifers rather than the ages of water is sought [135, 136]. Given the increasing need for better understanding of the potential change of water resources as a result of global climate change, more frequent use of ^{36}Cl combined with other nuclides in groundwater investigations is envisioned.

Generally, over 90% of ^{129}I observed in the environment is due to intended or unintended activities of humans with ^{235}U or ^{239}Pu. Anthropogenic ^{129}I can be used as an effective global tracer for studying natural environmental cycles. One example is the use of the ^{129}I released into the North Atlantic in large recorded quantities by the two biggest nuclear fuel reprocessing facilities in France and the United Kingdom. The stepwisely released ^{129}I from these facilities have provided a unique opportunity for studying ocean circulation from the North Atlantic into the Arctic and the South Atlantic [137–139]. The anthropogenic ^{129}I can also be employed as a proxy for the more dangerous radioactive ^{131}I isotope, released accidentally into the environment [23]. In addition, naturally occurring ^{129}I could in principle be used to date groundwater and other geological processes on a 100 Ma timescale. In practice, the difficulty lies in the very low levels of ^{129}I and high ^{127}I concentration; a typical value of the natural ^{129}I/^{127}I ratio would be close to the AMS detection limit.

4.3. *Archeology*

^{14}C dating with the beta-counting method has been used in archeology for a long time, before the invention of AMS. The important advantage offered by AMS is the great reduction in the sample size (carbon content) required for ^{14}C measurements, hence extending the ability of radiocarbon dating and opening up many new possibilities. There have been numerous archeology-related studies using AMS during the past 30 years, but only a few examples are outlined below to highlight the importance of AMS radiocarbon dating for archeological applications.

The sample will have to be destroyed after it is submitted for radiocarbon dating. For some precious cultural relics, only small pieces are allowed to be taken. AMS therefore offers a unique tool for determining the age of this kind of relics.

The Shroud of Turin is a linen cloth kept in the royal chapel of the Cathedral of Saint John the Baptist in Turin, northern Italy. It was believed that the Shroud wrapped the body of Jesus. AMS radiocarbon dating of the Shroud was performed in 1988 by AMS laboratories of Arizona, Oxford and Zurich. Each laboratory received a small piece of linen measuring about $2\,cm^2$. The results from three laboratories were in very good agreement, 1290 AD–1360 AD at 90% confidence, obviously different with the expected time around Jesus's death [140]. Due to the sensitivity of the dating of the Shroud, the discussion about whether the sample was taken from the main part of the Shroud and whether there were serious contaminants continued for many years [141, 142].

Another example of dating precious cultural relics is concerned with oracle bones [143]. Oracle bones were discovered about 100 years ago and most of them were excavated around the Yin site in Henan province, China. Most of the oracle bones were made of ox shoulder blade and some tortoiseshell. There are inscriptions on many oracle bones which record the auspicious predictions and subsequent events in ancient Chinese history. The names of the nine kings of the late Shang Dynasty, from King Wuding to King Dixin, can be found in the oracle inscriptions. Many oracle bone samples were dated by AMS at Peking University. Only a small piece of about $1\,cm^2$ was cut from the part without inscriptions of each oracle bone (Fig. 7). The obtained AMS ^{14}C ages provided key data for establishing an accurate chronology for late Shang Dynasty in early Chinese history [143].

The lack of suitable materials is a common problem when people try to date the ancient pottery

Fig. 7. Oracle bones dated by AMS at Peking University [143].

and metal artifacts. AMS has made it possible to directly measure radiocarbon ages by isolating organic materials imbedded in the pottery and metal artifacts when they were manufactured. Pottery temper (short-living grass) was selected as dating material for determining the earliest pottery ages from the incipient Neolithic site, Amur River basin, Russian Far East. The age of 16,000 cal BP shows that it is one of the oldest pieces of pottery in East Asia, and in the Old World [144]. Another example is the dating of black pottery. The black color was formed by chaff (husks of rice or millet) which charred on the hot surface of fired raw pottery. The charred powder was taken from the black pottery excavated from the Lal-lo shell midden sites in Japan as dating material. The ages of around 2350–2095 cal BP have led to a more accurate chronology of these sites [145].

Recently, two bronze statues, of extraordinary importance for the history of the art of the Greek classical period, were dated by AMS and the dating materials were organic residues in the casting cores extracted from the inside of the statues, such as charred wood, vegetal remains and animal hairs [146]. These two bronze statues were found underwater along the Ionian coast in front of the town of Riace in Calabria, southern Italy. Overall, 17 samples were collected, and some of them produced an amount of graphite below 0.3 mg. The statistical analysis of the results obtained for the two

statues yields an age between 517 BC and 394 BC and between 509 BC and 394 BC at the two-sigma confidence level. The results lend support to the hypothesis that the ages of the two bronze statues should be in the 5th century BC and allow archeologists to reject the hypothesis that the two statues belong to the Roman period around the first century BC.

The radiocarbon ages obtained by dedicated AMS can be very precise, with an uncertainty as small as less than 30 years. However, the calibration of radiocarbon age to calendar age with the tree ring calibration curve will enlarge the error in most cases due to the wiggles on the calibration curve, and the specific error of calendar age depends on the shape of the curve. Figure 8 shows an example of calendar age calibration with OxCal: the upper one with a small error and the lower one with a large error, although the errors of their radiocarbon ages are the same. Therefore, the radiocarbon dating of individual samples may not always meet the requirement for an accurate and precise chronology. However, for a series

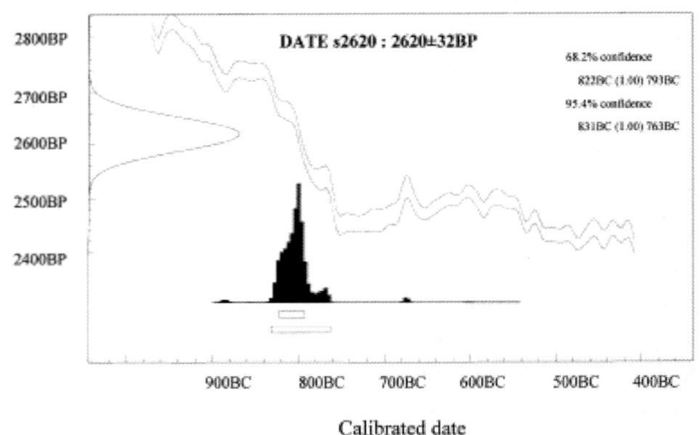

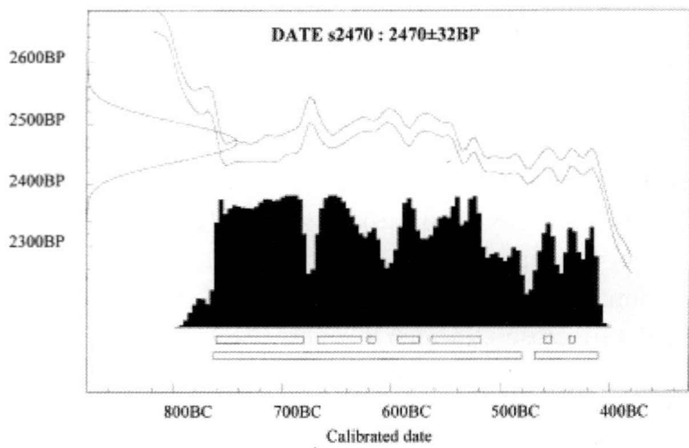

Fig. 8. Examples of calendar age calibration [143].

of samples with a certain sequence, the Bayesian method can be used for calibration and the errors of the calendar age can be substantially reduced [147]. Software for this is now available, such as OxCal, developed at Oxford University [148]. Using this method, the archeological samples unearthed from a site are grouped into different phases according to their cultural characteristics and a sufficient number of samples is needed for each phase. Usually, samples of small size have to be selected in a certain cultural phase and AMS is the only means of dating them.

One of the successful examples of using a series of samples is AMS radiocarbon dating for Xia–Shang–Zhou chronology [143]. Ancient Chinese civilization has lasted for several thousand years without interruption. It is well known that the earliest dynasties of China are Xia, Shang and Zhou. However, there have been many arguments about the particular dates of the Xia and Shang dynasties, because there was no complete chronology in well-recognized literatures before 841 BC. A radiocarbon dating campaign was carried out by archeologists and radiocarbon dating experts in China from 1996 and the objective was to establish a chronological framework for Xia, Shang and Zhou based on scientific evidences. More than 200 samples in series collected from seven archeological sites, such as Fengxi [149] and Xinzhai [150], were dated at Peking University and the results were calibrated with OxCal. A chronological framework of the Xia, Shang and Zhou dynasties was then established on the basis of the radiocarbon dating results and the astronomical analysis, which showed that the Xia dynasty lasted from about 2070 to 1600 cal BC, the Shang dynasty from 1600 to 1046 cal BC and the Zhou dynasty from 1046 to 771 cal BC.

Other recent applications of AMS dating using series samples with the Bayesian method include, for example, the chronological study of the Paracas and Nasca cultures in southern Peru [151] and the timing of some eruptions of Somma-Vesuvius volcano in Italy during the Copper–Middle Bronze Age [152].

4.4. *Art authentication and detection of art forgery*

AMS is also widely used for art authentication and detection of art forgery. Numerous artworks, including paintings, textiles, metal artifacts and others, from museums and private collections have been dated by AMS. The results for many of these samples are often crucially important in testing the authenticity of these works of art and artifacts. In addition, AMS can be used to detect 20th century forgeries of art and artifacts purportedly created before 1955. This is achieved by the detection of atomic-bomb-derived ^{14}C occurring after 1955, which can, in some cases, date objects at plus or minus one year's resolution. The bomb peak of ^{14}C caused by nuclear weapon tests is at around the year 1963.

The wooden support of a precious panel painting, which is privately owned and depicts the Madonna with Child, St. Joseph, St. Giovanino and another saint, was dated by AMS to clear any doubts about the attribution of the painting to the ancient painter Cesare da Sesto [153]. The AMS ^{14}C measurement dated the wooden support to 1593 AD ± 23 a. This means that the artwork cannot be attributed to da Sesto, who died in 1523 AD.

Another interesting example is the mystery of the Persian mummy [154]. In 2000, a mummy was confiscated at Kharan in Baluchistan, and handed over to the National Museum of Pakistan. The mummy was wearing a gold crown, and a golden cypress tree was embossed on the chest of the mummy, below a gold plate with two names. This strongly supported the suggestion that the mummy was Ruduuna, the daughter of Xerxes, the great Persian king, who lived from 518 BC to 465 BC. Since no other Persian mummy has been found so far, it was supposed to be the archeological find of the century. AMS measurements were performed on pieces of textile, a straw mat and charcoal. The ^{14}C contents of these samples were about 113 pMC, instead of an expected value of about 75 pMC. According to the bomb-pulse curve, and by comparing the ^{14}C contents of bone, skin and muscular tissue samples, the mummy died in 1994–1996 AD, and it was revealed to be a modern fake.

4.5. *Biomedical sciences*

Most of the AMS applications in the biomedical sciences are for tracing purposes. Long-lived radioisotopes are absorbed by experimental organisms in different ways, and then radioisotope contents of samples taken from blood, urine, tissue and organs are measured by AMS to study their distributions and metabolism kinetics. Tracing isotopes measured by AMS are mainly ^{14}C, ^{26}Al and ^{41}Ca.

4.5.1. ^{14}C

^{14}C is widely used as a tracer in biomedical research and drug testing. AMS provides the advantage that much smaller doses are required than for liquid scintillation counting, thereby reducing the amounts of labeled compounds. After AMS application in the biomedical sciences was demonstrated at the Lawrence Livermore National Laboratory in 1990 [155], it was developed very rapidly in the early years, as summarized by Vogel and Turteltaub [156] and Freeman and Vogel [157]. An AMS system dedicated to biomedical study was then set up; it was sponsored by large pharmaceutical companies, and many clinical studies for drug development were carried out [158, 159].

Although publication statistics showed a general decline in the ^{14}C-labeling method from the late 1990s as other analytical methods developed, the AMS ^{14}C method is continuously and critically needed in low level isotopic tracing such as molecular assays, human metabolism and kinetics, investigation of aging-related generation and regeneration of human tissues, and cellular metabolism [160]. For example, the potential molecular toxicity of acrylamide (AA) was evaluated [161]. In this study, the lowest doses taken by mice were 7.5×10^{-2} and $7.5 \times 10^{-1} \, \mu g/kg$ b.w., very close to the WHO-published data on the average daily intake of AA for the general population. The results of AA-protamine, AA-sperm DNA and AA-sperm head/tail adducts at low dose levels suggest that the quality of sperm might be damaged gradually by adducts of AA in the sperm compositions followed by cumulate daily AA exposure.

4.5.2. ^{26}Al

It is well known that aluminum is a highly neurotoxic element and can degenerate nerve cells in the brain. AMS measurement of ^{26}Al is used to study the absorption, distribution and excretion of aluminum in human volunteers or experimental animals. The use of ^{26}Al enables the study of Al pharmacokinetics at physiological Al concentrations without interference from ^{27}Al in the environment or the subject.

Early studies with human subjects included the biokinetics of aluminum [162], the distribution of aluminum among the various components of blood [163], absorption from the gut with different chemical forms of the ingested aluminum [164], influence of dissolved silicate on the physiological chemistry of aluminum [165] and so on. Gastrointestinal absorption of aluminum in patients associated with neuropathological changes was also investigated using ^{26}Al measured by AMS. Gastrointestinal absorption of ^{26}Al under normal dietary conditions in Down's syndrome subjects exceeded that of controls by a factor of 6 [166].

Studies using rats have been extended in many aspects. The oral aluminum bioavailability from drinking water and food has been estimated by serum-analyzing for ^{26}Al by AMS. The results showed that 0.3% of Al was orally absorbed from water [167] and 0.1–0.3% from different kinds of food [168, 169]. The presence of citrate, maltolate and fluoride would not be expected to greatly influence Al absorption from water [170]. Daily Al intake by humans from food is about 3–10 mg (mainly from food additives), much higher than the ~ 0.1 mg from water. Considering the oral bioavailability, food provides at least 25 times more Al to systemic circulation. A series of studies have been carried out to investigate Al incorporation into the rat brain. ^{26}Al incorporation into the brain of sucking rats through lactation was measured and the ^{26}Al amount in the brain was diminished only slightly up to 140 days after weaning [171]. Subsequent study investigated ^{26}Al incorporation into rat fetuses through the placental barrier. Approximately 15% of ^{26}Al incorporated into the brain of fetuses still remained in the brain 730 days (average lifetime of the rat) after birth [172].

4.5.3. ^{41}Ca

^{41}Ca is a promising radioisotope measured with AMS for biomedical studies. It has the potential to monitor the long term changes in bone metabolism directly. Most studies are focused on the bone loss problem of premenopausal women by monitoring changes in the urinary ^{41}Ca/^{40}Ca isotope ratios, which are expected to directly reflect changes in bone Ca metabolism, and using compartmental modeling [173–175]. An efficacy study using urinary ^{41}Ca excreted from prelabeled bone was performed to identify the effective dose of isoflavone for suppressing bone resorption [176]. The results showed that soy protein given to subjects with an isoflavone dose of up to 135.5 mg/d did not suppress bone resorption in postmenopausal women.

4.6. *Nuclear astrophysics and records of supernova explosions*

The understanding of the elemental abundances of the solar system is one of basic issues in nuclear astrophysics. The complex isotopic scenario can be understood as the interplay of principles of nuclear physics and the specific conditions in the stellar environment [177]. The synthesis of elements from carbon up to Fe and Ni is dominated by charged-particle-induced reactions during the stellar burning phases, while most of the elements heavier than iron are produced via neutron-induced reactions, which are triggered either by the slow neutron capture process (s-process) on timescales of months to years, or by the rapid neutron capture process (r-process), most likely along with the supernova (SN) explosions [178]. The ultrahigh sensitivity AMS is the best technique for precise measurements of ultralow microbar cross-sections related to the reactions such as ^{26}Mg (p, n)^{26}Al [179], ^{40}Ca$(\alpha, \gamma)^{44}$Ti [180], ^{62}Ni(n, $\gamma)^{63}$Ni [181, 182] and ^{78}Se(n, $\gamma)^{79}$Se [183]. In addition, there are some records of SN explosions on the Earth, such as ^{14}C variation in a tree ring [184] and ^{60}Fe enhancement in the Earth's crust [185]. In the following subsections, a few examples related to these studies will be given.

4.6.1. *Cross-section of $^{62}Ni(n, \gamma)^{63}Ni$ at stellar temperatures*

An accurate knowledge of the neutron capture cross-sections of ^{62,63}Ni is crucial, since both isotopes hold key positions which affect the whole reaction flow in the weak s-process up to A $\approx$ 90. However, the cross-section values for ^{62}Ni(n, $\gamma)^{63}$Ni at stellar temperature kT $=30$ keV measured by the TOF method in the 1990s were scattered between 12 and 37 mb. Nassar *et al.* [181] measured this cross-section with the activation-AMS method. A ^{62}Ni-enriched sample was activated, and the ^{63}Ni AMS measurement was performed at the linac ATLAS in ANL with a gas-filled magnet. The authors gave a

Maxwellian-averaged (kT $=25$ keV) cross-section of 28.4 ± 2.8 mb, which is equivalent to $(\sigma)_{30\,keV} = 26.1 \pm 2.6$ mb. Subsequently, Dillmann *et al.* [168] measured this cross-section at the Munich 14 MV MP tandem accelerator with a gas-filled analyzing magnet system using two activated samples. The resulting Maxwellian cross-section at kT $=30$ keV was determined to be $(\sigma)_{30\,keV} = 23.4 \pm 4.6$ mb, which is also in perfect agreement with a recent TOF measurement [186]. Combining the two AMS results and two other TOF results [186, 187], a weighted average of $(\sigma)_{30\,keV} = 26 \pm 2$ mb was recommended for future stellar modeling [182].

4.6.2. *Measurement of the ^{60}Fe remnant of a supernova explosion*

Normally, in the background of cosmic rays, the content of ^{60}Fe $(t_{1/2} = 1.49 \pm 0.27$ Ma) is very low. During an SN explosion, ^{60}Fe may reach the Earth with a surface density of about 4×10^9 atom/cm^2, assuming that the mass of the SN is about 1.5 M$_o$ (M$_o$ is the mass of the Sun), as theoretically predicted [188]. Hence, ^{60}Fe may be deposited on the Earth's crust and will be a good proxy for an SN explosion.

An interesting AMS measurement was performed at the Munich 14 MV tandem AMS facility for minute traces of ^{60}Fe in a sample obtained from a deep-sea ferromanganese crust in the Pacific Ocean at a depth of 4830 m. The results of the experiment indicate an anomalous increase in ^{60}Fe at ~ 3 million years ago [185].

Improved measurements in the subsequent years resulted in a more convincing picture, suggesting that this event may really have been caused by an SN explosion. The slow growth rate of the crust (2.5 mm/Ma) converts depth to the timescale in Fig. 9. It can be seen that a clear ^{60}Fe anomaly above the background level of 2.4×10^{-16} is at an age of 2.8 Ma. This figure indicates clearly a possible extraterrestrial event. According to astrophysical models, it is consistent with the deposition of a remnant from an SN a few tens of parsecs (~ 100 light years) away from the Earth [189].

Apart from ^{60}Fe there are more radionuclides, such as ^{182}Hf (8.9×10^6 a) [190] and ^{244}Pu (8.1×10^7 a) [191], which are produced in similar cosmic events and are to be studied by AMS technology [192, 193].

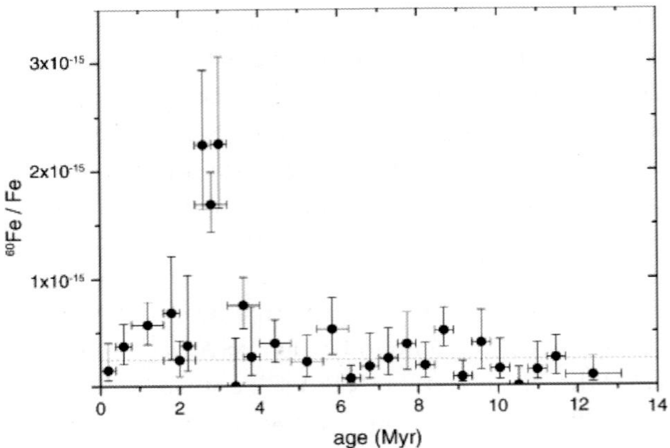

Fig. 9. Depth profile of ^{60}Fe/Fe ratio measurements of a deep-sea ferromanganese crust collected from the Pacific Ocean (4830 m water depth) [189].

4.7. *Nuclear environmental safeguards*

Since the 1990s, the International Atomic Energy Agency (IAEA) has used environmental sampling as part of its nuclear safeguard program. Sampling environmental materials such as water, air, soil or vegetation taken from within and around nuclear facilities could in principle be done to identify undeclared nuclear activities by measuring trace levels of uranium, plutonium and other indicators. Among them, ^{236}U and ^{239,240}Pu are the most significant tracers.

In the disaster of the Fukushima nuclear accident which occurred in Japan in March 2011, a variety of long-lived radionuclides, such as ^{137}Cs, ^{90}Sr, ^{129}I, ^{239}Pu, ^{236}U and other radioisotopes, have leaked into the environment from the accident site and have spread over almost the entire Northern Hemisphere in the atmosphere, oceans and groundwater. Even ^{131}I contained in milk samples was detected in Phoenix and Los Angeles in North America. This disaster reminds us that the operating status of a nuclear power plant, the storage and transfer of used fuel, as well as the processing of nuclear waste, have to be under stricter inspection by quantifying the leakage, and tracing the subsequent dispersion of all these nuclides. As an extremely sensitive tool for measuring the relative abundance of the long-lived radionuclides, AMS is playing an important role in the environmental monitoring for nuclear safeguards. Much can be learnt using such a sensitive technique toward establishing safer practices

in nuclear environmental safeguards in normal situations, ahead of any serious leak events.

Currently, a number of AMS laboratories are working on actinide measurements. Among about 80 AMS facilities in the world, there are 15 or more working on nuclear safeguards. Most of them are operating at a terminal voltage of 3–5 MV; a few have used higher energy tandem accelerators, such as 14 UD or 14 MV MP. The small facilities with a 0.2–1 MV terminal voltage can also be employed. A few examples are outlined below.

4.7.1. *Measurements of* ^{236}U

^{236}U is a long-lived radioactive isotope produced principally by thermal neutron capture on ^{235}U. Natural ^{236}U is produced at very low levels in uranium ores, due to the low neutron flux existing in ore bodies. AMS measurements of these nuclides were first presented by the IsoTrace facility of the Toronto group in 1994 [194]; the ratio of ^{236}U/^{238}U is lower than 5×10^{-10} in a natural uranium ore from the Cigar Lake deposit in Canada. Subsequently, ^{236}U measurements were performed on groundwater samples from the US Idaho Chemical Processing Plant [195]. The natural ratio ^{236}U/^{238}U has also been measured by several other groups, such as VERA [196], ANU [17], ETH [197], Weizmann Institute [198], CIAE [199], LLNL [200] and ANSTO [201]. These low ratios contrast sharply with that for nuclear fuel, where the ^{236}U/^{238}U ratio builds up to 10^{-4}–10^{-2}. Such high ratios relative to the naturally occurring ratio make its measurement an important signature of contamination by irradiated uranium in the environment. ^{236}U is also present in depleted uranium as a result of the use of recycled uranium in enrichment plants.

^{236}U have been measured both by α-particle spectroscopy and by conventional mass spectrometry (MS), using either thermal ionization (TIMS) or inductively coupled plasma (ICP-MS) with positive ion sources. Sensitivities at the femtogram (fg) level have been reported for modern high resolution ICP-MS systems that employ a secondary electron multiplier to count the measured ions [202]. Nevertheless, AMS offers advantages in terms of eliminating molecular interferences which might not be recognized in a conventional MS system, and sensitivities at the 0.1 fg level have been achieved [200, 203, 204]. For the ^{236}U/^{238}U ratio measurement, modern ICP-MS reports a sensitivity of 1×10^{-10} [205] while AMS allows measurements of the isotope ratios as low as $\sim 10^{-12}$. However, due to the fact that ^{235}U and ^{238}U are abundant in the sample and form an intense source of background, the detection system is more complex. Besides the energy measurement to separate the ^{236}U ions from the lower energy ions with the same mass-to-charge ratio, a high resolution velocity measurement by TOF must be incorporated [206, 207]. In the nuclear safeguard area, only very small amounts of uranium are available. In these cases, the sample must be bulked with iron oxide [201]. Beam currents of $100\,\mathrm{nA}$ of UO$^-$ ions (6×10^{11} ions/s) are readily extracted from a uranium oxide sample.

Apart from monitoring the leakage of relevant nuclides from the nuclear power plant and from the transfer of used fuel as well as the processing of nuclear waste, a potential application is the identification of clandestine nuclear activities by measuring ^{236}U in small amounts. These may have been collected as aerosols or from the leaves of plants, and an increase in the ^{236}U/^{238}U ratios would provide explicit evidence of material that had passed through a nuclear reactor [208]. Another closely related application is the identification of areas contaminated by depleted uranium weaponry. Armor-piercing shells made of depleted uranium were used in the recent Iraq and Kosovo wars, and there are concerns about possible health effects on both military personnel and the civilian population of the affected areas. Depleted uranium is a byproduct of the uranium enrichment process, and the feedstock often contains some recycled reactor fuel with high levels of ^{236}U. Hence, some ^{236}U finds its way into the depleted uranium and can serve as a very sensitive fingerprint for the presence of depleted uranium in an area [209].

4.7.2. *Measurement of plutonium isotopes*

The cosmogenic ^{244}Pu largely decayed away over the first 3 billion years of the Earth's history, leaving ^{239}Pu as the predominant isotope at a constant base level in the Earth's crust [210]. The anthropogenic plutonium isotope of ^{239}Pu is produced when ^{238}U absorbs one or two neutrons in the fission process of ^{235}U, while ^{240}Pu is the second-most-common isotope, formed by occasional neutron capture of ^{239}Pu.

Since ^{240}Pu does not undergo fission in the same way as ^{239}Pu, its concentration in nuclear fuel builds up steadily. The ^{240}Pu/^{239}Pu ratio is a useful indicator of the source of the plutonium. Human activity since 1945, especially in the 1950s and 1960s, has led to several tons of plutonium being released into our biosphere as a legacy of atmospheric weapon testing. Alpha-ray-emitting plutonium can get lodged in people's bones and lungs after being inhaled and could become highly carcinogenic.

For the AMS measurement of plutonium, there is actually no stable isotope. Hence, it is necessary to add a "spike" of a known amount of one of the long-lived isotopes, generally ^{242}Pu, to the sample in order to quantify the concentrations of the isotopes of interest, which may be either ^{239,240}Pu or ^{244}Pu. Both cost and radiological issues dictate that the amount of the spike added is in the range of 1–10 picograms rather than the quantities of milligrams that are commonly used for the Cs sputter source of conventional AMS. Consequently, the extra bulk is achieved by dispersing the plutonium in an iron oxide matrix, and mixing it with aluminum or silver powder to ensure electrical and thermal conductivity. Both the isotopes of interest and the ^{242}Pu spike must be measured by ion counting [203, 204]. In traditional AMS measurements, PuO$^-$ molecular ions rather than Pu$^-$ are selected for the injection magnet, because their intensities are about two orders higher than that of Pu$^-$. For Pu, backgrounds from uranium can be reduced to very low levels with appropriate chemistry. However, ^{238}U remains as the primary background. Child *et al.* have explored this ^{238}U background as a function of the amount of uranium added to a blank sample, and found that even 1 ng of uranium leads to an apparent ^{239}Pu signal of 10 fg [210]. Fortunately, at ANU, the excellent high vacuum in the high energy acceleration tube close to the high voltage terminal helps reduce the effect, where routine blank operation exhibits an apparent ^{239}Pu signal of $< 0.2\,\mathrm{fg}$ [211], while for ^{240}Pu measurement uranium backgrounds can be higher than at ^{239}Pu because the ^{238}U^{18}O$^-$ ion is five times richer than the ^{238}U^{17}O$^-$ ion. Consequently, the ^{238}U background at ^{240}Pu might be approximately five times higher than at ^{239}Pu.

At ANU, transmission in the 5+ charge state is $\sim 3\%$; correspondingly, the overall efficiency of detecting a plutonium ion is $\sim 2 \times 10^{-4}$. Hence, sensitivities of $\sim 10^5$ atoms of Pu are achievable with AMS, while the daily urinary excretion of Pu for one person (in the general population) is $\sim 10^6$ atoms [212]. On the other hand, it has been shown at Zurich that AMS of plutonium can be performed very well on a small accelerator operating at only 0.3 MV [197, 213]. It appears that this system may even be suitable for high sensitivity ^{236}U measurements [197]. The superb performance of this small system was unexpected, but will prove highly useful as it has a transmission of 15%, which is higher than for any traditional AMS system for actinides.

AMS has been applied to the measurement of ^{240}Pu/^{239}Pu ratios to detect leaked fuel from some types of accidents. This ratio is the most sensitive indicator of small traces of the weapons-grade material from the accident, allowing it to be distinguished from global fallout [211]. Other examples of possible plutonium isotope applications are the use of ^{239}Pu as a tracer of soil erosion and sediment transport instead of ^{137}Cs [214], and the use of ^{244}Pu as a tracer of biomedical studies on human subjects, such as lungs [215].

5. Trends and Challenges

Currently, there are more than 80 AMS facilities around the world [216]. Highly diversified and innovative AMS applications are contributing to almost all domains of science and technology. Many of these applications are closely associated with research designed to quantify the spatial and temporal scales of processes and interactions which shape our planet [217]. While it is hard to predict any specific breakthrough or improvement, some trends of AMS developments can be noted, as discussed below.

5.1. *Trend toward small size AMS instruments*

In the last decade, there has been a continual drive to move to sub–1 MV smaller machines so as to reduce substantially the size, complexity and costs of the facility, while maintaining high quality of measurement, volume of output, sensitivity and background, as mentioned in Subsec. 3.1. The use of lower voltages and simpler technology has resulted in a boom in such new AMS facilities, with more than 20 installations over the past several years [216], and this expansion is likely to continue. With some

modifications and extensions the sub–1 MV accelerators would be able to measure not only ^{14}C but also many of the relevant isotopes, such as ^{10}Be, ^{26}Al, ^{129}I, ^{236}U and the Pu isotopes, with acceptable sensitivity, and for ^{41}Ca with a sensitivity of 5×10^{-12} [218]. Although the terminal voltage of compact AMS machines has been remarkably reduced to 200 kV, the possibility of further reduction still exists. An investigation into the dissociation of ^{13}CH and ^{12}CH$_2$ molecules in He and N$_2$ at low beam energies which was carried out at ETH indicated that the dissociation cross-sections in He were constant over the range of 80–150 kV and sufficient molecule dissociation can be obtained with gas densities of $\sim 0.4\,\mu\text{g/cm}^2$, for which angular straggling is relatively small. In addition, the 1+ charge state fraction using a He stripper is shown to increase at lower stripping energies. Thus, the usage of He for stripping and molecule dissociation might allow the development of even smaller ^{14}C-AMS systems than those available today [219].

5.2. *Role of larger AMS facilities*

As smaller AMS instruments are installed in new laboratories, the existing larger facilities are used to analyze isotopes with severe interference from isobars, or to detect as-yet-unexplored isotopes.

Recent developments for multipurpose AMS systems based on 5–6 MV accelerators show quite a renaissance. These systems have the capability to detect ^{36}Cl at natural levels, which requires quite efficient suppression of the ^{36}S isobar background with elaborate ion detection procedures. Remarkable progress has been made with these facilities [220] and also at the 3 MV tandem-based VERA system, where detection levels in the low 10^{-14} range have been demonstrated [221].

For more exotic AMS nuclides like ^{32}Si, ^{53}Mn, ^{59}Ni, ^{55}Fe, ^{60}Fe, ^{63}Ni and ^{99}Te, large tandem accelerators with the terminal voltage up to 14 MV are necessary in order to provide a high-enough energy (around 3 MeV/amu) for isobar separation and to achieve high sensitivity [222]. Fifield [15] has reported that the ANU 14UD facility is able to perform ^{36}Cl measurements routinely with the background even below 10^{-16}. With the 25 MV Pelletron of the Radioactive Ion Beam Facility (RIBF) in ORNL, the reported sensitivity of ^{36}Cl measurement was a few times of 10^{-16} [223].

5.3. *Challenges*

5.3.1. ^{14}C measurements

The small uncertainty, low level background and tiny sample size have been the goals of ^{14}C measurements all along.

In the early days, AMS results with 1% uncertainties were regarded as very precise measurements, and even 10 years ago 3‰ results were state-of-the-art. Recently, several attempts have been made to demonstrate capabilities below 2‰, and in the near future the challenge will be the 1‰ level of uncertainty. A number of experiments have been performed to demonstrate the increased quality of AMS ^{14}C measurements [224, 225]; a decrease in the overall uncertainty will not only result in more precise dates but also make possible new promising applications [226]. However, measurements at the 1‰ level require counting statistics beyond 1 million ^{14}C events per sample. Therefore, stable and reproducible conditions for measurements over a quite long time range must be guaranteed. In addition to the measurement procedure itself, there are a number of different factors preventing reliable measurements at this uncertainty level — in particular, fractionation effects and uncontrollable introduction of contamination during sample preparation as well as in the measurement procedure itself, which can cause shifts in isotopic ratios well above the 1‰ level [224]. A set of standard samples measured on MICADAS at ETH has shown a reproduced uncertainty level of 1.12–1.35‰ [216]. Accuracy reproducible to ±0.1% from laboratory to laboratory is difficult to obtain, but near ±0.25% seems to be attainable currently [23]. The critical requirements for low energy ion beam optics, stripper and apertures in the design of high precision low energy compact AMS systems have been emphasized by Suter [227], and special attention should be paid to the beam transport process, such as small angle scattering and energy straggling.

The limit of ^{14}C dating by AMS mainly depends on the contamination control in the process of sampling in the field, sample preparation in the chemical laboratory, and the memory effect in the ion source. Nowadays, several laboratories have a reported machine background as low as 70–80 ka BP [14, 228], which corresponds to the $^{14}C/^{12}C$ ratios of 2.5×10^{-16} and 7×10^{-17}, respectively. At

ISOTRACE several samples of interstratified peats associated with paleosols have been dated and the ages are 55–60 ka BP [13]. With enriched natural gas samples, a $^{14}C/^{12}C$ ratio of 10^{-18} has been measured [229].

Reliable measurement of ^{14}C samples with a tiny amount of carbon is another challenge for AMS. Actually, many AMS facilities are now engaged in the "1 μg race" [217]. A variety of approaches and techniques are being used, such as micrographitization by laser heating [230], CO_2 gas sources [29] and Zn [72] or porous Fe [231] graphitization. Apart from issues of fractionation differences when using larger-mass ^{14}C standards, for merely ~ 1–2 μg of effective "modern ^{14}C" introduced during sample preparation, a detailed understanding and control of contamination is most important for these ultralow C masses. One of the more challenging and exciting issues is related to on-line continuous flow ^{14}C AMS analyses of compound-specific organic factions chemically separated from bulk samples. The currently available AMS systems with gas-fed ion sources would need the integration and automation of the combustion and analysis stages. This would save on both the time and the cost required for graphitization. The system at ISOTRACE [232] includes an elemental analyzer used for sample combustion, and an HVEE SO-110 hybrid gas/solid sample ion source provides the ion beam from the CO_2. The connections from elemental analyzer to transfer line and to ion source and finally to AMS system are controlled by PC units. Although it is still under development, the system looks highly promising.

Such technical advancement will inevitably lead to wider applications of AMS in many fields. Compound-specific radiocarbon analyses of particular biomarkers in aerosol, lake/ocean sediment or soil samples would be among the first to benefit [74, 233]. New applications in association with the development of new energy sources would also immediately benefit from the improved capacity of small scale ^{14}C analysis, as shown in the recent use of ^{14}C measurements for testing the accuracy of biodiesel blending [234].

5.3.2. *High precision ^{10}Be measurement*

As outlined in Subsec. 4.2, the investigation into the timings of interhemispheric climate changes during the Holocene based on the glacier fluctuations over the past 7000 years, including at least five events during the last millennium in New Zealand's Southern Alps, presents a new challenge for the AMS ^{10}Be methodology. High precision measurement is required with small sample sizes and in particular for low ^{10}Be concentrations. Currently, the benchmark for ^{10}Be is the CAMS AMS facility at LLNL [12], which delivers ^{9}Be^{3+} (analyzed) beams up to $30\,\mu$A with very low backgrounds (ratio of ^{10}Be/^{9}Be $\sim 5 \times 10^{-17}$). Cathodes prepared with only $\sim 80\,\mu$g of ^{9}Be as BeO and with ^{10}Be chemistry blanks of $\sim 5 \times 10^{-16}$ ($\pm 15\%$). The total transmission efficiency for the CAMS FN mass spectrometer is $\sim 34\%$ for 7.5 MV Be^{3+} and the detection limit is estimated to be ~ 1000 ^{10}Be atoms. With this efficiency, background and detection limit, the CAMS system can routinely measure Be samples with $<1\%$ precision. These results demonstrate that Little Ice Age exposure ages of 200–400 years (with 10% errors) are now attainable [235]. This is the challenge for both the new compact 1 MV and 3–6 MV facilities in ^{10}Be AMS, because young ages, high throughput, equally enhanced ^{26}Al statistical errors and throughput are what research on cosmogenic exposure dating is now demanding [217].

5.3.3. *Isobar suppression*

Effective suppression of the isobar is essential for enabling low energy compact AMS to extend to all major AMS radionuclides. At present, isobar–radionuclides pairs, such as ^{41}Ca–^{41}K, ^{36}Cl–^{36}S and ^{59}Ni–^{59}Co, are still in the domain of TV > 5 MV AMS systems. However, there are already a number of ways to suppress the radionuclide isobar by using basic nuclear and atomic physics. For instance, laser-induced photodetachment of selected anions in the low energy AMS beam line was first reported for ^{59}Ni and ^{36}Cl at the 14UD Rehovot Pelletron in 1990 [236]. To improve the efficiency and cross-section, similar arrangements with lasers using RF quadrupole ion coolers were presented [237].

Another improved technology is isobar separation for anions (ISA) using RFQ reaction cells, at ISOTRACE [238]. An impressive suppression $\sim 10^{-6}$ of S to Cl$^-$ has been demonstrated (Fig. 10) using a three-stage setup of retardation to eV energies via RFQ cooling, isobar suppression in a NO$_2$ reaction

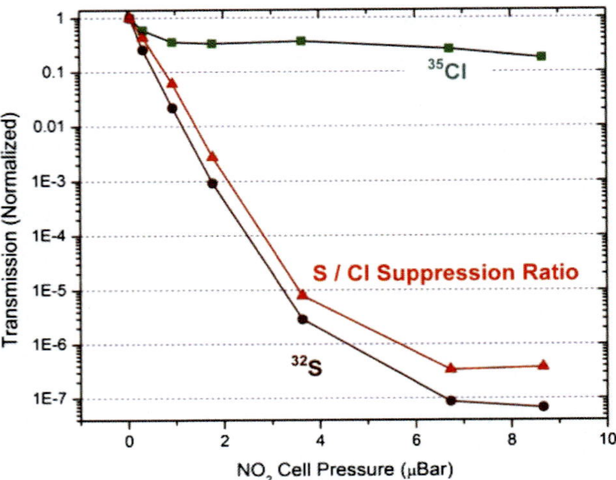

Fig. 10.   ISA in the reaction cell [238].

cell with extraction via a small longitudinal electric field, and final RFQ mass–energy analysis.

Furthermore, the fact that the slow anions travel along the axis of a solenoid magnetic field would permit laser suppression of the unwanted isobar. For 5 keV S ions, a theoretical attenuation of 3.2×10^{-6} can be reached with a total laser power of 50 W. An advantage of the laser method is that the ions do not have to be slowed down to low eV energies; the separation of atomic anions of keV energies may become possible. Both methods are being actively developed at the moment and look very promising [23].

6. Summary

During the past few decades, AMS has developed successfully in instrumentation, methodology and applications.

The most important progress of AMS instrumentation is the continuous efforts to lower the terminal voltage and reduce the facility size. After the commercial 0.5 MV compact AMS became available about 10 years ago, the 200 kV tandem AMS and 250 kV single stage AMS appeared in succession. The latest study indicates that even smaller ^{14}C AMS systems with a terminal voltage as low as 80 kV might be available in the future. Although the AMS systems based on 5–6 MV accelerators showed quite good performance for a number of radionuclides, the compact AMS is now able to measure ^{10}Be, ^{26}Al, ^{129}I, ^{236}U and the Pu isotopes with acceptable sensitivity.

The AMS methodology developments are mainly the increase in the number of measured radionuclides, the new sample preparation techniques and isobar-suppressing methods. More than 30 radionuclides with half-lives from about 1 a to 100 Ma have been detected by AMS so far, and more than 50% of them have been reported after AMS-8 in 1999. The compound-specific sample preparation combined with the μg carbon-analyzing technique will open new fields of AMS applications. To measure some radionuclides, large tandem AMSs still have to be used due to the limitation of existing isobar-suppressing techniques. However, some new techniques of isobar suppression under development may change this situation.

AMS has been successfully applied to many different fields. So far ^{14}C is still the most widely used radionuclide. Earth and environmental sciences are the most important application fields, and global carbon cycling, climate change and landscape evolution are currently hot topics of wide interest. Archeology and biomedical sciences are also taking up quite a large proportion of AMS applications. Other important application fields are nuclear physics and nuclear chemistry, including the determination of the half-life of radionuclides and the nuclear reaction cross-section, the search for exotic particles and the study of nuclear astrophysics and nuclear safeguards. There are also some very special applications, such as detection of art forgery and nuclear forensic medicine.

In summary, over 30 years' development in AMS has resulted in great advancement in the technology and instrumentation, which has in turn led to innovative applications of AMS in many disciplines. There is no doubt that more progress in AMS technology will come along with even wider applications in the future.

Acknowledgments

The authors would like to thank Alexander W. Chao and Weiren Chou for their encouragement. We are thankful to M. Suter, A. E. Litherland, W. Kutschera, A. J. T. Jull and L. K. Fifield for sharing their experience and thoughts over the years. Partial support from the National Natural Science foundation of China and the Ministry of Science and Technology of China is acknowledged.

References

[1] R. C. Finkel and M. Suter, *Adv. Anal. Geochem.* **1**, 1 (1993).

[2] W. Kutschera, *Int. J. Mass Spectrum.* **242**, 145 (2005).

[3] L. W. Alvarez and R. Cornog, *Phys. Rev.* **56**, 379 (1939).

[4] D. E. Nelson, R. G. Korteling and W. R. Stott, *Science* **198**, 507 (1977).

[5] C. L. Bennet *et al.*, *Science* **198**, 508 (1977).

[6] K. H. Purser, R. B. Liebert and C. J. Russo, *Radiocarbon* **27**, 794 (1980).

[7] D. Elmore *et al.*, *Nature* **277**, 22 (1979).

[8] D. Elmore *et al.*, *Nature* **286**, 138 (1980).

[9] J. Klein, R. Middleton and H. Tang, *Nucl. Instrum. Methods* **193**, 601 (1982).

[10] R. Middleton *et al.*, *Nucl. Instrum. Methods* **218**, 430 (1983).

[11] P. W. Kubik *et al.*, *Nature* **319**, 568 (1986).

[12] D. H. Rood *et al.*, *Nucl. Instrum. Methods B* **268**, 730 (2010).

[13] W. C. Mahaney *et al.*, *Geoderma* **104**, 215 (2001).

[14] S. Freeman *et al.*, *Nucl. Instrum. Methods B* **259**, 66 (2007).

[15] L. K. Fifield *et al.*, *Nucl. Instrum. Methods B* **268**, 858 (2010).

[16] D. Fink *et al.*, *Nucl. Instrum. Methods B* **47**, 79 (1990).

[17] K. M. Wilcken *et al.*, *Nucl. Instrum. Methods B* **266**, 3614 (2008).

[18] D. Elmore and F. M. Philips, *Science* **236**, 543 (1987).

[19] L. K. Fifield, *Rep. Prog. Phys.* **62**, 1223 (1999).

[20] M. Suter, *Nucl. Instrum. Methods B* **223–224**, 139 (2004).

[21] A. J. T. Jull and G. S. Burr, *Earth Planet. Sci. Lett.* **243**, 305 (2006).

[22] C. Tuniz and G. Norton, *Nucl. Instrum. Methods B* **266**, 1837 (2008).

[23] A. E. Litherland, X.-L. Zhao and W. E. Keiser, *Mass Spectrum. Rev.* DOI: 10.1002/mas.20311 (2010).

[24] R. Middleton, *Nucl. Instrum. Methods B* **5**, 193 (1984).

[25] S. P. H. T. Freeman *et al.*, *Nucl. Instrum. Methods B* **99**, 557 (1995).

[26] R. Middleton, J. Klein and D. Fink, *Nucl. Instrum. Methods B* **43**, 231 (1989).

[27] C. B. Ramsey, P. Ditchfield and M. Humm, *Radiocarbon* **46**, 25 (2004).

[28] S. Xu *et al.*, *Nucl. Instrum. Methods B* **259**, 76 (2007).

[29] M. Ruff *et al.*, *Nucl. Instrum. Methods B* **268**, 790 (2010).

[30] A. E. Litherland, *Nucl. Instrum. Methods B* **5**, 100 (1984).

[31] M. Suter *et al.*, *Nucl. Instrum. Methods B* **5**, 242 (1984).

[32] K. H. Purser, T. H. Smick and R. K. Purser, *Nucl. Instrum. Methods B* **52**, 263 (1990).

[33] G. Bonani *et al.*, *Nucl. Instrum. Methods B* **52**, 338 (1990).

[34] P. W. Kubik *et al.*, *Nucl. Instrum. Methods B* **52**, 238 (1990).

[35] P. Steier *et al.*, *Nucl. Instrum. Methods B* **223–224**, 67 (2004).

[36] A. Alfimov and H.-A. Synal, *Nucl. Instrum. Methods B* **268**, 769 (2010).

[37] M. Suter, *Nucl. Instrum. Methods B* **52**, 211 (1990).

[38] H. Reithmeier *et al.*, *Nucl. Instrum. Methods B* **239**, 273 (2005).

[39] P. Steier *et al.*, *Radiocarbon* **46**, 5 (2004).

[40] M. Stuiver and H. Polach, *Radiocarbon* **19**, 355 (1977).

[41] R. A. Muller, *Science* **196**, 489 (1977).

[42] M. B. Chen *et al.*, *Radiocarbon* **37**, 675 (1995).

[43] M. L. Roberts *et al.*, *Nucl. Instrum. Methods B* **172**, 262 (2000).

[44] Z. Y. Guo *et al.*, *Nucl. Instrum. Methods B* **259**, 204 (2007).

[45] P. Collon, W. Kutschera and Z.-T. Lu, *Annu. Rev. Nucl. Part. Sci.* **54**, 39 (2004).

[46] M. Suter, S. Jacob and H. A. Synal, *Nucl. Instrum. Methods B* **123**, 148 (1997).

[47] H.-A. Synal, S. Jacob and M. Suter, *Nucl. Instrum. Methods B* **161–163**, 29 (2000).

[48] H.-A. Synal, S. Jacob and M. Suter, *Nucl. Instrum. Methods B* **172**, 1 (2000).

[49] K. Liu *et al.*, *Nucl. Instrum. Methods B* **259**, 23 (2007).

[50] M. Suter *et al.*, *Nucl. Instrum. Methods B* **259**, 165 (2007).

[51] A. M. Muller *et al.*, *Nucl. Instrum. Methods B* **268**, 2801 (2010).

[52] M. G. Klein, D. J. W. Mous and A. Gottdang, *Nucl. Instrum. Methods B* **249**, 764 (2006).

[53] E. Chamizo *et al.*, *Nucl. Instrum. Methods B* **266**, 2217 (2008).

[54] H. Hernández-Mendoza *et al.*, *Nucl. Instrum. Methods B* **268**, 1331 (2010).

[55] H.-A. Synal *et al.*, *Nucl. Instrum. Methods B* **223–224**, 339 (2004).

[56] T. Schulze-König *et al.*, *Nucl. Instrum. Methods B* **268**, 891 (2010).

[57] J. B. Schroeder *et al.*, *Radiocarbon* **46**, 1 (2004).

[58] G. Skog, *Nucl. Instrum. Methods B* **259**, 1 (2007).

[59] S. P. H. T. Freeman *et al.*, *Nucl. Instrum. Methods B* **268**, 715 (2010).

[60] S. J. Fallon, L. K. Fifield and J. M. Chappell, *Nucl. Instrum. Methods B* **268**, 898 (2010).

[61] G. M. Raisbeck and F. Yiou, *Nucl. Instrum. Methods B* **5**, 91 (1984).

[62] G. M. Raisbeck and F. Yiou, *Earth Planet. Sci. Lett.* **89**, 103 (1988).

[63] M. Paul, *Nucl. Instrum. Methods B* **52**, 315 (1990).

[64] H. Artigalas *et al.*, *Nucl. Instrum. Methods B* **92**, 227 (1994).

[65] W. Kutschera *et al.*, *Nucl. Instrum. Methods B* **42**, 101 (1989).

[66] I. Hajdas, *Eiszeitalter und Gegenwart Quaternary Science Journal* **57**(1–2), 2 (2008).

[67] I. Hajdas, *Radiocarbon* **51**(1), 79 (2009).

[68] M. I. Bird, L. K. Ayliffe and L. K. Fifield, *Radiocarbon* **41**, 127 (1999).

[69] I. Hajdas *et al.*, *Nucl. Instrum. Methods B* **223–224**, 267 (2004).

[70] X. M. Xu *et al.*, *Nucl. Instrum. Methods B* **259**, 320 (2007).

[71] J. S. Vogel, *Radiocarbon* **34**(3), 344 (1992).

[72] M. S. Khosh, X. M. Xu and S. E. Trumbore, *Nucl. Instrum. Methods B* **268**, 927 (2010).

[73] L. A. Curie *et al.*, *Nucl. Instrum. Methods B* **123**, 475 (1997).

[74] A. E. Ingalls and A. Pearson, *Oceanography* **18**, 18 (2005).

[75] S. R. Beaupré, E. R. M. Druffel and S. Griffin, *Limnol. Oceanogr.* **5**, 174 (2007).

[76] S. Griffin, S. R. Beaupré and E. R. M. Druffel, *Radiocarbon* **52**(2–3), 1224 (2010).

[77] J. C. Gosse and F. M. Phillips, *Quaternary Sci. Rev.* **20**, 1475 (2001).

[78] G. Shen *et al.*, *Nature* **458**, 198 (2009).

[79] https://c14.arch.ox.ac.uk/embed.php?File=oxcal.html

[80] P. J. Reimer *et al.*, *Radiocarbon* **46**(3), 1029 (2004).

[81] G. M. Raisbeck and F. Yiou, *Nuclear Methods of Dating*, eds. E. Roth and B. Poty (CEA, Paris, 1989), pp. 353–378.

[82] I. Levin and V. Hesshaimer, *Radiocarbon* **42**(1), 69 (2000).

[83] E. M. Wild *et al.*, *Nucl. Instrum. Methods B* **172**, 944 (2000).

[84] D. Lal, *Scientific World J.* **2**, 1 (2002).

[85] A. J. T. Jull *et al.*, *Global Planet. Change* **41**, 309 (2004).

[86] S. Ivy-Ochs and F. Kober, *Quaternary Sci. J. (Eiszeitalter und Gegenwart)* **57**(1–2), 179 (2008).

[87] W. S. Broecker, *Radiocarbon* **51**(1), 109 (2009).

[88] M. Frank *et al.*, *Geochim. Cosmochim. Acta* **73**(20), 6114 (2009).

[89] I. U. Olsson, *Radiocarbon* **51**(1), 1 (2009).

[90] G. M. Raisbeck and F. Yiou, *Nucl. Instrum. Methods B* **5**, 91 (1984).

[91] G. Bond *et al.*, *Nature* **360**, 245 (1992).

[92] W. S. Broecker and S. Hemming, *Science* **294**, 2308 (2001).

[93] L. Wang *et al.*, *Mar. Geol.* **156**, 245 (1999).

[94] Y. H. Wu, S. M. Wang and L. P. Zhou, *Radiocarbon* **53**(2), 359 (2011).

[95] J. Z. Hou *et al.*, *Anal. Chem.* **82**, 7119 (2010).

[96] L. C. Skinner and N. J. Shackleton, *Paleoceanography* **19**, 2005 (2004).

[97] M. Sarnthein, *Science* **331**, 156 (2011).

[98] N. Z. Jiao et al., *Nat. Rev. Microbiol.* **8**, 593 (2010).

[99] I. B. Levin et al., *Geophys. Res. Lett.* **30**, 2194 (2003).

[100] J. C. Turnbull et al., *J. Geophys. Res.* **112**, D11310 [doi:10.1029/2006JD008184 (2007)].

[101] S. Szidat et al., *Atmos. Environ.* **38**, 4035 (2004).

[102] C. W. Lewis and D. C. Stiles, *Aerosol Sci. Technol.* **40**, 189 (2006).

[103] M. Jacobson et al., *Rev. Geophys.* **38**, 267 (2000).

[104] S. H. Chung and J. H. Seinfeld, *J. Geophys. Res.* **107**, 4407 (2002).

[105] I. Levin, K. O. Münnich and W. Weiss, *Radiocarbon* **22**, 379 (1980).

[106] D. Y. Hsueh et al., *Geophys. Res. Lett.* **34**, L02816 (2007).

[107] W. W. Wang and D. E. Pataki, *Landscape Ecol.* **25**, 35 (2010).

[108] X. T. Xi et al., *Chin. Sci. Bull.* **56**, 1721 (2011).

[109] S. E. Trumbore, *Annu. Rev. Earth Planet. Sci.* **37**, 47 (2009).

[110] M. S. Torn et al., *Biophysico-Chemical Processes Involving Natural Nonliving Organic Matter in Environmental Systems*, eds. P. M. Huang and N. Senesi (Wiley, Hoboken, 2009), pp. 219–272.

[111] S. E. Trumbore and C. I. Czimczik, *Science* **321**, 1455 (2008).

[112] D. Lal and B. Peters, *Handb. Phys.* **46**, 551 (1967).

[113] L. R. Mchargue and P. E. Damon, *Rev. Geophys.* **29**(2), 141 (1991).

[114] G. M. Raisbeck et al., *Nature* **444**, 82 (2006).

[115] J. Beer, W. Mende and R. Stellmacher, *Quaternary Sci. Rev.* **19**, 403 (2000).

[116] R. Muscheler et al., *Quaternary Sci. Rev.* **26**, 82 (2007).

[117] M. Frank et al., *Paleoceanography* **23**(1), 1S02 (2008).

[118] E. Sellén et al., *Global Planet. Change* **68**, 38 (2009).

[119] T. J. Dunai, *Cosmogenic Nuclides: Principles, Concepts and Applications in the Earth Surface Sciences* (Cambridge University Press, 2010).

[120] P. R. Bierman and K. K. Nichols, *Annu. Rev. Earth Planet. Sci.* **32**, 215 (2004).

[121] S. Ivy-Ochs, *Quaternary Sci. J.* (*Eiszeitalter und Gegenwart*) **57**(1–2), 179 (2008).

[122] B. Lemieux-Dudon et al., *Quaternary Sci. Rev.* **29**, 8 (2010).

[123] A. E. Putnam et al., *Nat. Geosci.* **3**, 700 (2010).

[124] M. E. Mann, R. S. Bradley and M. K. Hughes, *Nature* **392**, 779 (1998).

[125] J. M. Schaefer et al., *Science* **324**, 622 (2009).

[126] D. E. Granger and P. F. Muzikar, *Earth Planet. Sci. Lett.* **188**, 269 (2001).

[127] A. Dehnert and C. Schlüchter, *Quaternary Sci. J.* (*Eiszeitalter und Gegenwart*) **57**(1–2), 210 (2008).

[128] P. Häuselmann et al., *Quaternary Int.* **164–165**, 33 (2007).

[129] N. J. Shackleton, *Nature* **215**, 15 (1967).

[130] J. Zachos et al., *Science* **292**, 686 (2001).

[131] G. Balco and C. W. Rovey II, *Geology* **38**, 795 (2010).

[132] M. Zreda and J. Noller, *Science* **292**, 1097 (1998).

[133] L. Benedetti et al., *Geophys. Res. Lett.* **29**, 871 (2002).

[134] L. Palumbo et al., *Earth Planet. Sci. Lett.* **225**, 163 (2004).

[135] A. J. Love et al., *Water Resour. Res.* **36**(6), 1561 (2000).

[136] Y. Tosaki et al., *Nucl. Instrum. Methods B* **259**, 479 (2007).

[137] F. Yiou et al., *Nucl. Instrum. Methods B* **92**, 436 (1994).

[138] J.-C. Gascard et al., *Geophys. Res. Lett.* **31**, L08302 (2004).

[139] J. N. Smith et al., *J. Geophys. Res.* **110**, C05006 (2005).

[140] P. E. Damon et al., *Nature* **337**, 611 (1989).

[141] B. Philip, *Nat. Mater.* **7**, 349 (2008).

[142] R. A Freer-Waters and A. J. T. Jull, *Radiocarbon* **52**(4), 1521 (2010).

[143] Z. Guo et al., *Nucl. Instrum. Methods B* **172**, 724 (2000).

[144] A. P. Derevianko et al., *Nucl. Instrum. Methods B* **223–224**, 735 (2004).

[145] S. Mihara et al., *Radiocarbon* **46**(1), 407 (2004).

[146] L. Calcagnile et al., *Nucl. Instrum. Methods B* **268**, 1030 (2010).

[147] C. E. Buck et al., *Antiquity* **65**, 808 (1991).

[148] C. B. Ramsey, *Radiocarbon* **37**, 425 (1995).

[149] Z. Guo et al., *Radiocarbon* **47**(2), 221 (2005).

[150] K. Liu et al., *Radiocarbon* **47**(1), 21 (2005).

[151] I. Unkel et al., *Radiocarbon* **49**(2), 551 (2007).

[152] I. Passariello et al., *Nucl. Instrum. Methods B* **268**, 1008 (2010).

[153] M. Bernabei et al., *J. Cult. Herit.* **8**, 202 (2007).

[154] W. Kretschmer et al., *Nucl. Instrum. Methods B* **223–224**, 672 (2004).

[155] J. S. Vogel et al., *Nucl. Instrum. Methods B* **52**, 524 (1990).

[156] J. S. Vogel and K. W. Turteltaub, *Nucl. Instrum. Methods B* **92**, 445 (1994).

[157] S. P. H. T. Freeman and J. S. Vogel, *Int. J. Mass Spectrom. Ion Process* **143**, 247 (1995).

[158] R. C. Garner and D. Long, *Nucl. Instrum. Methods B* **172**, 892 (2000).

[159] G. C. Young and W. J. Ellis, *Nucl. Instrum. Methods B* **259**, 752 (2007).

[160] J. S. Vogel et al., *Nucl. Instrum. Methods B* **259**, 745 (2007).

[161] Q. Xie et al., *Toxicol. Lett.* **163**, 101 (2006).

[162] R. J. Talbot et al., *Hum. Exp. Toxicol.* **14**, 595 (1995).

[163] J. P. Day *et al.*, *Nucl. Instrum. Methods B* **92**, 463 (1994).

[164] N. D. Priest *et al.*, *Biometals* **9**, 221 (1996).

[165] S. J. King *et al.*, *Nucl. Instrum. Methods B* **123**, 254 (1997).

[166] P. B. Moore *et al.*, *Biol. Phychiatry* **41**, 488 (1997).

[167] R. A. Yokel *et al.*, *Toxicology* **161**, 93 (2001).

[168] R. A. Yokel and R. L. Florence, *Toxicology* **227**, 86 (2006).

[169] R. A. Yokel, C. L. Hicks and R. L. Florence, *Toxicology* **46**, 2261 (2008).

[170] Y. Zhou, W. R. Harris and R. A. Yokel, *J. Inorg. Biochem.* **102**, 798 (2008).

[171] S. Yumoto *et al.*, *Nucl. Instrum. Methods B* **223–224**, 754 (2004).

[172] S. Yumato *et al.*, *Nucl. Instrum. Methods B* **268**, 1328 (2010).

[173] E. Denk *et al.*, *Anal. Bioanal. Chem.* **386**, 1587 (2006).

[174] S. K. Hui *et al.*, *Nucl. Instrum. Methods B* **259**, 796 (2007).

[175] W.-H. Lee *et al.*, *Anal. Bioanal. Chem.* **399**, 1613 (2011).

[176] J. M. K. Cheong *et al.*, *J. Clin. Endocrinol. Metab.* **92**(2), 577 (2007).

[177] G. Wallerstein *et al.*, *Rev. Mod. Phys.* **69**, 995 (1997).

[178] A. Wallner, *Nucl. Instrum. Methods B* **268**, 1277 (2010).

[179] M. Paul *et al.*, *Phys. Lett. B* **94**, 303 (1980).

[180] H. Nassar *et al.*, *Phys. Rev. Lett.* **96**, 041102 (2006).

[181] H. Nassar *et al.*, *Phys. Rev. Lett.* **94**, 092504 (2005).

[182] I. Dillmann *et al.*, *Nucl. Instrum. Methods B* **268**, 1283 (2010).

[183] G. Rugel *et al.*, *Nucl. Instrum Methods B* **259**, 683 (2007).

[184] P. E. Damon and A. N. Peristykh, *Radiocarbon* **42**, 137 (2000).

[185] K. Knie *et al.*, *Phys. Rev. Lett.* **83**, 18 (1999).

[186] A. M. Alpizar-Vicente *et al.*, *Phys. Rev. C* **77**, 015806 (2008).

[187] H. Beer and R. Spencer, *Nucl. Phys. A* **240**, 29 (1975).

[188] Z. Guo and C. Zhang, *Nucl. Phys. Rev.* **19**, 395 (2002).

[189] K. Knie *et al.*, *Phys. Rev. Lett.* **93**, 171103 (2004).

[190] C. Vockenhuber *et al.*, *New Astron. Rev.* **48**, 161 (2004).

[191] M. Paul *et al.*, *Astrophys. J.* **558**, L133 (2001).

[192] C. Wallner *et al.*, *New Astron. Rev.* **48**, 145 (2004).

[193] S. Winkler *et al.*, *New Astron. Rev.* **48**, 151 (2004).

[194] X. L. Zhao, L. R. Kilius and A. E. Litherland, *Nucl. Instrum. Methods B* **92**, 249 (1994).

[195] X. L. Zhao *et al.*, *Nucl. Instrum. Methods B* **126**, 297 (1997).

[196] P. Steier *et al.*, *Nucl. Instrum. Methods B* **188**, 283 (2002).

[197] L. Wacker *et al.*, *Nucl. Instrum. Methods B* **240**, 452 (2005).

[198] D. Berkovits *et al.*, *Nucl. Instrum. Methods B* **172**, 372 (2000).

[199] X. Wang *et al.*, *Nucl. Instrum. Methods B* **268**, 2295 (2010).

[200] T. A. Brown *et al.*, *Nucl. Instrum. Methods B* **223–224**, 788 (2004).

[201] M. A. C. Hotchkis *et al.*, *Nucl. Instrum. Methods B* **172**, 659 (2000).

[202] E. J. Wyse *et al.*, *J. Anal. At. Spectrom.* **16**, 1107 (2001).

[203] L. K. Fifield *et al.*, *Nucl. Instrum. Methods B* **117**, 295 (1996).

[204] L. K. Fifield *et al.*, *Nucl. Instrum. Methods B* **123**, 400 (1997).

[205] S. Richter *et al.*, *Int. J. Mass Spectrom.* **193**, 9 (1999).

[206] X. Wang *et al.*, *Nucl. Instrum. Methods B* **268**, 2295 (2010).

[207] O. J. Marsden *et al.*, *Analyst* **126**, 633 (2001).

[208] M. Hotchkis *et al.*, *Appl. Rad. Isot.* **53**(1–2), 31 (2000).

[209] P. R. Danesi *et al.*, *J. Environ. Radioact.* **64**, 121 (2003).

[210] D. P. Child *et al.*, *Nucl. Instrum. Methods B* **223–224**, 788 (2004).

[211] L. K. Fifield, *Quater. Geochron.* **3**, 276 (2008).

[212] R. Hellborg and G. Skog, (Wiley Inter-Science, 2008), doi:10.1002/mas.20172.

[213] L. K. Fifield *et al.*, *Nucl. Instrum. Methods B* **223–224**, 802 (2004).

[214] S. G. Tims *et al.*, *Nucl. Instrum. Methods B* **268**, 1150 (2010).

[215] G. Etherington *et al.*, *Radiat. Prot. Dosim.* **105**, 321 (2003).

[216] H.-A. Synal and L. Wacker, *Nucl. Instrum. Methods B* **268**, 701 (2010).

[217] D. Fink, *Nucl. Instrum. Methods B* **268**, 1334 (2010).

[218] T. Schulze-König *et al.*, *Nucl. Instrum. Methods B* **268**, 752 (2010).

[219] T. Schulze-König *et al.*, *Nucl. Instrum. Methods B* **269**, 34 (2011).

[220] K. M. Wilcken *et al.*, *Nucl. Instrum. Methods B* **268**, 748 (2010).

[221] P. Steier *et al.*, *Nucl. Instrum. Methods B* **268**, 744 (2010).

[222] K. Knie *et al.*, *Nucl. Instrum. Methods B* **172**, 717 (2000).

[223] A. Galindo-Uribarri *et al.*, *Nucl. Instrum. Methods B* **259**, 123 (2007).

[224] C. B. Ramsey, T. Higham and P. Leach, *Radiocarbon* **46**, 17 (2004).

[225] F. Terrasi *et al.*, *Nucl. Instrum. Methods B* **266**, 2221 (2008).

[226] H. D. Graven, T. P. Guilderson and R. F. Keeling, *Radiocarbon* **49**, 349 (2007).

[227] M. Suter *et al.*, *Nucl. Instrum. Methods B* **268**, 722 (2010).

[228] R. E. Tylor and J. Southon, *Nucl. Instrum. Methods B* **259**, 282 (2007).

[229] R. P. Beukens, *Nucl. Instrum. Methods B* **79**, 620 (1993).

[230] A. M. Smith *et al.*, *Nucl. Instrum. Methods B* **268**, 919 (2010).

[231] M. de Rooij, H. van der Plicht and H. Meijer, *Nucl. Instrum. Methods B* **268**, 947 (2010).

[232] W. E. Kieser *et al.*, *Nucl. Instrum. Methods B* **268**, 784 (2010).

[233] E. R. M. Druffel *et al.*, *Radiocarbon* **52**, 1215 (2010).

[234] C. M. Reddy *et al.*, *Environ. Sci. Technol.* **42**, 2476 (2008).

[235] J. Schaefer *et al.*, *Science* **324**, 622 (2009).

[236] D. Berkovits *et al.*, *Nucl. Instrum. Methods B* **52**, 378 (1990).

[237] A. Galindo-Uribarri *et al.*, *Nucl. Instrum. Methods B* **268**, 834 (2010).

[238] J. Eliades *et al.*, *Nucl. Instrum. Methods B* **268**, 839 (2010).

Jiaer Chen is a professor of physics at Peking University. His main interest is in the physics and applications of low energy particle accelerators (including tandem-based AMSs), various types of RFQ accelerators, and SRF-cavity-based energy recovery linacs. He was elected as an academician of the Chinese Academy of Sciences and of the Third World Academy of Sciences in 1993 and 2001, respectively. He is now the senior advisor to the NSF of China and a member of the PAC of the ILC.

Zhiyu Guo is a professor at Peking University. His main research interests are in accelerator physics and technology, including RFQ accelerators and ion sources, accelerator mass spectrometry and its applications, as well as neutron imaging and its applications. He is now a board member of the International Society for Neutron Radiology, the Vice-President of the Chinese Particle Accelerator Society and an honorary council member of the Chinese Nuclear Society.

Kexin Liu has been working at Peking University for more than 20 years on electrostatic accelerators, mainly on accelerator mass spectrometry and its applications. He worked at the Australian National University on the 14UD-based AMS from 1996 to 1998, as a visiting scholar. Since 2005, he has extended his work to superconducting RF accelerators and energy recovery linacs.

Liping Zhou is a geoscientist. He is Cheung Kong Professor at Peking University. He has mainly worked on long term environmental change and geochronology of geological records in arid and semi-arid regions in the northern midlatitudes. His current research involves extensive use of ^{14}C and ^{10}Be, e.g. compound-specific radiocarbon analysis of lacustrine and soil samples, carbon dynamics on land and in the coastal and deep oceans, and past changes in the production rate of cosmogenic nuclides. He serves as Vice-President of the Commission on Stratigraphy and Chronology, International Union for Quaternary Research.

Reviews of Accelerator Science and Technology
Vol. 4 (2011) 147–159
© World Scientific Publishing Company
DOI: 10.1142/S1793626811000501

Electron Accelerators for Environmental Protection

Andrzej G. Chmielewski

Institute of Nuclear Chemistry and Technology,
Dorodna str. 16, 03-195 Warsaw, Poland

Department of Chemical and Process Engineering,
University of Technology, Warynskiego str. 1,
00-645 Warszawa, Poland
a.chmielewski@ichtj.waw.pl

This article gives an overview of existing and possible electron accelerator applications for environmental pollution control. Laboratory and pilot plant tests and industrial applications have illustrated the possibility of applying this technology for purification and treatment of gaseous, liquid, and solid wastes. Examples of ionizing radiation application to protect the environment and human health are discussed.

Keywords: Electron accelerator; flue gas treatment; wastewater treatments; biological sludge disinfection.

1. Introduction

The powerful tools of ionizing radiation, electron accelerators, have been used for radiation processing of materials for more than half a century [1–4]. However, the possibility of radiation applications for environmental pollution control was realized in the 1970s, when environmental protection agencies were established and standards for pollutant emission limits were set. The pioneer in these applications was the Japan Atomic Energy Research Institute, Takasaki [5]. The special input for application of the technology was the development of new high power electron accelerators which can be used for on-line processing of huge flow streams of liquid or gaseous pollutants. The accelerators were employed for off-gas and wastewater treatment [6–8] and biological sludge from wastewater treatment plant disinfection testing [9], and have higher throughput in comparison with gamma sources applied to the last-mentioned technology as well [10]. Technologies which apply particle accelerators are considered important for further high-tech processes in different fields of national economies: material processing, sterilization of medical products, environmental protection, medicine (patient treatment and diagnosis; manufacturing of radiopharmaceuticals), cargo inspection, chemical analysis, nuclear power (ADS and transmutation), and so on [11]. Electron

beam processing of materials was introduced during the 1950s, and this technology has been continually evolving since then. A variety of industrial electron accelerators can now provide electron energies from 0.3 MeV to more than 10 MeV, with average beam power capabilities of up to 300 kW [12]. Nowadays higher power accelerators are available as well.

2. Industrial Off-Gas Purification

Pollutants are emitted to the atmosphere along with off-gases from industry, power stations, residential heating systems, and municipal waste incinerators. Fossil fuels, which include coal, natural gas, petroleum, shale oil, and bitumen, are the main source of heat and electrical energy. Recently, biomass has also been a main fuel for renewable energy production in heat boilers. Besides the major constituents (carbon, hydrogen, oxygen), all these fuels contain metal, sulfur, and nitrogen compounds.

During the combustion process, different pollutants, such as fly ash, sulfur oxides (SO_2 and SO_3), nitrogen oxides ($NO_x = NO_2 + NO$), hydrochloride (HCl), and volatile organic compounds, including chlorinated species, are emitted. Ninety-five percent of emitted NO_x is NO, an insoluble and nonreactive compound that is difficult to remove. Fly ash contains different trace elements (heavy metals). Mercury is emitted in adsorbed or free forms.

Gross emission of pollutants is tremendous in most countries, all over the world. These pollutants are present in the atmosphere in conditions in which they can affect man and his environment. Air pollution caused by particulate matter and other pollutants acts not only directly on the environment but also by contamination of water and soil, leading to their degradation. Wet and dry deposition of inorganic pollutants leads to acidification of the environment. These phenomena affect the health of people, increase corrosion, and destroy forest and plants.

Different air pollution control technologies are sought. The conventional technologies most often used for air pollution control are: wet FGD (flue gas desulphurization), based on SO_2 absorption in lime or limestone slurry; and SCR (selective catalytic reduction), based on NO_x reduction over a catalyst to atmospheric nitrogen with ammonia as a reductor. However, technologies which treat different pollutants in one step are of special interest. Electron beam treatment technology (EBFGT) is such a process.

2.1. *Interaction of electrons with flue gas, components*

After irradiation of polluted gas, fast electrons interact with gas, creating various ions and radicals, and the primary species formed include e^-, N_2^+, N^+, O_2^+, O^+, H_2O^+, OH^+, H^+, CO_2^+, CO^+, N_2^*, O_2^*, N, O, H, OH, and CO. In the case of high water vapor concentration, the oxidizing radicals $^\bullet OH$ and $HO_2^\bullet$ and excited ions such as $O(^3P)$ are the most important products. The SO_2, NO, NO_2, and NH_3 present cannot compete with the reactions because of very low concentrations, but react with N, O, OH, and HO_2 radicals. Ammonia, as mentioned above, is added to the gas to neutralize acids formed in reactions, with aerosol of ammonium sulfate and nitrate being the final products of the reaction. The interaction of electrons with gas forms visible cold plasma (Fig. 1).

2.2. *SO_x and NO_x removal from fossil fuel combustion flue gases*

The method of sulfur and nitrogen oxide removal is based on the oxidation of both pollutants and their reaction with water to form acids. The acids

Fig. 1. The visible glow indicates that cold plasma is formed inside the process vessel, in which gas is irradiated.

are neutralized with gaseous ammonia to form the solid aerosol, a mixture of ammonium nitrate and sulfate, which is the popular nitrogen-bearing component of NPK (nitrogen, phosphor, potassium) fertilizer. There are several pathways of NO oxidation known. In the case of EBFGT the most common are as follows [13]:

$$NO + O(^3P) + M \rightarrow NO_2 + M,$$

$$O(^3P) + O_2 + M \rightarrow O_3 + M,$$

$$O(^3P) + O_2 + M \rightarrow O_3 + M,$$

$$NO + O_3 + M \rightarrow NO_2 + O_2 + M,$$

$$NO + HO_2^\bullet + M \rightarrow NO_2 + {}^\bullet OH + M.$$

After the oxidation NO_2 is converted to nitric acid in the reaction with $^\bullet OH$, and HNO_3 aerosol reacts with NH_3, giving ammonium nitrate. NO is partly reduced to atmospheric nitrogen.

$$NO_2 + {}^\bullet OH + M \rightarrow HNO_3 + M,$$

$$HNO_3 + NH_3 \rightarrow NH_4NO_3.$$

There can also be several pathways of SO_2 oxidation, depending on the conditions. In the EBFGT process the most important are radiothermal and thermal reactions. Radiothermal reactions proceed through radical oxidation of SO_2 and HSO_3, which creates ammonium sulfate in the following steps [14]:

$$SO_2 + {}^\bullet OH + M \rightarrow HSO_3 + M,$$

$$HSO_3 + O_2 \rightarrow SO_3 + HO_2^\bullet,$$

$$SO_3 + H_2O \rightarrow H_2SO_4,$$

$$H_2SO_4 + 2NH_3 \rightarrow (NH_4)_2SO_4.$$

The thermal reaction is based on the following process:

$$SO_2 + 2NH_3 \rightarrow (NH_3)_2SO_2,$$

$$(NH_3)_2SO_2 \xrightarrow{O_2, H_2O} (NH_4)_2SO_4,$$

The total yield of SO_2 removal consists of the yields of thermal and radiothermal reactions, and can be written as follows [15]:

$$\eta_{SO_2} = \eta_1(\phi, T) + \eta_2(D, \alpha_{NH_3}, T),$$

where η, ϕ, T, D, and α_{NH_3} are process efficiency, gas humidity, gas temperature, dose deposited (amount of energy transferred to gas by means of irradiation), and ammonia stoichiometry (NH_3 concentration in relation to the stoichiometric value), respectively. The yield of the thermal reaction (η_1) depends on the temperature and humidity, and decreases with the temperature increase. The yield of the radiothermal reaction (η_2) depends on the dose, temperature, and ammonia stoichiometry. The main parameter in NO_x removal is the dose. The rest of the parameters play a minor role in the process. The high dose is required for high concentrations of NO_x removal, while SO_x is removed under proper conditions at low energy consumption. SO_x removal efficiency equal to 95% is easily achieved, while at the nitrogen oxide concentrations observed in coal- or oil-fired boilers a removal efficiency of 70–80% is observed. This level of pollutant removal is requested by power plant operators due to the existing standards of air pollution control. The scheme of the flue gas treatment process is presented in Fig. 2.

Recent tests have illustrated the possibility of using process applications to treat the polyaromatic

(PAH) and volatile organic hydrocarbons (VOC) present in off-gases in trace concentrations [16], and the process has some features which may allow its application to mercury control as well [17]. The applications of this technology to treatment of municipal and medical flue gases, where the gas flow rate is not so high, would be very economical and feasible from the environmental point of view [18, 19].

2.3. Technical applications of the process

The above mechanism of the process, studied in laboratory conditions, was a basis for the technical implementation of the technology. However, in real, industrial conditions, dose distribution and gas flow patterns are important from the technological point of view [20]. These parameters influence the electrons' energy transfer, mass, and heat transfer before, after, and in the process vessel. After humidification and lowering of the temperature, flue gases are guided to the reaction chamber, where irradiation by electron beam takes place. The electrons are introduced into the process vessel via thin $50\,\mu$m titanium foil. NH_3 is injected upstream of the irradiation chamber. The ammonia is used to neutralize the sulfuric and nitric acids, and to form the solid particle aerosol. The size of the aerosol particles is about one micron and the by-product is sticky; therefore high efficiency dust collectors have to be applied downstream of the chemical reactor. Electrostatic precipitators (ESPs) are equipped in the screw conveyers installed at the heated bottom and hammering systems at the electrodes and other filter components. The insulators are protected by air jets. The solid by-product is a high class fertilizer.

In 1970–71, Japanese scientists [5] demonstrated the removal of SO_2 using an electron beam from a linear accelerator (2–12 MeV, 1.2 kW). A dose of 50 kGy at 100°C led to the conversion of SO_2 to an aerosol of sulfuric acid droplets, which were easily removed. Ebara Co. employed an electron accelerator (0.75 MeV, 45 kW) to convert SO_2 and NO_x into a dry product containing $(NH_4)_2SO_4$ and NH_4NO_3 which could be used as a fertilizer. By the "Ebara process," two larger scale pilot plants were constructed in Indianapolis, USA, and Karlsruhe, Germany. The Indianapolis plant was equipped with two electron beam accelerators (0.8 MeV, 160 kW)

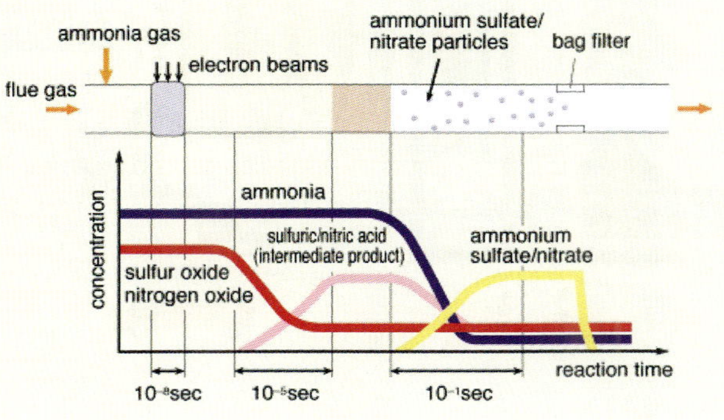

Fig. 2. Scheme presenting sequences of the physicochemical reactions which lead to acidic pollutant removal and solid fertilizer particle formation.

and had a capacity of 1.6–$3.2 \times 10^{+4}\,\mathrm{m^3/h}$, with gas containing 1000 ppm SO_2 and 400 ppm NO_x. In Karlsruhe, two electron accelerators (0.3 MeV, total power 180 kW) were used to treat 1–$2 \times 10^{+4}\,\mathrm{m^3/h}$ flue gas containing 50–500 pm SO_2 and 300–500 ppm NO_x.

However, the final engineering design technology for industrial applications was achieved at the pilot plants operated in Nagoya, Japan [21], and Kaweczyn, Poland [22]. In the case of the latter, new engineering solutions were applied: double-longitudinal gas irradiation, an air curtain separating the secondary window from corrosive flue gases, and modifications of the humidification/ammonia system (high enthalpy water or steam injection, ammonia water injection), and others. The obtained results have confirmed the physicochemistry of the process discussed earlier. In Fig. 3 the applied process vessel is presented. The double window was applied to protect the window of the accelerator from the corrosive

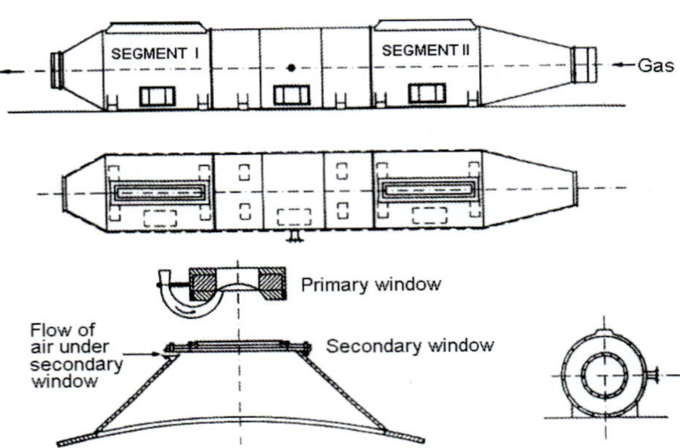

Fig. 3. Scheme of the flue gas irradiation vessel applied at the EPS Kaweczyn, Poland, pilot plant and the EPS Pomorzany, Poland, industrial plant.

flue gas atmosphere, and the air curtain protects a secondary window from such effects as well.

These new solutions led to improvements in economic and technical feasibility and final industrial scale plant construction.

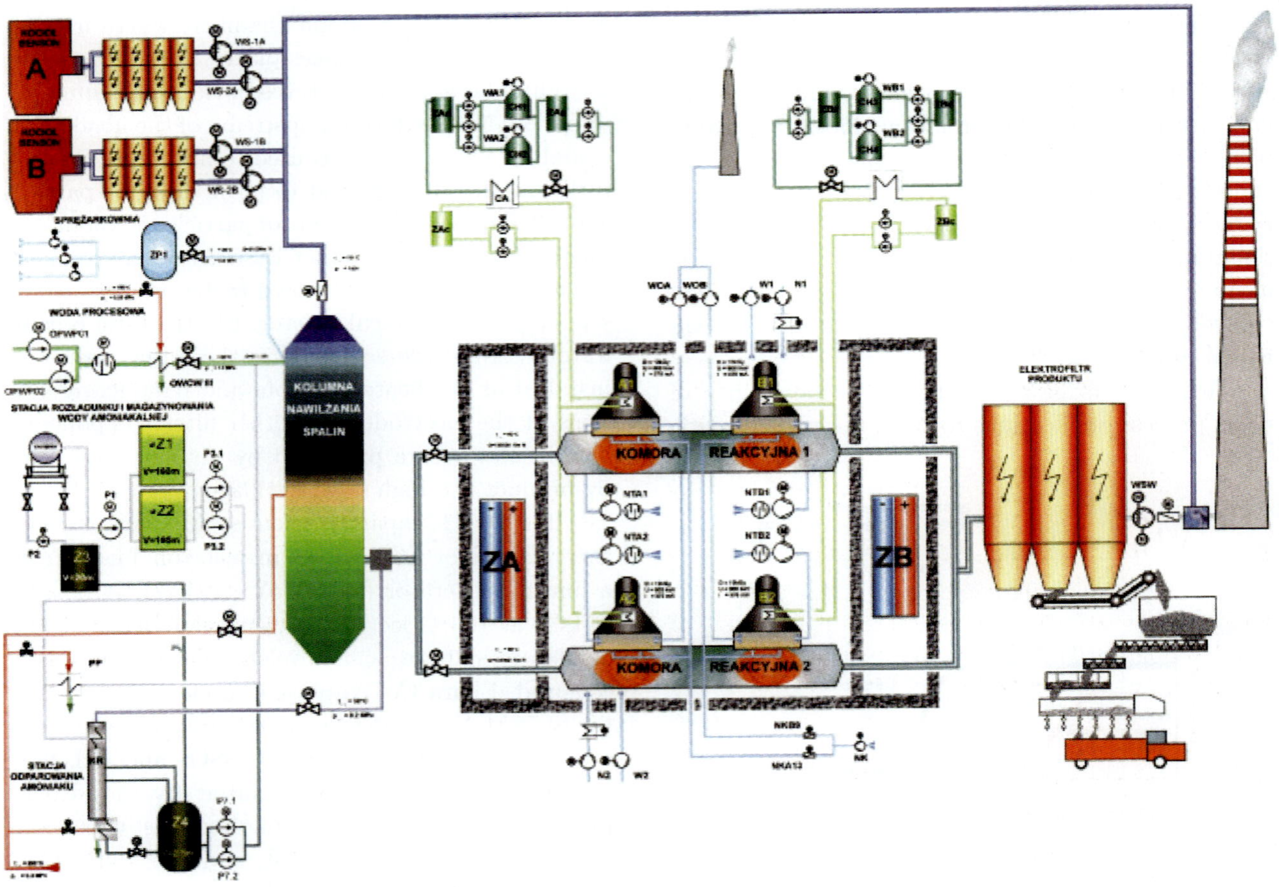

Fig. 4. Scheme of an industrial plant for electron beam flue gas treatment: EPS Pomorzany, Szczecin, Poland.

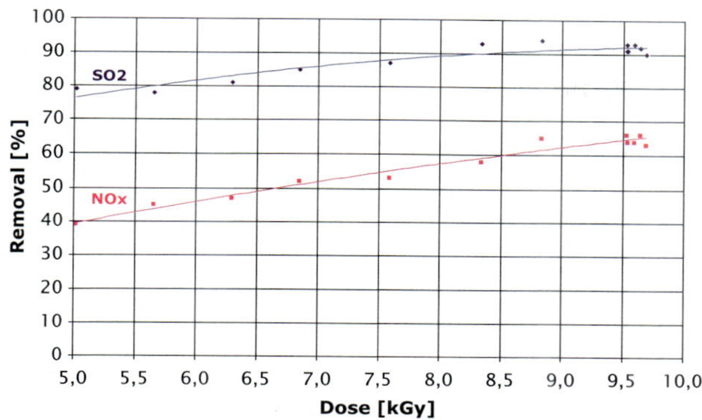

Fig. 5.   Removal efficiency of SO_2 and NO_x as a function of energy absorbed in gas.

Ebara Corporation has constructed a full scale plant in Chengdu, China, mostly for SO_x removal, and therefore the power of accelerators applied is 320 kW for treatment of 270,000 Nm^3/h of the flue gas. The reported efficiency is 80% for SO_x and 20% for NO_x [23].

The flue gas treatment industrial installation is located in EPS Pomorzany in Szczecin, in the north of Poland (Fig. 4) [24]. The installation purifies flue gases from two Benson boilers of 65 MW_e and 100 MW_{th} each. The maximum flow rate of the gases is 270,000 Nm^3/h and the total beam power exceeds 1 MW. There are two reaction chambers with nominal flow gas rates of 135,000 Nm^3/h. Each chamber is irradiated by two accelerators (260 kW, 700 keV) installed in series. The applied dose is in the range of 7–12 kGy. The removal of SO_2 approaches 80–90% in this dose range, and that of NO_x is 50–60% (Fig. 5). The by-product is collected by the electrostatic precipitator and shipped to the fertilizer plant.

Other laboratory tests have proven the applicability of the technology to the treatment of flue gases from oil-fired boilers [25] and the feasibility of applying the process for the treatment of mercury (in flue gas). A review of the process vessel construction for accelerator-based continuous processing flow systems was given by Berejka [26].

2.4. *Electron accelerators applied for flue gas treatment*

In Table 1 accelerators applied in different laboratory, pilot, and industrial size installations are listed; early project lists are taken from the paper

Table 1.   List of accelerators applied at test sites.

Site	Accelerator (MeV; kW)	System (Nm^3/h)
Tokyo 1970–71	2–12; 1.2 linear	20 l, batch
Takasaki 1972–74	–; 15 Cocroft–Walton	60
Ebara 1974–77	0.3–0.75; 30	1000
Univ. Tokyo 1974–78	1; 90–120 Dynamitron	36–84
Ebara 1977–78	0.6–0.75; 2×10–45	3000–10,000
JAERI 1981	1.5; 30	0.09
Res. Cottrell 1984–85	2×0.8; 80	5300
Indianapolis 1984–88	2×0.88; 160	8000–24,000
Karlsruhe 1984	190–220; 22	100–1000
Karlsruhe 1984	150–300; 36	60–1000
Badenwerk 1985	260–300; 2×90	10,000–20,000
Warsaw 1990–now	1; 20	1–400
Kawęczyn 1990–94	700; 2×50	20,000
Fujisawa 1991	500; 15	1500
Matsudo 1992	900; 15	1000
Nagoya 1992	800; 3×36	12,000
Tokyo 1992	500; 2×12.5	50,000
Mianyang 1999	800; 36	3000–12,000
Chengdu 1999–2004	800; 2×320	270,000
Bejing 2000		
Pomorzany 2002–now	700; 4×268	270,000
Maritza East 2003–05	700; 3×12.5	10,000

by Frank [27]. The small units were widely used in R&D and industry accelerators with low and medium power. The power supplies had power of up to 100 kW and appropriate window dimensions (due to the cooling requirements).

The most popular accelerators, due to the electron energy requirements (up to 1 MeV), were transformer accelerators like the one presented in Fig. 6. These are the most economical units with high energy efficiency, and their preferability for applications is connected with the fact that the density of flue gas is close to 1.25 kg/Nm^3, which assures good penetration of the medium treated by electrons, much deeper in comparison with the liquid or solid phases.

The series of accelerators presented in the picture is produced by the Budker Institute of Nuclear Physics in Novosibirsk and similar ones are produced by the D. V. Efremov Scientific Research Institute of Electrophysical Apparatus in Saint Petersburg, Russia. The ELV units are available from EB Tech Co., Republic of Korea, as well. Vivirad S.A., France,

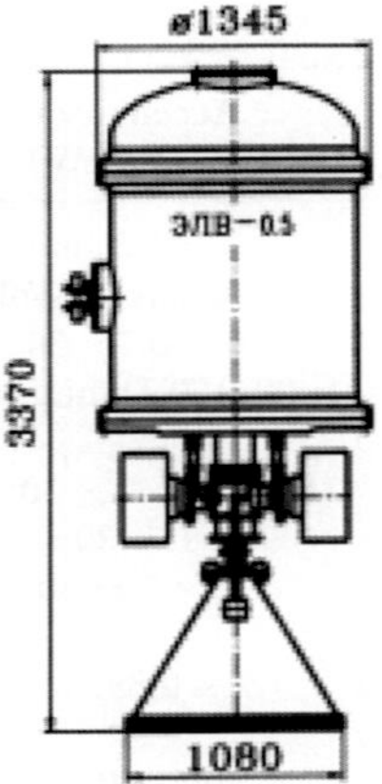

Fig. 6. An example for an accelerator used for EBFGT. An ELV-3a accelerator (two units) has been applied at the EPS Kaweczyn pilot plant.

offers insulated core transformer accelerators with a cable-connected arrangement (500–750 keV) and integral tank (0.8–5 MeV). In the above-presented design, a pencil-shaped beam of electrons from a point source is accelerated through a multistage accelerator tube and then scanned by an electromagnetic field through the window into the gas. This type of equipment is normally used for applications requiring accelerating voltages above 300 kV. The second type of EB equipment uses one or more filaments arranged inside a vacuum chamber across the product or the web to be treated or several filaments in parallel to the web direction. These filaments generate a curtain of electrons over the entire product width, without the need for scanning of the electron beam. Its small dimensions make this type of EB equipment especially suited for installation on small mobile units. Modern "low energy EB systems" are made for electron energies in the range of 70–300 kV. However, in this case the single window has to be used in the system, which causes difficulties in the application of these units for treatment of corrosive media. Accelerators of this type are produced by Energy Sciences Inc., USA, and Nissin High Voltage Corporation, Japan.

Very low energy, self-shielded accelerators which can be applied for VOC treatment are manufactured by Advanced Electron Beams (AEB), USA, and the BroadBeam division of PCT Engineering, Davenport, USA. The typical parameters of accelerators which may be applied in emission control technologies are given in Table 2 [28].

The lessons learned from pilot and industrial installations have shown that the technology itself is superior and very competitive with more conventional flue gas treatments.

However, problems reported at industrial plants in China were connected with the failure of accelerators, and similar problems occurred at the installation in Poland. The technology requires application of accelerators of very high power. Therefore the accelerators applied in the mentioned installations were of power higher than 250 kW and the power to two of them was supplied from a single high power supply. This was a breakthrough in the technology, since the required availability of the system is equal to 92% of boiler operation time (7000–8000 h a year). The problems related to the accelerators can be avoided by proper system design, manufacturing, and quality control. One reason for some of the quality issues may be that the equipment manufacturers themselves have not performed sufficient research in the development of very high power accelerators.

Table 2. Parameters of selected electrons accelerators.

Accelerator type parameter	EPS-800-375	Dynamitron	ELV-12
Nominal energy	800 keV	1–5 MeV	0.6–1.0 MeV
Energy stability	–	±2%	±1%
Nominal beam current	375 mA	50 mA	500 mA
Beam current stability	–	±2%	±2%
Beam power	300 kW × 2	250 kW	400 kW (three heads)
Scan width	225 cm	200 cm	200 cm
Dose uniformity	±5%	< ±5%	< ±5%
Mode of operation	Continuous	Continuous	Continuous
Number of scanners	Two heads	One head	Three heads
Total beam power	600 kW	250 kW	400 kW
Power consumption	682 kW	350 kW	500 kW
Electrical efficiency	88%	71%	80%
Manufacturer	NHV, Japan	IBA(RDI), USA	BINP, Russia; EB Tech., Rep. of Korea

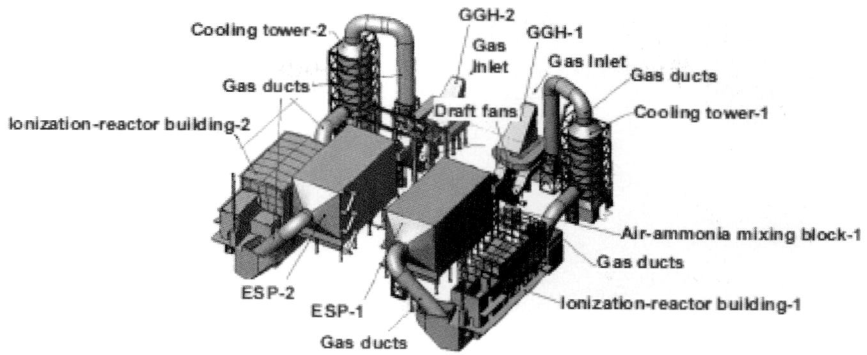

Fig. 7. General view of the EBFGT plant for the Sviloza Power Plant, Bulgaria [31].

New developments in the field of electron beam linear accelerators have been reported by PAVAC Industries, Canada. Future use of the electron beam treatment of flue gases technology is predicted. Reliable and moderately priced accelerators are a key factor for progress in the field. Guidelines for feasibility studies have been elaborated [29]. The scheme of a new planned unit at the Svilosa Power Plant, Bulgaria, is presented in Fig. 7 [30].

The plant capacity is $580\,\mathrm{MW_t}$ (four coal-fired boilers). The flow rate of flue gases will be $600{,}000\,\mathrm{Nm^3/h}$ and the estimated installation construction cost is equal to 26 million euros.

3. Wastewater Treatment

Because of the increasing levels and complexity of polluted effluents from municipalities and industry, current wastewater treatment technologies are often not successful for the remediation of polluted waters and disinfection. Development and implementation of alternative technologies for the cleanup of industrial wastewater, municipal water, groundwater, and drinking water are critical to sustainability in many countries [32]. Very important R&D work on a large scale wastewater treatment application has been performed at the Miami Electron Beam Research Facility [33].

The water purification process uses a product of water radiolysis (Fig. 8) in pollutants degradation [34].

Some of these free radicals formed are oxidative species ($^{\bullet}$OH), and the others are reductive (H, e_{aq}^-) ones. Thus there is competition between oxidation and reduction processes in the system, and the application of synergy with ozone may improve

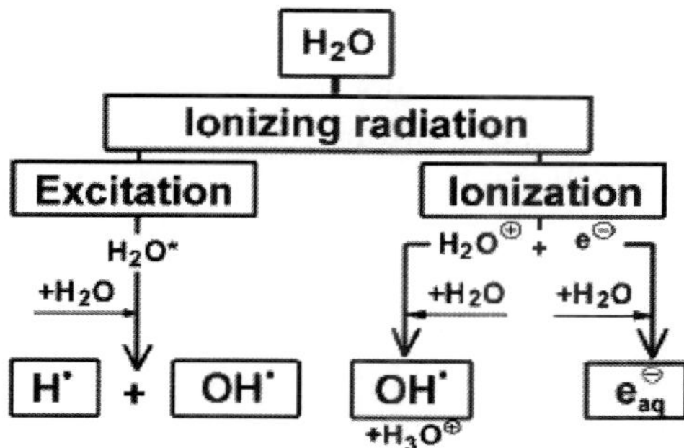

Fig. 8. Water radiolysis; the main active species formed.

the overall efficiency of destruction of organic pollutants. In this case the ozone reaction with the strong reductive species leads to the formation of hydroxyl radicals (Fig. 9) [35].

Aqueous effluents that have been treated by irradiation include polluted drinking water and liquid industrial and agricultural wastes. However, attention must be paid to the toxicity of the by-products formed in the process, which is the main limitation on its implementation. An industrial plant has been constructed in the Republic of Korea. Based on

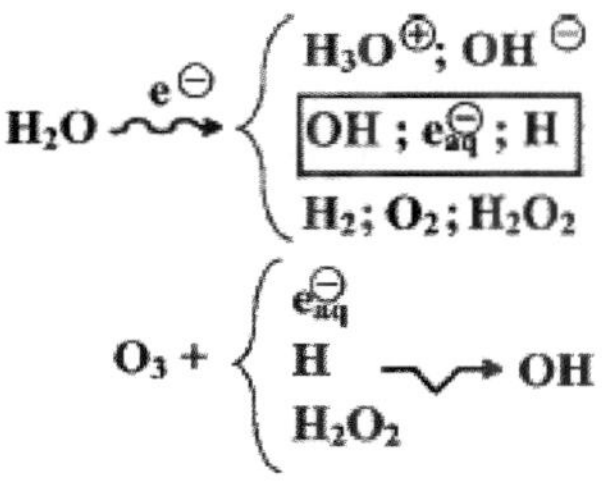

Fig. 9. Reactive species formation by radiation in the presence of ozone.

Fig. 10. Three scanners of the accelerator over jet type of wastewater treatment system (left) and the scheme of the accelerator (right) [38].

the data obtained in the laboratory and pilot plant experiments, suitable doses were determined to be around 0.2 kGy for the flow rate of 10,000 m^3 of effluent per day [36, 37]. In this case a high power accelerator (1 MeV, 400 kW) manufactured by EB Tech Co., Republic of Korea, has been applied (Fig. 10).

High energy electron disinfection of sewage wastewater in flow systems was proposed and tested very early [39] and the hybrid application (COD and microbiological load) is the most promising for future implementations.

A review of the different usages of radiation for wastewater treatment has been given by in Ref. 40. Decoloration and degradation of aqueous solution of reactive azo dye, namely Reactive Red-120 (RR-120), was carried out by electron beam irradiation. The BOD$_5$/COD ratio increased upon irradiation and it indicated the transformation of non-biodegradable dye solution into biodegradable solution. This study showed that electron beam irradiation could be a promising method for treatment of textile wastewater containing RR-120 dye [41]. Ionizing radiation such as electron beam irradiation was utilized in the decoloration and degradation of the remaining dye waste after dyeing cotton fabrics with four different dyestuffs based on azo and anthraquinone structures. The results showed that higher decoloration and degradation of dyes was obtained when using an electron beam than when using gamma irradiation under the same conditions. The TOC and COD reductions for all dye solutions were approximately 72–91% and 71–93%, respectively [42]. Other applications of ionizing radiation concern removal of heavy metals from water [43].

In developing wastewater systems it is necessary to consider the penetration range of electrons in the

Table 3. Penetration of electrons of different energies in water (cm) [44].

Energy (MeV)	1.5	3.0	5.0
R(opt)*	0.45	1.05	1.86
R(50e)†	0.53	1.21	2.13
2 R(50e)	1.06	2.42	4.26

*R(opt) — exit dose equals entrance dose; †R(50e) — exit dose equals half entrance dose.

medium, and data for different energies are given in Table 3 [44].

Due to the limited penetration range of electrons in a medium of density equal to 1000 kg/m^3, special construction of the irradiation vessel has to be carried out (Fig. 11).

These types of solutions assure the treatment of the entire water stream.

New developments in environmental applications of accelerator-generated electron beams concern degradation of antibiotics and leftover drugs released in liquid effluents. Regarding the use of antibiotics for animal husbandry, administered drugs, metabolites, or degradation products penetrate the ecosystem via the application of manure or slurry to areas used for agricultural purposes or from pasture-reared animals that excrete directly on the land. Degradation of ampicillin in pig manure slurry and an aqueous ampicillin solution has been studied using electron beam irradiation. The results demonstrate that the technology is an effective means of removing antibiotics from manure and bodies of water [45].

Ion exchangers and other substances used in the wastewater treatment process are synthesized employing electron beam grafting or polymerization. Liquid phase polymerization of acrylamide–acrylic acid to form polyelectrolytes used in wastewater

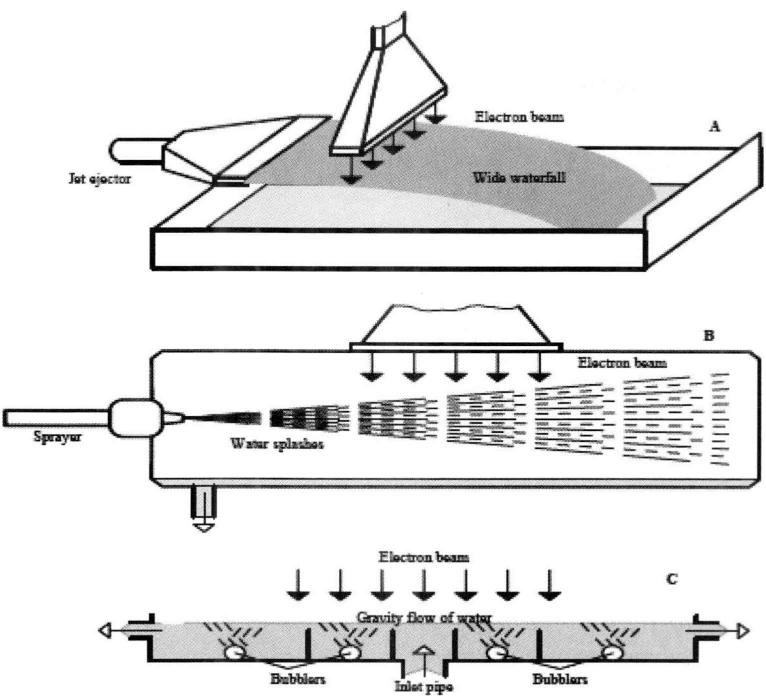

Fig. 11.   Process vessels for wastewater irradiation; jet, spray, and upflow.

cleaning was examined employing accelerated electron beam and microwave irradiation methods. Quality indicators such as total suspended matter (TSM), chemical oxygen demand (COD), biological oxygen demand (BOD), and fat, oils, and grease (FOG) were measured before and after the treatment with polymeric flocculants and compared with the results obtained from classical treatment with $Al_2(SO_4)_3$. It was found that the combined treatment with polymers and $Al_2(SO_4)_3$ increases the degree of purification of both wastewaters up to 99% [46].

4. Biological Sludge Disinfection

The problem of water contamination by chemical and biological matter is well known. Due to the fact that in many regions deficits of water for municipal, agricultural, and industrial use are observed, water from reservoirs, mainly rivers, is reused many times. Therefore perfect purification and disinfection are necessary for protecting the health of consumers; even so, bottled water and household filters are very popular as a source of good quality drinking water.

The most popular and efficient wastewater purification systems are biological treatment plants. As a result of the process, these plants are a source of biological sludge, which is a waste (which contains approximately 3% solids; to obtain a higher concentration of solids, a dewatering process is applied). Unfortunately, the sludge of municipal wastewater origin is biologically contaminated by viruses, bacteria, and eggs of parasites. In the case of landfill storage these contaminants survive for many years, due to the fact that even in regions with severe winters the sludge undergoes continuous fermentation and the temperature is much higher than the freezing point of water. Some years ago, different countries solved the problem by dumping at sea, which is prohibited nowadays. The sludge is a good organic fertilizer and is especially good for sandy soil applications, so some countries are applying injection under the soil level, which is not so safe from a health point of view if the field is used for cultivation of food industry crops. Therefore, in the EU, sludge incineration is the main direction taken to solve the problem; however, all combustion processes emit pollutants and greenhouse gases to the atmosphere.

Different methods of disinfection are proposed: heat pasteurization, mixing with lime, and ionizing radiation treatment. Radiation treatment is used

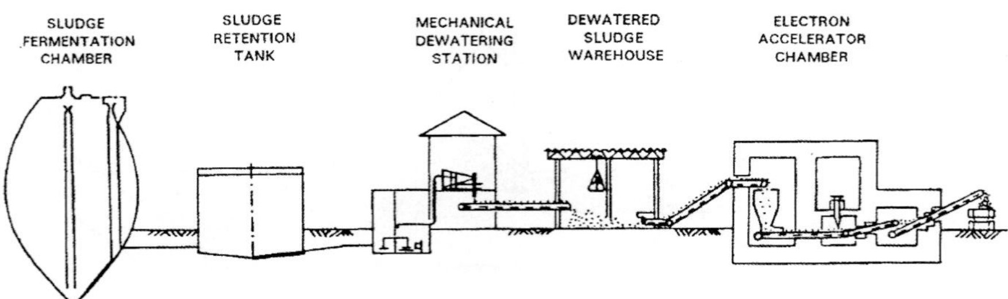

Fig. 12.   Scheme of the installation for electron beam sludge treatment.

for hygienization of food items [47] and sterilization of medical products [48]. In the first case low doses are applied to control microbiological contamination of the consumable product (e.g. spices) and in the second case much higher doses are applied to obtain microbe-free products. The destruction of the microbes is achieved by direct and indirect DNA double and single strand breaks and other damage to cell components. Again, due to the high concentration of water in a living organism, the free radicals formed play a most important role in the indirect damage to the living organism's structure. The indirect action of ionizing radiation, which is very similar to that discussed earlier for nonliving physicochemical matter, is connected with water radiolysis and the effect of active species on the DNA strain. Research has shown that sewage sludge can be disinfected successfully by exposure to high energy radiation. Doses of 2–3 kGy destroy more than 99.9% of bacteria present in sewage sludge [10]. Higher doses (up to 10 kGy) are required to inactivate more radiation-resistant organisms. Both gamma sources (Co-60, Cs-137) and electron accelerators can be used for the irradiation of sewage sludge. Gamma source radiation has better penetration, allowing thicker layers of sludge to be irradiated, although they are less powerful and require a longer irradiation time than electron sources. The irradiated sludge, being pathogen-free, can be beneficially used as manure in agricultural fields as it is rich in nutrients required for the soil. Initial field trials of sludge as manure in agricultural fields of winter wheat crops as well as summer green gram crops have been very encouraging. Since the irradiated sludge is free from bacteria, it can also be employed as a medium for growing bacteria useful for soil like rhizobium and azotobacter produce biofertilizers, which can be used to enhance crop yields. In the case of sludge or soil irradiation,

the high energy accelerators are preferable [49]. The efficiency of sludge disinfection by irradiation was investigated using an electron beam accelerator, with *Ascaris ovum* as a model. The D_{10} values obtained for irradiation of residual sludge contaminated with ova depended on the source of the ova: the D_{10} values were 788 ± 172 Gy for suspensions of ova extracted from slaughterhouse sludge and 1125 ± 145 Gy for suspensions freshly prepared by dissection. Ovum suspensions freshly prepared by dissection were more irradiation-proof. Similarly, the D_{10} value was affected by storage: it was 1125 ± 145 Gy for freshly produced ovum suspensions and 661 ± 45 Gy for suspensions of ova stored for two months at $48°C$ in deionized water [50]. An accelerator of 10 MeV at 10 kW is able to irradiate 70 tonnes of sludge a day at a dose of 5–6 kGy, and the concept of such a plant is presented in Fig. 12. The estimated cost of installation is US$ 4 million [9].

The feasibility study was performed for an Elektronika accelerator (10 MeV, 10 kW) NPO "Torij," Moscow, Russia. In the USA, high power linacs for radiation processing are manufactured by IBA Industrial, USA (formerly known as Radiation Dynamics, Inc.). The company has the longest record of continuous operation in the business of making industrial electron accelerators.

EB Tech Co. Ltd., Republic of Korea, which is very active in the field of accelerator applications for environmental pollution control, tested and prepared a feasibility study for 1 MeV beam application in sludge irradiation (Fig. 13).

Application of this technology may play a very important role in the reclamation of desert land to increase food production and ensure food security. For example, cultivated areas in Egypt existing around the Nile valley and delta represent only 4% of the total area, with the remaining 96% being barren

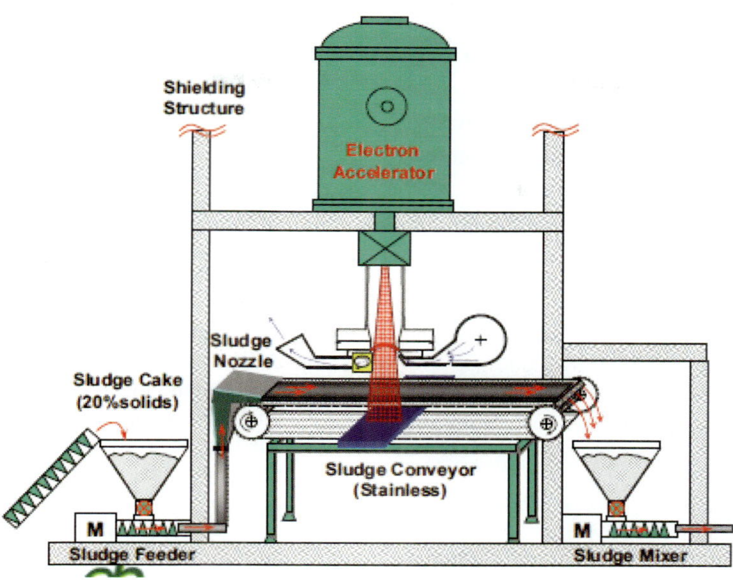

Fig. 13.   Technical solution for sludge treatment on a conveyer [51].

deserts. Reclamation of such desert lands requires water and fertilizer input [52]. The electron beam was studied to enhance the biodegradability of sewage sludge. Changes in the physicochemical characteristics of the sludge were examined with various irradiation doses, sludge thicknesses, and exposure times. Irradiation thickness was suggested as the key factor in the efficiency of solubility of solid organic matter, whereas exposure time would be the most critical parameter in inducing cell lysis in sewage sludge. In addition, biogas production was improved by as much as 22% when the sludge was irradiated by EB, cake thickness was 0.5 cm with a dose of 7 kGy [53].

Economics is a driving force in accelerator technology environmental applications, and the economic feasibility of irradiation-composting plants for sewage sludge was presented by Hashimoto *et al.* [54]. Some aspects of comparison of feasibility of electron accelerators versus a gamma source are discussed in Ref. 55.

5. Conclusions

The electron beam is an important technique for environmental protection applications. Thousands of electron accelerators based on different principles have been constructed and used in the field of radiation chemistry and radiation processing. The progress in accelerator technology means that not only a growing number of units but also lower cost,

higher dose rate, more compact size suitable for production lines, beams shaped adequately for the process, reliability, and other parameters that are important in radiation-processing application can be realized. Modern industrial accelerators can provide electron beams with an average power up to several hundred kilowatts, with an energy range suitable for radiation processing (0.15–15 MeV) [56]. The major producers of accelerators are located mainly in the USA, Japan, Russia, Korea, France, and Germany. Several other countries are capable of producing accelerators, among them China, Poland, Canada, and Sweden, but their instruments are usually prototype constructions and are used rather as pilot and R&D installations. Although the present level of accelerator development can satisfy most commercial requirements, this field continues to expand and stimulate radiation-processing activity. On the other hand, the specific demands of the growing field of industrial applications including environmental protection have a strong impact on R&D in accelerator technology.

Acknowledgments

Work partially supported by EU grant Baltic Sea Region Program–Project PlasTEP: "Dissemination and fostering of plasma-based technological innovation for environment protection in the Baltic Sea region" (electron beam gas treatment) and a grant of the National Centre for Research and Development — SP/E/4/65786/10: Task 4, "Development of integrated technologies for production of fuels and energy from biomass, agricultural wastes and other substrates" (waste and sludge electron beam treatment).

References

[1] A. G. Chmielewski, M. Haji-Saeid and Sh. Ahmed, *Nucl. Instrum. Methods B* **236**, 328 (2005).

[2] A. G. Chmielewski, *Nukleonika* **51**(S1), S3 (2006).

[3] A. G. Chmielewski, Ch. M. Kang, Ch. S. Kang and J. L. Vuijc, *Radiation Technology* (Seoul National University Press, 2006).

[4] A. G. Chmielewski, *Russ. J. Phys. Chem.* **81**(9), 1488 (2007).

[5] S. Machi, *Rad. Phys. Chem.* **22**(1–2), 91 (1983).

[6] W. J. Cooper, R. D. Curry and L. E. O'Shea (eds.), *Environmental Applications of Ionizing Radiation* (John Wiley & Sons, New York, 1998).

[7] IAEA, *Radiation Processing: Environmental Applications* (IAEA–RPEA, Vienna, Austria, 2007).

[8] A. G. Chmielewski, *Nukleonika* **50** (Supp. 1), S17 (2005).

[9] A. G. Chmielewski, Z. Zimek, T. Bryl-Sandelewska, W. Kosmal, L. Kalisz and M. Kazmierczuk, *Rad. Phys. Chem.* **46**(46), 1071 (1995).

[10] S. Gautam, M. R. Shah, S. Sabharwal and A. Sharma, *Water Environ. Res.* **77**, 472 (2005).

[11] DOE, *Accelerators for America's Future* (Washington, 2010).

[12] M. R. Cleland and L. A. Parks, *Nucl. Instrum. Methods B* **208**, 74 (2003).

[13] O. Tokunaga and N. Suzuki, *Rad. Phys. Chem.* **24**(1), 145 (1984).

[14] H. Mätzing, H. Namba and O. Tokunaga, *Rad. Phys. Chem.* **42**(4–6), 673 (1993).

[15] A. G. Chmielewski, *Rad. Phys. Chem.* **46**, 1057 (1995).

[16] A. G. Chmielewski, A. Ostapczuk, Z. Zimek, J. Licki and K. Kubica, *Rad. Phys. Chem.* **63**(3–6), 653 (2002).

[17] J.-C. Kim, K.-H. Kim, A. Armendariz and M. Al-Sheikhly, *J. Environ. Eng.* **136**(5), 554 (2010).

[18] H.-R. Paur, W. Baumann, H. Mätzing and K. Jay, *Rad. Phys. Chem.* **52**(1–6), 355 (1998).

[19] K. Hirota, T. Hakoda, M. Taguchi, M. Takigami, H. Kim and T. Kojima, *Environ. Sci. Technol.* **37**(14), 3164 (2003).

[20] A. G. Chmielewski, B. Tymiñski, A. Dobrowolski, S. Sato, O. Tokunaga and S. Machi, *Rad. Phys. Chem.* **52**(1–6), 339 (1998).

[21] H. Namba, O. Tokunaga, S. Hashimoto, T. Tanaka, Y. Ogura, Y. Doi, S. Aoki and M. Izutsu, *Rad. Phys. Chem.* **46**(6), 1103 (1995).

[22] A. G. Chmielewski, *Nukleonika* **50** (Supp. 1), S17 (2005).

[23] Y. Doi, I. Nakanishi and Y. Konno, *Rad. Phys. Chem.* **57**(3–6), 495 (2000).

[24] A. G. Chmielewski, J. Licki, A. Pawelec, B. Tyminski and Z. Zimek, *Rad. Phys. Chem.* **71**, 439 (2004).

[25] A. A. Basfar, O. I. Fageeha, N. Kunnummal, A. G. Chmielewski, A. Pawelec, J. Licki and Z. Zimek, *Fuel* **87**(8–9), 1446 (2008).

[26] A. J. Berejka, *Rad. Phys. Chem.* **71**, 309 (2004).

[27] N. W. Frank, *Rad. Phys. Chem.* **45**(6), 989 (1995).

[28] Z. Zimek, *Rad. Phys. Chem.* **45**(6), 1014 (1995).

[29] IAEA, Radiation processing of flue gases: guidelines for feasibility studies (TECDOC–1189, Vienna, Austria, 2000).

[30] Energoproekt, Feasibility study for installation of elektron-beam desulfurization and denitration plant in TPP "Sviloza" (Sofia, Bulgaria, 2006).

[31] N. Doutzkinov and K. Nikolov, *The Possibility for Implementation of E-Beam Technology in TPS "Svilosa," Bulgaria; IAEA Meeting on Electron Beam Flue Gas Treatment* (Warsaw, Poland; 14–18 May 2007).

[32] A. K. Pikaev, *Rad. Phys. Chem.* **65**(4–5), 515 (2002).

[33] Ch. N. Kurucz, Th. D. Waite and W. J. Cooper, *Rad. Phys. Chem.* **45**(2), 299 (1995).

[34] N. Getoff, *Rad. Phys. Chem.* **47**(4), 581 (1996).

[35] P. Gehringer and H. Eschweiler, *Rad. Phys. Chem.* **5**, 379 (2002).

[36] B. Han, J. Ko, J. Kim, Y. Kim, W. Chung, I. E. Makarov, A. V. Ponomarev and A. K. Pikaev, *Rad. Phys. Chem.* **64**(1), 53 (2002).

[37] B. Han, J. Kim, I. E. Makarov and A. V. Ponomarev, *Water Sci. Tech.* **52**(10–11), 317 (2005).

[38] B. Han, S. M. Kim, R. A. Salimov and N. K. Kuksanov, High power accelerators for environmental application, in *Proc. 8th International Topical Meeting on Nuclear Applications and Utilization of Accelerators* (Pocatello, Idaho; July 2007).

[39] T. Miyata, M. Kondoh, T. Minemura, H. Arai, M. Hosono, A. Nakao, Y. Seike, O. Tokunaga and S. Machi, *Int. J. Rad. Appl. Instrum. Part C. Rad. Phys. Chem.* **35**(1–3), 440 (1990).

[40] M. H. Sampa *et al.*, *Nukleonika* **52**(4), 137 (2007).

[41] P. Jhimli, K. P. Rawat, K. S. S. Sarma and S. Sabharwal, *Appl. Rad. Isotop.* **69**, 982 (2011).

[42] L. A. W. Abdou, O. A. Hakeima, M. S. Mahmouda and A. M. El-Naggar, *Chem. Eng. J.* **168**, 752 (2011).

[43] M. Chaychian, M. Al-Sheikhly, J. Silverman and W. L. McLaughlin, *Rad. Phys. Chem.* **53**(2), 145 (1998).

[44] M. R. Cleland, The use of ionizing radiation to protect the environment (TIS-01736, IBA Industrial, Inc., 2007).

[45] B. Y. Chung, J.-S. Kim, M. H. Lee, K. S. Lee, S. A. Hwang and J. Y. Cho, *Rad. Phys. Chem.* **78**, 711 (2009).

[46] M. T. Radoiu, D. I. Martin, I. Calinescu and H. Iovu, *J. Hazard. Mater. B* **106**, 27 (2004).

[47] A. G. Chmielewski and W. Migdał, *Nukleonika* **50**(4), 179 (2005).

[48] IAEA, *Trends in Radiation Sterilization of Health Care Products* (IAEA-TCS-31, Vienna, Austria, 2008).

[49] J. McKeown, M. R. Cleland, C. B. Lawrence and A. Singh, Engineering studies for sludge and sludge disinfection with IMPELA accelerator, in Ref. 6 (p. 537).

[50] S. Capizzi-Banas and J. Schwartzbrod, *Water Res.* **35**(9), 2256 (2001).

[51] J. Kim, B. Han, Y. Kim and N. Ben Yaacov, Economics of electron beam sludge hygenization plant, in *International Meeting on Radiation Processing*, London, UK (2008).

[52] R. A. El-Motaium, *Alleviation of Environmental Pollution Using Nuclear Techniques Recycling of Sewage Water and Sludge in Agriculture: A Case Study*, ICEHM2000 (Cairo University, Egypt, 2000), pp. 323–332.

[53] W. Park, M. H. Hwang, T. H. Kim, M. J. Lee and I. S. Kim, *Rad. Phys. Chem.* **78**, 124 (2009).

[54] S. Hashimoto, K. Nishimura and S. Machi, *Int. J. Rad. Appl. Instrum. Part C. Rad. Phys. Chem.* **31**(1–3), 109 (1988).

[55] L. Chu, J. Wang and B. Wang, *Biochem. Eng. J.* **54**, 34 (2011).

[56] Z. Zimek and A. G. Chmielewski, *Nukleonika* **38**(2), 3 (1993).

Andrzej G. Chmielewski, Ph.D., D.Sc., is Director of the Institute of Nuclear Chemistry and Technology, Warsaw, Poland, and a professor at the Department of Chemical and Process Engineering, Warsaw University of Technology. In 1976–77 he was employed by the University of Tennessee, Knoxville, USA, and in 2003–05 by IAEA, Vienna, Austria. He then lectured at the universities in Sao Paulo, Brazil; Hefei, China; and Pavia, Italy. He is a member of the Academy of Engineering and in 1999 he was nominated a "Gold Engineer of the Year" by the Polish Federation of Engineering Associations. He has authored more than 150 publications and books, and over 60 international patents — some of these inventions were the bases for construction of the big installations for air (accelerator-based) and water pollution control built in industry. Prof. Chmielewski supervised the teams constructing the accelerator plants for health care product sterilization and food irradiation. As a UN IAEA expert he undertook missions to more than 30 countries and served as a DOE expert on the panel elaborating the report "Accelerators for America's Future." Eleven Ph.D. and over 50 M.Sc. theses have been written under his supervision. He is Editor-in-Chief of *Nukleonika*.

Reviews of Accelerator Science and Technology
Vol. 4 (2011) 161–182
© World Scientific Publishing Company
DOI: 10.1142/S1793626811000513

Studying Radiation Damage in Structural Materials by Using Ion Accelerators

Peter Hosemann

Department of Nuclear Engineering,
University of California, Berkeley,
4169 Etcheverry Hall, CA 94720, USA
peterh@berkeley.edu

Radiation damage in structural materials is of major concern and a limiting factor for a wide range of engineering and scientific applications, including nuclear power production, medical applications, or components for scientific radiation sources. The usefulness of these applications is largely limited by the damage a material can sustain in the extreme environments of radiation, temperature, stress, and fatigue, over long periods of time. Although a wide range of materials has been extensively studied in nuclear reactors and neutron spallation sources since the beginning of the nuclear age, ion beam irradiations using particle accelerators are a more cost-effective alternative to study radiation damage in materials in a rather short period of time, allowing researchers to gain fundamental insights into the damage processes and to estimate the property changes due to irradiation. However, the comparison of results gained from ion beam irradiation, large-scale neutron irradiation, and a variety of experimental setups is not straightforward, and several effects have to be taken into account. It is the intention of this article to introduce the reader to the basic phenomena taking place and to point out the differences between classic reactor irradiations and ion irradiations. It will also provide an assessment of how accelerator-based ion beam irradiation is used today to gain insight into the damage in structural materials for large-scale engineering applications.

Keywords: Ion beam irradiation; structural materials; radiation damage; small-scale materials testing; postirradiation examinations (PIE).

1. Motivation

Today's industrial societies make wide use of radiation-producing equipment and facilities, in areas such as energy production, medicine, and science. Since the early days of the nuclear age [1], it has been realized that radiation can produce significant effects on materials. This fact leads to major concerns regarding product output, structure lifetime, and development and reliability of conventional and new, advanced nuclear facilities. Advanced fission reactor concepts, such as fast reactors [2] as well as fusion reactors [3], are limited by the materials that can withstand the combination of radiation dose, temperature, long lifetime, corrosive environments, stresses, etc. on the structure of the facility. The same is true for medical isotope production facilities [4], such as the 100 MeV proton facility at Los Alamos National Laboratory, where the beam window can have a very limited lifetime. High-power neutron sources, such as the SINQ spallation neutron

source in Switzerland which is used for research applications, also experience limitations and frequent target exchanges due to radiation damage in the materials of the target [5, 6].

In order to improve the safety, reliability, and efficiency of nuclear facilities or to develop future applications, it is essential to learn more about the nature of radiation damage in materials and the fundamental processes taking place [7, 8] and to gain engineering data on materials degradation due to irradiation so that structures can be designed appropriately [9, 10].

While, for most engineering applications, structural materials degradation is caused by the combination of time, stress, temperature, and environment, the application of radiation adds the new dimension of "dose" to it, which makes it a scientifically interesting challenge.

Studying radiation damage in materials and its impact on nuclear structures and the environment is a truly multidisciplinary area, involving many fields of basic technical sciences and all fields of engineering sciences, and can even involve the life sciences when the question of failed nuclear components due to radiation damage and the effect on people are discussed and their costs for replacement and construction might involve economics. This multidisciplinary nature requires bringing together specialists from a wide range of topics, in order to gain a comprehensive view of the effects and their impact on all of us.

Because materials science investigates the behavior of almost any type of matter on all length and time scales, it can be stated that materials science is a truly multiscale discipline. Effects taking place on the smallest scales and shortest time frames influence strongly the behavior of large and long-lasting structures of any kind. Therefore, studies investigating the phenomena need to be performed likewise with a multiscale approach, as outlined by B. D. Wirth [11]. However, from an experimental point of view, the information gained from a specific irradiation experiment is only as good as the diagnostics and the subsequent postirradiation characterization, which has to be performed after the irradiation itself to study the effects of the irradiation on a specific material. It is the intention of this article to give the reader an overview of why and how ion beam irradiations are used today to study radiation effects on structural materials, pointing

out the difficulties and challenges, and the successes achieved so far. In the following sections, the basic effects that radiation can produce on the properties of structural materials are discussed. While the article cannot achieve the depth that a focused, original research article or a textbook can, it should give the reader a basic understanding of the effects taking place and introduce him or her to the field of ion beam irradiation to study structural materials for nuclear application.

2. Radiation Damage in Structural Materials

At low energy (typically below 1 MeV for protons), ion irradiation does not transmute elements or produce He in most materials. With low-energy ion irradiation, therefore, the induced radiation damage is based primarily on the displacement damage caused by the incoming particle. Modern understanding of radiation damage is based on the idea that the incident radiation particle creates a primary knock-on atom (PKA) with energy T, which then causes a displacement cascade by knocking further atoms from their original lattice sites. Brinkman depicted this concept with a cartoon for the first time, suggesting a high vacancy concentration surrounded by interstitials [12]. Later, Seeger created a more realistic image that considers effects of the crystallinity of materials [13] (see Fig. 1).

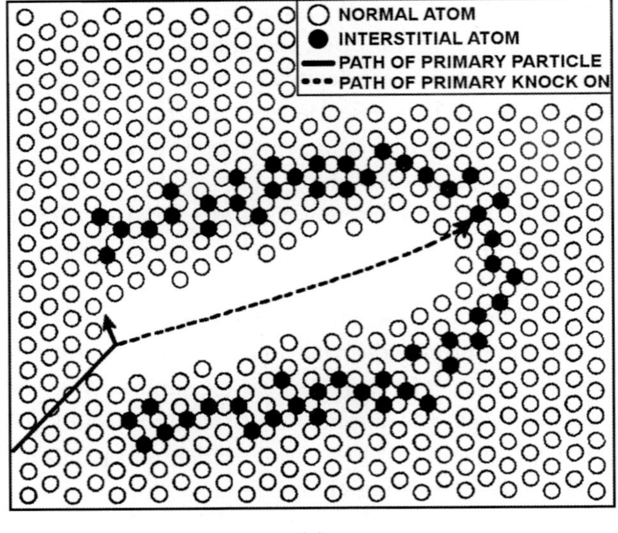

(a)

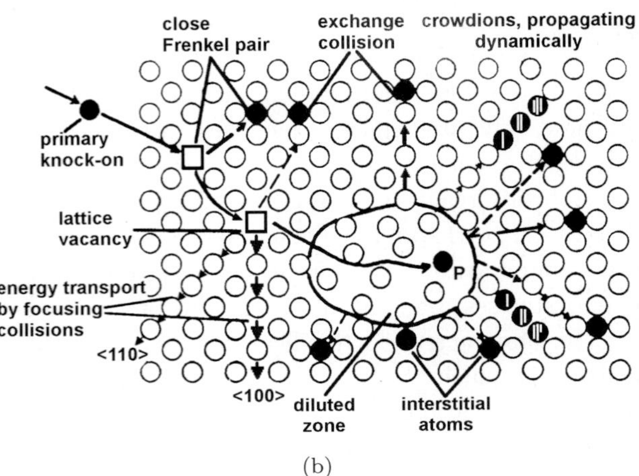

(b)

Fig. 1.   (a) First schematic sketch of a knockon cascade (after Brinkman [12]), and (b) the advanced model considering crystallographic effects (after Seeger [13]).

The Kinchin–Pease (KP) model [14, 15] describes the number of displaced atoms in the cascade as a function of the original PKA energy, T. The number of displacements in a collision cascade, $\nu(T)$, depends on T. If $T < E_d$ (E_d is the threshold displacement energy necessary to displace an atom away from its stable lattice site), no displacement will take place. If $E_d < T < 2E_d$, the number of displacements is 1, since only enough energy is available to displace one atom before T becomes less than E_d. For $2E_d < T < E_c$, $\nu(T) = T/2E_d$. E_c is the electronic stopping energy, above which only electronic stopping of the ions occurs. Energies of $T > E_c$ cause a fixed number of displacements equal to $E_c/2E_d$, since the ions slow down via electronic stopping until $T < E_c$. However, it should be pointed out that the KP model significantly overestimates the amount of displacements by a factor of 2–10 [16–18]. A more comprehensive and accurate view is given by Lindhard, Scharff, and Schioett [19], commonly called the LSS theory; this view has also been discussed by others [20–27]. In the LSS model, a clear deviation to the KP model is found.

Today J. F. Ziegler's computer code, the Stopping and Range of Ions in Matter (SRIM) [28], implements stochastic methods and is widely used by scientists studying radiation damage in materials by ion accelerators. It allows quick calculations of the range of ions, the number of vacancies produced, the energy lost, and several other parameters that are needed to plan and conduct an ion beam experiment. One of the reasons for the widespread use is its user-friendliness and simple application, which makes it a tool employed by a wide range of students and scientists.

However, none of the models above acknowledge crystal structure effects, which can take place in a real material: E_d depends on the crystallographic structure, as demonstrated by Fig. 2 [29].

Molecular dynamics (MD) and kinetic Monte Carlo (KMC) methods allow more accurate calculations of the displaced atoms but require intense computing time.

One measure of quantifying dose is through calculating "displacements per atom" (dpa) describing the total number of initially displaced atoms due to the incoming particles. The unit does not, however, describe the immediate effect on specific materials, since it does not acknowledge the number of atoms

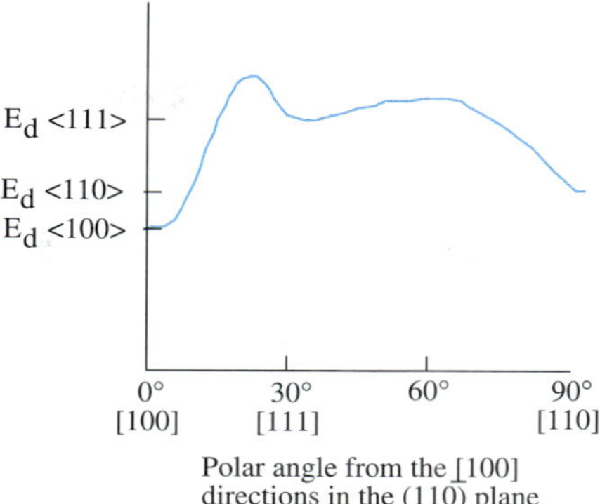

Fig. 2.   Displacement energy as a function of crystallographic direction in Cu [29].

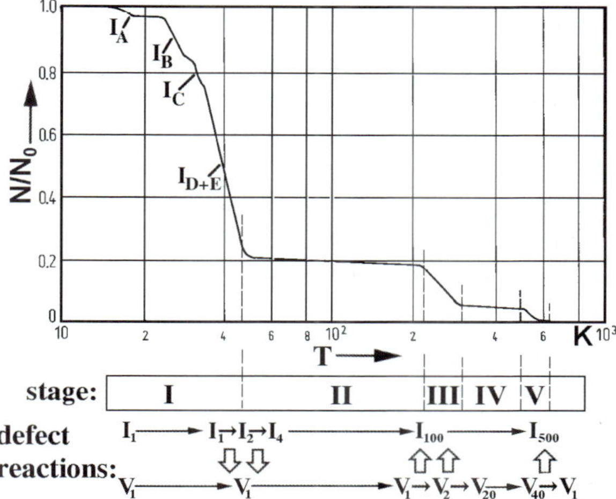

Fig. 3.   Five recovery stages in electron-irradiated Cu (after Ref. 32). I and V denote single interstitials and vacancies, respectively; and I_2 and V_2, di-interstitials and di-vacancies. Stage I: The lower temperature substages I_A, I_B, and I_C are due to collapse of close Frenkel pairs. I_D is due to correlated recombination and I_E due to uncorrelated recombination. Stage II: Interstitial clusters grow, leading to identifiable small interstitial loops. Stage III: Vacancies migrate and adulate at interstitial clusters. Stage IV: The vacancy clusters surviving stage III grow in size. The vacancy clusters dissociate thermally.

recombining with the vacancies. In the literature, oftentimes different stages of defect production in a cascade are discussed: (1) the initial displacing collisions, (2) the formation of a thermal spike, (3) the quenching of the thermal spike, and (4) the annealing within the cascade [30].

In the past electron radiation was used to evaluate the recovery stages of radiation on a material, as shown in Fig. 3. There, five stages of recovery can be distinguished [31, 32].

However, it is the fate of the defects remaining at longer time frames that determines the changes of engineering materials properties due to irradiation. Therefore, extra parameters such as time, temperature, stress, and microstructure are needed in addition to the dose to evaluate the effect and impact of radiation on materials.

Studying radiation damage in materials for reactor applications using ion accelerators is a difficult and controversial task. Several artifacts, as discussed in later sections, have to be taken into account. One of the first issues in planning and conducting an ion beam irradiation experiment for structural materials is to select the proper ion type and ion energy for the anticipated postirradiation test. Oftentimes it is a competition between maximizing the penetration depth of the ion, by keeping a reasonable damage rate, and at the same time avoiding going to energies that are so high that the sample material becomes activated.

Figure 4 shows a comparison of different radiation particles, which have different cascades and penetration depth in materials [31–36].

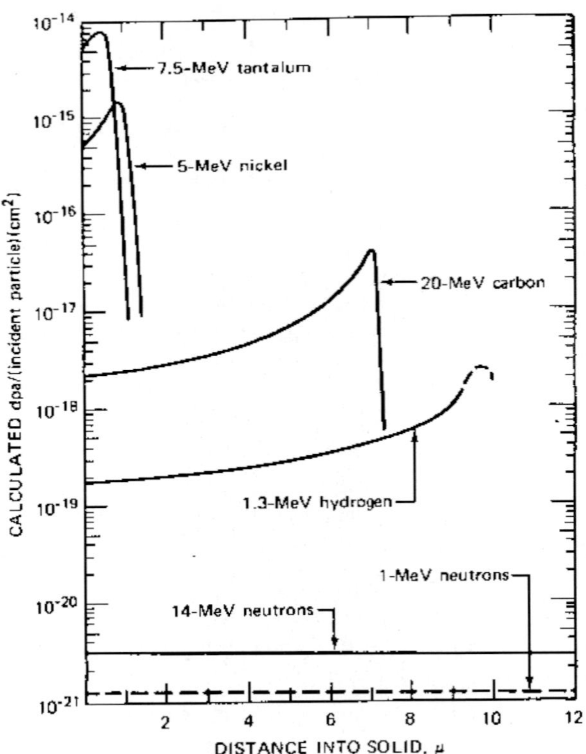

Fig. 4. Displacement damage (dpa) as a function of beam penetration and particle type (note the log scale) [31, 36].

Because of the significantly different cascade shapes that the various particles produce, the dose rate depends on the type of particles used. This variance in the dose rate is particularly obvious in the comparison between ion irradiation and neutron irradiation, but similar phenomena are observed among reactors of different neutron spectra, as has been widely reported [37, 38]. Mansur developed a model that can account for the different dose rates by using an adjusted irradiation temperature [39, 40]. This approach will be described in more detail in a later section.

The following subsections, though, will describe, in general, the different effects that radiation can produce on structural materials due to displacement cascades.

2.1. *Radiation-induced hardening/embrittlement*

As described above, many of the atoms that are initially displaced eventually recombine, and the material is left with a certain amount of point defects. These excess point defects can combine to form a large amount of dislocations loops or voids, can enhance diffusion via the large concentration of vacancies, and can allow intermetallic precipitates to form that would not otherwise form without radiation, due to too-slow kinetics.

Good examples of this last phenomenon are Cu precipitates in reactor pressure vessel steels [41–43] and alpha prime formation in HT-9 (a ferritic–martensitic steel with 12 wt% Cr) [44, 45]. These intermetallic precipitates can contribute to hardening and, more importantly, to embrittlement of the materials, which in turn can cause sudden failure of a component. In general, it is the sudden failure of a component which is especially worrisome, since it does not obey the "leaking before breaking" concept. Figure 5 shows an image of a stainless steel pipe which broke in a hot cell by simple clamping in a vice and a stress–strain curve illustrating the change in the mechanical behavior of a material exposed to a spallation source. The behavior shown in Fig. 5 clearly does not obey the "leaking before breaking concept" [46, 10].

The increased number of dislocation loops also significantly contributes to the radiation-induced hardening, by hindering the movement of dislocations; this movement is responsible for the

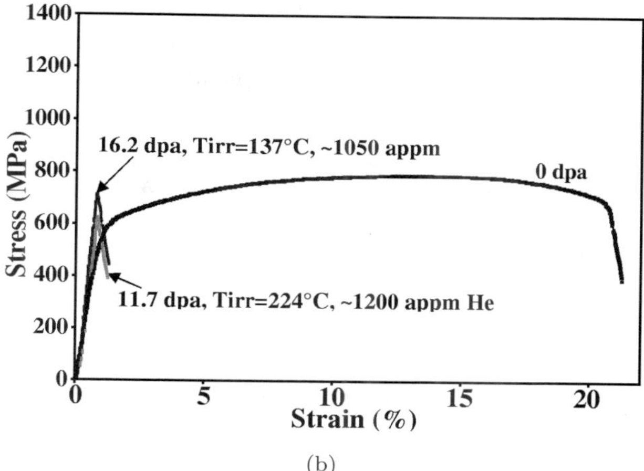

(a) (b)

Fig. 5. (a) A 316L tube that fractured during handling in a hot cell [46]. (b) Stress–strain tensile test results from samples irradiated in SINQ Target Irradiation Program (STIP) [10].

ductility of a material. In FCC materials, stacking fault defects also contribute to an increase in hardness.

While hindering the dislocation movement does increase the yield strength (YS), it is the development of defect-free channels that leads to localized sudden failure without warning. Defect-free channels are created when moving dislocations clear all the defects from a path through the material, as discussed in Refs. 47 and 48. The phenomenon of radiation-induced hardening and embrittlement has been studied on reactor-irradiated and spallation-irradiated materials for a while and has more recently also been studied on ion beam-irradiated materials since the advent of small-scale mechanical testing [49–52].

2.2. *Radiation-induced segregation*

As described above, radiation causes the formation of an excessive amount of vacancies and interstitials, leading to fast diffusion. This enhanced diffusion can cause local segregation of elements around grain boundaries. Past research has shown that radiation-induced segregation (RIS) often occurs on grain boundaries, with the enrichment of a specific element

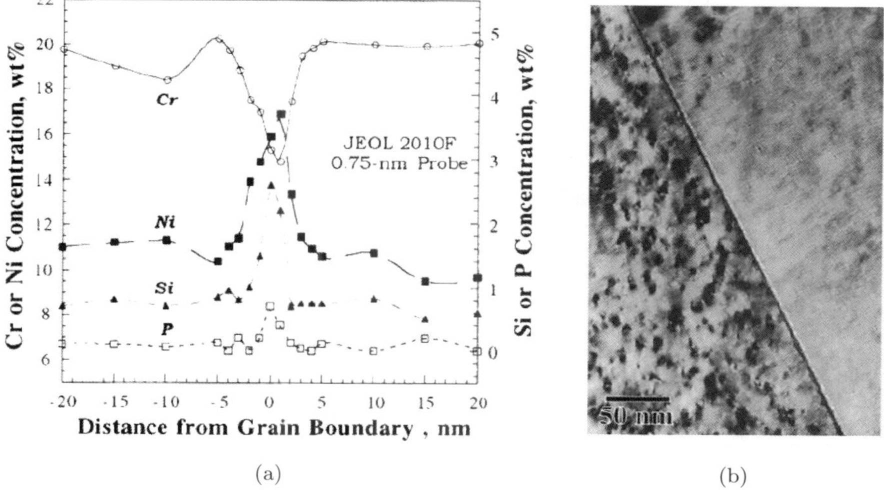

(a) (b)

Fig. 6. (a) Segregation at a grain boundary in neutron-irradiated, 300-series stainless steel, as measured quantitatively with TEM. (b) TEM picture of the grain boundary (after Ref. 52).

on the grain boundary and depletion around it, and vice versa for the other elements. Thus, a line scan of the elemental distribution across the grain boundary shows a Cr-distribution in the typical "W" shape. Figure 6 shows this typical element concentration on a grain boundary in a 304 stainless steel [52].

Local electrode atom probe (LEAP) measurements, as presented in Ref. 53, show similar results at the grain boundaries of ion-beam-irradiated and cold-worked 316L stainless steel. Recently, Stergar has reported segregation also at dislocations, in highly neutron-irradiated ferritic–martensitic (F/M) steels [54].

2.3. *Radiation-induced swelling*

Radiation-induced swelling [55–58] is a widely discussed phenomenon, with direct implications for the application of a specific component. It is caused by the agglomeration of the excess amount of vacancies and interstitials, which increases the volume of a material while decreasing its density. At certain doses and temperatures, the vacancies recombine into nanometer-sized voids, as has been observed in TEM images. Stress can also have an effect [59]. Austenitic stainless steels have been found to be more susceptible to swelling than F/M stainless steels. However, in recent times it has been realized that even BCC materials can swell significantly, even though they have a larger transient regime. (See, for example, the work of Garner [60].) Thus, great attention needs to be paid to high-dose irradiation, which is difficult to access experimentally. Swelling is also studied in ion accelerators, but the significantly higher dose rate from ion accelerators has to be taken into account. It is known that a higher dose rate causes a shift of the onset of swelling to a higher

dose, which can lead to a dangerous underestimation of swelling rates when one is analyzing data from accelerators. For example, Fig. 7 shows a shift to a higher dose on neutron-irradiated austenitic materials due to higher dose rates [61].

2.4. *Effects of helium*

Neutron irradiation and high-energy proton irradiation can cause the production of helium (He) in materials; a good example is the He produced in potential fusion applications. In fact, it is suggested that He is responsible for void formation by stabilizing a void embryo and thereby assisting the nucleation process [62]. Depending on the nuclear cross section of a specific nuclear reaction, a large amount of He can be produced especially in spallation neutron sources or fast reactors due to n, alpha reactions. Helium has a low solubility in metals [63] and therefore can lead to the formation of bubbles, the size of which depends on the temperature to which the material is exposed. However, He also segregates to interfaces. For example, He agglomerates at grain boundaries, leaving He-denuded zones around grain boundaries [64]. Other interfaces, such as W–Fe multilayers [65] or Cu–Nb, also attract He.

In general, He bubble growth is governed by the dose, He content, microstructure, and especially temperature, which allows the He atoms to diffuse far enough to form the bubbles within a grain or at the grain boundary. It is believed that crack propagation through the array of voids located at grain boundaries leads to intergranular fracture and a brittle failure of the material.

2.5. *Radiation and environment interactions*

2.5.1. *Stress*

While irradiation by itself causes several physical changes to materials (e.g. changes in density, changes in strength, and swelling, in combination with stress), it can also lead to phenomena such as radiation-enhanced creep. In general, creep is governed by dislocation glide, dislocation climb, and diffusion; the last two mechanisms are greatly influenced by the diffusion of vacancies [66]. Radiation changes the amount of defects present, and their motion leads to macroscopic deformation by diffusion, dislocation climb, and slide. Two mechanisms

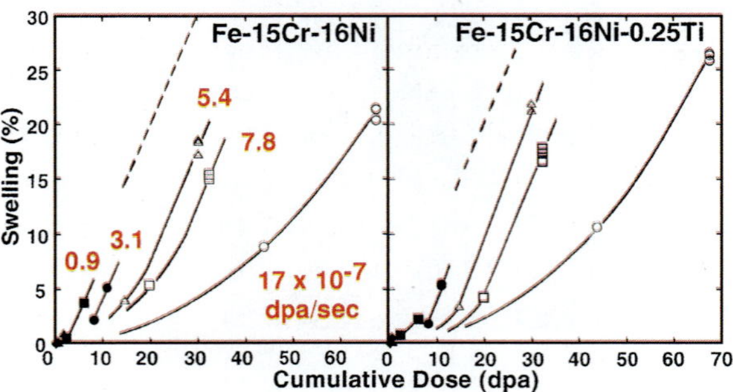

Fig. 7. Effect of the dose rate on the swelling of an austenitic stainless steel [61].

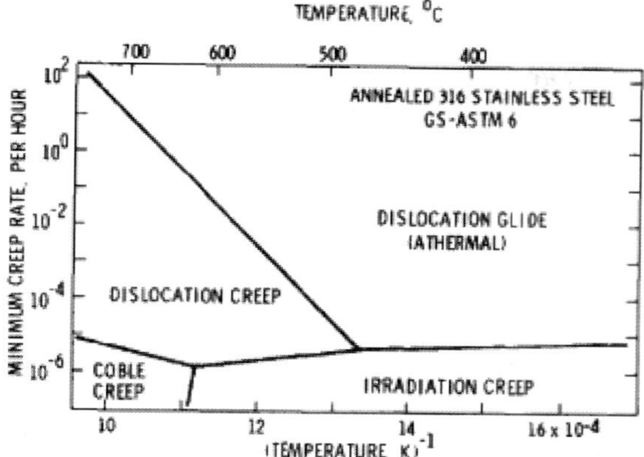

Fig. 8. Creep map (after Ref. 71). The different creep regimes are shown in a temperature vs. creep rate plot. It can be seen that irradiation-induced creep exists over a wide temperature range, starting at rather low temperatures. At high temperatures thermally activated creep is dominant.

of irradiation-induced creep are discussed in the literature: stress-induced, preferred absorption (SIPA) creep and swelling-driven creep [29, 67, 68].

However, all theories are governed by the idea of enhancing the ability of dislocations to overcome glide obstacles by climbing, due to increased point defect action. It is known that the effect of radiation is dominant at temperatures where thermal creep usually does not occur. At higher temperatures where thermal creep is expected, the effect of radiation can be neglected, since the vacancies produced by the thermal activation process become dominant. Figure 8 presents a creep map (from Ref. 71) that shows the temperature regime in which irradiation creep becomes an issue. In Ref. 72 it is clearly stated that the dose rate has an effect on the creep rates. It was found that as the dose rate increases the creep rate decreases. Therefore, the data from accelerator ion beam studies (which have high dose rates) need to be used carefully when they are compared to data from neutron studies (which have low dose rates). Historically, thermal creep and radiation-enhanced creep have been studied using pressurized tubes. Accelerators can allow the use of tensile-type samples. Currently, several groups are working on introducing *in situ* creep stages in low-energy accelerators [6].

2.5.2. *Corrosion*

In every engineering application, a specific material is surrounded by a specific medium, which can cause environmental degradation of the material. Therefore, the effects of radiation on environmental degradation (corrosion) have to be taken into account, not only for light-water reactors but also for other media in other applications [73, 74]. Several models have been introduced for modeling corrosion in aqueous systems; they take into account the radiolysis of water [75]. But little is known about the effects of radiation on materials in other environments, such as liquid metals or molten salts, which are also considered as reactor coolants for advanced reactor concepts. Especially for reactors cooled by heavy liquid metal, corrosion is the limiting factor and cannot be neglected. Since most metal elements are soluble in lead-based alloys, a passivation layer is the key to developing a corrosion-resistant system [76]. Although such systems have been proposed for decades, little is known about the effects of radiation on the passive film growth and its stability. The Swiss project MEGAPIE has looked at these issues for T91 steel at spallation source operation temperatures [77], but very little work has been done at advanced-reactor temperatures (greater than 450°C) or on other materials, such as HT-9 or 316L.

As discussed in the later sections of this article, *in situ* ion beam irradiation experiments are currently being developed and conducted in order to address these issues on a larger variety of materials, temperatures, and media.

3. How Radiation Damage Is Studied Today

Radiation damage in structural materials can be studied experimentally in several different ways, each method having its advantages and disadvantages.

Ultimately, a specific material has to be tested in the actual environment in which it is going to be used. However, oftentimes a prototypic environment for materials testing does not exist, since the material for the use in a prototype application is under development, leading to a catch-22 situation. Therefore, other approaches are taken, such as irradiating in thermal reactors instead of fast reactors, neglecting the He effects while spallation sources overestimate the He effects compared to reactors, and irradiating with ions instead of neutrons to cause radiation damage.

Table 1. Different options of how radiation damage can be caused in a structural material and their benefits or disadvantages.

Issue	Reactor	Spallation source	Ion beams
Comparable to thermal neutron irradiation	Good	Limited	Difficult
Comparable to fast neutron irradiation	Limited	Limited	Difficult
Temperature control	Limited	Difficult	Good
Temperature monitoring	Good	Good	Good
In situ measurements	Difficult	Difficult	Good
Bulk samples	Good	Good	Limited
Cost-effectiveness	Poor	Poor	Good
Sample turnaround	Poor	Poor	Good
Safety; hot samples	Difficult	Difficult	Good
Flexible irradiation conditions	Limited	Limited	Good

3.1. *Sources of radiation damage*

Currently, three principal sources can be used to irradiate materials: nuclear reactors, spallation neutron sources, and ion beams (accelerators). These are discussed below. The main points of studying structural materials with various sources of radiation are briefly summarized in Table 1. It has to be noted that this article does not cover all the effects taking place but rather gives an introduction to the issues to consider when looking into irradiation sources to study structural materials.

3.1.1. *Reactor irradiation campaigns*

Since the beginning of the nuclear age, materials have been studied by irradiating them in reactors. Actual components are investigated (cladding tubes, ducts, or fuels) by preparing samples after irradiation which have been exposed to the reactor operation history and condition, as shown in Ref. 44.

Specific locations in reactors are made available for preshaped specimens, allowing proper postirradiation examination (PIE). However, in this type of irradiation, one is usually tied to the reactor's neutron spectrum and the reactor temperatures. Recent sample designs as introduced by G. R. Odette allow one to investigate specific He/dpa ratios by adding a NiAl layer to the sample surface, leading to the

production of He, and further He is injected in the near-surface areas. The temperature is controlled by controlling the thermal transport from the sample to the coolant by appropriate sample design, allowing for irradiation temperatures other than the reactor temperature. However, one is always tied to the spectrum in a specific reactor. One of the highest possible dose that can be achieved today is ∼20 dpa per year, in BOR-60, a Russian experimental fast reactor. *In situ* studies are difficult to perform due to limited access, with the exception of pressurized tubes for creep studies. In addition, materials irradiation in reactors usually requires long-term planning and the capability of handling radioactive samples in a safe manner. A good overview of the capabilities of different reactors for materials irradiation is given in Table 2 [78, 79].

3.1.2. *Spallation source irradiations*

The target design from the SINQ (Schweizer Spallations Neutronen Quelle) facility allows materials irradiation in the target itself within STIP (SINQ Target Irradiation Program) [80, 81]. In such irradiations, the temperatures can be monitored but are tied to the proton beam power and also to its fluctuations. The spectrum is usually a mixed proton and neutron spectrum, leading to high He/dpa ratios that are not typical for light-water reactors (LWRs) but are very interesting for the fusion and fast reactor communities to get a good understanding of the influence of He and dose on the materials properties.

As in the reactor case in this setting, *in situ* experiments are difficult to conduct. Also, the samples are going to be highly radioactive, making it necessary to handle them in a controlled or hot laboratory, requiring long-term planning, and leading to high costs.

3.1.3. *Ion beam irradiations*

Ion beam irradiations can cause radiation damage in materials in a time- and cost-efficient way in order to gain basic knowledge about the interaction of materials and radiation. Of course, relating the data gained by ion beam irradiations to those gained by reactor or spallation source irradiations is not a straightforward matter, due to the significant differences among the sources. Also, no

Table 2. Neutron sources available for materials irradiation.

Facility	Country	Fast flux $E_a > 1\,\mathrm{MeV}$ $(10^{18}/\mathrm{m}^2\mathrm{s})$	Displace m. damage in steel (dpa/yr)	Useful vol. (cm^3)	Temp. range $(^\circ\mathrm{C})$	Comments
BR2						
Core	Belgium	1.5–3.0	<3/yr	90	50–1000	• ~ 105 days/yr
reflector		0.05–1.0	<1/yr	250	50–1000	• Caps Ø: 50–200 mm
						• *In situ* fatigue rigs
OSIRIS	France	2.5	few/year	230	50–1000	
HFR						
Core	Netherlands	2.5	<7/yr	1540	50–1100	• 275 days/yr
						• *In situ* experiments
ATR						
A and H,	USA	2.3		240	50–>150 $^\circ$C	• Caps Ø: <127 mm
B, I-positions		0.8	6–10/yr	1390		• Large irrad. volume
flux traps						
		0.03	6–8/yr	5560		• Versatile facility
		2.2		5560		
HFIR						
Tgt. pos. 37	USA	11	18/yr	100	300–1500	• Very high peak flux
RB pos. 8		5.3	5–7/yr	720		• Accelerated testing in smaller volumes
JOYO	Japan	5.7	~ 30/yr		300–700	• Temp. control $+4\,\mathrm{K}$
BN600	Russia	6.5	20–52	350	375–750	• Very high dose rate
						• Only passive instrumentation
BOR-60	Russia	3.0	~ 20	358	300–700+	• Only passive instrumentation
						• High level PIE

fission products are created during low-energy ion beam irradiations, making *in situ* dual or triple ion beam irradiations necessary in order to produce an environment that is more similar to the one experienced by nuclear fuels in fast reactors. However, in recent years, due to the lack of reactor irradiations and funding restrictions, ion beam irradiation has become a popular tool for academic research.

If one intends that the samples should not contain any remaining radioactivity after ion irradiation, then the beam penetration is very limited, which is one of the main restrictions as compared to the other options (see Fig. 4). This restriction makes it necessary to develop new PIE techniques or improve existing ones. It has to be noted that ion beam irradiation and *in situ* or postirradiation examination are equally important for conducting a successful experiment, and both need to be carried out with great care, keeping in mind the specific question which is going to be answered. It is important to acknowledge that, if the high dose and fast sample turnaround advantages of ion beam irradiation are to be utilized, then one has to consider the orders-of-magnitude higher dose rates from using ion beams, which can potentially lead to delayed effects such as swelling (see Fig. 7).

3.2. *The benefits and issues in the use of ion beam irradiation*

It is clear that ion beam irradiation can provide additional basic information on the processes of materials under radiation. The fast and efficient sample turnaround, high-dose regimes, low material activation, well-defined temperature and dose, ion content, and simple accessibility that allow *in situ* experiments are obvious benefits of this type of radiation damage study. However, because of the limited penetration depth, significantly higher dose rate, and different spectra, the data have to be carefully examined when compared to reactor data. Table 3 provides an overview of the advantages and disadvantages of various particle types that can be selected for an experiment [82].

Table 3. Advantages and disadvantages of different particle types (after Ref. 82).

Advantages	Disadvantages
Electrons	
Relatively "simple" source — TEM	Energy limited to $\sim 1\,\text{MeV}$
Uses standard TEM sample	No cascades
High dose rate — short irradiation time	Very high beam current (high dpa rate) requires high temperature
	Poor control of sample temperature
	Strong "Gaussian" shape (nonuniform intensity profile) to beam
	No transmutation
Heavy Ions	
High dose rate — short irradiation times	Very limited depth of penetration
High T_{avg}	Strongly peaked damage profile
Cascade production	Very high beam current (high dpa rate) requires high temperature
	No transmutation
	Potential for composition changes at high dose via implanted ion
Protons	
Accelerated dose rate — moderate irradiation times	Minor sample activation
Modest ΔT required	Smaller, widely separated cascades
Good depth of penetration	No transmutations
Flat damage profile over 10 s of μm	

4. Ion Beam Experiments

4.1. *Selecting the right accelerator*

From the perspectives of the materials and the user, the way the ions are produced is secondary, as long as the proper beam condition on the target is achieved in a simple, effective, and reliable way. Greater focus needs to be given to the beam end station, where the samples are located. Most of the time, the needed ion species and energy preselect the accelerator. However, various accelerator types have different advantages and disadvantages for a particular experiment. Maybe the most flexible and widely used accelerators today are electrostatic linear accelerators, since they can be easily built in a laboratory setting and also allow selection of a wide range of ions with a wide range of energies. A lot of experience has been gained from using these machines for ion beam irradiations. As scientists usually want to get samples quickly, and as large irradiation areas are desirable for irradiating more samples simultaneously, high beam currents are desired. But high maintenance and source degradation make older accelerators difficult to use for long-term irradiation campaigns. Maybe modern cyclotrons can address this issue, but they can be rather limited in flexibility. In this article, no recommendation for one accelerator versus another can be given, because the choice truly depends on the specific beam condition and the availability of the resources for which one is looking.

4.2. *Conducting ion beam irradiations on structural materials*

In order to conduct a successful ion beam irradiation experiment, all parameters need to be chosen and monitored carefully. Dose, dose rate, temperature, beam profile, spot size, and raster condition (if rastered) are equally important and are discussed below in brief paragraphs. It is realized that this short review article cannot address each point to the fullest extent needed but can encourage consideration when planning an experiment.

Dose. It is essential to know the actual dose on a given sample at all times, which makes frequent beam sampling and/or charge collection on the target necessary. Because the beam is, in general, not homogeneous but has a Gaussian-like profile, it is necessary to condition it. Beam rastering clearly achieves a homogeneous dose profile and is widely conducted but can increase significantly the irradiation time if the raster area is large and the beam current is low. However, because not every laboratory has a beam rastering option or needs to irradiate a large area for a large sample irradiation, spreading the beam and cutting off the ends of the Gaussian-shaped beam spot are possibilities. The beam current on the samples is usually monitored continuously, but obtaining an accurate charge measurement through the beam current integration directly from the target requires a proper bias to the target and careful secondary electron suppressions. Alternatively, Faraday cups have to be inserted in the beam path periodically during the irradiation to measure the total beam current. Ideally, an accurate reading of beam current as a function of position in the beam tube is desired, but such a reliable reading is often difficult to gain.

Beam locating and conditioning (shaping) is often conducted by optical observations using scintillators [83]; thus, cameras that are suitable for use in vacuum are desirable. It is essential to keep the beam properly aligned and focused during an experiment, and oftentimes, in older machines, the beam has to be reconditioned frequently.

Temperature. Controlling and measuring the irradiation temperature accurately is essential to achieving reproducible and meaningful results. Some groups [84, 85] use indium in between the sample and the sample stage to obtain good thermal contact, and great effort is put into developing heating/cooling stages for ion beam irradiation experiments. Depending on the chosen irradiation condition, the stage may need to be cooled or heated. However, the energy deposited on the sample due to the beam hitting the sample also needs to be taken into account. Usually several watts per square centimeter (W/cm^2) can be deposited on the sample just by the beam. It has to be kept in mind that most metals have phase transformations at certain temperatures — a fact that is used to heat-treat the metals. Therefore, the irradiation temperature has to be kept well below this phase transformation temperature in order not to anneal the damage or change the heat treatment state unintentionally. In recent times, thermal transport measurements in various sample-mounting conditions have been conducted to ensure accurate irradiation conditions [86]. Accurate temperature monitoring is required in order to adjust the heating/cooling power of the sample stage and to be able to report the temperature in publications. Spot welding of thermocouples on the sample is done, as well as IR camera temperature monitoring. The IR monitoring also allows one to visualize any potential inhomogeneities in the beam heating and adjust the irradiation parameters to take it into account. However, IR monitoring can make the setup rather difficult, due to complex sample chamber designs, and costly, due to the high price of IR cameras. The IR cameras also need to be cross-calibrated with the thermocouple, because the emissivity of the sample can change because of surface effects on the sample during irradiation.

If samples are preshaped (by micromachining), the thermal transport from the preshaped geometry to the sample holder or stage can change, which might result in local overheating. The same is true for the cases of free-standing samples and thin foils, in which all the beam energy is deposited in the specimen, leading to potential hotspots.

Utilizing today's data acquisition allows continuous monitoring of temperatures and doses, leading to more accurate data and better samples. Nevertheless, computerized systems for data acquisition and control are not substitutes for adjustments of the beam parameters by knowledgeable operators during the irradiation.

4.3. *Performing postirradiation examination*

While a wide range of PIE can be conducted using various techniques, PIE can be categorized into two areas: (1) microstructural and chemical changes; (2) mechanical property changes.

Microstructural and chemical changes. Various techniques, including ion beam analysis techniques such as proton-induced x-ray emission (PIXE), ion channeling, and Rutherford backscattering spectrometry (RBS), can be used to measure changes in composition on or in a material's surface, and also damage profile if the target is a single crystal [87–89]. For most structural materials, like steels, changes occurring within a given microstructure are of great interest, and measurements need to be performed locally. FIB-based sample preparation techniques allow the investigation of specifically chosen locations, such as grain or phase boundaries, or of specific sample depths, such as for determining the effect of a changing dose throughout the sample depth. Transmission electron microscopy (TEM) and local electrode atom probe (LEAP) tomography techniques are heavily utilized today to detect microstructural or chemical changes due to ion beam irradiation. TEM samples can be prepared using either FIB-based techniques or conventional backside dimpling techniques; the conventional preparation avoids changing the sample surface due to damage from the gallium (Ga) ions of the FIB. While TEM

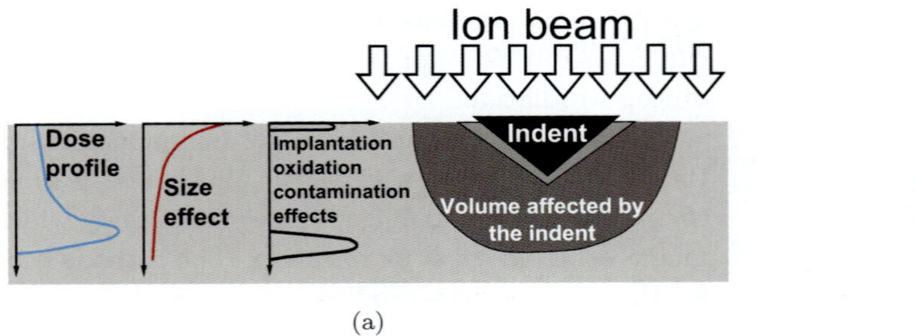
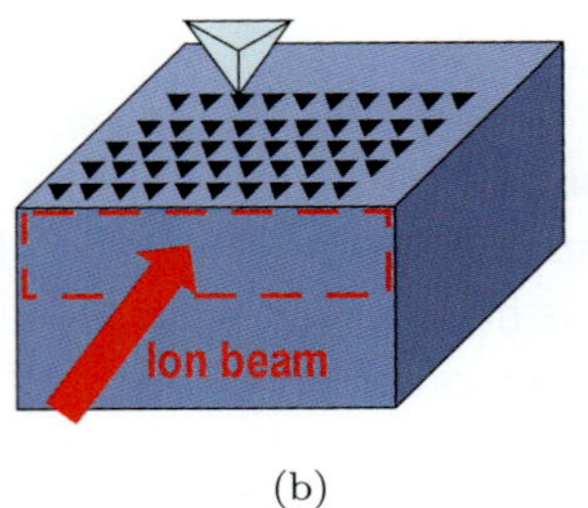

(a)   (b)

Fig. 9.   (a) Summary of the effects to consider when performing postirradiation surface indentation. (b) Schematic of cross-section nanoindentation on ion-beam-irradiated materials [49].

is often performed without FIB-based sample preparation techniques using conventional backside dimpling in order to avoid sample surface changes due to the Ga beam damage from the FIB, it is rather difficult to manufacture LEAP samples without an FIB instrument on an ion-beam-irradiated sample unless the samples are pre-etched [90]. Changing the sample and the radiation damage structure by preparing the sample is always a valid concern, especially for TEM investigation. The main question is whether the structure that one observes has been caused by the original ion beam damage or by the postirradiation sample preparation using the FIB. Today's new FIB instruments and plasma cleaners, however, have the option to use a very low beam energy (2000 eV) to minimize the risk of FIB damage. Kiener *et al.* and several others have studied the influence of Ga damage on FIB-produced samples [91]. Also, flash electropolishing techniques can be used to clean the potential FIB damage from the surface of an FIB-produced sample [92]. However, such a technique is rather difficult to master, due to the difficulty of finding the proper etching parameters for a given material. On some complicated structural materials, it can become so difficult as to be impractical.

Mechanical property changes. Since one of the main purposes of structural materials in nuclear applications is to hold certain stresses and to ensure the structural integrity of a component or machine, scientists are always interested in the changes of mechanical properties due to irradiation. Since the beginning of ion beam irradiation studies, hardness testing has always been an essential tool for evaluating changes in mechanical properties. Several studies have been conducted utilizing microhardness testing on ion-beam-irradiated materials

[93–95]. While it is essential to know what the change in hardness is, the ultimate goal is to determine the full tensile curves and the fracture toughness, which hardness data alone do not provide. The main limiting factor is the shallow irradiation depth in typical ion-beam-irradiated samples. Due to the development of instrumented small-scale hardness measurements (also called nanoindentation), the small region of irradiated material has become accessible for testing. In recent times, nanoindentation has become a popular tool for investigating mechanical property changes of ion-beam-irradiated materials [49, 83, 96, 97].

While surface indentation (i.e. indentation in the same direction as the incident ion beam) has been applied in the past [65, 98], cross-section indentation became available with more accurate indenter positioning. If surface indentation is performed, great care has to be taken in how the measurement is performed to avoid an overlap of different effects. These effects are displayed in Fig. 9, which shows the overlaying of a varying dose profile, the indentation size effect, surface sputtering or deposition, and end-of-ion-range implantation [99]. Therefore, one has to be careful about what the indentation experiment is actually sampling, considering the size of the plastic zone around the indenter.

Cross-section indentation (i.e. indentation that is 90° to the irradiation direction) avoids several of these effects. It also might lead to a way of estimating a yield strength (YS) value from the hardness value, by using a factor of approximately 3 [100].

However, when one is performing cross-section indentation, polishing artifacts have to be considered, and sample preparation can be an issue. Figure 10 presents a TEM image of indents performed in cross-section on an ion-beam-irradiated

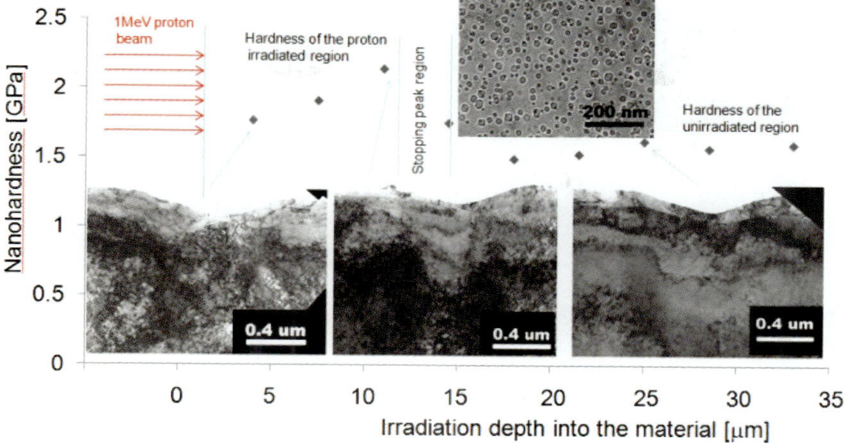

Fig. 10.   Hardness measurements on ion-beam-irradiated Cu and TEM image of the indents. The inset shows the voids in the ion beam stopping area.

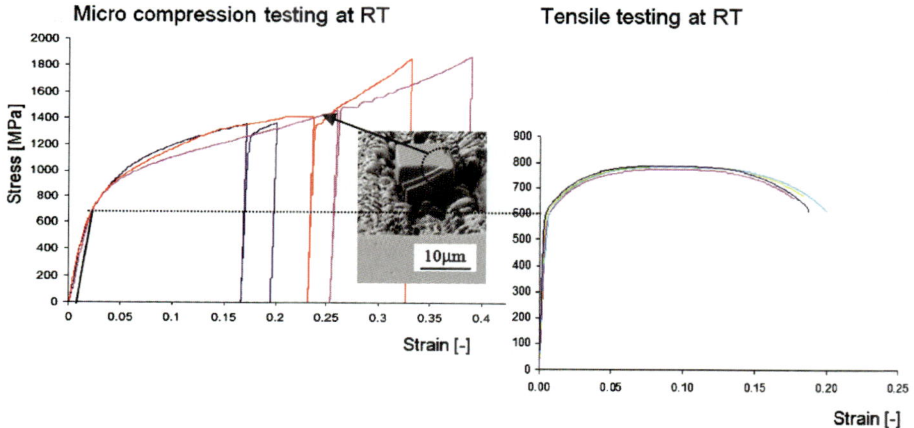

Fig. 11.   Microcompression testing on an F/M steel (HT-9) in direct comparison with a regular tensile test (after Ref. 51).

material. The hardness change due to ion beam irradiation can be seen clearly in the data plot, while the TEM pictures reveal the defects causing the hardness. However, in most cases, one is interested more in the change of a property due to irradiation than in its absolute, numerical value. Spherical indentations can also be used to evaluate YS. A more direct measurement of YS after irradiation can be achieved with microscale mechanical testing, such as microcompression testing or microtensile testing, as shown in Fig. 11 [51].

More recently, tensile testing on small samples has become available, and it is currently being used to investigate ion-beam-irradiated materials. While small-scale mechanical test devices are built for compression, it is rather difficult to modify them for tensile testing. Two approaches being developed

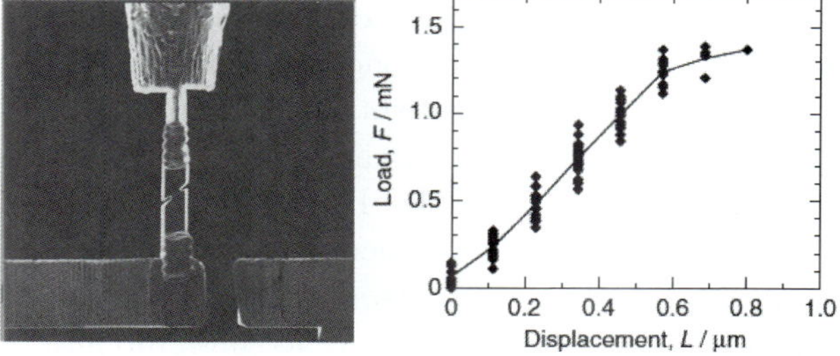

Fig. 12.   Microtensile testing and the resulting tensile curves (after Ref. 103).

are push-to-pull devices and direct tensile testing (Fig. 12) [101–103].

However, when performing these types of measurements, it is essential to keep in mind what is the property that one is after. Absolute engineering YS numbers obtained from using these techniques

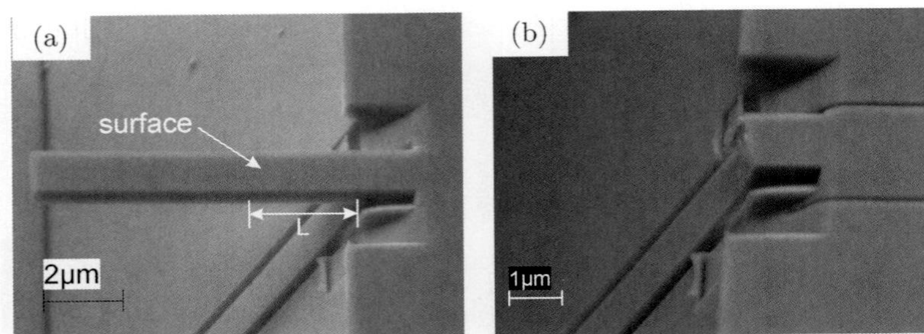

Fig. 13. Manufactured FIB bend bars for testing fracture toughness on interfaces: (a) before testing; (b) after testing [105].

are oftentimes questionable, since one needs to keep size effect and "strength-determining" length scales in mind.

It has been found that nanostructured materials have little to no size effect in compression testing as long as the strength-determining features are significantly smaller than the sample size.

Therefore, ODS alloys [50] are prime candidates for testing with small-scale techniques, while large-grained materials are more difficult. Fracture measurements are significantly more difficult to access on these scales. However, in recent times, scientists have attempted to perform also this measurement with small-scale materials testing on rather brittle materials or interfaces, as shown in Fig. 13 [104, 105]. In this work performed by Matoy, the bend bars were manufactured to measure fracture toughness on amorphous silicon [104, 105]. It will only be a matter of time before scientists also use this method on irradiated materials.

Dimensional changes of ion-beam-irradiated samples: as mentioned in Sec. 2, radiation-induced swelling is of significant concern in nuclear applications. Naturally, scientists attempt to measure swelling using ion beam irradiations. However, evaluating swelling during the PIE is complex, since volumetric changes are difficult to measure. Some groups attempt to measure swelling using atomic force microscopy or other surface-profiling methods on masked and irradiated samples, as shown in Ref. 106. While these types of measurements are possible, special care has to be taken on what is measured using AFM on the ion-beam-irradiated samples. Since AFM is only a surface-sensitive technique, the measurement of the step-height increase due to swelling can be overlapped by measurements of the step-height increases due to surface deposition

or other phenomena. Since surface sputtering or the deposition of carbon can occur, these measurements have to be cross-calibrated with TEM investigations in which the void density is observed and the resulting volume changes are calculated [107].

5. Previous Ion Beam Experiments and Data Evaluation

In the past, a wide range of comparative experiments were conducted in an attempt to prove that data from ion beam irradiations and from reactor irradiations can be compared. In particular, G. Was's and T. Allen's groups did a significant amount of work on microstructural and chemistry comparisons. Also, L. Mansur developed the invariant equations used to compare different dose rates to each other — one of the significant differences between neutron and ion beam irradiations.

In Mansur's work [39, 40], it is suggested that the N_s and N_r invariant equations (shown below) can be used to adjust the irradiation temperatures so that the resulting defects are similar despite the different dose rates. The principal idea behind this is that if one radiation variable is changed, the others can be adapted accordingly to preserve an integral quantity, such as the number of defects absorbed at sinks, N_S, or the number of defects per unit volume that have recombined, N_R.

N_R invariant is shown below, where ϕ_i denotes the flux, E_m^v the vacancy migration energy, E_f^v the vacancy formation energy, and T_i the irradiation temperatures:

$$\Delta T = T_2 - T_1 = \frac{\left(\frac{kT_1^2}{E_m^v + 2E_f^v}\right)\ln\left(\frac{\phi_2}{\phi_1}\right)}{1 - \left(\frac{kT_1}{E_m^v + 2E_f^v}\right)\ln\left(\frac{\phi_2}{\phi_1}\right)}.$$

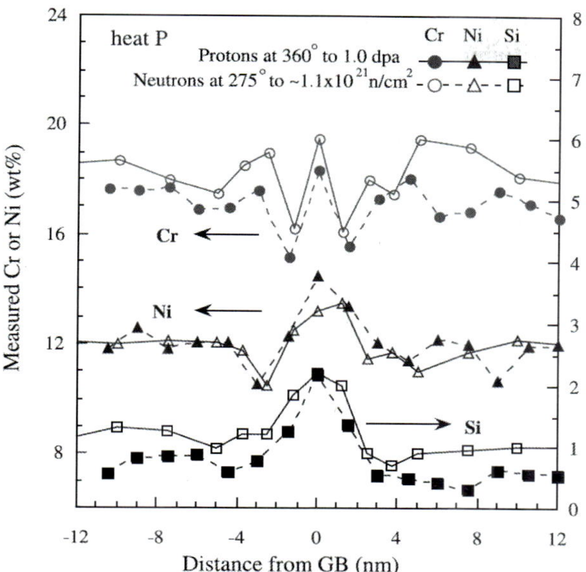

Fig. 14.　Direct comparison between the radiation-induced segregation caused by ion beam irradiation and that caused by neutron irradiation [108].

N_S invariant [39, 40]:

$$\Delta T = T_2 - T_1 = \frac{\left(\dfrac{kT_1^2}{E_m^v}\right)\ln\left(\dfrac{\phi_2}{\phi_1}\right)}{1 - \left(\dfrac{kT_1}{E_m^v}\right)\ln\left(\dfrac{\phi_2}{\phi_1}\right)}.$$

Using the N_R and N_S invariant equations, a dose rate relationship can be calculated.

Several comparison studies have been performed considering these equations, measuring and comparing defect densities, such as dislocations loops, or local segregations, utilizing TEM [108, 109] between neutron and proton irradiation.

Also, studies investigating the change in hardness were performed, showing comparability between proton-irradiated and neutron-irradiated samples of HT-9, a ferritic–martensitic material; see Fig. 15. The red line represents data gained from FFTF irradiation (including He generation), and the blue data points are from the combination of proton irradiation in an ion accelerator and nanoindentation measurements [109]. It is noted that the ion beam irradiation was preformed at 450°C and the reactor irradiation was preformed at 400°C based on Mansur's model described above. The green data point was obtained from two individual microcompression tests.

While great care must still be taken in how proton and neutron data are compared to each other, it appears that protons can be used to estimate changes in microstructure, chemical composition, and basic mechanical properties. A recent paper on single crystal Cu showed the limit of downscaling mechanical testing. It appears that there is a fundamental limit to how small one can go in the mechanical testing of compression samples. Apparently, for Cu, the minimum pillar diameter is about 400 nm for measuring a difference between nonirradiated samples and irradiated samples. If a sample is smaller than 400 nm, the deformation of the sample is source-controlled; while a sample larger than 400 nm in diameter is defect-controlled. In the latter case, radiation-induced defects and their consequences for mechanical property changes can be measured [110]. Also, in Ref. 110, proton-irradiated samples were compared to samples irradiated in a reactor. It was found that the change in YS of pillars larger than 400 nm is comparable to the change of YS in reactor-irradiated samples. This *in situ* work clearly shows the benefit of combining small-scale mechanical testing, TEM and ion beam irradiation to gain new and comprehensive insight into materials property changes due to irradiation.

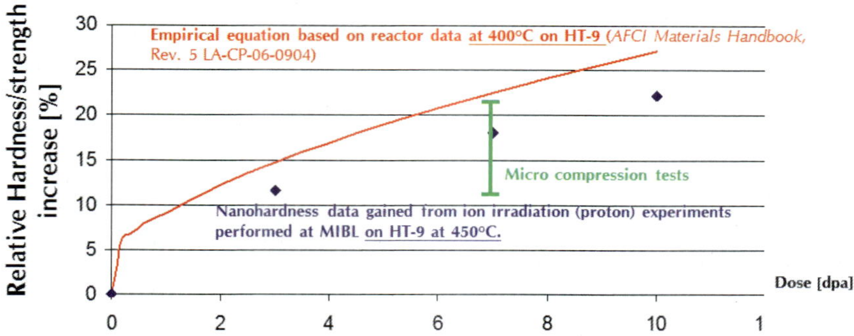

Fig. 15.　Comparison between the change in YS due to irradiation in a reactor (FFTF irradiation; red data) and that due to irradiation with an ion beam (blue and green data) [109].

6. The Benefits of *In Situ* Ion Beam

While radiation damage alone causes significant changes to a material's microstructure (dislocation densities, SFT, segregation, phase formation, etc), the combination of irradiation and an external second environment can lead to additional phenomena, as described below. This complication has resulted in the development of various *in situ* ion beam irradiation experiments, whether environmental (corrosion) stress–related (irradiation-induced creep) or multibeam. In the following subsections, various *in situ* testing techniques that have been described in the literature are presented. It has to be noted that most of these techniques were developed to study the behavior of materials in a reactor-type environment. Therefore, these techniques can have the same issues as those described above, such as issues caused by differences in dose rates or cascade shapes, leading to the question of how comparable the data are to true reactor environments.

6.1. *In situ TEM*

While most ion beam irradiations allow the study of the change of microstructure or mechanical properties due to radiation damage, the actual development of the defects cannot be observed directly by doing PIE after the irradiation. Therefore, ion beam irradiation in a TEM was developed to observe *in situ* the change of a material due to irradiation damage. The first experiments utilized an oxygen beam in a TEM more than 50 years ago in Great Britain, while today sophisticated tandem accelerators and sometimes dual beam setups are combined with a TEM [111]. Maybe the most difficult task in combining the TEM with an ion beam is the beam–TEM interface. As pointed out in Refs. 111 and 112, there are three main ways to combine a TEM with anion accelerator. First, one can bring in the beam in between the gap of the upper and lower pole pieces of the objective lenses — a place where several models of TEMs have a port.

The ion beam can also be brought in at a steep angle, which can require significant TEM modification. And the third method is the use of a deflection, as demonstrated by the facility previously located in Salford [113] and now at the University of Huddersford, UK; this allows a very high angle of incidence on the sample but requires electrostatic devices inside

the TEM. During the design stage, special care has to be taken of the fact that the high magnetic fields necessary to bend a beam in a desired direction cannot be close to the TEM, since it would disturb the image.

Obviously, it is desirable to achieve a truly *in situ* irradiation condition so that the sample does not need to be tilted in order to get irradiated. But this is sometimes difficult to achieve, due to geometric restriction in the facility available.

One thing that has to be kept in mind is that, although many scientists are interested in studying bulk effects, surface effects can dominate the results from the 100 nm thin foils used for a TEM *in situ* irradiation experiment. Therefore, the results obtained in this way can be difficult to interpret regarding their implications for a bulk material.

Figure 16 shows an image of the facility at Hokkaido University in Japan, which has a JEOL JEM-ARM1300 with two beams (20–300 kV and 20–400 kV). In this case, two different ion species can be used simultaneously on one sample.

Fig. 16.   JEOL JEM-ARM1300 with two beams (20–300 kV and 20–400 kV) [111].

Table 4. TEM ion beam facilities operating in 2009, after Ref. 112. An asterisk indicates uncertain current status due to the recent earthquake in Japan.

Facility	TEM	Ion source	Degrees
Sandia Nat. Lab, USA	JOEL, 200 keV	MeV heavy ions; keV D^{2+}, He^+	90 20
Kuyushu Univ., Japan	JEOL, 200 keV	0.1–10 kV	
JAEA Takasaki, Japan	JEOL, 400 keV	2–40 kV; 20–400 kV	30
JAEA Tokaimura, Japan	JEOL, 200 keV	40 keV	30
NIMS, Tsukuba, Japan*	JEOL, 200 keV	5–25 kV FIB	35
NIMS, Tsukuba, Japan*	JEOL, 1000 keV	30/200 kV	
Hokkaido Univ, Japan	JEOL, 1300 keV	20–300 kV / 20–400 kV	44
Shimane Univ., Japan	JEOL, 200 keV	1–20 kV	17
IVEM, ANL, USA	Hitachi, 300 keV	0.65–2 MV	30
JANNus, CSNSM, France	FEI, 200 keV	0.19–2 MV	68
Wuhan Univ., China	Hitachi 200 keV	0.2–3.4 MV	90
Salford Univ., UK	JEOL, 200 keV	1–100 kV	25

Table 4 shows a summary of all facilities operating today around the world [112] and their basic features. The newly developed facility at Sandia National Laboratory in the USA is not listed.

6.2. *In situ corrosion*

Almost every nuclear facility has some sort of cooling integrated within it in order to keep a desired temperature. While commercial power reactors are all water-cooled, liquid metal, gas, and molten salt have also been used as coolants or are being considered as coolants. Since structural materials and fuels are exposed to these environments, it is necessary to gain information about the effect of radiation on the corrosion mechanism. Several groups have attempted in the past to study radiation-assisted corrosion by using ion beam accelerators. Lewis and Hunn [114] built a setup for irradiating a stainless steel foil in a liquid (mercury or water). It used protons with energies up to 2 MeV and also collected the gas produced during the irradiation. A schematic of this setup is presented in Fig. 17 [114].

A later experiment targeting only liquid-metal corrosion was conducted by P. Hosemann. This experiment was conducted at temperatures as high as 450°C; a sketch of it is shown in Fig. 18. In the experiment, the beam window was the sample; and, as in the Lewis experiment, the sample was the only barrier separating the liquid from the high vacuum

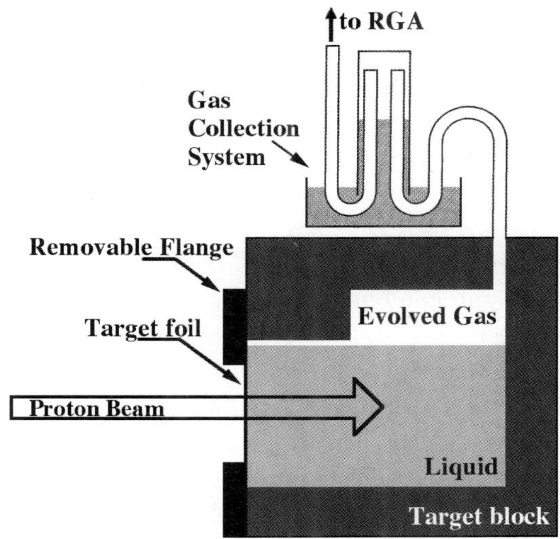

Fig. 17. Schematic after Ref. 114, from the irradiation and corrosion experiment to study liquid–solid interfaces.

in the accelerator. This configuration was a particularly difficult situation, since a sample failure would have resulted in significant damage to the accelerator. Therefore, great care had to be taken on the sample design. A concave-shaped sample geometry was chosen in order to achieve a maximum of physical stability while also allowing a wide dose range of the backside of the sample due to the changing sample thickness.

In aqueous corrosion experiments, parameters such as the pH value or gas production (as in

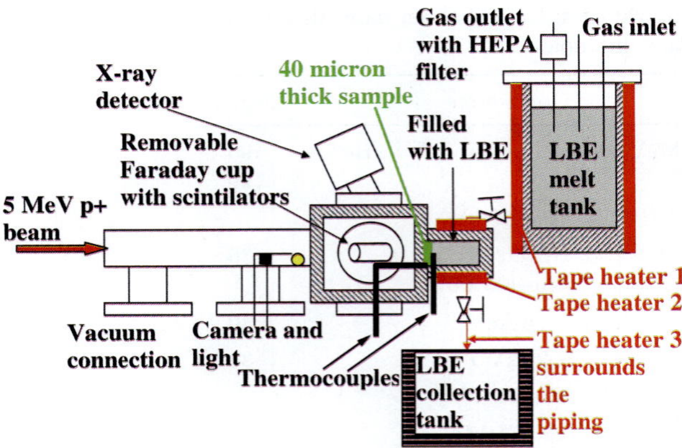

Fig. 18.   Sketch of the irradiation and corrosion experiment as shown in Ref. 73.

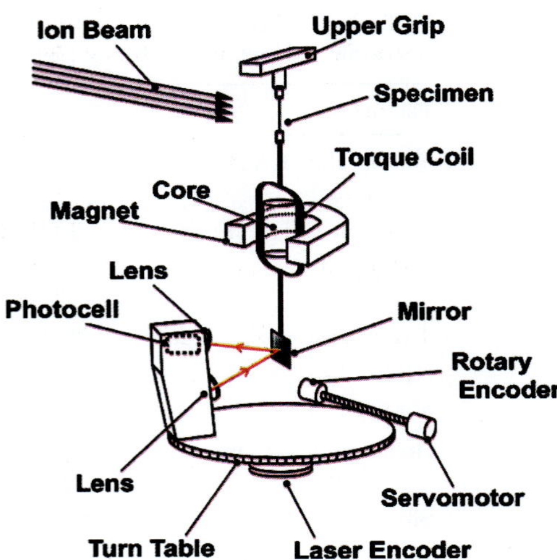

Fig. 19.   Torsion cell implemented in an ion accelerator [121].

the case of the experiment by Lewis and Hunn) can be recorded over time. In the case of liquid-metal corrosion, it would be rather difficult to implement an *in situ* corrosion rate measurement such as impedance spectroscopy [115].

However, with the increasing interest in studying materials in extreme environments, *in situ* corrosion testing clearly has the potential to be expanded in the future.

6.3. *In situ mechanical testing*

As with the combination of environment and radiation described above, stress in combination with radiation damage can result in new phenomena. As described above, radiation leads to an excessive amount of defects, which can help dislocations climb over obstacles, resulting in radiation-enhanced creep. This phenomenon is usually observed at temperatures lower than those at which regular, thermal creep is observed, due to the fact that the amount of defects created during radiation is independent of the temperature at which the sample is being held. However, at higher temperatures, radiation-enhanced creep is no longer dominant, because the climbing of dislocations becomes activated mainly by the temperature. Since the 1970s, scientists have built stressing devices in ion beam accelerators. Most facilities and experiments employ a uniaxial stress stage, as first reported by Ref. 116, but later also used by Refs. 117–119. Great effort is spent for accurate temperature control of the specimen, since the heat input due to the actual ion beam needs to be considered. Combined heating/cooling stages are

built as in Ref. 117. In this particular setup, purified He is used on the sample so as to be able to actively cool it when the beam heating becomes too high. On the other hand, dc heating can be applied if the heating is not enough to hold the desired temperature. In Ref. 120, the sample is connected to a Cu heat sink, with a liquid indium film in between the sample and the heat sink that allows good thermal transport to the temperature-controlled Cu block. However, some other ideas, such as a torsion cell [121], have been introduced in accelerators as well. Such a setup is schematically shown in Fig. 19. Overall, it can be said that, while the instrumental setup for performing such an experiment appears to be simple, it is rather difficult to manufacture a homogeneous sample with micrometer dimensions and to provide a controlled temperature, stress, and displacement system on such thin specimens for an extended period of time in an accelerator environment. Therefore, great care has to be taken to provide the relevant controls in order to guarantee comparable, repeatable, and good results.

6.4. *Multibeam irradiations*

When displacement damage in a material caused by various ion beams is studied, the additional effects of other fission products are often neglected. Spallation sources can produce as much as 100 appm of He per dpa, and fusion neutron spectra can produce 10s of appm of He per dpa. Such He production leads to He bubble formation due to the insolubility of

Table 5.   List of multi-ion-beam facilities around the world. It does not claim to list all the facilities available today [79].

Location	Facility	Applications
Kalpakkan, India	1.7 MV Tandetron	Nuclear alloys
Tokyo, Japan	3.7 MV V. d. Graaff 1 MV Tandetron	Nuclear alloys and ceramics
Nagoya, Japan	2 MV V. d. Graaff 200 kV implanter	Nuclear alloys and ceramics
Rossendorf, Germany	3 MV Tandetron 500 kV implanter	Material synthesis
Jena, Germany	3 MV Tandetron 400 kV implanter	Material synthesis, radiation damage
Kyoto, Japan	1.7 MV Tandetron 1 MV V. d. Graaff 1 MV Singeltron	Material synthesis, radiation damage
JAEA Takasaki, Japan	3 MV Tandem 3 MV V. d. Graaff 400 kV implanter	Material synthesis
Saclay, France	3 MV Pelletron 2.5 MV V. d. Graaff 2.25 MV Tandetron	Radiation damage, modifications
Kharkov, Ukraine	2 MV ESU 50 kV proton 50 kV helium	Radiation damage
Los Alamos, USA	3 MV Tandem 200 kV ion implanter	Radiation damage, modification

He in the material's matrix; the bubbles, in turn, result in additional effects on the mechanical properties of the material. While He effects are of particular importance, other fission products are also of interest, especially in nuclear fuel research.

In order to address the issue of displacement damage and issues such as He buildup, multi-ion-beam facilities have been constructed and are in operation around the world.

In Ref. 79, a detailed description of facilities with dual and triple ion beams is given, based on a workshop hosted by Lawrence Livermore National Laboratory in 2009. Table 5 presents a list of dual- and triple-ion-beam facilities based on this document.

The Joint Accelerators for Nano-science and Nuclear Simulation (JANNuS) and the Takasaki Ion Accelerators for Advanced Radiation Applications (TIARA) facility is a triple-ion-beam facility that should be mentioned specifically, since it is their aim to simulate the synergies of ion displacements in the presence of hydrogen and helium production as well as other transmuted elements. In Ref. 122, a clear example is presented of how synergetic effects can affect the development of larger damage. Void

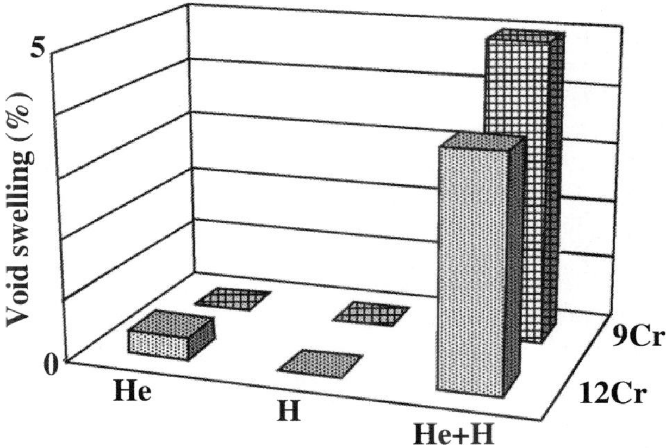

Fig. 20.   Results of a triple-ion-beam irradiation experiment using He, H, and heavy ions (Fe3+), after Ref. 122.

swelling seems to be dependent on the presence of simultaneous H and He and displacement damage, as shown in Fig. 20.

This and other experiments demonstrate the need for multi-ion-beam facilities in order to study radiation damage in materials in a comprehensive manner.

7. Summary

In this review article, the main issues concerning ion beam irradiations for studying mainly structural materials for neutron irradiation application have been listed, along with the advantages that scientists can gain by using these techniques.

While the article does not claim to discuss every aspect of the application of ion beams to materials for nuclear applications, it gives an overview of both the capabilities and the limitations of these types of experiments. The list of references provided is not complete on all the important work performed in the past, but it does give the interested reader a starting point for deeper literature research.

Acknowledgments

Erich Stergar, Yongqiang Wang and Alan Bolind are thanked for their editorial suggestions and support in the extensive literature research. S. A. Maloy and F. A. Garner are acknowledged for their advice and help with older articles not easily available today. B. Doyle is thanked for the opportunity to introduce this topic into RAST.

References

[1] E. P. Wigner, Report for Month Ending December 15, *U.S. Atomic Energy Commision Report CP387* (1942).

[2] K. L. Murty and I. Charit, *J. Nucl. Mater.* **383**(1–2), 189 (2008).

[3] K. Ehrlich, *Philos. Trans. R. Soc. London, Ser. A* **357**(1752), 595 (1999).

[4] F. M. Nortier, J. W. Lenz and P. A. Smith, Large-scale isotope production with an intense 100 MeV proton beam: recent target performance experience, in *Cyclotrons and Their Applications 2007: 18th Int. Conf.* (2007), pp. 257–259.

[5] E. Lehmann, P. Vontobel and M. Estermann, *Appl. Radiat. Isot.* **61**(4), 603 (2004).

[6] S. A. Maloy, *J. Nucl. Mater.* **343**(1–3), 367 (2005).

[7] B. D. Wirth, *Science* **318**(5852), 923 (2007).

[8] X.-M. Bai, A. F. Voter, R. G. Hoagland, M. Nastasi and B. P. Uberuaga, *Science* **327**(5973), 1631 (2010).

[9] T. S. Byun and S. A. Maloy, *J. Nucl. Mater.* **377**(1), 72 (2008).

[10] S. A. Maloy, T. Romero, M. R. James and Y. Dai, *J. Nucl. Mater.* **356**(1–3), 56 (2006).

[11] B. D. Wirth *et al.*, *J. Nucl. Mater.* **329–333**, Part 1, 103 (2004).

[12] J. A. Brinkman, *Am. J. Phys.* **24**(4), 246 (1956).

[13] A. Seeger, On the theory of radiation damage and radiation hardening, in *Proc. Second United Nations International Conference on the Peaceful Uses of Atomic Energy* (1958), p. 250.

[14] G. H. Kinchin and R. S. Pease, *Rep. Prog. Phys.* **18**, 1 (1955).

[15] G. H. Kinchin and R. S. Pease, *J. Nucl. Energy* **1**(3–4), 200 (1955).

[16] W. Köhler and W. Schilling, *Nukleonik* **7**, 389 (1965).

[17] P. Sigmund, *Appl. Phys. Lett.* **14**(3), 114 (1969).

[18] D. A. Thompson and J. E. Robinson, *Nucl. Instrum. Methods B* **132**, 261 (1976).

[19] J. Lindhard, M. Scharff and H. E. Schioett, Range concepts and heavy ion ranges (notes on atomic collisions, II), *Kgl. Danske Videnskab. Selskab Mat.-fys. Medd.* **33**(14), 1 (1963).

[20] C. Lehmann, Zur Bildung von Defekt-Kaskaden in Kristallen beim Beschuss mit energiereichen Teilchen, *Nukleonik* **3**, Dissertation (1961).

[21] J. Lindhard and P. V. Thomson, Sharing of energy dissipation between electronic and atomic motion, in *Proc. Radiation Damage in Solids Vienna: International Atomic Energy Agency*, (1962), Vol. 1, pp. 66–67.

[22] O. S. Oen and M. T. Robinson, *Appl. Phys. Lett.* **2**(4), 83 (1963).

[23] P. Sigmund, *Phys. Lett.* **6**(3), 251 (1963).

[24] M. T. Robinson, *Philos. Mag.* **12**(118), 741 (1965).

[25] W. S. Snyder and J. Neufeld, *Phys. Rev.* **97**(6), 1636 (1955).

[26] J. Lindhard, V. Nielsen, M. Scharff and P. V. Thomsen, Integral equations governing radiation effects (notes on atomic collisions, III), *Kgl. Danske Videnskab. Selskab Mat.-fys. Medd.* **33**, 1 (1963).

[27] J. Lindhard, V. Nielsen and M. Scharff, Approximation method in classical scattering by screened Coulomb fields, *Kgl. Danske Videnskab. Selskab Mat.-fys. Medd.* **36**, 1 (1968).

[28] J. F. Ziegler, J. P. Biersack and U. Littmark, *SRIM: The Stopping and Range of Ions in Matter* (Pergamon, 1996).

[29] D. R. Olander, *Fundamental Aspects of Nuclear Reactor Fuel Elements* (National Technical Information Service, 1976).

[30] H. L. Heinisch, *Nucl. Instrum. Methods B* **102**, 47 (1995).

[31] G. S. Was, *Fundamentals of Radiation in Materials* (Springer, Berlin, Heidelberg, New York, 2007).

[32] W. Schilling, P. Ehrhart and K. Sonnenberg, *Fundamental Aspects of Radiation Damage in Metals: in Proc. Int. Conf. Gatlinburg*, eds. M. T. Robinson and F. W. Young (1975).

[33] W. Schilling, *J. Nucl. Mater.* **72**, 1 (1978).

[34] W. Schilling and K. Sonnenberg, *J. Phys. F* **3**, 322 (1973).

[35] J. W. Corbett, R. B. Smith and R. M. Walker, *Phys. Rev.* **114**, 1452 (1959).

[36] G. L. Kulcinski, J. L. Brimhall and H. E. Kissinger, Production of voids in pure metals by high energy heavy ion bombardment, in *Int. Conf. Radiation-Induced Voids in Metals* (Albany, New York, 9 June 1971), UNCL, p. 449.

[37] L. K. Mansur and K. Farrell, *J. Nucl. Mater.* **170**, 236 (1990).

[38] R. K. Nanstad, K. Farrell, D. N. Braski and W. R. Corwin, *J. Nucl. Mater.* **158**, 1 (1988).

[39] L. K. Mansur, *J. Nucl. Mater.* **206**, 306 (1993).

[40] L. K. Mansur, *J. Nucl. Mater.* **216**, 97 (1994).

[41] M. K. Miller and K. F. Russel, *J. Nucl. Mater.* **371**, 145 (2007).

[42] G. R. Odette and G. E. Lucas, *JOM* **53**, 18 (2001).

[43] G. R. Odette and G. E. Lucas, *Radiat. Eff. Defects Solids* **144**, 189 (1998).

[44] B. H. Sencer, J. R. Kennedy, J. I. Cole, S. A. Maloy and F. A. Garner, *J. Nucl. Mater.* **393**, 235 (2009).

[45] B. H. Sencer, J. R. Kennedy, J. I. Cole, S. A. Maloy and F. A. Garner, *J. Nucl. Mater.* [doi:10.1016/j.jnucmat.2011.03.050].

[46] D. L. Porter and F. A. Garner, *J. Nucl. Mater.* **159**, 114 (1988).

[47] M. Victoria *et al.*, *J. Nucl. Mater.* **276**, 114 (2000).

[48] Y. Dai *et al.*, *J. Nucl. Mater.* **296**, 174 (2001).

[49] P. Hosemann *et al.*, *J. Nucl. Mater.* **375**, 135 (2008).

[50] M. A. Pouchon, J. Chen, R. Ghisleni, J. Michler and W. Hoffelner, *Exp. Mech.* 79 (2010).

[51] P. Hosemann *et al.*, *Exp. Mech.* [doi:10.1007/s11340-010-9419-2, 2010].

[52] S. M. Bruemmer *et al.*, *J. Nucl. Mater.* **274**, 299 (1999).

[53] A. Etienne, B. Radiguet, N. J. Cunningham, G. R. Odette and P. Pareige, *J. Nucl. Mater.* **406**, 244 (2010).

[54] E. Stergar, personal communication.

[55] F. A. Garner and D. L. Porter, *J. Nucl. Mater.* 1006 (1988).

[56] F. A. Garner, *J. Nucl. Mater.* **122**, 459 (1984).

[57] E. A. Little, *J. Nucl. Mater.* **87**, 11 (1979).

[58] E. A. Little and D. A. Stow, *J. Nucl. Mater.* **87**, 25 (1979).

[59] F. A. Garner, E. R. Gilbert and D. L. Porter, Stress-enhanced swelling of metals during irradiation, in *ASTM Special Technical Publication* (1981), Vol. 725, pp. 680–697.

[60] F. A. Garner, M. B. Toloczko and B. H. Sencer, *J. Nucl. Mater.* **276**, 123 (2000).

[61] T. Okita *et al.*, Neutron-induced microstructural evolution of Fe–15Cr–16Ni alloys at $\sim 400°C$ during neutron irradiation in the FFTF fast reactor, in *Proc. 10th Int. Conf. Environment Degradation of Materials in Nuclear Power Systems–Water Reactors* (2001).

[62] Y. Hidaka, S. Ohnuki, H. Takahashi and S. Watanabe, *J. Nucl. Mater.* **212**, 330 (1994).

[63] E. V. Kornelsen and A. A. Van Gorkum, *J. Nucl. Mater.* **92**, 79 (1980).

[64] B. N. Singh, T. Leffers, W. V. Green and M. Victoria, *J. Nucl. Mater.* **125**, 287 (1984).

[65] N. Li *et al.*, *J. Nucl. Mater.* **389**, 233 (2009).

[66] G. E. Dieter, *Mechanical Metallurgy* (McGraw-Hill, 1961).

[67] R. Bullough and M. H. Wood, *J. Nucl. Mater.* **90**, 1 (1980).

[68] R. Bullough and M. R. Hayns, *J. Nucl. Mater.* **65**, 184 (1977).

[69] V. Fidleris, *J. Nucl. Mater.* **159**, 22 (1988).

[70] L. K. Mansur and T. C. Reiley, *J. Nucl. Mater.* **90**, 60 (1980).

[71] E. R. Gilbert, J. L. Straalsund and G. L. Wire, *J. Nucl. Mater.* **65**, 266 (1977).

[72] G. W. Lewthwaite and D. Mosedale, *J. Nucl. Mater.* **90**, 205 (1980).

[73] P. Hosemann *et al.*, *J. Nucl. Mater.* **376**, 392 (2008).

[74] R. S. Lillard, M. Paciotti and V. Tcharnotskaia, *J. Nucl. Mater.* **335**, 487 (2004).

[75] W. G. Burns, *Nature* **339**, 515 (1989).

[76] J. Zhang, P. Hosemann and S. Maloy, *J. Nucl. Mater.* **404**, 82 (2010).

[77] H. Glasbrenner, Y. Dai and F. Gröschel, *J. Nucl. Mater.* **343**, 267 (2005).

[78] A. Möslang, Existing and future irradiation facilities: their availability and access, in *Workshop Proceedings: Structural Materials for Innovative Nuclear Systems (SMINS)* (2007).

[79] G. Bench, V. Bulatov and M. Serrano De, Workshop on Science Applications of a Triple Beam Capability for Advanced Nuclear Energy Materials, *Report* (2009).

[80] Y. Dai and G. S. Bauer, *J. Nucl. Mater.* **296**, 43 (2001).

[81] Y. Dai *et al.*, *J. Nucl. Mater.* **343**, 33 (2005).

[82] G. S. Was and T. R. Allen, *Radiation Damage from Different Particle Types*, eds. K. E. Sickafus, E. A. Kotomin and B. P. Uberuaga (Springer, 2007), Vol. 235.

[83] P. Hosemann, S. A. Maloy, R. R. Greco, J. G. Swadener and T. Romero, *J. Nucl. Mater.* **384**(1), 25 (2009).

[84] D. L. Damcott, J. M. Cookson, V. H. Rotberg and G. S. Was, *Nucl. Instrum. Methods B* **99**, 780 (1995).

[85] J. M. Cookson, R. D. Carter Jr., D. L. Damcott, M. Atzmon and G. S. Was, *J. Nucl. Mater.* **202**, 104 (1993).

[86] A. T. Nelson and S. A. Hosemann, in *Proc. CARRIE Conference in 2010: Dallas, Fort Worth* (2010).

[87] T. L. Alford, D. Adams and J. W. Mayer, *Encyclopedia of Condensed Matter Physics*, eds. F. Bassani, G. L. Liedl and P. Wyder (Elsevier, 2005).

[88] G. W. Grime, High energy ion beam analysis, in *Encyclopedia of Spectroscopy and Spectrometry*, ed. John Lindon (Academic, 1999), pp. 844–853.

[89] Y. Q. Wang and M. A. Nastasi, *Handbook of Modern Ion Beam Materials Analysis*, 2nd Edition (MRS Publisher, Warrendale, Pennsylvania, 2009).

[90] P. Pareige *et al.*, *J. Nucl. Mater.* **360**, 136 (2007).

[91] D. Kiener, C. Motz, M. Rester, M. Jenko and G. Dehm, *Mater. Sci. Eng. A* **459**, 262 (2007).

[92] L. Veleva, R. Schäublin, A. Ramar, Z. Oksiuta and N. Baluc, Focused ion beam application on the investigation of tungsten-based materials for fusion application, in *Proc. EMC 2008: 14th European Microscopy Congress* (Aachen, Germany, 1–5 Sept. 2008), pp. 503–504.

[93] A. D. Pogrebnjak *et al.*, *Vacuum* **58**, 45 (2000).

[94] G. Gupta, Z. Jiao, A. N. Ham, J. T. Busby and G. S. Was, *J. Nucl. Mater.* **351**, 162 (2006).

[95] T. Miura, K. Fujii, K. Fukuya and Y. Ito, *J. Nucl. Mater.* **386–388**, 210 (2009).

[96] P. M. Rice and R. E. Stoller, *J. Nucl. Mater.* **244**, 219 (1997).

[97] P. Hosemann *et al.*, *J. Nucl. Mater.* **389**, 239 (2009).

[98] N. Li *et al.*, *J. Appl. Phys.* **105**, 123522–1 (2009).

[99] P. Hosemann, D. Kiener and S. A. Maloy, *J. Nucl. Mater.* Vol. submitted (2011).

[100] J. T. Busby, M. C. Hash and G. S. Was, *J. Nucl. Mater.* **336**, 267 (2005).

[101] M. Haque and M. Saif, *Exp. Mech.* **42**, 123 (2002).

[102] D. Kiener, W. Grosinger, G. Dehm and R. Pippan, *Acta Mater.* **56**, 580 (2008).

[103] K. Fujii and K. Fukuya, *Mater. Trans.* **52**, 20 (2011).

[104] K. Matoy, T. Detzel, M. Müller, C. Motz and G. Dehm, *Surf. Coat. Technol.* **204**, 878 (2009).

[105] K. Matoy *et al.*, *Thin Solid Films* **518**, 247 (2009).

[106] M. Terasawa, T. Mitamura, L. Liu, H. Tsubakino and M. Niibe, *Nucl. Instrum. Methods B* **193**, 329 (2002).

[107] W. G. Johnston, J. H. Rosolowski, A. M. Turkalo and T. Lauritzen, *J. Nucl. Mater.* **54**, 24 (1974).

[108] G. S. Was *et al.*, *J. Nucl. Mater.* **300**, 198 (2002).

[109] P. Hosemann, *Material Studies for Lead Bismuth Eutectic Cooled Nuclear Applications* (University of Leoben, 2008).

[110] D. Kiener, P. Hosemann, S. A. Maloy and A. M. Minor, *Nat. Mater.*, advance online publication, [doi:10.1038/NMAT3055, 2011].

[111] M. A. Kirk *et al.*, *Microsc. Res. Tech.* **72**, 182 (2009).

[112] J. A. Hinks, *Nucl. Instrum. Methods B* **267**, 3652 (2009).

[113] C. Kinoshita, H. Abe, K. Fukumoto, K. Nakai and K. Shinohara, *Ultramicroscopy* **39**, 205 (1991).

[114] M. B. Lewis and J. D. Hunn, *J. Nucl. Mater.* **265**, 325 (1999).

[115] X. Chen, J. F. Stubbins, P. Hosemann and A. M. Bolind, *J. Nucl. Mater.* **398**, 172 (2010).

[116] J. A. Hudson, R. S. Nelson and R. J. McElroy, *J. Nucl. Mater.* **65**, 279 (1977).

[117] P. Jung, J. Viehweg and C. Schwaiger, *Nucl. Instrum. Methods B* **154**, 207 (1978).

[118] P. Jung, C. Schwaiger and H. Ullmaier, *J. Nucl. Mater.* **85–86**, 867 (1979).

[119] P. Jung, *J. Appl. Phys.* **86**, 4876 (1999).

[120] P. L. Hendrick, A. L. Bement Jr. and O. K. Harling, *Nucl. Instrum. Methods B* **124**, 389 (1975).

[121] J. Nagakawa, S. Uchio, Y. Murase, N. Yamamoto and K. Shiba, *J. Nucl. Mater.* **386–388**, 264 (2009).

[122] T. Tanaka *et al.*, *J. Nucl. Mater.* **329–333**, 294 (2004).

Peter Hosemann is currently an assistant professor at the University of California, Berkeley, in the Department of Nuclear Engineering. Previously he was at Los Alamos National Laboratory, working on radiation damage in structural materials for nuclear application and liquid metal corrosion for advanced fast reactors. He built the only irradiation and corrosion experiment (ICE) for low-energy ion beam accelerators to study the changes in corrosion phenomena due to irradiation and applied small-scale mechanical testing on ion-beam- and spallation-source-irradiated materials. He graduated in 2008 from the Montanuniversitaet in Leoben, Austria with a Ph.D. in Materials Science.

Reviews of Accelerator Science and Technology
Vol. 4 (2011) 183–212
© World Scientific Publishing Company
DOI: 10.1142/S1793626811000525

Direct Current Accelerators for Industrial Applications

Ragnar Hellborg

*Department of Physics, Lund University,
Sölvegatan 14, SE-223 62 Lund, Sweden
ragnar.hellborg@nuclear.lu.se*

Harry J. Whitlow

*Department of Physics, P. O. Box 35 (YFL),
FIN-40014, University of Jyväskylä, Finland
harry.j.whitlow@jyu.fi*

Direct current accelerators form the basis of many front-line industrial processes. They have many advantages that have kept them at the forefront of technology for many decades, such as a small and easily managed environmental footprint. In this article, the basic principles of the different subsystems (ion and electron sources, high voltage generation, control, etc.) are overviewed. Some well-known (ion implantation and polymer processing) and lesser-known (electron beam lithography and particle-induced X-ray aerosol mapping) applications are reviewed.

Keywords: Particle accelerators; ion sources; electron sources; electron microscopy; high voltage terminal; ion beam analysis; ion implantation; electron beam polymer processing; electron beam sterilization; neutron generators; electron beam lithography; isotope production; sterilization.

1. Introduction and History

Accelerators can be classified according to the different principles by which they work, such as direct current accelerators (with the cascade and electrostatic subgroups), cyclotrons, linacs, synchrocyclotrons and synchrotrons. All principles are based on the only known method to accelerate a particle: to charge it and subject it to an electric field. This occurs either in one large step or in a number of smaller steps.

Direct current accelerators were the first type to open nuclear physics to extensive experimentation. They continued to play an important role in basic nuclear and atomic physics for 40 years, before these research fields moved on to larger accelerators. At the same time their uses for applied physics began to expand rapidly in many fields. This type of accelerators became the new tools for the new techniques delivering both analysis and modification in many branches of technology. Improvements in the design of the DC accelerators during those years gave them simplicity, precision, reliability and versatility — i.e. they became ready for industrial use!

Today DC accelerators and other types are used in very diverse fields (Table 1) such as radiotherapy, isotope production, ion implantation in the manufacture of semiconductor devices, synchrotron light production, spallation, neutron production, radiography, sterilization, inertial fusion, age determination of samples that are thousands and even millions of years old, or nondestructive analysis. High power electron beams are used for enhancing insulation on wire, producing shrink tubing and sterilizing disposable medical products. More than 30,000 accelerators have been built worldwide over the decades. More than half of them have been used for industrial purposes. Ion implanters dominate this group by some 10,000 [1]. Today more than 15,000 accelerators are in operation around the world; only a handful are used in elementary particle physics research, a few hundred in physics and applied physics research, while one third are involved in medical applications, such as therapy, imaging and the production of short-lived isotopes. Some — such as electron microscopes and lithographic systems — are often not registered as accelerators. The vast majority are used for industrial applications ranging from semiconductor manufacture, electron beam processing and micro-machining to food sterilization and national security applications, which include X-ray inspection of cargo containers and nuclear stockpile stewardship.

The years around 1930 can be taken as the beginning of the accelerator era, when people at different laboratories started development work employing different principles. Robert Van de Graaff (see Fig. 1), Ray Herb (see Fig. 2) and others developed the electrostatic accelerator, Cockcroft and Walton the cascade accelerator.

The tremendous progress in the construction of accelerators since the 1930s is illustrated in Fig. 3 with an exponential increase — about an order of magnitude in beam energy per seven years! This graph is called a "Livingston plot," after Stanley Livingston, the accelerator physicist who first constructed such a plot in the 1960s. In the same time period, the cost per eV beam energy has been drastically reduced, roughly by a factor of 1000. The similarity to Moore's law in electronics is striking (see for example the chapter "High Speed Electronics" in Ref. 2).

Fig. 2.   Ray Herb (*on the left*) and the "Long Tank" machine in 1936. (Courtesy of Mrs. Ann Herb.)

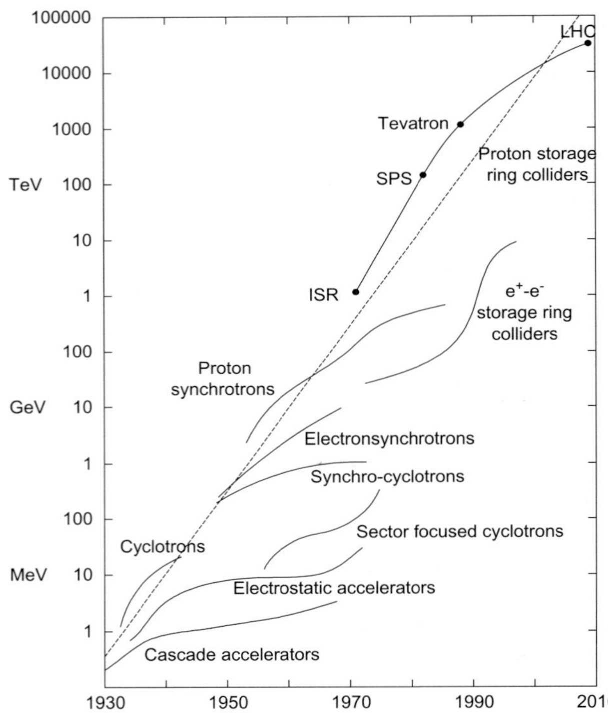

Fig. 3.   A modified Livingston diagram showing the exponential growth of accelerator beam energy. The energy plotted is limited to single-charged ions. The energy of colliders is plotted in terms of the laboratory energy of particles colliding with a proton at rest to reach the same center-of-mass energy.

Fig. 1.   Robert Van de Graaff (*on the left*) and Carl Compton with Van de Graaff's test generator in 1931. (Reprinted from Ref. 3, copyright 1974, with permission from Elsevier.)

Industrial use of accelerators places special demands on the accelerator itself and the associated plant. Obviously, where an accelerator is part of a batch or a continuous production process, *reliability* is a key factor, as well as *compactness*. Further on the accelerator must be available for on-demand use 24 hours a day, seven days a week. The *environment* where the accelerator is to operate is also a key factor. Special consideration will be necessary where the accelerator is to operate in a clean room environment for electronic or pharmaceutical production or for security applications.

Generally, the facility must be suitable for use by operators with limited technical skills, although in some cases the supervisory staff, such as at a semiconductor plant, may have extensive engineering training and are able to carry out maintenance in-house. Compared to university and research environments, a characteristic of industrial accelerator facilities is that they are adapted for handling a large number of samples/work-pieces. An example is for quality control (QC) where a large number of standardized samples are used. Computerized records are often generated (such as image records for security scanners and logs for traceability in QC applications). Another difference is that, in contrast to research accelerators, industrial machines are constructed or set up to produce beams of electrons or a single ion species and energy. If a beam of another species is needed, it is often more efficatious in a production environment to procure a new machine.

In this article an overview of DC accelerators is briefly presented in Sec. 4. In Sec. 2 the environmental stewardship related to industrial accelerators is discussed. In Sec. 3 the main types of ion sources used for DC accelerators are described. Section 5 deals with ion optics, beam diagnosis and computer control. In Sec. 6 some specific details about DC accelerators used for different industrially applications are discussed. Finally, in Sec. 7, a perspective and outlook is given. A detailed technical description of DC accelerators and especially electrostatic accelerators can be found in Ref. 4.

2. Industrial Accelerators and Environmental Stewardship

Today there is increasing focus on environmental stewardship in industry. Electricity provides the energy to operate accelerators. Generally, only a small amount of energy is deposited in the sample or work-piece; the vast majority is converted to low grade heat, which is generally removed by cooling water. In modern plants a closed-circulation cooling system is employed and the waste heat is recycled for heating offices, etc. Compared to a continuous flow this conserves water, and also reduces corrosion which deposits metallic ions into the environment, besides decreasing maintenance costs and downtime.

Some of the materials used, either as feedstock (As, Sb, Hg, Pb) or construction materials (Be, Ni), are toxic. Other toxic materials can be the oil used in (some) vacuum pumps. However, because of the efficiency of accelerators, only gram quantities of feedstock materials are needed, and because the material is confined inside the accelerator, the probability of a major release to the environment is very small, compared to conventional chemical-based engineering approaches.

SF_6, which is often used as an insulating gas in the tank of DC accelerators and the products (e.g. SF_4, S_2F_2, SO_2F_2) resulting from electrical breakdown (e.g. corona voltage stabilization), deserves special mention, because it is a very potent greenhouse gas compared to CO_2. In accelerators the breakdown product production in SF_6 is much lower than for its main use in electrical switchgear, and this can easily be taken care of using gas circulation through an activated alumina filter [5]. Legislation in many countries requires that SF_6 be recycled and losses controlled. In any case the economic cost of procuring the gas makes this effectively mandatory. This can be done either using an in-house unit (Fig. 4) or through specialist companies that take care of SF_6 handling for the electrical power generation and distribution industries. For accelerator use, where the volumes are large (2–$10\,m^3$ and even larger for the biggest accelerators) and the pressures low (0.3–$0.6\,MPa$), it is important that the unit be fitted with a vacuum pump capable of removing the appreciable amount of the tank SF_6 inventory that remains at compressor inlet pressures below $50\,kPa$, below which compressor efficiency falls off.

Where external electron and ion beams are used, irradiation of the air produces NO_x and O_3, both of which are reactive and toxic. This can be removed by powerful air exhaust that may be supplemented with a scrubber to prevent release to the environment.

3. Ion Sources and Electron Guns

In this section positive and negative ion sources used for ion beam formation, as well as electron guns for electron beam formation for industrial applications, are described in some detail. Readers interested in

Fig. 4.   An SF$_6$ recycle unit suitable for use with a small accelerator. The gas is stored as a liquid in standard gas bottles (*left*) at a pressure of 2.16 MPa. The unit contains a dry vacuum pump which allows removal of SF$_6$ from the accelerator tank even below atmospheric pressure.

more details about positive and negative sources are referred to the chapter "Positive-Ion Sources" and the chapter "Negative-Ion Formation Processes" respectively in Ref. 4. Another valuable piece of literature is the ion source book edited by Ian Brown [6].

3.1. *Positive ion sources*

There are many types of ion sources designed to address different needs and for application to different types of accelerators. However, there are relatively few that have been widely adopted for commercially produced accelerators. These can be classified broadly into three groups: "electron bombardment sources," "radio frequency (RF) sources" and "microwave sources." The ion currents produced by these sources are from a few μA to several hundred mA. The design of a source depends highly on the specific use.

A few typical positive ion sources of different design principles are described below in some more detail.

3.1.1. *Electron bombardment sources*

This type of source utilizes a discharge in the range 40–150 V. The beam current is controlled by the discharge current — often in the range 1–2 A — from a heated cathode. The electrons emitted from the cathode pass through a gas or vapor consisting of the type of material desired for the beam. This material is fed into the discharge chamber at a pressure of ~ 1 Pa. The electrons are accelerated by the discharge voltage and confined by a magnetic field. The confinement increases the number of electron — atom collisions and therefore increases the plasma density. A number of sources have been designed using this principle.

An example is the *Freeman source*, which has been widely adopted by the semiconductor industry. The Freeman source has a slit aperture for producing a laminar beam. The narrow dimension of the beam should be aligned with the dispersion plane of the analyzing system. In this way a very high mass resolution can be obtained. The source can be designed for use with gases, oven/chemical synthesis variants and sputter versions. In Fig. 5 a Freeman source adapted with an oven assembly is shown. When adapted in sputter mode, the oven will be replaced by a sputter electrode. This electrode projects into the arc chamber through an insulator in the oven port.

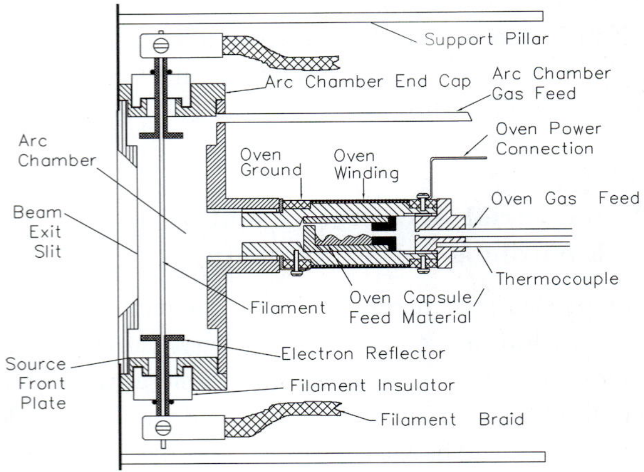

Fig. 5.   A sectioned view of the main components of a Freeman source in the oven mode. (Reprinted from Ref. 2, copyright 2009, with permission from Springer.)

The Freeman source has been used in high current ion implanters because of its simplicity and consistent performance. Example of species used for implantation are Ar, As, B, Cd, O, P, Sb, Se, Si, Te and Zn.

The duoplasmatron is another type of electron bombardment source. The plasma is composed of two regions (hence "duo") inside the source: a lower density cathode plasma maintained at a relatively high pressure (10 Pa) between the cathode and an intermediate electrode, and a high density ($\sim 10^{14}/\text{cm}^3$) plasma at a much lower pressure (about 0.1 Pa) between the intermediate electrode and the anode. Duoplasmatrons are typically used for production of singly charged gaseous ions from materials like H_2, D_2, He, C, N_2, O_2 and SF_6. The left part of Fig. 6 shows the principal design of a duoplasmatron.

A third type of source belonging to this group is the *Penning source*, named after the inventor of the Penning or Philips ionization vacuum gauge, F. M. Penning. Penning sources (also called PIG sources when in multiply charged mode) are arc discharge sources with electrons oscillating between a hot and a cold cathode or between two cold cathodes through a hollow anode, in a magnetic field. The cathode and the anticathode are at the same negative potential with respect to the anode. All of these elements are placed on a common axis, parallel to the magnetic field. The electrons attracted by the anode move along an expanding helical orbit and

collide with molecules producing ions. Examples of odd species are xenon and mercury.

A variant of this source is the *Kaufman source*, where the anode is composed of narrow metal strips positioned between two adjacent magnet pole pieces. This minimizes the power requirement.

Electron bombardment sources produce mostly singly charged ions and relatively few multiply charged ions. The number of multiply charged ions can be increased by tuning and by running the source aggressively. Unfortunately, the stability will then suffer and the source lifetime will be badly affected.

3.1.2. *RF ion sources*

The radio frequency (RF) or "high frequency" (HF) source produces ions from gaseous materials (like H_2, D_2, He, N_2 and CO_2). They are used on a wide range of accelerators, from low voltage (20–30 kV) to several tens of MV or higher. The source is constructed from a glass tube which is closed at one end with a metal extraction electrode. The other end of the tube consists of a plate with an extraction canal of 1–2 mm diameter. The gas is bled into the tube at a pressure of 0.1 Pa. An RF field of 10 to a few hundred MHz is applied to the tube and is tuned by a matching network. The RF power is 50–400 W. The beam current is controlled by the extraction voltage and the gas pressure. A few-mA beam current can be obtained. It is mostly used for light ions, for which the electron bombardment sources do not perform efficiently. In Fig. 7 a typical RF source for gaseous ions used on many single-ended accelerators is schematically presented. In Fig. 8 a photo of an RF source is shown.

3.1.3. *ECR ion sources*

The electron cyclotron resonance (ECR) source is a microwave source. A gas feed delivers the source material. This is excited by a microwave in the range 5–15 GHz. Electrons moved by the field are constrained by an axial and transverse multipole magnetic fields. The magnetic fields are arranged so that the electrons pass many times through the plasma. An ECR source produces a high yield of highly charged ions. The reason is the high temperature. The heating of the electrons by the ECR heating gives a very small energy spread of the ions. Beams of ions from solid material can be obtained by direct insertion of the material into the plasma.

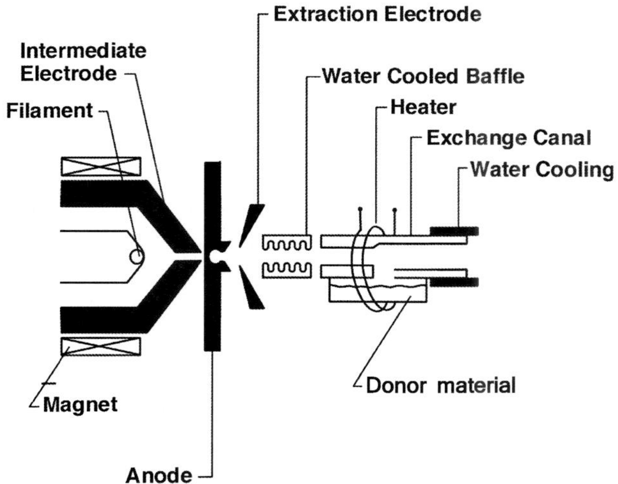

Fig. 6. A schematic representation of a close coupled positive duoplasmatron and a charge exchange cell.

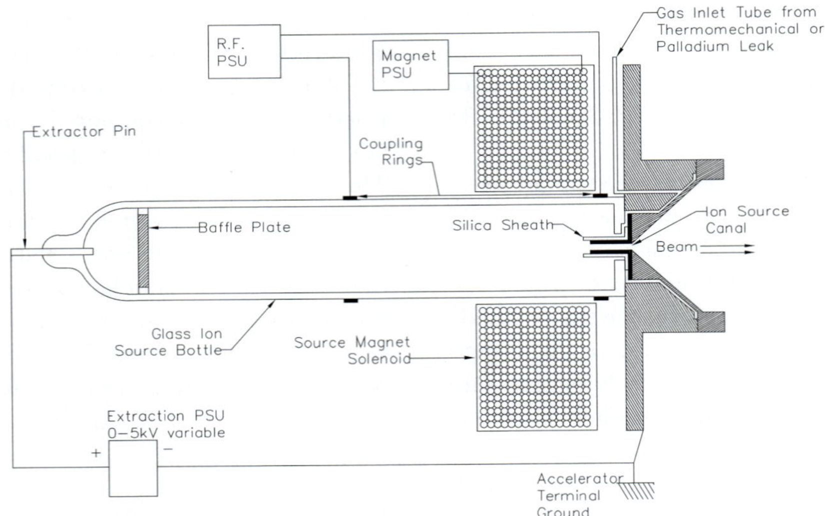

Fig. 7.   An RF source for gaseous ions. (Reprinted from Ref. 2, copyright 2009, with permission from Springer.)

Fig. 8.   Photo of an RF source in the terminal of an HVEC CN accelerator. The red plasma is clearly seen and indicates that the source is fed with H_2 gas. (Photo by R. Hellborg.)

3.2. *Negative ion sources*

Many elements and molecules can form negative ions with lifetimes long enough to be used in accelerators by adding an extra electron. The negative ions can be formed through a number of processes. These processes are all exothermic. The binding energy of the additional attached electron is the parameter that characterizes the negative ion state. It is called the electron affinity and must have a positive value. About 80% of the elements have a positive electron affinity. Unfortunately, a few have negative electron affinity, such as Be, N, Mg, Mn, the noble gas atoms and a few others.

One type of negative ion formation is in a plasma environment. Those sources employ electron impact attachment methods to form negative ions directly in a plasma and are often called direct extraction sources. The sources belonging to this group are the duoplasmatron type source, diode negative ion source and Penning discharge source or electron bombardment sources (similar to the sources described in Subsec. 3.1.1). The materials to be ionized are fed into the plasma as gases, vaporized from an oven or sublimed from a solid rod of the material.

A second type of negative ion formation is based on charge exchange. During a collision between an ion or a neutral atom/molecule and another ion or neutral atom/molecule, an electron can be transferred from one particle to the other. The probability depends on the collision energy, ionization potentials and electron affinities of the colliding partners. If a positive beam will be converted to negative, each ion has to undergo two collisions, first to become neutral and then to become negative. The practical arrangement is often a positive ion source followed by a charge exchange cell. The source can be a gas source such as a duoplasmatron or an RF source or, if needed, a more universal source capable of producing positive ions from solid material. Several materials, including Li, Na, Mg, K, Ca, Rb and Cs, have been utilized as charge exchange media. Efficiencies up to over 90% have been obtained for suitable combinations between beam and charge exchange media.

A third type of negative ion formation is through thermodynamic equilibrium surface ionization. Atoms (and molecules) impinging on a hot metal surface can be emitted as positive or negative ions. Unfortunately, negative surface ionization has not been utilized frequently for generation of an ion beam, owing to the lack of chemically stable low-work-function material for use as ionizers. However, for positive surface ionization this technique is employed frequently, as several high-work-function metals may be chosen for that purpose. LaB_6 is normally used for negative surface ionization.

A fourth type of negative ion formation is through non-thermodynamic-equilibrium surface ionization. This is the most widely used technique today to obtain a negative beam. A surface covered with a thin layer of a highly electropositive absorber like Cs is sputtered by a primary beam. The sputtering process is measured in terms of the yield of ejected surface atoms per incident primary particle. The positive ions in the primary beam are formed either by direct surface ionization of a group IA element (Cs, Rb, etc.) or in a Cs-rich noble gas plasma discharge. The primary beam is accelerated to between a few hundred eV and several keV, to sputter a sample containing the element of interest. A small proportion of the sputtered atoms are negatively charged and are thus accelerated from the negatively biased sample through an anode aperture.

A few typical negative ion sources of different design principles are described below in some more detail. For each negative source the phase space emittance area of the beam is given. This is an important figure when one is trying to fit the negative beam into the space charge acceptance area of a tandem accelerator (see Subsecs. 4.3 and 5.1).

3.2.1. *The duoplasmatron*

This type of source belongs to the group "negative ion formation in a plasma environment."

Duoplasmatron sources for production of positive ions were in use long before their capability for negative production was discovered. It was found that negative ions are more abundant in the periphery of the plasma and by offsetting the extraction electrode with respect to the center of the plasma the yield can be enhanced. To extract the negative ions from the source, it needs to be operated in reverse polarity. A duoplasmatron can produce > 10 μA of H^- and D^- from H_2 and D_2 gases, respectively. A few heavy ions (e.g. $C^-, CN^-, O^-, F^-, S^-, Cl^-$) can be produced from gaseous or liquid species and with currents $> 1\mu$A. When producing heavy ions, the duoplasmatron often becomes contaminated in a short time and the lifetime of the filament is decreased dramatically. This happens even if the gas feeding the source contains only a few percent of the gas/liquid (like CO_2) added to the H_2 to obtain the heavy ions (like C^- or O^-). A negative duoplasmatron source producing a hydrogen beam has a phase space emittance area of $(1-2)\pi$ mm·mrad·MeV$^{1/2}$ for nearly 100% of the beam. The left part of Fig. 6 shows the principal design of a duoplasmatron.

3.2.2. *Charge exchange cell*

The combination of a positive ion source and a charge exchange cell for converting the positive ions to negative can, as mentioned above, have a very high efficiency. For suitable combinations of ion type–donor material–ion energy, it can be over 90%, i.e. nearly all positive ions will be converted to negative. For heavy ions in particular, a very high efficiency can be obtained. To obtain negative helium ions, the charge exchange method is the only one available (as He^- is a metastable state [7]). Unfortunately, the efficiency for He^- production (for all donor materials) is very low, around 1%. To maximize the negative current and to increase the operational lifetime, special attention has been given to the cell design. By accurately modeling the thermal gradients in relation to the physical design and the construction material, the lifetime of the cell can today be several months. The cell should be designed with temperature gradients along the entrance and exit pipes to ensure that the donor vapor condenses, liquefies and drains back into the reservoir. The main problem with earlier cells was the condensation of the donor material in the entrance and exit openings. The negative beam currents from a cell are strongly dependent on the type of positive beam and donor material, as well as on the beam energy. The phase space emittance area of the negative ion beam from the charge exchange cell is in the range of $(1.6–2.8)\pi$ mm·mrad·MeV$^{1/2}$, from Ref. 8. In Fig. 6 a schematic drawing of a positive duoplasmatron and a charge exchange cell is shown.

3.2.3. *Spherical geometry ionizer sputtering source*

This sputtering source belongs to the group "non-thermodynamic-equilibrium surface ionization." The main development of sputtering sources started around 1970 and has gone through several stages. The original investigator was Roy Middleton at the University of Pennsilvania [9]. Most tandem accelerators around the world are today equipped with sputtering sources and many of them have an ionizer (to obtain the Cs$^+$ ions) with spherical geometry. Both computer simulations and experimental experience show that the sputter sample is optimally positioned at the focal point of the ionizer. For applications, such as accelerator mass spectroscopy (AMS) [10], where high efficiency and/or high frequency sample changes are desirable, the source has been designed with a multisample wheel. Typical beam currents are tens of μA with exceptions for some ions having higher/lower currents. A phase space emittance area of $(5–7)\pi$ mm·mrad·MeV$^{1/2}$ is typical for this type of source. In Fig. 9 a schematic drawing of a spherical geometry sputtering source is shown.

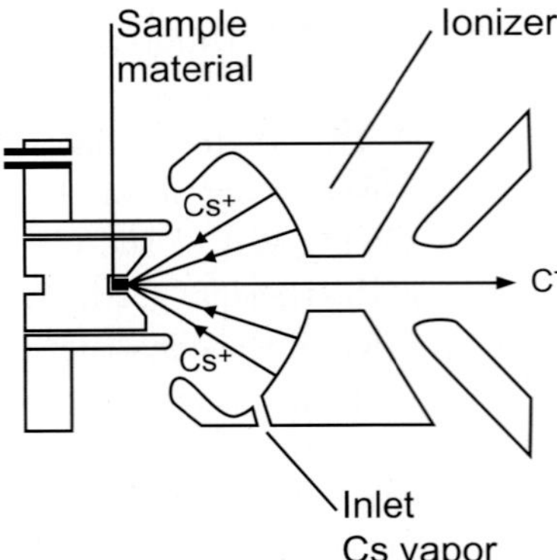

Fig. 9. A Cs-sputtering ion source equipped with a spherical geometry surface ionizer and provided with a graphite sample for C$^-$ production.

3.3. *Electron guns*

In electron accelerators a rather simple type of electron gun is used to obtain the electron beam. It can consist of a tungsten filament adapted into a cathode. Beam currents from $\sim$10 μA or lower up to several hundreds of mA or higher can be obtained. Electron accelerators are nearly exclusively used in industry for many different applications. Several thousands of different products are processed daily throughout the world. Worldwide there are over 1400 high current industrial electron beam accelerators in commercial use. An overview can be found in the chapter "Industrial Electron Accelerators" in Ref. 4. Electron accelerator use in a wide variety of processes is also overviewed in Ref. 11. An electron gun with beams in the ampere region is employed for a free electron laser based on a tandem electrostatic accelerator. The very high beam current needs to be balanced by an equally high current to and from the high voltage terminal of the tandem. An overview of this technique can be found in the chapter "Electrostatic-Accelerator Free-Electron Lasers" in Ref. 4.

4. DC Accelerators

Low energy accelerators (a few MeV or less) are nearly exclusively of the DC type. In accelerators utilizing this principle (other names are "potential drop" and "high voltage DC accelerators," because the current is DC, unlike all other accelerator types), the particle (after ionization) is accelerated through an accelerator tube in one step. The tube is constructed as a long rectilinear drift tube with a number of electrodes along the axis with a controlled voltage for each electrode, partly to aid in focusing the beam and partly to distribute the voltage gradient uniformly along the insulation surfaces. The potential drop from the high voltage terminal to ground potential accelerates the ions to an energy E equal to the charge on the ion q times the terminal voltage V (i.e. $E = qV$). The positive ions (or electrons) to be accelerated are generated in an ion source (or electron gun) located at high voltage (except for tandem accelerators — see Subsec. 4.3 — in which negative ions are generated at ground potential). DC accelerators are often identified by the type of high voltage generator used. The high voltage can be generated by rectifying AC voltage (often called a cascade generator) or using electrostatic charging in which a mechanical system carries the charge to the high voltage terminal (called an electrostatic generator). The open air accelerator fails above a few MV, mainly because electrical breakdown of the air results in sparking. To overcome this the accelerator can be enclosed in a steel tank with a suitable insulating gas at high pressure. This will significantly reduce the size of these accelerators and is required for reliable operation above 1–2 MV. SF_6 is the most widely used insulating gas for accelerators today (see Sec. 2.) The electric strength of SF_6 is nearly three times that of air (and nitrogen) at normal atmospheric pressure. It rises approximately linearly with pressure up to 0.6–0.7 MPa [12]. At this high pressure SF_6 sustains fields of about 200 kV/cm. For stable operation at high voltages, the insulators of the support structure (and the accelerator tube, as mentioned below) must be subdivided by metal electrodes which are connected to a voltage divider for voltage grading from terminal to ground. The voltage available today is up to a few tens of MV. Normally, for industrial applications, much more compact accelerators with voltages of 2 MV or less are used.

The tube for acceleration of the ions fulfills two functions: as a high vacuum compatible insulator it is required to support the highest possible fields, and as a beam transport element it must transport and focus the ions with minimum degradation and beam loss. An ideally designed tube would sustain the same voltage gradient as the gas that surrounds the terminal and column. Further, it should transmit intense ion beams of a certain emittance area without loss and the focusing must be predictable, and scattering should be small. It should also have a high vacuum conductance. Modern accelerating tubes are made up of annular insulators bound to metal electrodes across which the high voltage is applied to accelerate ions through a series of central apertures. The electrodes are bound to the insulators either by a thin layer of thermoplastic adhesive such as polyvinyl acetate or by a diffusion bond of aluminum foil formed under sustained high pressure and temperature. For more details about tubes see the chapter "Accelerator Tubes" in Ref. 4.

The two main technologies for grading the voltage of accelerators are resistors and corona systems. The ability to hold the voltage of the accelerator depends on management of the distribution of electric field stress. This relies on the voltage distribution system. Resistor systems are today the norm. Corona grading provided an adequate solution during the development of resistors. The large amount of energy stored in the electric field is distributed during a spark and caused earlier resistors to be destroyed. These failures motivated the improvement of resistors and techniques to protect them. To protect the resistors, spark gaps intrinsic to the accelerator, antenna effects, local shielding and the structure of the resistors themselves have been developed. A corona grading system is an inexpensive and spark-tolerant alternative. Another advantage of a corona system over one based on resistors is that the voltage across a corona discharge is much less sensitive to changes in the current due to beam loading. A disadvantage of the corona point system is that the corona is extinguished when the gap voltage is reduced to below a certain threshold. This would strand the accelerator without any reliable grading at all. Shorting out some sections of the accelerator to preserve a gradient sufficient to keep the remaining corona points lit eliminates this problem. For more details about voltage distribution, see the chapter "Voltage Distribution Systems — Resistors and Corona Points" in Ref. 4.

The DC principle has been used in thousands of accelerators around the world. The reasons why this type became so popular are:

- All types of ions can be accelerated;
- the ion energy can be changed continuously;
- the high voltage stability is extremely good and therefore:
- the ion energy has a very low energy spread.

An overview of different types of accelerators, including the DC type, can be found in the introduction to a book on electrostatic accelerators edited by Hellborg [4].

Each type of accelerator described in the subsections below has advantages for specific industrial applications. Each is serving needs in the growing range of industrial uses of accelerators. For analytical purposes and for voltages above 5–6 MV, the electrostatic accelerator is preferred. For implantation to a few hundred keV, for neutron-induced reactions and for electron beam applications, the cascade type is most convenient. Industrial use of very intense beams of ions or electrons is most easily met by the insulating core transformer (see Subsec. 4.1.4) and parallel-driven accelerators (see Subsec. 4.1.3).

4.1. *Cascade accelerators*

The high voltage unit of a cascade accelerator is a multiplying rectifier–condenser system. In this subsection the principal design of the power supplies for different types of cascade accelerators (or "rectifying AC voltage accelerators") will be briefly described.

4.1.1. *Asymmetrical circuit*

The original design of a cascade accelerator — employed by Cockcroft and Walton [13] to produce the first accelerator-based nuclear reaction — can be seen schematically in Fig. 10. The circuit is asymmetrical, and often the word "asymmetrical" is used for this type of accelerator. Sometimes it is called the "Cockcroft–Walton type" or a "Cockcroft–Walton voltage multiplier." The circuit was originally designed to transfer AC into high voltage DC already more than ten years before Cockcroft and Walton used it in their accelerator. It was known in the electrical engineering community as the *Greinacher doubling voltage circuit*, after Heinrich Greinacher — a professor of physics at the University

of Bern, Switzerland — who developed this circuit around 1920 [14].

The circuit includes n identical stages (in Fig. 10 three stages are shown) called cascades. It uses two stacks of series-connected capacitors C. The right capacitor stack in Fig. 10 is connected at one end to the ground and at the other to the high voltage (HV) end of the accelerating tube. The voltage across each capacitor in this stack is constant except for the ripple. One end of the left capacitor stack in the figure is connected to the transformer giving peak voltages of $\pm U$. Voltages at all points along this stack oscillate over a range of $2U$. Series-connected rectifiers R link the two stacks. As the voltage on the transformer oscillates, charge is transferred stepwise through the rectifiers from the ground to the HV terminal. The terminal voltage will be $2nU$. The chain can be extended to higher potentials, limited only by the ability of the HV terminal to hold its potential without spark discharge to the surroundings.

A few commercial producers of this type of accelerator were available for a number of years.

4.1.2. *Symmetrical circuit*

In practice, the asymmetrical circuit was soon replaced by a symmetrical circuit, as seen in Fig. 11.

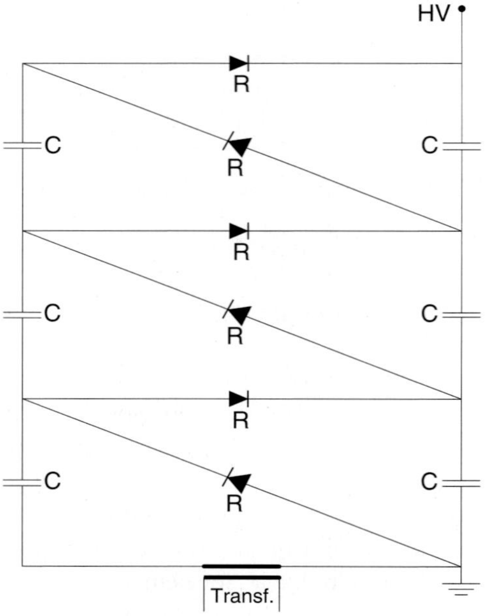

Fig. 10.   An asymmetrical circuit cascade generator consisting of capacitors C and rectifiers R. (Reprinted from Ref. 4, copyright 2005, with permission from Springer.)

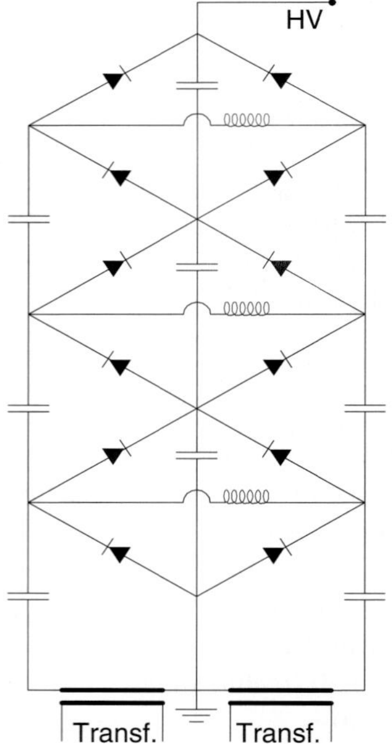

Fig. 11.   A symmetrical circuit cascade generator. (Reprinted from Ref. 4, copyright 2005, with permission from Springer.)

This type is sometimes called a "symmetric cascade voltage multiplier accelerator." It employs two transformers and two capacitor stacks (the outer two stacks in Fig. 11) that oscillate in voltage. Both oscillating stacks feed one fixed voltage capacitor stack (the central stack in the figure). The advantage of the symmetrical circuit compared to the asymmetrical circuit for voltages above 400–500 kV can be seen from the voltage drop ΔU and the voltage ripple δU when loaded with a current I. For the asymmetrical circuit these are given as [15]

$$\Delta U_{\text{as}} = \frac{I}{fC}\frac{n}{3}\left(2n^2 + \frac{3}{2}n - \frac{1}{2}\right), \qquad (1)$$

$$\delta U_{\text{as}} = \frac{I}{fC}\frac{n}{2}(n+1). \qquad (2)$$

For the symmetrical circuit the corresponding equations are [15]:

$$\Delta U_{\text{s}} = \frac{I}{fC}\frac{n}{3}\left(n^2 + \frac{3}{2}\right), \qquad (3)$$

$$\delta U_{\text{s}} = \frac{I}{fC}\frac{n}{2}. \qquad (4)$$

f is the frequency of the AC supply, C the capacitance of a given stage and n the number of stages. For an increasing number of stages n, both the voltage drop and the ripple become considerably lower for the symmetrical circuit than for the asymmetrical. As both the voltage drop and the ripple vary inversely with the frequency f, a high frequency of the primary sinusoidal voltage is of importance.

Accelerators for up to several MV have been constructed. With currents of several hundred mA, they give a total beam power of several hundred kW. These generators have often been employed in injectors to high energy machines and they are commonly used as power supplies in electron microscopy. The asymmetrical and symmetrical accelerators are often open and not enclosed in a pressure tank. However, size and space requirements for open air accelerators increase rapidly with voltage, and a practical upper limit is found at about 0.5–1.0 MV.

High Voltage Engineering Europe (HVEE) and National Electrostatics Corp. (NEC) provide open air symmetrical systems for ion implantation and ion beam analysis. A few other suppliers can be found. The symmetrical type is also utilized as injectors for almost all MeV and GeV accelerators. An example

Fig. 12. A symmetrical cascade system for an injector of 500 kV, designed by NEC for the 25 MV folded Pelletron tandem at the Holifield Radioactive Ion Beam Facility at Oak Ridge National Laboratory. (Courtesy of NEC.)

can be seen in Fig. 12, showing the 500 kV injector of the Oak Ridge National Laboratory's 25 MV tandem.

4.1.3. *Parallel-driven circuit*

A third principle, sometimes called a "parallel-driven cascade rectifier," is shown in Fig. 13. This way to obtain the high voltage for a cascade accelerator was introduced by Radiation Dynamics Inc. Their product is called the Dynamitron. Inside the accelerator tank two large semicylindrical RF electrodes are mounted near the wall of the tank and surrounding the column. These electrodes are supplied with power from an RF oscillator at 100 kHz and they will form the tuning capacitance of an LC resonant circuit. The high voltage column is enclosed by half-rings with smooth exterior surfaces to inhibit corona and spark discharges. In these segments along the accelerator column, secondary voltages are induced by capacitive coupling. These segments are coupled to rectifiers, and the rectified voltage from each segment

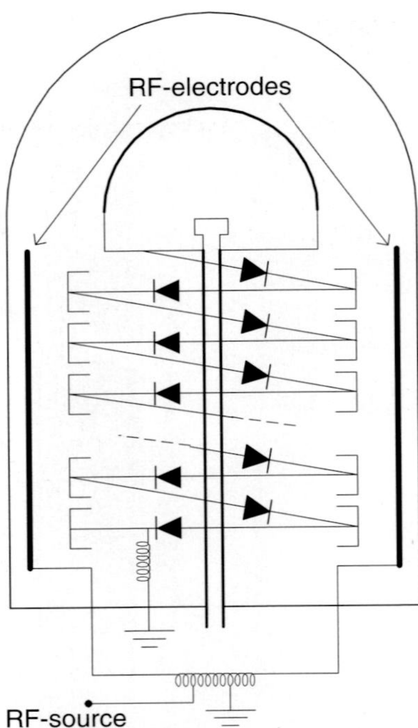

Fig. 13. A parallel-driven circuit. (Reprinted from Ref. 4, copyright 2005, with permission from Springer.)

is added up in two rows on opposite sides to supply the terminal with a high voltage. The accelerator tank is filled in the normal way with spark-protecting gas. The DC voltage produced by each segment is 50 kV.

An analog of the Dynamitron relying on magnetic parallel coupling was developed by the Budker Institute of Nuclear Physics in Novosibirsk.

At the end of the 1970s, Purser and coworkers [16] designed a similar type of parallel-driven circuit to be used for the — at that time — new Tandetrons produced by the General Ionex Corporation. The Tandetron (today produced by HVEE) is a compact tandem for material analysis, accelerator mass spectrometry, ion implantation, etc. It has a 50 kHz driver, delivering several mA with very high stability. For accessibility, the high voltage stack in Tandetrons up to 3 MV is at a right angle to the accelerator column. In the 5 MV Tandetron delivered in 2001–2002 to the Centro de Micro-Analisis de Materiales in Madrid [17], the high voltage stack is parallel to the high energy column. A photo of the Madrid machine is shown in Fig. 14. The solid state power supply is wrapped around the high energy accelerator tube (to the left in the photo). The RF field is supplied by a pair of large electrodes, called dynodes. The end of the upper dynode can be seen at the top of the accelerator tank opening. The RF voltage present on the dynodes is capacitively coupled to the power supply diode stack via the corona rings, also seen in the photo. A very high beam current can be accelerated in a parallel-driven accelerator with a total power up to 200 kW. Driving the stages in parallel rather than in series reduces the stored energy to levels comparable with those of electrostatic accelerators. Minimizing stored energy is important, especially in MV accelerators, because high stored energy released in a discharge can damage capacitors, rectifiers, column components, etc. Parallel-driven accelerators are ordinarily designed

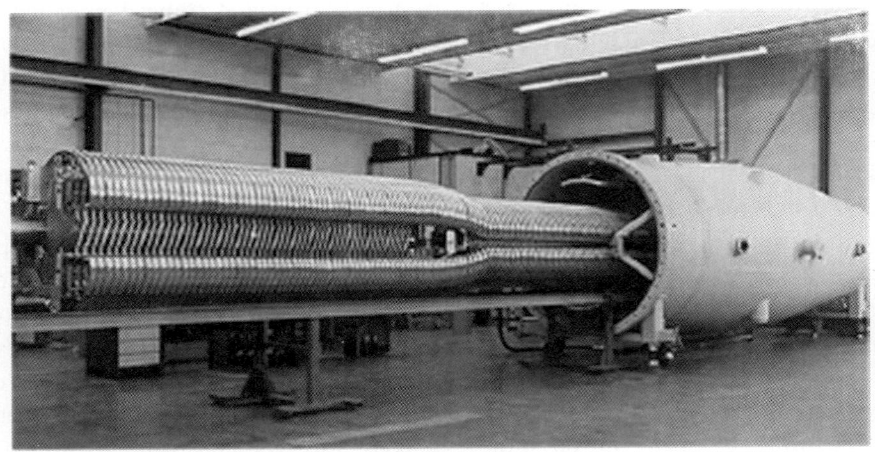

Fig. 14. The 5 MV Tandetron for the Universidad Autonoma de Madrid, under test at the HVEE plant in Amersfoort, The Netherlands. (Courtesy of HVEE.)

for higher voltages than series accelerators and are enclosed in a tank. A second Tandetron with the high voltage stack in parallel to the high energy column (i.e. coaxial) was recently installed at Helmholtz Zentrum Dresden-Rosendorf. The design voltage of this machine is 6 MV.

Another main producer of commercial cascade accelerators today is Kobe Steel Ltd., Japan.

Most Dynamitrons have been built for electron acceleration to be used for industrial polymerization and sterilization. These applications require electron energies up to a few MeV and with a very high beam power up to 200 kW at 5 MeV.

4.1.4. *Insulating core transformer*

The High Voltage Engineering Corporation (HVEC) developed the insulating core transformer (often abbreviated to "ICT") circuit for industrial high current electron beam processing systems. The design of an ICT is shown in Fig. 15 and a photo is found in Fig. 16. The core is divided into sections separated by spacers of insulating material. It is excited through the primary windings (using a three-phase, 400 V system at 50 or 60 Hz). Input power is magnetically coupled to secondary coils (three per deck) by a three-phase iron core electrically insulated between each deck. Each of all the secondary sections is coupled

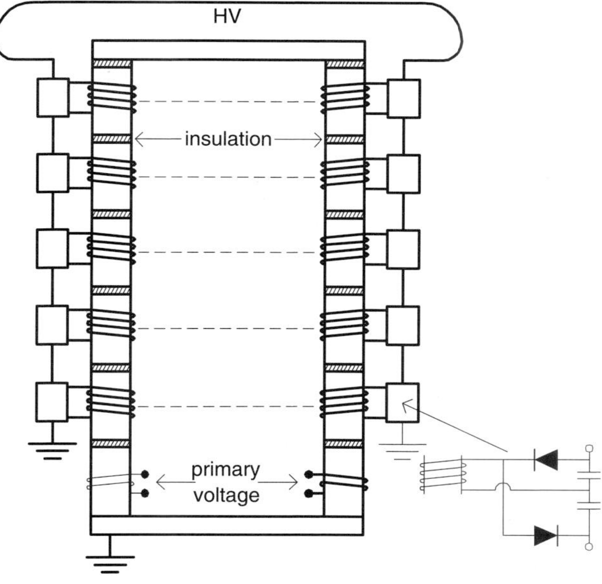

Fig. 15.   An insulating core transformer. Two phases of the three-phase system are shown in the drawing. (Reprinted from Ref. 4, copyright 2005, with permission from Springer.)

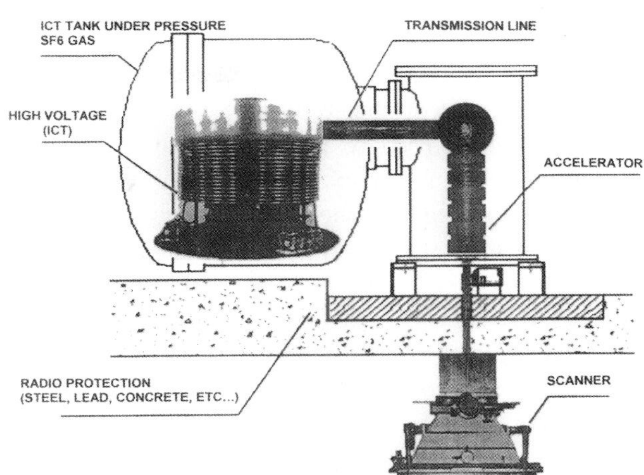

Fig. 16.   An insulating-core transformer for industrial applications. (Reprinted from Ref. 4, copyright 2005, with permission from Springer.)

to a rectifier operating as a voltage divider and is an independent 50 kV unit. The rectifier outputs are connected in series to produce the high voltage.

This type of voltage supply is housed in a tank filled with gas and can be built to be very compact. Units up to several MV and tens of mA giving beam power up to several hundred kW are available. The accelerating column may be directly connected to the high voltage terminal or may be physically separated from the transformer and connected to it by a high voltage shield cable. In 1983, HVEC withdrew from the accelerator business and the ICT technology was adopted by other companies like Vivirad-High Voltage, the Cryovac division of the Sealed Air Corporation, Nissin-High Voltage and Wasic Associated. ICTs are today produced by HVEE and VIVIRAD with energies up to 3 MeV and beam power up to 100 kW.

4.2. *Electrostatic accelerators*

The high voltage unit of an electrostatic accelerator is a charge transport system moving charge from the ground to an insulated unit. In this subsection the principal design of the different systems available (belt-driven, chain-driven and drum-driven) will be briefly described. A disadvantage of an electrostatic accelerator is the low beam current output compared to the cascade accelerator. More details about different types of accelerators and especially electrostatic accelerators can be found in the extensive textbook Ref. 4.

4.2.1. *Belt-driven*

In 1929 Robert Van de Graaff demonstrated the first generator model of this type [18]. An electrostatic charging belt is used to produce the high voltage. Two rollers are provided, one at ground potential driven by a motor and the other located in the high voltage terminal, well insulated from the ground. An endless belt of insulating material passes over the rollers. A metal screen or thin shim is pressed lightly onto the surface of the belt which is in contact with the (grounded) roller. The screen is mounted on an insulating holder and up to several tens of kV can be supplied to the screen. The polarity can be positive or negative, depending on what polarity is needed on the high voltage terminal.

The belt conveys the charge to the insulated high voltage terminal. In the high voltage terminal a similar screen connected to the terminal removes the charge from the moving belt. The voltage required is 5–10 kV, to obtain 100–200 μA of charging current. Belt-driven accelerators can carry up to 1 mA of current on their outer surface. For most applications this is more than enough.

The voltage on the terminal rises until it is discharged by a spark to the ground or limited by the corona or an electrical load, such as a resistor string along the insulating support or an ion current through the accelerating tube. Conventional charging belts are made of a cotton multilayer carcass, which has been coated with a vulcanized rubber material on the inside and outside. Belts have been produced in narrow (150 mm wide) widths for accelerators up to 1 MV and in wide (520 mm) widths for accelerators up to 20 MV. The primary reason for variations in the current delivered to the high voltage terminal is the inhomogeneity of the belt surface. The belts are handmade and the rubber coating is vulcanized in sections, producing a variation in thickness at the overlap. The principle of a belt-driven electrostatic generator is illustrated in Fig. 17.

4.2.2. *Chain(ladder)-driven*

The insulating belt for transport of the charge has been replaced almost entirely in many modern electrostatic accelerators by a chain (or "ladder"; see below) of metal cylinders (often called pellets). The individual cylinders are connected to each other

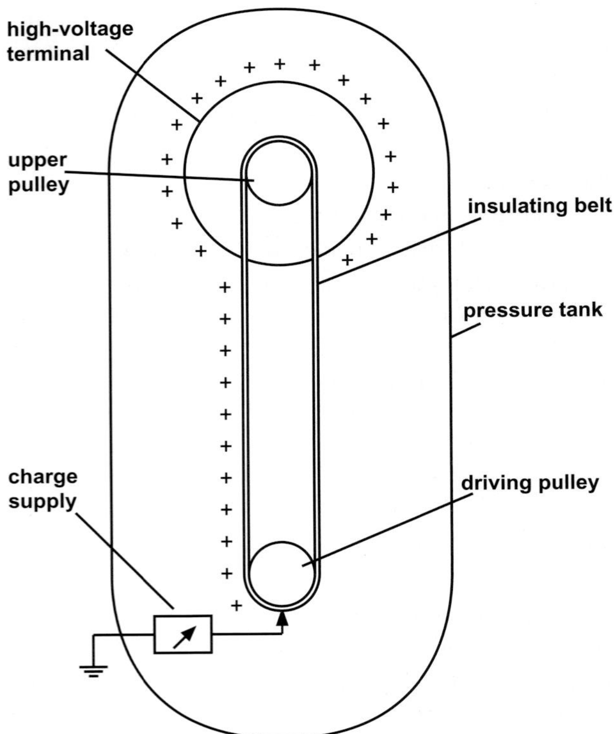

Fig. 17. The principle of an electrostatic generator. By including an ion source and an accelerator tube, the generator is converted to an accelerator. (Reprinted from Ref. 2, copyright 2009, with permission from Springer.)

by insulating links. Each metal cylinder is charged inductively. In this way a more robust transport system with much more well-defined charge transport and hence better voltage stability is obtained, compared to using an insulating belt. Another principal disadvantage is belt flexure, which leads to mechanical oscillations and thereby a terminal ripple. A further disadvantage with a belt is the large amount of dust built up in the tank after some period of operation since the last cleaning. This dust can initiate sparks between the tank and high voltage parts. A chain is capable of carrying only up to 150 μA each. Multiply chains are therefore installed in accelerators requiring greater charging capacity. In Fig. 18 a chain system containing four chains with a total capacity of 600 μA is shown.

The first charging chain system was developed by Jim Ferry and Ray Herb in the early 1960s [19]. In 1965 NEC was founded by Herb, and the chain system has been included in all electrostatic accelerators produced by NEC. Accelerators produced by NEC are sold under the trademark Pelletron.

Fig. 18. A Pelletron provided with four charging chains. (Courtesy of NEC.)

The NEC charging chain is constructed of plated steel tubing with the diameter and length equal (31.75 mm). Bushings are mounted between the inner pellet wall and the nylon link, which prevents the nylon link from moving from side to side and keeps it centered in the pellet. The rolled ends of the pellet act as spark gaps to protect the nylon links from spark damage. The chain rotates on two wheels and travels at a speed of about 13 m/s. To overcome the large mechanical strength on the chain when starting the chain motor, some laboratories have installed an electronic system so as to have a smooth start and a reduced stress on the chain during the startup procedure [20]. The same system can also be used as a simple way to adjust the chain current. Charge is induced on the chain as it leaves the grounded end by a negatively (to obtain a positive high voltage terminal) charged electrode called an inductor. This is biased up to 50 kV. The induced charge on the chain corresponds to 3–4 μA/kV of inductor voltage at full chain speed. Positive charge flows from the ground through the wheel to the metal link on the chain. As the wheel rotates, the contact between the link and the wheel is broken and the positive charge is trapped by the insulating nylon connecting links. The charge is in this way mechanically transported to the high voltage terminal.

NEC is the main producer of chain-driven commercial electrostatic accelerators today. Several photographs showing different sections of a Pelletron facility appear in Fig. 19.

Researchers at Daresbury Laboratory in the UK developed a different type of charging chain in the 1970s and HVEC commercialized it. It consisted of two chains running side by side, with metal crossbars connecting adjacent pellets. It was termed a "ladder." The ladder was guaranteed to carry 250 μA of current.

Fig. 19. The 1.7 MV Pelletron tandem accelerator facility in Jyväskylä, Finland, used for industrial consulting and materials research and development. *Left photo*: The low energy end of the accelerator, with two different ion sources. One is for production of ions of gaseous elements (H, He, O), while the other is for production of ions from solid materials (C, Cl, Cu, Br, I, etc.). *Middle photo*: The accelerator tank. *Right photo*: The different materials measurement and modification stations. An RBS measurement station used for industrial consulting work is on the far right. To the left of this is a ToF–E ERDA system. On the far left is a MeV ion beam lithography system used for materials modification.

4.2.3. *Drum-driven*

A method similar to the use of a belt or chain has been employed by a French company to obtain up to a few hundred kV in a compact machine. In this accelerator, often used as a neutron generator to obtain fast neutrons through a suitable nuclear reaction, a cylinder is rotating and transporting charges at a high speed around another fixed cylinder, all enclosed in a tank with gas at high pressure; see Fig. 20. Such accelerators are often designed for a few hundred kV and up to 0.5 MV, and have a stability of $\Delta E/E = 10^{-3}$–10^{-4} and a beam current of up to 10 mA. The speed of the rotating cylinder is a few thousand revolutions per minute. The distance between the fixed and the rotating cylinder is a few tenths of a mm. The rotating cylinder is made of an insulating material.

Industrial applications of drum-driven accelerators include extensive use of many techniques, such as:

- Study of neutron diffusion;
- Study of neutron-induced reactions;
- Production of radioisotopes;
- Activation analysis;
- Moderator and shielding studies;
- Subcritical assembly studies;
- Neutron radiography;
- Radiation effects on materials;
- Metallurgical analysis;
- Ecological investigations.

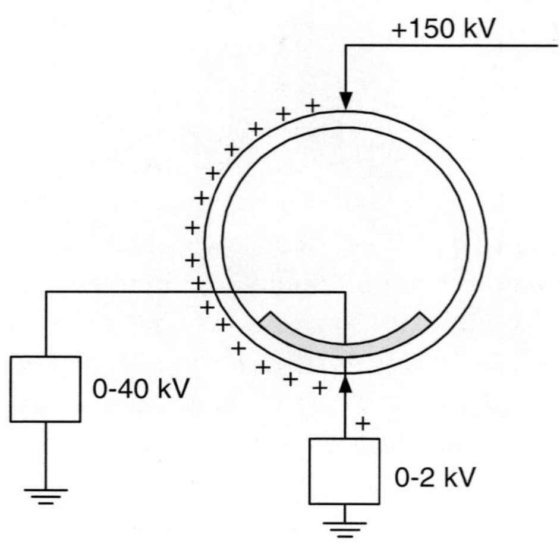

Fig. 20. An electrostatic generator using a rotating cylinder to obtain the high voltage. (Reprinted from Ref. 4, copyright 2005, with permission from Springer.)

Sealed tube type neutron generator. In connection with the drum-driven type accelerator, the sealed tube type accelerator should also be mentioned, as it is used for the same purpose, i.e. to produce fast neutrons (see Sec. 6). The sealed tube type provides a source of fast neutrons of up to around 10^9 n·s^{-1}, compared with 10^{13} n·s^{-1} from the drum type accelerator (distributed over 4π steradians). The maximum useful flux available (dependent on the sample size and physical arrangement) is a factor of 100 lower in units of n·cm^{-2}·s^{-1}. This system is relatively compact, allowing for a degree of portability. Because of its degree of simplicity, it requires less monitoring and supervision than the larger drum type.

A sealed tube contains an ion source, an accelerator system (usually 100–200 kV), a replenisher to maintain a constant pressure within the tube and a self-replenishing tritium gaseous or metallic target. The tubes are frequently 300–650 mm in length and interconnected with the required power supplies and control.

The metallic targets are usually tritium targets containing 350–750 GBq of tritium. The target is usually water- or oil-cooled.

The tube lifetime for continuous operation for neutron outputs of 10^{10} n·s^{-1} can be expected to exceed 500 h.

4.3. *Tandem vs. single stage accelerators*

During the 1950s, negative ion sources were developed, i.e. beams of ions with an extra electron added to the neutral atom became available. This development made it possible to build two-stage (or tandem) accelerators. In a tandem the high voltage is utilized twice. Negative ions are formed at ground potential and injected into the first stage, where acceleration to the positive high voltage terminal takes place. The energy gain is $e\,U_T$ eV, where e is the elementary charge and U_T the terminal voltage. In a stripper system located in the high voltage terminal, the negative ions lose a few electrons and become positive ions. In the second stage, the positive ions once more gain energy. Now the energy gain is $q\,e\,U_T$ eV, where q is the charge state of the positive ion. Thus a total energy gain of $(q+1)\,e\,U_T$ eV is obtained. For heavy ions and high voltages, i.e. high speed ions undergoing stripping, q can be quite high and therefore the final energy of the ions can be hundreds of MeV.

Changing the injected negative ions into positive ions is the defining feature of acceleration in a tandem accelerator. This process is called "stripping." It results in loss of beam transmission, efficiency and voltage-holding ability. The charge state fraction and scattering leads to a reduced beam intensity. The pressure in the tubes will also affect the transmission, and therefore methods for achieving a low tube pressure have been developed. This often includes terminal pumping, i.e. pumps installed in the high voltage terminal. An obvious advantage of a tandem over a single stage accelerator is that the ion source is more accessible, making service work easier.

As a stripping medium gas or foils can be used. If a stable transmission has high priority, one should choose gas stripping. If the highest beam energy has high priority, foil stripping is preferred. Unfortunately, by foil stripping the beam intensity will vary and nonoptimal transmission will be obtained. A short foil lifetime for heavy ions limits the application of foils to low and medium mass ions and to low intensity applications.

For more details about stripping, see the chapter "Stripper Systems" in Ref. 4. Two boxes in that reference deal with "Charge Exchange and Electron Stripping" and "Carbon Stripper Foils — Preparation and Quality."

A drawing illustrating the principle of a tandem accelerator is shown in Fig. 21.

4.4. *Radiation protection at an accelerator laboratory*

The phenomenon called "radiation" is the transport of energy in the form of a stream of atomic particles or electromagnetic quanta (photons). No supporting medium is required. Radiation can be divided into ionizing and nonionizing. Ionizing radiation has a higher energy than nonionizing. (As a rule of thumb, ionizing radiation has an energy of the order of atomic or molecular binding energy, i.e. 10 eV or higher, while nonionizing radiation has an energy below 10 eV.) Ionizing radiation can — as the name suggests — ionize material when interacting. "Ionizing" means that electrons are removed from the atoms/molecules in the material by the radiation.

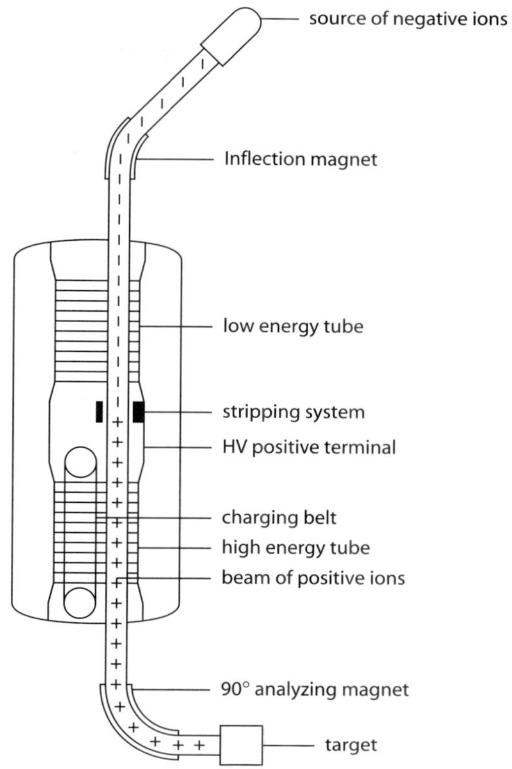

Fig. 21. The principle of a two-stage (tandem) accelerator. (Reprinted from Ref. 4, copyright 2005, with permission from Springer.)

In this way, charged particles, i.e. ions, are produced. If this happens in a human body, radiation injuries can result.

Ionizing radiation is not a new phenomenon connected with human activity. It has always been available and is present throughout the environment. However, it is only during the last century that we have learned to detect ionizing radiation and to produce artificial ionizing radiation.

Many accelerators are used with low intensities and therefore normally have a low radiation level. However, it should be remembered that in general any accelerator can produce hazardous levels. Even if the ion (or electron) source is switched off, stray electrons can be accelerated over the high voltage gap, producing bremsstrahlung when hitting material.

For standard use with low intensities, only minimal shielding is normally required in the accelerator laboratory. This could be a risk, as personal may become careless if one considers the radiation hazard as negligible. It is therefore important to measure the radiation level whenever uncertainty exists.

Readers interested in more details about fields like radiation and its interaction with matter including living material, quantities and units, dose measurements, radiation hazards and safety considerations, are referred to the chapter "Radiation Protection at an Accelerator Laboratory" in Ref. 4.

4.4.1. *Detecting ionizing radiation*

All detectors for ionizing radiation are based on the same fundamental principle — the transfer of energy to the detector. In the detector the energy is converted into some other form that can be registered. Modern detectors are essentially electrical — at some point, the information from the detector is transformed into electrical pulses that can be treated by electronic means.

It should be recognized that all instruments are limited to measuring certain types of radiation within a fixed range of energies. Outside these limits, instrument readings are not to be trusted. The detectors can be simple counters, spectrometers or more sophisticated 1D and 2D imaging detectors. The latter are commonly met with in security applications.

X-ray, β-ray and γ-ray detection. Ionization detectors were the first electrical devices to be developed for radiation detection. These instruments are based on the direct collection of the ionization electrons and ions produced in a gas by passing radiation. Three basic types of detectors have been developed: the ionization chamber, the proportional counter and the Geiger–Müller counter. These types of detectors are today mostly used as radiation monitors, as they are cheap, simple to operate and easy to maintain.

The *scintillation detector* makes use of the fact that certain materials when irradiated emit small flashes of light, i.e. they scintillate. When coupled to an amplifying device such as a photomultiplier, this scintillation light is transmitted through a shaped light pipe to the photocathode of a photomultiplier. There, photons release electrons, which are accelerated and focused onto the first dynode. For each primary electron hitting a dynode, between two and five secondary electrons are released. Up to 15 multiplying stages can be used, reaching overall multiplying factors of up to 10^9. A few incident photons therefore produce a measurable electrical pulse, which can then be analyzed and counted electronically.

A recent development is the silicon avalanche photodiode array ("Si photomultiplier"). This is geometrically much smaller than conventional photodiodes and has similar gain and quantum efficiency but is faster, requires 20–50 times smaller voltages and is insensitive to magnetic fields.

Semiconductor detectors are based on crystalline semiconductor materials, most notably silicon and germanium. The basic operating principle of semiconductor detectors is analogous to that of gas ionization devices. However, the medium is now a solid material. The passage of radiation through the solid material creates electron–hole pairs (instead of electron–ion pairs). The advantage of a semiconductor is that the average energy required per electron–hole pair is some 10 times smaller than that required for gas ionization. Thus, the amount of ionization produced for a given particle energy is an order of magnitude greater, resulting in superior energy resolution. Moreover, semiconductor detectors have a greater density and, therefore, a greater stopping power than gas detectors. As a result, they are more compact in size and can have a very fast response time.

Tritium detection. In accelerator laboratories that use tritium (e.g. for neutron generation) it is important that proper ventilation techniques are used to ensure that any release of tritium is adequately exhausted from the laboratory. It is also necessary to provide air monitoring for realization of tritium gas. A common device for tritium monitoring is an ionization detector through which air is drawn at a fixed rate. The ionization chamber may be preceded by an ion collector and/or filter in order to remove other ions or radioactive material. Tritium contamination can be detected and analyzed through the use of samples collected with filter paper, or other type of traps, and then transferred to a liquid scintillation solution. Liquid scintillation counters are also the instrument of choice for monitoring tritium in urine samples.

Neutron detection. Neutron radiation may be accompanied by relatively high levels of γ-radiation. Consequently, in order to measure the neutron level adequately, it is necessary to have an instrument insensitive to γ-radiation. Ionization instruments are in general not satisfactory for measuring neutrons, since they are also sensitive to γ-radiation.

The boron trifluoride (BF_3) proportional counter provides a sensitive detector for a neutron survey instrument and can be relatively insensitive to γ-radiation. This instrument uses BF_3 gas enriched to more than 95% in the isotope ^{10}B, and typically the detector has about a 100 mm active length. The BF_3 counter is sensitive to thermal neutrons. It is also possible to detect intermediate and fast neutrons by enclosing the detector in polyethylene or paraffin wax, which brings down the neutron energy before the neutrons enter the BF_3 gas. By a suitable moderator configuration it is possible to achieve a count rate which approximates the dose equivalent rate for neutrons in the energy range below 10 MeV.

The ^{3}He neutron detector is more sensitive than the BF_3 counter. It is based on the reaction ^{3}He(n, p)T. The proton and the tritium will ionize the gas, and an electrical pulse will be obtained for every absorbed neutron. A disadvantage is that the ^{3}He gas is very expensive and to have a sensitive detector it needs to be large. The sensitivity is strongly dependent on the ^{3}He gas pressure, volume and the enrichment, and can be up to several hundred pulses per incoming neutron per cm^2.

Cerium-doped lithium-silicate glasses are widely used as neutron detectors. Recent developments incorporate this scintillator into plastic fibers acting as light wave guides [21]. Both flat and bent large area neutron detectors can be built by this technique. The light output is monitored at the end of the glass fibers.

5. Ion Optics, Beam Diagnosis, Computer Control

5.1. *Ion optics*

The ion beam from the ion source of a cascade or electrostatic accelerator is divergent. The source is therefore frequently followed by an accelerating gap (often the ion source extractor or a downstream gap lens). The next optical unit is the accelerating tube. The tube can be described as consisting of an (strong) entrance lens, a field region where the acceleration takes place, and a weak exit lens. The accelerator is mostly followed by a quadrupole doublet or triplet. This lens will focus the beam at the object point of the magnetic dipole. The magnetic dipole acts as a strong lens to refocus onto the image slits. The dipole is also used to convert variations in beam momentum (magnetic dipole) or energy (electric dipole) into spatial displacements that are detected at slits and fed back to a control system which corrects for errors in the accelerating voltage. The first dipole magnet can be followed by additional lenses. The ability to change the beam direction is a valuable one, and multiport magnets facilitate directing the beam to more than one target station. Between different dipoles, focusing elements like quadrupoles may be necessary.

For low energy (< 300 keV) accelerators, the magnetic analysis of the beam often takes place after the extraction acceleration (at ~ 20–50 kV) analysis, and is then followed by a second acceleration (or, for very low energy, deceleration) stage. The advantage of this arrangement is that the secondary acceleration supply then only needs to supply enough current for the analyzed beam and not the full output of the source, which may contain different charge states, molecular beams or beams of ions from ion source components (such as W from filaments).

The negative beam from the source of a tandem accelerator is very divergent and has a low energy. It has to be focused by a lens and accelerated in a first accelerating step. This acceleration can be from a few tens of kV to several hundred kV, depending on the size of the tandem. After this first acceleration, the beam is generally analyzed in a magnetic or electrostatic analyzer before being fed to the accelerator. The entrance lens of the accelerating tube behaves like a very strong lens. To avoid the loss of part of the beam, the beam phase space emittance area when entering the tube must be within the phase space acceptance area of the accelerator. The emittance area of the beam is mainly determined by the design of the source in use. The emittance areas of a few different negative ion sources are given in Subsec. 3.2. The phase space acceptance area of a tandem accelerator is governed by the dimensions of the stripper canal and the entrance of the low energy beam tube, in conjunction with the strong lens effect at the beginning of the tube. As an example, the phase space acceptance area of the 3 MV Pelletron in Lund (in use between 1976 and 2005) was calculated to be 27π mm·mrad·MeV$^{1/2}$ at a terminal voltage of 3.0 MV and 13π mm·mrad·MeV$^{1/2}$ at a terminal voltage of 0.5 MV [22]. This value is well above the emittance area values for different negative sources given in Subsec. 3.2. For a well-designed system, the beam will have a focus in the stripper canal at the center of the tandem. In the high energy accelerating tube, the beam is divergent and must, after the accelerator, be handled similarly to the beam from a single stage accelerator.

Focusing of the ion beam along the experimental beam line down to submicrometer dimension is described in detail in the chapter "Focusing keV and MeV Ion Beams" in Ref. 2.

5.2. *Beam diagnosis*

Beam envelope and trajectory calculations can today be made using a number of general purpose beam transport codes. They will, in most cases, describe the real situation in a very satisfactory way.

A number of parameters are related to an ion beam. The main parameters are the type of isotope, charge state and energy. These parameters are assumed to be determined by the operation of the source and the setting of the analyzing magnets and of the different acceleration voltages. As has been mentioned earlier, the emittance and brightness of the beam are given from the setting of the source and acceleration voltage. In the case of a tandem accelerator, the stripping process will also influence these parameters. Very few accelerator laboratories actually measure these parameters continuously. Other parameters — such as beam current, beam diameter, position and intensity distribution, and energy spread — are often measured during operation by different types of diagnostic instruments. Information about the beam current is needed for optimizing the operation and, in some types of experiments, for calculating the implanted fluence. The current can be measured by a destructive or a nondestructive method. Examples of destructive methods are total beam stopping (with a Faraday cup) and partial beam stopping (with a scanning wire or a rotating sector disk). An example of a nondestructive method is measurement of the residual-gas ionization. Monitoring the beam diameter, position and stability can be achieved through a screen which emits light under beam irradiation. The screen can be a plate coated with a luminescent material like ZnS, MgO or Al_2O_3. Another possibility is a quartz disk. The lifetime of a coated screen is limited and it has to be replaced regularly. To suppress discharge effects when using quartz, the irradiated surface can be covered with a metallic net or pencil marks in the case of a ceramic

screen. The emitted light can be observed directly through a glass window or in a remote mode by the use of a TV camera. The position and intensity distribution of the beam can be measured by a beam profile monitor. Such a device can consist of one or more insulated, thin, moving wires. The current signals from the wires give information about the intensity distribution in the horizontal and vertical directions. The profile monitor produced by NEC has a different design, in that it has a grounded wire surrounded by a collection cylinder. The secondary electrons hitting the cylinder are measured. The voltage stability of an electrostatic accelerator (and therefore also, indirectly, the energy stability of the beam) can be monitored by a capacitive pickup. This is an electrode at the tank wall connected to a current integrator. Since the electrode current is proportional to the derivative of the terminal voltage, the output of the integrator represents the AC component of the terminal voltage. The signal is usually displayed on an oscilloscope and may even be fed back into the accelerator stabilization system to improve stability.

5.3. *Computer control and data acquisition*

Some decades ago, the control of an accelerator and associated elements was carried through a central console containing a large number of knobs, switches, meters, etc. This console was often located some distance from the accelerator and with long and expensive cables. Each element of the system was controlled by an often-homemade chassis in the console. The startup of the system or major changes of the settings could require help from a skilled operator. Often it needed hours of tuning and retuning of the system. In most modern accelerator systems, a properly designed computer system has limited most of these issues. Computerized control systems are less expensive to implement, more reliable and precise, easier to modify, and even allow remote operation over the Internet. The change from one set of parameters to another is rapid and more flexible.

A large number of software and hardware solutions are available for such a control system. Interested readers are referred to, for example, the chapter "Computer Control" in Ref. 4. A principal design of a control system consists of a computer, a device interface and the different devices. The operator interacts with software in the computer. The computer communicates with the device interface. This unit contains all the analog and digital inputs and outputs needed to control the different devices. The devices can be power supplies, vacuum valves, beam stops, profile monitors, etc. In large accelerator systems, several or many device interfaces can be needed and may require multiple computers distributed throughout the whole accelerator complex. The control system is organized in a hierarchical way, with subsystems for vacuum, high voltage, RF supplies, etc. Many accelerator control systems have been based on the CAMAC and VME standards. However, newer, more economical and more powerful interface standards are evolving all the time, such as the industry standard PXI interface, which is based on the industry standard PCIe computer bus, and these new standards will certainly be implemented at accelerator facilities. Accelerators, from a control theory viewpoint, are multi-input, multi-output systems and generally present a mixture of on/off, threshold, PID and fuzzy logic control.[a] Several software packages are available. LabVIEW™ is perhaps the most widely used among small and medium accelerators. On larger facilities, which often use a distributed UNIX environment, C and C++ are generally preferred. The communication between the computer and the interface is usually based on standard computer protocols and interfaces with wireless, copper cables and fiber-optic physical links with standards such as UTP, FTP, TCP/IP RS-232 and RS-422. Fiber-optic and wireless data links are not only useful for communicating across large potential differences, such as a high voltage terminal or an ion source platform, but also they provide a convenient way to eradicate ground loops. This is particularly valuable, because in accelerator facilities one often wants to measure extremely small signals (e.g. from a high resolution detector) in close electrical proximity to a high current plant such as magnet power supplies, transformers and vacuum pump motors. Figure 22 illustrates an antonymous industrial computer for LabVIEW computer control and monitoring applications (National Instruments™ c-RIO system) during development.

Fig. 22. A National Instruments™ c-RIO system under development for web-based control of a Pelletron accelerator.

The unit has a standalone computer running a field-programmable gate array (FPGA) that interfaces with standard slot-in multi-input, multi-output and relay controller interfaces in the crate. Using standard library routines, this unit can handle the control functions (such PID control of magnets and terminal voltage) of an industrial accelerator. One of the most useful features of computer control of accelerators is the possibility of displaying real-time and historical data on machine parameters. Figure 23 shows a web-based display from a

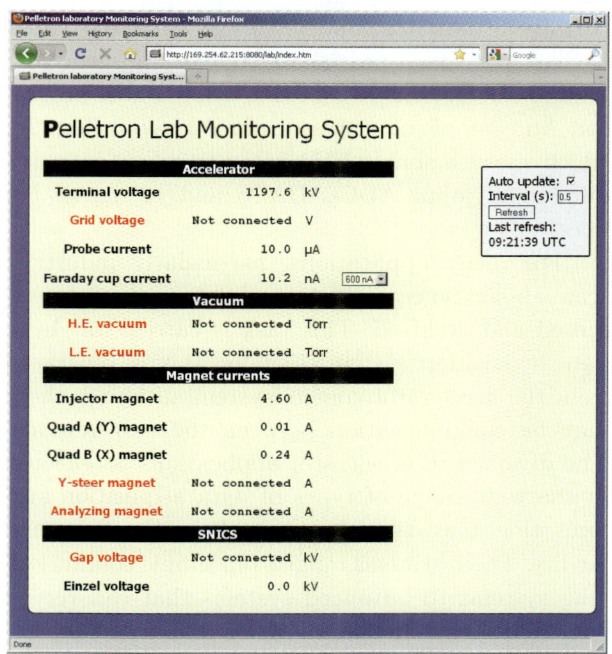

Fig. 23. Screen shot of the webpage for monitoring a small Pelletron using the system in Fig. 22. (Courtesy of Jaakko Julin.)

---

[a]The term "fuzzy logic" presumably originated from poor translation of "expert control logic" from Japanese. Fuzzy logic is in fact as precise as other control methods.

Fig. 24. A National Instruments^{TM} USB-6509 computer interface suitable for controlling accelerator subsystems such as the ion source, vacuum systems and magnets.

parameter logger implemented using the controller shown in Fig. 22 at a small Pelletron accelerator. For control of small subsystems (such as an injector vacuum system), low cost solutions based on peripheral interface controllers may be used. These devices have small processors built-in, but generally require calculations to be made using a host processor. An example of such a device is given in Fig. 24, which shows a simple USB-based interface with 4/8 differential input ADCs, DACS and 12 digital i/o lines.

For many applications, particularly industrial scale applications, data must also be collected, presented and archived. This task is carried out by a data acquisition system. Usually, this is separated from the accelerator control system, although there may be communication between the two systems. The diversity of accelerator applications is reflected in the wide range of types of data acquisition systems that may be encountered, even in the same facility. These systems range from simple counterlogging systems to advanced systems that can record signals from many thousands of detectors simultaneously. For some industrial applications the data acquisition system may also control hardware to perform sequences of programmed tasks, such as changing samples. The measured spectra, images and key accelerator operational parameters are then recorded in files for later analysis.

6. Application-Specific Accelerators

Table 1 presents an overview of industrial applications of accelerators. The wide span of industrial applications of accelerators is clearly seen from this table. The list is by no means exclusive and "industrial applications" has been taken rather loosely to cover applications that are associated with industrial production, including the use of accelerators for governmental agencies to control pollution. A common denominator for the applications is that for all the applications the work-piece handling is highly automated. This may be a package conveyor to allow entire medical disposables to be irradiated within the outer package in electron beam sterilization to automated handlers for the PIXE analysis of aerosol sampler disks.

The industrial applications can be broadly classified into materials modification and characterization. The former includes electron beam processing and ion implantation. The latter is generally used in QC. As industrial production migrates toward nanotechnology-based products, there is an increasing demand for methods for QC that can work as small dimensions where conventional characterization methods are at, or beyond their limits. Accelerator-based techniques such as an RBS with magnetic spectrometers are able to analyze at the nm or sub-nm level and are therefore finding increased application for industrial QC.

A well-developed example of an industrial accelerator for industrial application is the ion implanter used for doping semiconductors. These have evolved from the early calutron electromagnetic separators which were used for industrial scale $^{234/235/238}$U isotope separation. The implantation machine consists of: (i) the accelerator, which provides the energetic ion beam; (ii) the scanning system, which distributes the ion beam irradiation uniformly over the surface of the semiconductor wafer; and (iii) the wafer handler, which transports the wafers from the input cassette and transfers them after irradiation to the output cassette. The ion source in the accelerator part is typically of the Freeman or Nielsen type, or a derivative of these. Usually, for dopant implantation in Si, gaseous feedstock source materials such as diborane (B_2H_6), phosphine (PH_3) and arsine (AsH_3) are used. Although these are highly toxic, the hazards associated with them are easily managed because

Table 1. Overview of industrial DC accelerators.

Application	Particle	Beam energy	Beam current	Accelerator type	Characteristics	Environment
Electron microscopy	e^-	60–250 keV	0.1–1 pA	Cascade	QC applications	Test lab
Electron beam processing	e^-	80–300 keV (surface curing) 300–800 keV (shrink film) 0.4–3 MeV (wire and cable) 3–10 MeV (sterilization)	Up to several mA and even higher	Cascade, linear	Package belt	Factory
Environmental monitoring (PIXE)	^{1}H	1–2 MeV	50 pA–30 nA	Cascade, electrostatic	Automated sampler analysis	Central laboratory
Ion implantation (semiconductors)	H^+, B^+ As^+, P^+	5 keV–2 MeV	1–100 mA	Cascade	Automated work-piece change	Clean room
(metals)	N^+, C^+, Ti^+	10–100 keV	1–500 mA	Cascade	Large work-piece change	Mechanical workshop
Isotope production	p, d, α	10–30 MeV	50 μA–2 mA	Cyclotron linear, electrostatic	Production of ^{11}C, ^{18}F, ^{123}I	Pharmaceutical clean room
Neutron generators	^{2}H	200 keV	1–20 μA	Cascade	QC applications Hydrocarbon Prospecting n-hardness	Test lab Security control facility Drilling rig Test lab
Ion beam analysis (IBA)	^{1}H, ^{4}He	300 keV–2.5 MeV	100 nA–1 μA	Cascade, electrostatic	RBS QC	Factory

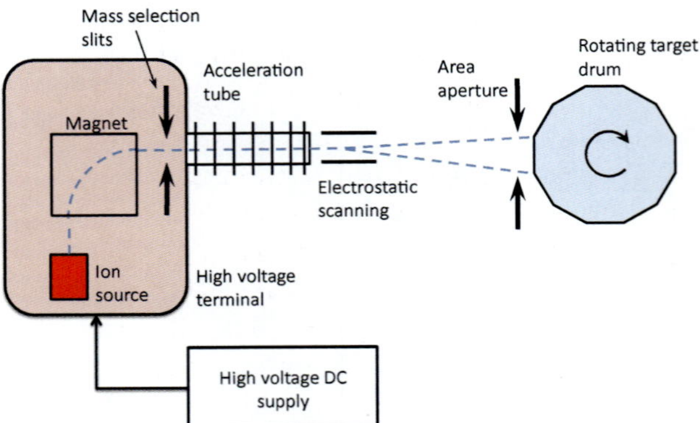

Fig. 25.   Schematic illustration of a typical ion implanter for small scale industrial semiconductor processing. The ions are produced by the ion source and accelerated to a few tens of keV. The 90° magnet filters the ions of the desired mass, which are then accelerated/decelerated to 1 keV to a few hundred keV by the accelerator tube. A combined electrostatic and mechanical scanning is then used to distribute the ion beam uniformly over the surface.

the quantities are small. Often, two acceleration stages are used; in the first the ions from the ion source are accelerated to 10–30 keV, and then the ions with the desired mass are selected using a magnetic sector, and subsequently the ions are accelerated/decelerated to their final energy. This is illustrated in Fig. 25, which shows a schematic of an ion implantation system for small batch industrial use. By placing the analyzing magnet in the terminal before the main acceleration stage rather than after, a smaller and hence lower cost magnet can be used.

Electrostatic lenses are used to focus the beam. The scanning system that distributes the beam fluence uniformly over the sample can be either electrostatic or mechanical, or a mixture. The former is well suited to small wafer sizes, which are often encountered in research and development applications. The largest Si wafer sizes are currently 300 mm diameter (in production) and 450 mm diameter (in development), and they present a challenge for electrostatic scanning because of defocusing. For this reason mechanical scanning is often preferred. The loading and unloading of wafers is time-consuming and poses a risk of introducing mechanical damage. For this reason cassette-to-cassette robotic handling is often used and this may be integrated into the scanning system. The entire system is computer-controlled and this may allow implantation of the

same ion species at different energies. Where different ions of p- and n-type dopants are to be implanted, separate machines are generally used in production because of the downtime associated with changing over the feedstock material, which usually involves stripping down and cleaning the ion source. Clean room environments are needed for industrial processing of semiconductors, in order to maintain economically high yields. For this reason the machine must be designed for clean room operation. For example, in order to minimize the detrimental effects of dust particles from ion source cleaning, etc., the ion source and accelerator end may be placed in a service area so that only the target station protrudes into the clean room. In this way the wafers are maintained in a clean room environment, while maintenance does not affect other clean room processing.

Another example where accelerators are used for materials modification is electron degradation polymerization. Degradation of linear chain polymers, such as poly(methyl methacrylate) (PMMA) by electron-induced main chain scission, renders them more susceptible to dissolution in selective solvents (developer). These polymers are termed "resists," because of their analogy to the classic photoresists used in the printing industry. By using an electron beam of 5–200 keV energy that is finely focused to ~10 nm spot size and scanned over the sample patterns with features of the order of 10 nm can be written in thin (20–180 nm) PMMA films. This electron beam lithography (EBL) method is a key technology for producing the many masks required for nanoelectronic devices, such as those described in the article by Gordon Moore. Figure 26 shows a schematic illustration of an EBL tool.

Figure 27 shows an EBL system in a clean room environment that is used for lithographic mask making.

In order to achieve high beam currents in a small beam spot, field emission electron sources are used. The small divergence and emittance area of these sources implies that the electron beam has a small divergence and source area giving a high brightness which allows small spot sizes to be achieved. The lenses in this case are cylindrical lenses closely similar to those used in electron microscopes. They act as condenser lenses which create a small spot by producing a demagnified image of the small aperture through which the electrons pass, on the surface

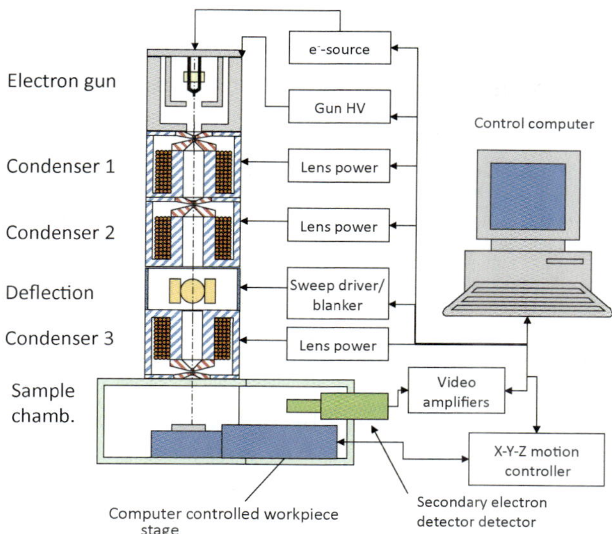

Fig. 26. Schematic illustration of an electron beam lithography (EBL) tool.

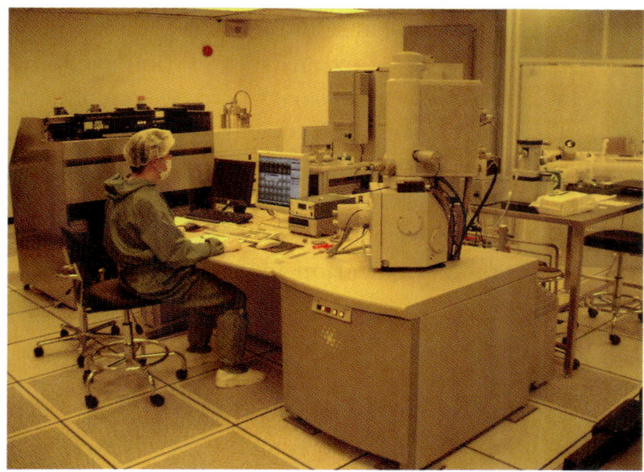

Fig. 27. An electron beam lithography (EBL) system for research and development application. This instrument is capable of writing features down to 10–20 nm or so. The tool is installed in a clean room at the National University of Singapore with yellow light to facilitate handling of blue-light-sensitive polymer resists. (Photo: H. J. Whitlow.)

of the resist material. The wafer (work-piece) is mounted on a precision stage. A secondary electron microscope is generally integrated into the tool to allow registration between different exposures on the same work-piece. The best stages with optical-interferometer-based position feedback allow registration with an accuracy of 50 nm or better. The entire EBL tool is controlled by a computer that translates the high level pattern layout instructions

into low level instructions for moving and turning the beam on and off, as well as controlling the stage and vacuum and focus, etc. The electron energy defines the penetration depth and the smearing of the dose (energy deposited per unit mass) in the resist material. The energy of the beam must be kept within tight limits ($1{:}10^5$), because otherwise the dispersion in the magnetic lenses will cause defocussing that will smear the pattern written. This requires voltage feedback regulator control of the AC supply voltage to the cascade generator.

Not only can bonds be broken in polymers, but new bonds can also be formed (cross-linking). Electron accelerators are frequently used in this way to modify polymers. For example, heat-shrink polymers are made by converting linear chain polymers to network polymers by creating cross-linking bonds. Linear chain polymers melt and flow under stress (e.g. like polyethylene); however, network polymers do not melt but become elastomers that return to their original shape when the stress is removed. Then, by heating the electron irradiation cross-linked polymer and allowing it to cool under stress, it will become glasslike and retain its shape after the stress is released. If this polymer is subsequently heated, it will relax back to its original shape, giving rise to its heat-shrink properties. A further application of this process is the production of graft polymers where two or more different polymer chains are grafted together. This can be used to modify polymer properties such as the glass transition point, improve solubility for use in drug delivery and include elastomers in crystalline polymers to improve toughness. The electron accelerators for this purpose need to produce high currents (10–40 mA) of 100 keV to 5 MeV electrons. The X-ray production is substantial and extensive shielding and safety interlocks are needed. Radiolysis of air leading to NO_x and O_3, as noted above, is often mitigated by maintaining the work-piece in a N_2 gas atmosphere. The demands on the electron gun are moderate and a thermionic emission source can be used with lifetimes of thousands of hours. Likewise, the energy stability requirement is moderate. The beam exiting from the accelerator may be formed into a broad beam to accommodate irradiation of wide work-pieces by electrostatic scanning or by the use of focussing magnets. The work-piece can be transported through the electron beam continuously (such as a film or cable) or in batches using a conveyor system. The beam current, which determines the dose rate, can be determined by sampling the beam using a Faraday cup, or by employing an ionization chamber.

As an illustration of the accelerators used for characterization, we take the example of environmental monitoring using particle-induced X-ray emission (PIXE) to measure the transport of heavy metals in the environment. While not strictly an industrial method, it has been used by legislative bodies to map out and control release of environmental pollutants from industries. The aerosol is collected in a sampler and deposited on a filter substrate. These are advanced several times a day, so that the variations with time can be followed by measuring different positions on the filter substrate. The sampler itself may separate the aerosol particles into different size fractions by, for example, a differential particle size analyzer. The filter substrates from a number of samplers at different locations are collected. The sample chambers used for aerosol analysis are generally different from normal research and development PIXE, in that they are able to measure a large number of filter substrates, each at a large number of positions (times when the aerosol was collected). The accelerator is either a single-ended or a tandem accelerator with an ion source capable of producing protons. This is often of the RF or duoplasmatron type. Although each measurement is fast (< 5 min), the sheer number of measurements means that the data collection system must be automated. Since protons of moderate energies and low beam currents (0.1–$1\,$nA) are used, the radiation shielding requirements are modest and, generally, modern machines can be placed in the laboratory without the need for shielding walls.

Neutrons have the property that, unlike X- and γ-rays, they interact strongly with hydrogen and other light elements (H, D, Li, Be, B). They penetrate 5–30 cm into many common construction materials (metals, polymers). These properties have led to significant industrial application, such as for neutron activation analysis and hydrocarbon detection. A more recent interest had been in detection of contraband explosives, narcotics, etc. in packages, because neutrons can induce specific reactions with H, C, N and O. There the relative intensity of the reactions gives a signature for the type of material. Neutrons can be produced using small DC accelerators (known as *neutron generators*) or, more commonly, radioactive sources and research reactors. While the latter can provide high neutron fluxes (10^{12}–10^{13} n·cm^{-2}·s^{-1}), their use is restricted. This is because they are available only at a few fixed locations and security issues may restrict access. Accelerator bases sources (see Sec. 4.2.3) are in comparison small and can even be transportable, and the neutron fluxes are lower (10^{6}–10^{10} n · cm^{-2} · s^{-1}) but still greater than for radioactive neutron sources. Table 2 presents a number of nuclear reactions used to produce fast neutrons with the help of an accelerator. Thermal neutrons can of course be obtained by surrounding the target with a moderator, such as a water tank or graphite. The neutron generator is simply a DC generator, frequently of the cascade type or an integrated ion accelerator, often referred to as a neutron tube. Unlike reactor and radioactive sources, the neutrons produced by neutron generators are monoenergetic. Neutrons interact strongly with water in organisms, inducing secondary radiation. For this reason, extensive radiation shielding (0.5–1 m concrete) is necessary where personnel must be in close proximity to the neutron source.

Neutron radiography generally needs higher fluxes and brightness, and therefore uses reactor neutrons. The low fluxes available from neutron generators limit their application for radiography unless the levels of the absorber are high (such as radiography of cast explosives for the military and space industry).

Neutron activation analysis (NAA) was a widely used technique in the 1960s and '70 s. Typically, a 50–500 mg sample is sealed in a vial of polyethylene

Table 2. Fast-neutron-producing reactions suitable for use with an accelerator.

Ion	Reaction	Threshold/ MeV	Q-value/ MeV	Neutron energy at threshold/ MeV
d^+	$T(d, n)^4He$	None	$+17.586$	14.046
α	$^9Be(\alpha, n)^{12}C$	None	$+5.708$	5.266
d^+	$^9Be(d, n)^{10}B$	None	$+3.79$	—
d^+	$D(d, n)^3He$	None	$+3.266$	2.46
α	$^{13}C(\alpha, n)^{16}O$	None	$+2.201$	2.07
d^+	$^{12}C(d, n)^{13}N$	0.328	-0.281	0.0034
p	$T(p, n)^3He$	1.019	-0.764	0.0639
p	$^7Li(p, n)^7Be$	1.882	-1.646	0.0299

or high purity silica glass and is activated by neutrons. The decay γ-rays are subsequently measured with regard to the energy and decay constant in a Ge detector. This allows measurement of up to about 70 elements in the sample with varying instrument detection levels (IDLs) of 10–10^7 ng/g when a neutron generator is used as the source of neutrons. Often, the sample need not be removed from packaging but the analysis can be made with the packaging intact. The technique has been used extensively in geological prospecting, because it is multielemental and because of the ease with which it can be applied to rock and mineral samples as well as assaying the total contents of toxic elements such as Hg, Pb and Cd in industrial waste. The neutron energy spectrum plays a significant part in determining the sensitivity. Generally, thermal neutrons are used. The fast neutrons from the neutron tube can be converted to thermal neutrons by means of a moderator; this may significantly reduce the neutron intensity. However, for some measurements the capture of fast 14.1 MeV neutrons from $T(d, n)^4He$ is used. These methods have fallen somewhat out of general use and been replaced by PIXE and inductively coupled plasma spectroscopies. Presumably, this is a consequence of the need for massive shielding and legislative restrictions concerning dual use technology and the requirement to treat the samples as radioactive waste after measurement.

Neutron security scanning methods have been attracting increasing interest over the last decade or so. This is because these methods are able to identify the composition of material by identifying the components of the γ-ray spectrum induced by fast neutrons. A promising technique is pulsed fast neutron analysis (PFNA), which is capable of 3D detection of small amounts of contraband (explosives, chemical weapons, narcotics, etc.) from the levels of nitrogen and oxygen in the object, which may be hidden in a shipping container. A pulsed neutron source driven by a DC accelerator is used to excite characteristic γ-rays. Several companies are developing advanced security scanners based on PFNA. These can be combined in hybrid systems that merge conventional X-ray and γ-ray scanning and can be in a fixed facility, or mounted on a lorry or trailer for mobile analysis. As yet, relatively few facilities have been installed, and there is still uncertainty about the economic impact of the method. This is according to a report to the US Congress, owing to the high cost of the installation (US\$10 million per installation) and because the scan times are still long. However, the potential is enormous, since cargoes can be scanned and verified without unpacking, which saves on personnel costs — a factor of paramount importance to airline container handling and commerce.

Neutron well-logging (often termed "nuclear well-logging") has been widely used in geological prospecting as one of a range of well-logging methods. It is often used to measure the porosity and identity of the pore fluid in the geological formation surrounding a borehole. It is a wire-line logging technique where a compact neutron tube producing pulses of fast neutrons (usually 14.1 MeV; see Table 2) is mounted in a sond along with detectors that log the scattered neutrons and induced γ-rays. The sond is progressively lowered into the borehole (which can be up to 10 km deep) by a cable and the different signals logged with time (depth). A number of different signals may be measured, such as: the degree of neutron moderation which indicates the porosity of the rock from the scattering by hydrogen; the C/O ratio; and the neutron absorption, which gives the rock density and induced radioactivity associated with the elemental composition indicating minerals, such as shales which are indicative of hydrocarbon reserves. In practice, other sensors are included that measure the resistivity of the rock to allow differentiation of hydrocarbons from water. The technique can be employed on uncased boreholes as well as cased boreholes in wells previously used for production. Obviously, because of the high cost of setting up a production well, and the fact that the number of boreholes made each year worldwide is on the scale of tens of thousands, the economic impact of neutron well-logging is extremely significant. The rising cost of oil has even prompted neutron well-logging of depleted wells and low yield boreholes to identify economically viable sources that were previously discarded. In addition to oil prospecting, neutron well-logging has been widely used for prospecting uranium and other precious metal ores.

An overview of industrial use of different ion beam analysis methods — including PIXE — can be found in an article in an upcoming book by R. W. Hamm and M. E. Hamm Ref. [23].

7. Perspective and Outlook

DC accelerators, which were historically the first form of accelerators to be developed,[b] continue to play a major role for a wide range of industrial applications. This is because they are able to provide high currents and very high energy stability. The role they have directly played as an *enabling technology* in the development of a wide variety of modern technologies is enormous. In the Second World War, the Manhattan Project would been impossible without the calutron for separation of ^{235}U from less fissile isotopes. The development of ion implantation doping of semiconductors in the 1960s, which stemmed from the early ion accelerators, has been one of the key technologies underpinning the remarkable predictive power of Moore's law, which has held true over many decades of the computing and mobile communication revolution. DC accelerator science has also played several other key roles in this. One is EBL, which is based on an electron accelerator and is an essential technique for making the extremely fine line masks for the manufacture of processors and memory chips. Another is the role of ion beam analysis in the evolution of silicide contact technology (see the chapter "High Speed Electronics" in Ref. 2), which is key to modern high speed electronics. Almost in parallel with the electronics revolution, life science technology has undergone a similar revolution, starting with the structure determination of DNA through DNA sequencing and modern gene technology. Remarkably, ion implantation has found several key niche areas. The most visible today is ion beam cancer therapy, which, although not based on DC accelerators, draws upon much of the know-how in ion implantation. Other emerging areas are ion implantation mutogenesis of seeds for breeding of cash crops and sterilization. Today there are several applications at the research stage that have major future potential, such as the development of micro- and nanofluidic devices for biomedical applications like biochemical test kits (e.g. Fig. 28) and modification of biosurfaces.

The *socioeconomic value* of DC accelerator technology has truly been enormous. Modern cars rely on e-beam welding technology to construct precision components for gears. The same technology is

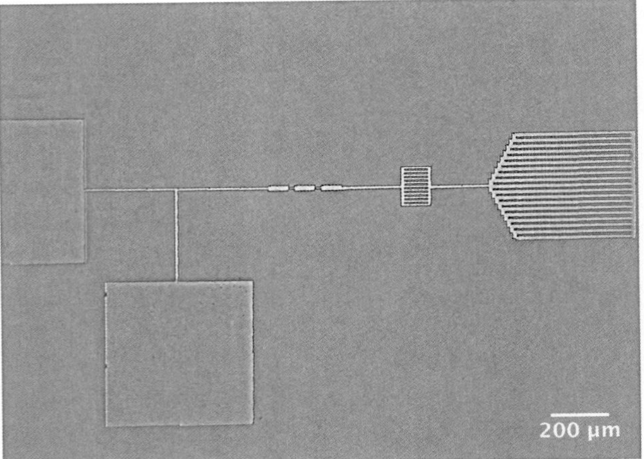

Fig. 28.   Optical microscope image of a prototype microfluidic analytical device for immunological use, fabricated using MeV ion beam lithography with 3 MeV ^{4}He ions in 12 μm PMMA/Si. (From Ref. 24.)

emerging as a cost-effective way to denature polluting toxins such as dioxins and PCB. It is even the only way to economically fabricate some essential products for society, such as shrink-wrap polymers. The direct effect on the tax bill of disposable consumables in medical and veterinary use that are sterilized by electron beam irradiation, reducing the need for troublesome and sometimes unreliable autoclave sterilization, is difficult to estimate but must be enormous. It is by no means unrealistic to claim that one particular DC accelerator, the CRT in 1950s–2000 TV receivers, has promoted world peace and humanistic actions, and in doing so has saved many lives by providing live news coverage of international developments. There is no obvious reason, such as a competing technology, why industrial applications of accelerators cannot continue to play an increasing role through new applications, particularly as enabling methods for nanoscience technology. Table 3 lists the (main) manufacturers of accelerators and the types of accelerators. Where AC accelerators are used, these are listed as well. According to a recent survey, over 1000 systems are shipped a year, with a sales value of US$2 billion a year [1]. The commercial value of products from accelerators is US$500 billion a year. DC accelerator systems are certainly a significant fraction of these figures.

---

[b]An interesting semifiction popular work that describes this is *The Fly in the Cathedral*, by Brian Cathcart.

Table 3.   Major producers of industrial accelerator equipment.

Application	Manufacturers	Accelerator type
Electron microscopy	Philips, FEI, JEOL, Hitachi	Cascade
Electron beam processing	Varian, Nissin-High Voltage, Radiation Dynamics*, Sealed Air Corporation, Vivirad, Wasik Ass., Efremov Inst., Budker	Cascade (ICT), electrostatic, linac
IBA (incl. PIXE, AMS, material anal.)	NEC, HVEE, Budker, Kobe Steel, Efremov Inst., Radiation Dynamics*	Cascade, electrostatic
Ion implantation	Varian, Nissin-High Voltage, HVEC, Axcelis, SEN, NEC	Cascade (rarely electrostatic)
Isotope production	HVEC, NEC, IBA	Electrostatic, cyclotron (IBA)
Neutron generators and equipment	Adelphi Technol., SODERN, Halliburton, Schlumberger, OSI Systems Security	Cascade
Injectors for large acc	HVEE, NEC	Cascade (sym.)

*Today IBA (Ion Beam Applications SA).

Acknowledgments

This work was supported by the Academy of Finland, Center of Excellence in Nuclear and Accelerator Based Physics, Ref. 213503. Max Strandberg, from the Faculty of Engineering, Lund University, kindly prepared Figs. 3, 10, 11, 13, 15, 17, 20 and 21.

References

[1] R. W. Hamm and M. E. Hamm, The beam business: accelerators in industry, *Phys. Today*, June 2011, pp. 46–51.

[2] R. Hellborg, H. J. Whitlow and Y. Zhang (eds.), *Ion Beams in Nanoscience and Technology* (Springer, Heidelberg, 2009).

[3] D. A. Bromley, *Nucl. Instrum. Methods* **122**, 1 (1974).

[4] R. Hellborg (ed.), *Electrostatic Accelerators: Fundamentals and Applications* (Springer, Heidelberg, 2005).

[5] R. Hellborg and K. Håkansson, *Nucl. Instrum. Methods A* **235**, 407 (1985).

[6] I. Brown (ed.), *The Physics and Technology of Ion Sources* (Wiley and Sons, 1989).

[7] A. Wolf, K. G. Bhushan, I. Ben-Izhak, N. Alstein, D. Zajfman, O. Heber and M. L. Rapaport, *Phys. Rev. A* **59**, 267 (1999).

[8] P. Tykesson, in *Symp. Northeastern Accelerator Personnel* (Oak Ridge, Tennessee, 23–25 Oct. 1978).

[9] R. Middleton and C. T. Adams, *Nucl. Instrum. Methods* **118**, 329 (1974).

[10] R. Hellborg and G. Skog, *Mass Spectrum. Rev.* **27**(5), 398 (2008).

[11] International Atomic Energy Agency, *Industrial Electron Beam Processing*, IAEA draft report, Dec. 2010; available at http://www.cirm.org/pdf/industrial-eb-processing-december-2010-revision4.pdf.

[12] R. Hellborg, *Nucl. Instrum. Methods A* **379**, 185 (1996).

[13] J. D. Cockcroft and E. T. S. Walton, *Proc. R. Soc. A* **129**, 477 (1930).

[14] H. Greinacher, *Z. Phys.* **4**, 195 (1921).

[15] M. Minovic and P. Schulze, *Hochspannungstechnik* (VDE-Verlag, Berlin, 1992).

[16] K. H. Purser, R. B. Liebert and C. J. Russon, *Radiocarbon* **27**, 794 (1980).

[17] G. G. Lopez *et al.*, The Centro de Micro-Analisis de Materiales equipped with a 5 MV Tandetron, in *Proc. Symp. North-Eastern Accelerator Personnel* (Strasbourg, Oct. 2003).

[18] R. J. Van de Graaff, *Phys. Rev.* **37**, 1919 (1931).

[19] G. A. Norton *et al.*, A retrospective of the career of Ray Herb, in *Proc. Eighth Int. Conf. Heavy Ion Accelerator Technology*, ed. K. W. Shephard (AIP, 1999), pp. 3–23.

[20] R. Hellborg and K. Håkansson, *Nucl. Instrum. Methods A* **268**, 408 (1988).

[21] M. J. Weber, M. Bliss, R. A. Craig and D. S. Sunberg, *Radiat. Eff. Defects Solids* **134**, 23 (1995).

[22] R. Hellborg, K. Håkansson and G. Skog, *Nucl. Instrum. Methods A* **287**, 161 (1990).

[23] R. W. Hamm and M. E. Hamm (eds.), *Industrial Accelerators and Their Applications* (World Scientific, to be published in 2012).

[24] H. J. Whitlow and M.-Q. Ren, Ion beam analysis in biomedicine, in *Ion Beam Analysis Principles and Applications*, eds. M. Natasi, J. W. Mayer and Y. Wang (Taylor and Francis, 2011), in press.

Ragnar Hellborg is Professor in applied nuclear physics. He received his Ph.D. in Nuclear Physics at Lund University in 1973. His research at different Direct Current Accelerators in Lund and elsewhere, has mainly been in the field of applied nuclear physics. That is he has used nuclear techniques and equipment to solve problems in other areas, mainly within radioecology and biomedicine. He has produced more than 100 publications and he has edited and co-edited four books and written a number of overview chapters for books edited by colleagues. The deep knowledge about direct current accelerators has been obtained by detailed technical development work, mainly with the 3 MV tandem Pelletron in Lund.

Harry J. Whitlow was appointed Professor of Experimental Materials Physics at the University of Jyväskylä, Finland, in 2004. He obtained a B.Sc. (Hons.) from the University of Bath, UK, in 1976 and a doctorate from the University of Sussex, UK, in 1981. Subsequently he has worked as a researcher at Aarhus University, Denmark; Helsinki University, Finland; The Royal Institute of Technology, Sweden; and Lund University, where he was appointed Professor in 2000. His research field is ion–matter interactions, with significant contributions in time-of-flight ERDA, detector technology and MeV ion beam lithography. He has supervised nine Ph.D. students, produced over 160 publications and coedited two books.

Reviews of Accelerator Science and Technology
Vol. 4 (2011) 213–218
© World Scientific Publishing Company
DOI: 10.1142/S1793626811000537

Radio-Frequency Electron Accelerators
for Industrial Applications

Marshall R. Cleland

*IBA Industrial, Inc., 151 Heartland Blvd.,
Edgewood, NY 11717, USA*
marshall.cleland@iba-group.com

Two types of radio-frequency electron accelerators are described in this article. They operate in the frequency range
of 100–200 MHz and are energized with triode or tetrode tubes instead of klystrons. They can provide more powerful
electron beams than typical microwave linear accelerators.

Keywords: ILU accelerators; Rhodotron accelerators; electron beam processing; X-ray processing.

1. Introduction

Radio-frequency (RF) electron accelerators use a single or several large resonant cavities to generate strong, alternating electric fields. In contrast to typical microwave linear accelerators (linacs), with many smaller cavities that resonate in the S-band range at about 3000 megahertz (MHz), or in the L-band range at about 1300 MHz, RF accelerators are designed to operate in the frequency range of about 100–200 MHz. These much lower frequencies allow them to be energized with triode or tetrode tubes, which are more efficient and less expensive than higher-frequency klystrons for providing the higher-beam-power capabilities that are needed for operating industrial electron accelerators.

There are several different types and models of RF electron accelerators that can generate kinetic energies ranging from 0.8 to 10 million electron volts (MeV) with average beam power ratings from 20 to 700 kilowatts (kW). These systems and some of their applications for radiation processing of materials and products are described in this article.

2. ILU Single-Pass Axial Electric
Field Accelerators

A description of this type of accelerator is given in the US patent [1] and in later papers [2–7]. It was designed, and has since been manufactured, by the Budker Institute of Nuclear Physics in Novosibirsk, Russia. More than 30 of these systems are installed in European and Asian countries, and are being used for radiation processing of materials and industrial products. Some examples of practical applications are cross-linking polymeric materials, like electrical wire insulation, plastic tubing and a variety of other commercial products.

2.1. *Single-cavity versions*

A diagram of an ILU-6 accelerator is shown in Fig. 1. It consists of a single resonant cavity energized with a triode tube. This tube is mounted on one end of the cavity and is closely coupled to the internal electric field to form a self-tuning oscillator circuit. A separate RF source is not needed to drive the tube. The frequency of oscillation is 115.4 MHz. The resonant cavity is formed with two open-ended copper cups which overlap each other to provide capacitive coupling between the cups. The upper cup supports the electron emitter. The lower cup is connected to a dc voltage supply of a few kilovolts to remove low-energy ions from the cavity, which could produce resonant discharges between the opposite ends. These cups are mounted inside a grounded metallic vacuum housing [2, 3].

The numbers in Fig. 1 identify the major components: (1) vacuum enclosure, (2) RF accelerating cavity, (3) inductive coil, (4) ion vacuum pumps, (5) electron emitter, (6) electron beam scanner, (7) RF measuring loop, (8) triode tube, (9) loop

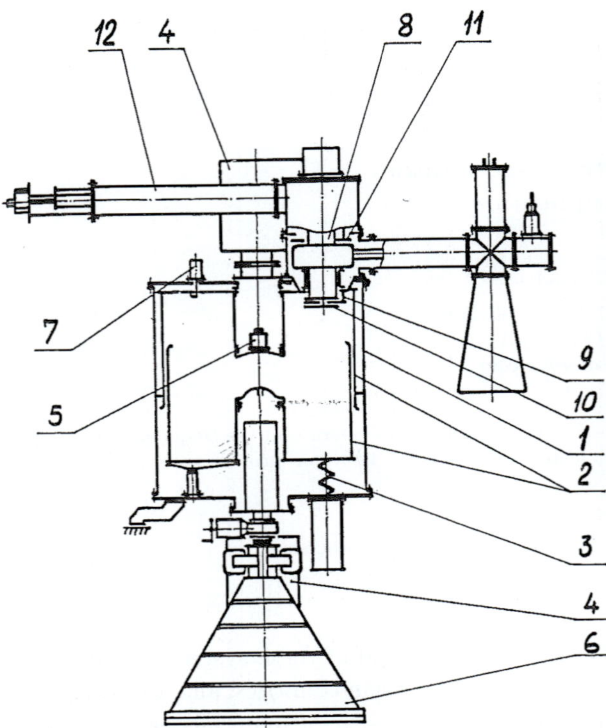

Fig. 1.   Diagram of a single-cavity ILU-6 electron accelerator.

Fig. 2.   Photograph of a single-cavity ILU-6 RF accelerator.

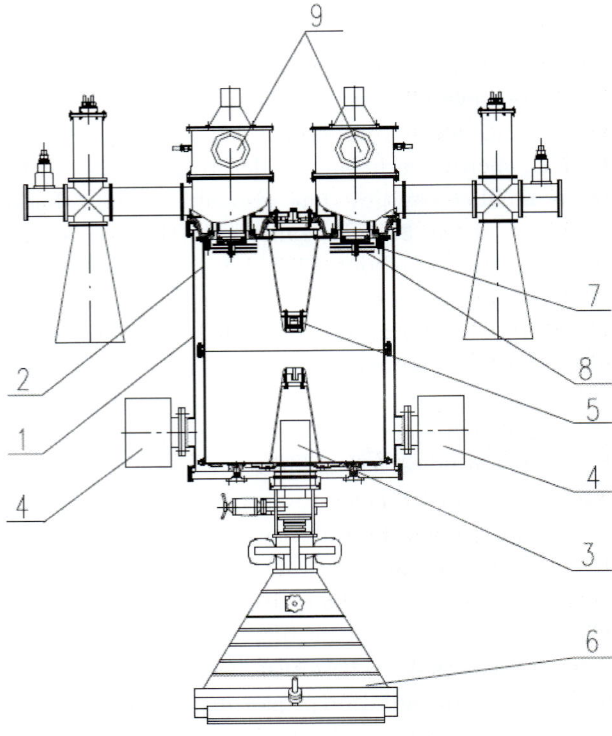

Fig. 3.   Diagram of a higher-energy, higher-power ILU-10 RF electron accelerator with two triode tubes.

for RF coupling, (10) vacuum capacitor of the coupling loop, (11) movable plate of the feedback capacitor, (12) resonant cathode stub for fine-tuning the system.

This type of RF accelerator has been made in several sizes for different applications. The ILU-6 provides electron energies from 1.7 to 2.5 MeV with an average beam power of 20 kW and an input power of 100 kW. The ILU-8 provides energies from 0.8 to 1.0 MeV with an average beam power of 20 kW and an input power of 80 kW.

For these performance ratings, the ILUs are operated in the pulsed mode, typically with a pulse length of about 400 μs and a repetition rate of 50 Hz. In the interval between pulses, the dc voltage on the triode tube is maintained at about 3–4% of the pulsed voltage. This low level of continuous excitation reduces the time for the RF field in the resonant cavity to build up to its maximum value when the full voltage is applied. A photograph of an ILU-6 system is shown in Fig. 2.

A diagram of an ILU-10 accelerator is shown in Fig. 3. It consists of a single resonant cavity with the upper and lower cups connected to each other and to the grounded enclosure. It is energized with two triode tubes operating in parallel, both of which use a self-tuning oscillator circuit like the ILU-6. The use of two tubes provides enough input power for the ILU-10 to accelerate electrons with energies from 4.0

Fig. 4. Photograph of an ILU-10 RF electron accelerator with two triode tubes.

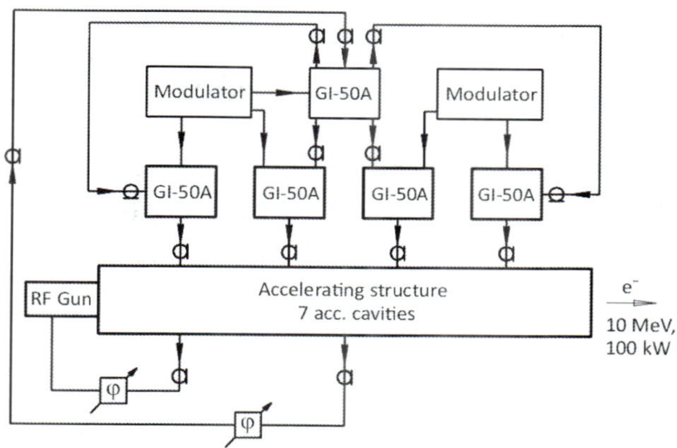

Fig. 5. Block diagram of an ILU-14 RF electron accelerator.

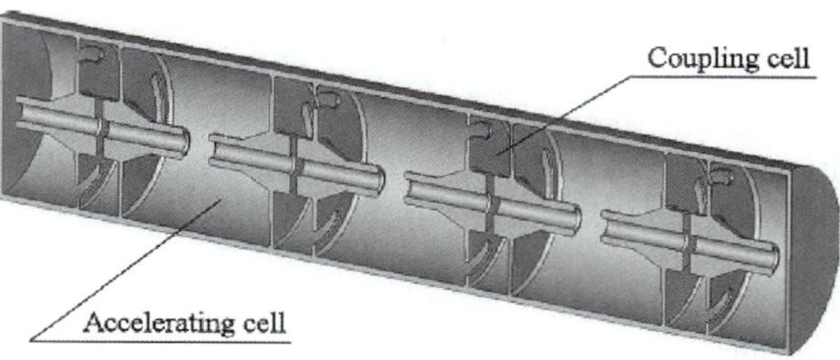

Fig. 6. Diagram of the internal cavities in an ILU-14 electron accelerator.

to 5.0 MeV. An input power of 150 kW gives an average beam power of 50 kW. The RF cavity is longer and the upper and lower internal stubs have smaller mutual capacitance, but the resonant frequency is nearly the same as for the ILU-6 — 115.9 MHz [4, 5].

The numbers in Fig. 3 identify the major components: (1) vacuum tank, (2) copper resonant cavity, (3) magnetic lens, (4) ion vacuum pumps, (5) electron gun, (6) beam scanner, (7) loop for RF coupling, (8) vacuum capacitor, (9) triode tubes. A photograph of the ILU-10 is shown in Fig. 4.

2.2. *Multiple-cavity versions*

This type of RF accelerator, which is designated as the ILU-14, consists of seven accelerating cavities with on-axis coupling cavities between the accelerating cavities to stabilize their phase relationships. The RF power is delivered by four triode amplifiers, distributed along the structure, which are all driven by a fifth triode tube. This more complex system is self-tuning by means of a feedback signal from the central cavity to the driving tube. The resonant frequency is 176 MHz. The electron energy range is from 7.5 to 10 MeV, and the average electron beam power rating is 100 kW [6, 7].

A block diagram of the ILU-14 system is shown in Fig. 5. A diagram of the cavity structure inside the accelerator is shown in Fig. 6 and a photograph of the prototype system is shown in Fig. 7.

Fig. 7. Photograph of an ILU-14 multiple-cavity accelerator.

3. Rhodotron Multiple-Pass Radial Electric Field Accelerators

Rhodotron™ accelerators employ a single, large, coaxial resonant cavity to accelerate electrons with an RF field. The electrons gain from 0.8 to 1.2 MeV of energy per pass through the cavity. Higher

216   *M. R. Cleland*

energies up to 10 MeV are produced by multiple passes through the same cavity. Different versions of these accelerators are designed with 6, 10 or 12 passes [8–12].

After each pass, the electron beam is deflected back into the cavity with dipole magnets which are placed outside of the cavity. The magnets also provide both horizontal and vertical focusing of the electron beam. Horizontal (transverse) focusing is due to variations in the electron trajectories within the magnets, and vertical focusing is obtained with external fringe fields from the edges of the magnet poles. These edges are slightly tilted from normal beam entrance and exit angles.

Beam bunching due to phase focusing occurs inherently within the dipole magnets because of variations in the paths of electrons that have gained slightly different energies by passing through the cavity at different times. This effect maintains a stable phase relationship between the electron beam bunch and the RF field in the cavity [9, 12].

A diagram of the beam trajectories for six passes through the cavity is shown in Fig. 8.

A cutaway drawing of a Rhodotron cavity that displays the internal structure, the beam deflecting magnets and the amplifier tube is shown in Fig. 9. The RF power is generated with a self-tuning tetrode amplifier at a frequency of 107.5 MHz. The central, coaxial structure produces a radial electric field that has its maximum value in the middle of the cavity, where the electrons are accelerated. The circular magnetic fields surrounding the central structure are

Operating Principle

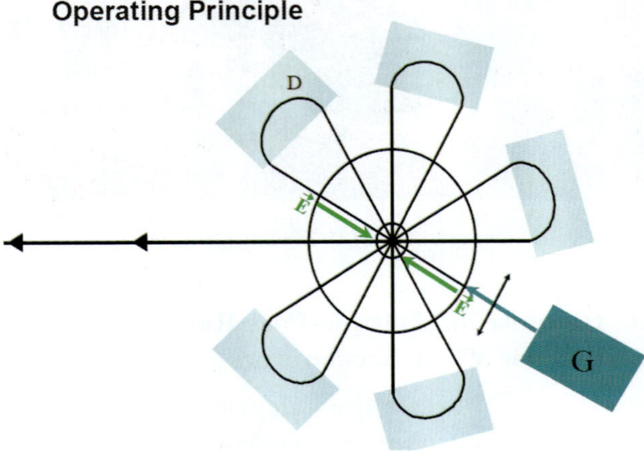

Fig. 8.   Diagram showing the beam trajectories in a six-pass Rhodotron accelerator.

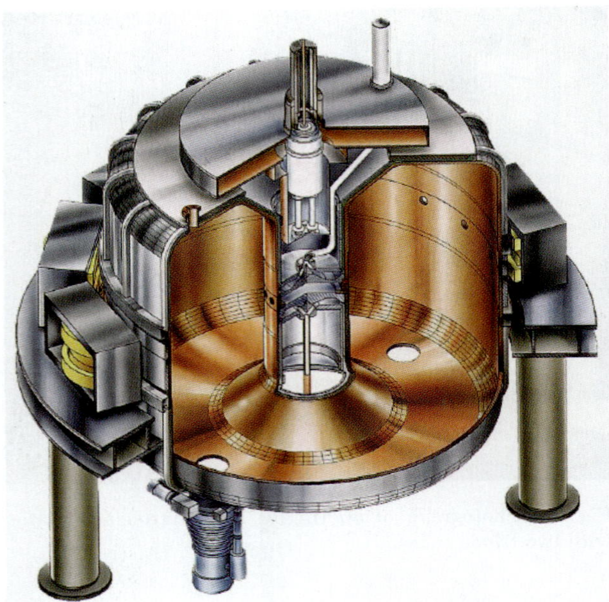

Fig. 9.   Cutaway drawing showing the main parts of a Rhodotron accelerator.

highest at its ends and vanish in the middle, so they do not interfere with the electron beam trajectories in the central plane.

The large size of a Rhodotron cavity, about 2 m in diameter, its high Q factor of 50,000 and the low energy gain of 1 MeV per pass allow the cavity to be energized in a continuous wave (cw) mode. This produces an average beam current that is bunched at the resonant frequency. The absence of macrobunching reduces the peak beam current, in comparison with an ILU RF accelerator or a microwave linear accelerator. The very short time interval (about 9 ns) between beam bunches allows the beam to be scanned with variable frequencies up to 200 Hz without producing gaps on a moving product conveyor.

A photograph showing the lower half of a Rhodotron cavity being assembled with its beam deflecting magnets is presented in Fig. 10. Another photograph of a complete 10 MeV Rhodotron system is shown in Fig. 11.

Several models of these accelerators with different electron energy and beam power ratings are made by selecting the appropriate number of passes and the power of the RF amplifier. The basic ten-pass electron beam model provides electron energies up to 10 MeV with beam powers from 40 to 200 kW [9]. The six-pass models are intended for high-power X-ray

Fig. 10. Photograph of the lower half of a Rhodotron cavity, showing the central structure and the beam-deflecting magnets.

Fig. 11. Photograph of a 10 MeV Rhodotron accelerator, showing the complete assembly with the RF amplifier on the top of the cavity and the external beam transport system.

Fig. 12. Photograph of a six-pass prototype of the very-high-power (7.0 MeV, 700 kW), Rhodotron for X-ray generation.

processing of materials. They are available with maximum electron beam power ratings of 200, 450 and 700 kW at 7.0 MeV [10]. A photograph of this system is shown in Fig. 12.

Electron energies above 7.5 MeV are not used in the X-ray mode, to avoid causing nuclear reactions in the X-ray target, which is made of tantalum. The main isotope of tantalum, Ta-181, has a photon-neutron threshold energy of 7.6 MeV. Tungsten should not be used for an X-ray target above 6.0 MeV, because one of its isotopes, W-183, has a threshold energy of 6.2 MeV. A gold target could be used up to 8.0 MeV, because the only isotope, Au-197, has a threshold energy of 8.1 MeV. High-power X-ray generators are now economically competitive with industrial gamma-ray facilities containing several megacuries of cobalt-60 [11].

Rhodotrons can provide lower beam energies by turning off one of the dipole magnets so that the beam is not deflected back into the cavity. Many Rhodotron facilities are equipped with two beam lines, one at 5 MeV and another at 10 MeV. One facility has three beam lines: two for X-rays at 5.0 and 7.0 MeV, and one for electrons at 10 MeV.

4. Conclusion

Both ILUs and Rhodotrons are employed extensively for industrial radiation processing applications. The ILUs have mainly been used for applications that can be done with electron energies below 5.0 MeV, while the Rhodotrons are mainly used for energies above 5.0 MeV. Because of such differences, these have not been competitive accelerator technologies.

References

[1] V. L. Auslender, G. I. Budker, G. B. Glagolev, A. A. Livshits, G. N. Ostreiko, A. D. Panfilov and V. A. Polyakov, Radio-frequency electron accelerator, US Patent **140**(4), 942 (1979).

[2] V. L. Auslender and I. N. Meshkov, *Radiat. Phys. Chem.* **35**(4–6), 627 (1990).

[3] V. L. Auslender, V. V. Bezuglov, A. A. Bryazgin, M. V. Korobeynikov, A. V. Sidorov and E. A. Shtarklev, Electron beam treatment line with ILU-6 machine for medicinal raw decontamination, in *Proc. RuPAC 2008* (Zvenigorod, Russia), pp. 333–335.

[4] V. L. Auslender, A. A. Bryazgin, B. L. Faktorovich, V. A. Gorbunov, E. N. Kokin, M. V. Korobeinikov, G. S. Krainov, A. N. Lukin, S. A. Maximov, V. E. Nekhaev, A. D. Panfilov, V. N. Radchenko, V. O. Tkachenko, A. A. Tuvik and L. A. Voronin, *Radiat. Phys. Chem.* **63**(3–6), 613 (2002).

[5] V. L. Auslender, V. V. Bezuglov, A. A. Brazgin, V. A. Gorbunov, B. L. Faktorovich, E. N. Kokin, M. V. Korobeynikov, A. N. Lukin, L. G. Makarov, M. A. Tiunov, V. E. Nekhaev, A. D. Panfilov, V. M. Radchenko, A. V Sidorov, E. A. Shtarklev, V. V. Tarnetski, V. O. Tkachenko and L. A. Voronin, Industrial high energy electron accelerators type ILU, in *Proc. RuPAC 2008* (Zvenigorod, Russia), pp. 367–369.

[6] V. L. Auslender, A. A. Bryazgin, K. N. Chernov, V. G. Cheskidov, B. L. Faktorovich, G. I. Kuznetsov, I. G. Makarov, G. N. Ostreiko, A. D. Panfilov, G. V. Serdobintsev, V. V. Tarnetsky and M. A. Tiunov, *Radiat. Phys. Chem.* **78**, 741 (2009).

[7] A. A. Brazgin, V. L. Auslender, K. N. Chernov, V. G. Cheskidov, B. L. Factorovich, V. A. Gorbunov, I. V. Gornakov, G. I. Kuznetsov, I. G. Makarov, N. V. Matyash, G. N. Ostreiko, A. D. Panfilov, G. V. Serdobintsev, V. V. Tarnetsky, M. A. Tiunov and A. A. Tuvik, 100 kW modular linear accelerator for industrial applications with electron energy of 7.5–10 MeV, in *Proc. RuPAC 2008* (Zvenigorod, Russia), pp. 379–381.

[8] A. N'Guyen and J. Pottier, Electron accelerator with Coaxial cavity, US Patent **107**(5), 221 (1992).

[9] Y. Jongen, M. Abs, F. Genin, A. Nguyen, J. M. Capdevilla and D. Defrise, *Nucl. Instrum. Methods B* **79**, 865 (1993).

[10] M. Abs, Y. Jongen, E. Poncelet and J.-L. Bol, *Radiat. Phys. Chem.* **71**(1–2), 287 (2004).

[11] J.-L. Bol, X-ray sterilization: review of available configurations for small to large capacity sterilization facilities, in *Proc. 2011 International Meeting on Radiation Processing (IMRP)* (Montreal, Canada), to be published: by Radiat. Phys. Chem. in 2011 or 2012.

[12] M. R. Cleland, Y. Jongen and M. Abs, Rhodotron, Handbook of Accelerator Physics and Engineering, 2nd edn., eds. A. Chao, M. Tigner, K. H. Mess and F. Zimmermann (World Scientific, Singapore, 2011).

Marshall R. Cleland received his Ph.D. in Nuclear Physics from Washington University in 1951. He cofounded Teleray Corporation in 1953, and then in 1958 also cofounded Radiation Dynamics, Inc. (RDI) to make particle accelerators. In 1994, he retired from RDI as its Chairman of the Board of Directors. Presently, he serves as Technical Advisor at IBA Industrial, Inc., which has acquired RDI. He has published more than 100 technical papers on topics including radiation effects, industrial electron beam and X-ray processing, ionizing radiation, radiation dose analysis, material curing, radiological safety, and industrial and commercial accelerator development and applications. Dr. Cleland holds 16 patents on particle accelerators and their applications.

Reviews of Accelerator Science and Technology
Vol. 4 (2011) 219–233
© World Scientific Publishing Company
DOI: 10.1142/S1793626811000549

Accelerators for Neutron Generation and Their Applications

Guenter Mank

Forschungszentrum Juelich, Project Management,
Basic Energy Research, 52425 Juelich, Germany
g.mank@fz-juelich.de

Guenter Bauer

79761 Waldshut-Tiengen, Germany
guenter@bauer-wt.de

Françoise Mulhauser

International Atomic Energy Agency,
Vienna International Center,
1400 Vienna, Austria
f.muelhauser@iaea.org

This article briefly reviews the most important points of neutron generation using accelerators in the small and medium energy ranges. The interest in neutron research is still growing, due to improvement of detectors and analyzing tools. Smaller accelerator-based neutron sources can serve for proof-of-principle experiments and for data generation. In developing and emerging economies they may replace underutilized research reactors, if there is a need for neutron research. The complementarity of neutron research to synchrotron and X-ray analysis will always keep some focus on effective and lower cost options for neutron production.

Keywords: Accelerators; neutron sources; spectrometers and spectroscopic techniques; neutron spectroscopy; neutron physics; spallation reactions.

1. Introduction

The interest in affordable accelerators for neutron research and applications of neutrons in industrial ranges is still high. In recent articles by Hamm [1, 2], the total number of neutron generators in use today, including sealed tube systems, is estimated to be around 1000, with about 50 systems sold every year. Compared to ion implantation, the number of neutron accelerators is an order of magnitude lower, but still they attract worldwide interest. This article will review the most important points, as to why neutron generation using accelerators on the small and medium energy scales is growing and of high interest in research and development not only for emerging and developing countries as there are new innovative approaches in materials research.

Many aspects as to why the development of these neutron sources is important have been reported in an IAEA publication on "Development Opportunities for Small and Medium Scale Accelerator Driven Neutron Sources" [3], which served as a background report for IAEA projects on neutron research and spallation systems.

In most of the industrial applications, radioisotope sources are used, including neutron radiographic testing [4, 5], as they are easily transportable and affordable. Due to increasing security concerns, it may get more and more difficult to transport radioisotopes. The backdrop of these sources, e.g. ^{252}Cf, is that they cannot be turned off and may be a burden in decommissioning of industrial and research equipment. Combined sources, e.g. Sb–Be, have the possibility of being turned off by removal of Be during transport or inspection [6] but have a permanent γ radiation background. Due to the limited half-life (e.g. about 2.6 years for ^{252}Cf), the availability of radioisotope sources is limited compared to other neutron sources.

The use of research reactors as neutron sources opens a new field in characterization of materials.

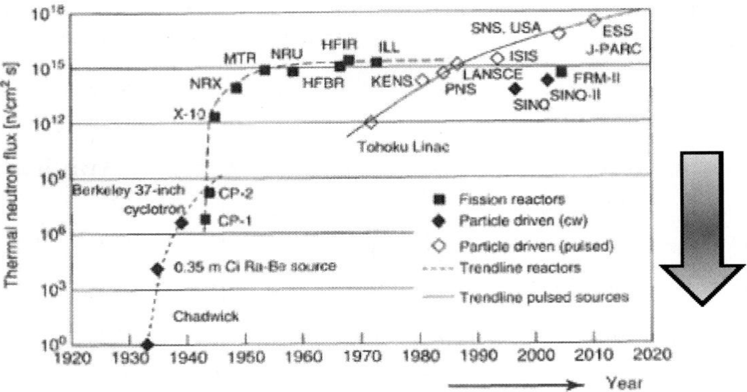

Fig. 1. Thermal neutron flux available from various neutron sources. (Courtesy of D. Filges and F. Goldenbaum [9].) The arrow indicates the range of the neutron sources discussed in this article.

New three-dimensional pictures providing analysis of products give insight into manufacturing processes and help to develop new and cheaper products. However, the requirement of a sophisticated experimental setup at the reactor and the increased costs for safety and operation have led to the shutdown of many research reactors worldwide. An overview of the dramatic change can be found in the IAEA's research reactor database [7]. Essentially, the developed countries shut down their less utilized reactors, with decreasing numbers from about 292 in 1975 to 143 in 2010, while in the developing world the reactors were still increasing to about 91 in 2010, but many of them are underutilized. One of the drawbacks of research reactors, besides high construction costs, licensing procedure and infrastructure requirements, compared to accelerator-based systems, is that the main system, i.e. the reactor core, cannot be easily turned off. The neutron beams can be chopped [6], and the neutron spectra modified using adequate moderators, but the flexibility is limited. Due to the high costs of a research reactor, new investments are expensive and are mainly made for new research perspectives, not mainly for industrial applications. In contrast to special and purpose-designed reactors, many commercial companies offer open vacuum system accelerators for neutron generation, e.g. AccSys Technology, Inc. (USA), Ion Beam Applications SA (Belgium), Sumitomo Heavy Industries (Japan), National Electrostatics Corporation (USA) and High Voltage Engineering Europe (The Netherlands) [1].

The source options for accelerator-driven neutron systems (ADNSs) are manifold. Due to the advances in neutron optics and detector technology, as e.g. developed at research reactors, ADNSs offer a real alternative to research reactors, as the costs are lower and the systems can be easily turned on and off. Further high power ADNSs, such as the American Spallation Neutron Sources (SNS) [8] and the Japanese Proton Accelerator Research Complex (J-PARC) [9], can have in their final design higher thermal neutron flux available in short pulses than even the best research reactors, such as ILL or FRM-2. Combined with adapted detectors and electronics for these pulses the final detection limit can be drastically increased. All these advantages draw the attention to smaller accelerators with power levels from the kW level to about 100 kW, as for these facilities the capital and operational costs will be much more feasible and attractive for industry and research in emerging and developing economies too [3], and for nondestructive testing.

Figure 1 provides an overview of neutron sources and the available thermal neutron flux [10]. Small and medium energy neutron sources have a neutron flux which is orders of magnitude lower, but offer the advantage of lower activation and much lower costs.

2. Low to Medium Energy Accelerators as Neutron Sources

Electrostatic tandem accelerators, cyclotrons and linear accelerators have been used for many years in industry [11]. Neutrons are generated through a variety of induced nuclear reactions, such as (p, n), (d, n), (t, n) and (α, n). Monoenergetic neutrons can be produced using, for example, the following reactions

Table 1. Example of reactions producing monoenergetic neutrons.

Incident particle	Reactions		
Proton	^{3}H(p, n)^{3}He	^{9}Be(p, n)^{9}B	^{10}B(p, n)^{10}C
	^{6}Li(p, n)^{6}Be	^{10}Be(p, n)^{10}B	^{11}B(p, n)^{11}C
	^{7}Li(p, n)^{7}Be	^{12}C(p, n)^{12}N	^{15}N(p, n)^{15}O
	^{36}Cl(p, n)^{36}Ar	^{13}C(p, n)^{13}N	^{14}C(p, n)^{14}N
		^{39}Ar(p, n)^{39}K	^{59}Co(p, n)^{59}N
Deuteron	^{2}H(d, n)^{3}He	^{3}H(d, n)^{4}He	^{24}Mg(d, n)^{25}Al
	^{13}C(d, n)^{14}N	^{7}Li(d, n)^{8}Be	^{18}O(d, n)^{19}F
	^{20}Ne(d, n)^{21}Na	^{15}N(d, n)^{16}O	^{32}S(d, n)^{33}Cl
			^{28}Si(d, n)^{29}P
Triton	^{1}H(t, n)^{3}He		

Table 2. Neutron yields and heat release for different nuclear reactions for neutron production [16].

Nuclear process	Example	Neutron yield	Heat release (MeV/n)
d-t in solid target	400 keV d on t in Ti	4×10^{-5} n/d	10,000
Deuteron stripping	40 MeV d on liquid Li	7×10^{-2} n/d	3500
Nuclear photo effect from e$^-$ bremsstrahlung	100 MeV e$^-$ on ^{238}U	5×10^{-2} n/e$^-$	2000
^{9}Be(d, n)^{10}Be	15 MeV d on Be	1 n/d	1000
^{9}Be(p, n; p, pn)	11 MeV p on Be	5×10^{-3} n/p	2000
Nuclear fission	Fission of ^{235}U by thermal neutrons	1n/fission	180
Spallation	800 MeV p on ^{238}U or Pb	27 n/p or 17 n/p	55 or 30

(Table 1), described in detail by Drosg and coworkers [12, 13].

High power ADNS facilities (SNS, J-PARC, ISIS, ESS) utilize about 1 GeV particle accelerators to produce neutrons through the nuclear spallation process — neutron evaporation from heavy nuclei (e.g. Ta, W, Pb or Hg) induced by collision with high energy particles — and consequently these facilities have high operation capital and operational costs. The neutron spectra vary, depending on the input energy and the extraction angle of the moderated neutrons [14]. Furthermore, a substantial amount of additional particles is produced, which influences the quality of the neutron production process in the spallation target [15]. The neutron yield can be increased by using fissile material such as ^{235}U or ^{238}U as target for the spallation process. An overview of neutron yields is provided in Table 2 [16]. All the processes above fission can be utilized for small neutron sources, but are not feasible for high flux facilities because of the huge amount of associated release of heat [17], as indicated in the last column.

The use of low and medium energy accelerators as drivers for neutron sources provides much more flexibility, together with reduced costs [3]. Commercially produced turnkey neutron sources are available with source strengths up to 10^{13} n/s, such as the LANSAR® model from AccSys Technology, Inc. [18].

2.1. *Neutron tubes*

Neutron tubes are the most commonly used neutron accelerators. They are commercially available, for example through the following manufacturers: Adelphi Technology (USA), EADS SODERN (France), Hotwell GmbH (Austria; only distributor), Thermo Fisher Scientific (USA) and VNIIA All Russia Research Institute of Automatics (Russia) [19]. Baker Hughes (USA), Halliburton (USA) and Schlumberger (USA) make them for internal use but do not sell them commercially.

The D-T and D-D reactions are used to produce 14.1 MeV and 2.45 MeV neutrons, respectively. The devices are very compact and are distributed mainly as sealed tubes with hydride deuterium, tritium or mixed targets. No vacuum pump is needed, so that these devices can be easily employed in field application. The amount of neutrons can be regulated using the accelerating voltage and ion source intensity. Neutron yields of sealed neutron tubes can reach up to about 1×10^{11} n/s in the case of D-T reactions. The tubes can be turned off and pulsed. Most of the applications are related to nondestructive testing (NDT) [5], such as for petroleum and uranium well-logging. For all smaller neutron sources, the range of applications is essentially the same; however, the measurement times are of course strongly dependent on the neutron yield, the angular distribution and the energy of the neutrons. Due to the compactness of the neutron tubes, they can be used in test or calibration setups, such as for the target of the European spallation source (see Fig. 2) [20].

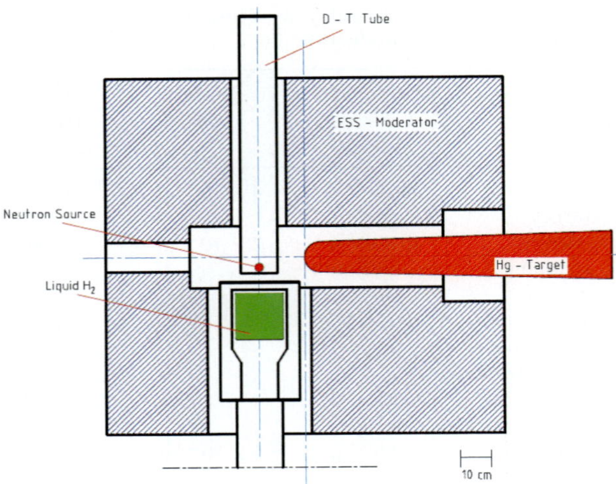

Fig. 2.   Use of a D-T tube at the ESS target station for calibration and MCNP calculation. (Courtesy of N. Paul.)

2.2. *Plasma accelerator gun, dense plasma focus*

A plasma focus device operates as a plasma accelerator gun, where the plasma is generated in a low pressure atmosphere by a pulsed powerful capacitive discharge between a pair of coaxial cylindrical electrodes [21]. The advantage of using a dense plasma focus (DPF) is that the hot plasma pinch which is created collapses in the order of about 100 ns. Intense pulses of neutrons, but also of X-rays, as well as energetic electron and ion beams, are created. Depending on the gas mixture, the kinetic energy of the neutrons is 14.1 MeV or 2.45 MeV, as for sealed neutron tubes. The DPFs are ideal setups for materials research, such as for controlled nuclear fusion. The capacity range determines the amount of neutrons [22]. Thus, very small neutron yields can be reached. The advantage of DPFs is that they can be pulsed and smaller devices are transportable. A recent coordinated research project of the IAEA displayed the variety of the DPFs and showed that DPFs are an ideal tool for starting neutron and materials research in developing countries with low budget sources [23].

2.3. *Inertial electrostatic confinement neutron sources*

Intense studies on inertial electrostatic confinement (IEC) reactors are being done at the University of Wisconsin–Madison [24]. The IEC reactor is a vacuum chamber filled with a fuel gas such as deuterium or a deuterium–tritium mixture at low pressure. There are inner and outer spherical or cylindrical grids centered inside the chamber. The outer grid is held at a nearly zero potential, and the inner, 90–99% transparent grid is held at a high negative potential, typically -100 kV. The potential difference between the grids accelerates ions inward to velocities of fusion relevance.

Modified versions which utilize the same principle can be employed as neutron generators, such as those sold through NSD-Fusion (Germany). In these devices fusion from D-D or D-T generates a neutron flux up to about 10^8 n/s (DD, 2.5 MeV).

2.4. *Low energy neutron sources*

Whereas most of the sealed sources, neutron tubes and DPFs are operated without moderators and provide monoenergetic fast neutron spectra, the neutrons created by higher energy accelerators are combined with moderators to produce cold or ultracold neutrons. The Low Energy Neutron Source (LENS) at the Indiana University Cyclotron Facility (IUCF) is an accelerator-based pulsed cold neutron source [25] (see Fig. 3). Fast neutrons (in the MeV energy range) are created via (p, xn) reactions in a Be target for proton energies of 13 MeV or less [26]. The LENS facility opens up the possibility of moderating the neutron spectra and including features of higher energetic neutron sources as spallation sources.

The beam is up to 30 kW of proton beam power and the beryllium target is surrounded by

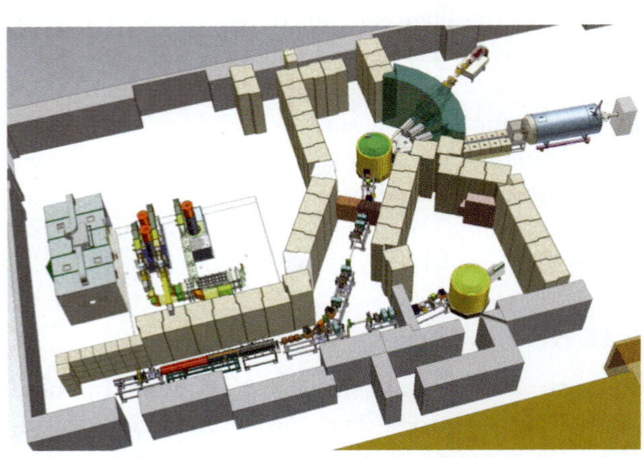

Fig. 3.   LENS layout. (Courtesy of D. Baxter.)

a moderator facility similar to spallation sources. Thus, the created cold or ultracold spectrum can be used as a proof-of-principle for bigger and much more expensive sources, but also opens up the possibility of doing top research at universities with lower budget options. Depending of the chosen setup, the spectrum can be broader than for spallation sources, thus providing a less stringent time of flight (TOF) resolution, but an acceptable neutron flux in a defined neutron tube. Neutron tubes have been used to test moderator setups such as for the ESS project [22, 27]; the lower energy proton accelerators like LENS allow moderating the neutrons from the water-cooled Be target effectively. The whole system consists of a proton linac (7 or 13 MeV for LENS), a water-cooled target, a room temperature reflector and moderator, a cryogenic moderator, beam tubes and a biological shielding [28]. The variety of additionally created reaction products increases with increasing proton energy through new reaction channels such as (p, t) and (p, α), and should be avoided or reduced by choosing the appropriate energy if possible.

2.5. *Van de Graaff accelerator*

The (d, n) reaction is ideal for use in producing neutrons employing a tandem Van de Graaff accelerator. Such a system can be coupled to a pulsed fast neutron analysis (PFNA) system, as used for air cargo inspection [29]. The system consists of an ion source, the low energy beam line with a chopper and buncher, and a 4 MV model 12 SDH Pelletron accelerator. The Van de Graaff accelerator operates at 3.5 MHz and provides a 140 μA deuterium beam pulse. The beam exits the Pelletron usually at 7.1 MeV and is directed to a 4.5-cm-thick deuterium gas target at about 608 kPa (see Fig. 4).

The advantage of this system is that the beam can be horizontally and vertically moved so that the final beam width at the inspected containers is about 9 cm wide and 12 cm high. Pelletrons are available worldwide [30], with a huge variety of applications in basic and applied physics. Due to the neutron beam requirements (determined by the experimental demand) and the energy needed for neutron production, lower beam energies are preferable (terminal voltages < 5 MV).

Fig. 4. Van de Graaff tandem accelerator. (Courtesy of E. Bakraji.)

2.6. *Electron accelerators*

High energy electrons can be injected into a preferably high Z-target producing gamma rays via bremsstrahlung. These gamma rays generate neutrons via photonuclear (γ, n) reactions. Medium energy electron accelerators between 25 MeV [31], 45 MeV [32] and 100 MeV [33] are used for this purpose. Basic neutron researches are ideal experiments to be conducted at the experimental facilities. In Ref. 31 electron pulses of 1.4 μs with a repetition between 12.5 and 100 Hz and an average current of about 25 μA are sufficient for producing neutrons from a water-cooled lead target. The arrangement offers the possibility of varying the moderator setup and modifying the moderator elements as well as the geometrical distribution. Thus, the production of subthermal, thermal or epithermal neutrons is possible. The pulsed setup is ideal for TOF measurements. Such an arrangement has high flexibility and shows the advantage of all smaller systems as described in this article: the neutron source is easily accessible and can be modified. New experiments can be done and deliver important input for less flexible and bigger neutron sources, which operate at a high radiation level. Experiments such as those performed in Argentina [31] can employ different moderators and geometries, with or without decoupler or 3-D inhomogeneous poisoning. Higher electron energies up to 100 MeV can be used together with water-cooled tantalum sheets, such as for the Pohang Neutron Facility (PAL) [33]. The neutron yield per kW beam power at the target is about 2×10^{12} n/s, i.e. about 2.5% lower than that based on Swanson's formula [34]. Due to the low pulse width of 1.5 μs and the repetition rate of 12 Hz, such beams can be used for TOF measurements.

2.7. *Improved moderators*

Moderated neutron beams are produced by a slowing-down and thermalization process, which suffers from very low efficiency. Indeed, only a few neutrons which enter the moderator will appear in the useful neutron beam direction. The most inefficient step in this process occurs when neutrons are emitted from the face of the neutron moderator uniformly in all directions, and only the small fraction, which happens to be going along the neutron beamline, makes it to the sample position. One exception to this generalization is the "grooved" moderator concept [35, 36], implemented at IPNS, ANL (USA), KENS, J-PARC (Japan) and TS2, ISIS (UK), in which neutrons leaving the bright "groove" surfaces in the wrong direction are scattered back into the moderator from the "fins" and get another chance to be usefully directed (see Fig. 5). However, even grooved moderators provide only a modest increase in the useful neutrons reaching the sample position.

At several facilities, research is underway to develop moderators which focus neutron beams in a preferred direction, as opposed to emitting neutrons isotropically. For example, recently the use of diamond nanoparticles for cold, very cold and ultra-cold neutrons produced very promising results at ILL (France), where significant increase in the neutron brightness was observed for wavelengths above 7 Å, reaching a factor of 5 at 20 Å. Similar studies are ongoing at SNS (USA), J-PARC (Japan), ISIS (UK) and elsewhere. However, it appears that little communication is occurring among the involved teams.

Small scale accelerator facilities at Sapporo (Japan), Bariloche (Argentina) and LENS (USA) are collaboratively developing new moderator experiments and simulations to improve available cold neutron fluxes. The results are encouraging and need to be shared with larger spallation neutron sources and research reactors.

The IAEA, in collaboration with many major and minor institutions, has organized meetings on directionally focused moderators for enhanced neutron beam intensities to support materials research and applications, and will provide a platform for additional cooperation between the interested parties.

2.8. *International fusion materials irradiation facility*

In the case of materials research for fusion experiments, a neutron source with a neutron energy of about 14 MeV is desirable. This requirement is fulfilled by the design of the International Fusion Materials Irradiation Facility (IFMIF), under development in Japan [37, 38]. The following reactions will be used for production of neutrons, through a stripping reaction between deuterium and lithium: ^{7}Li(d, 2n)^{7}Be and ^{6}Li(d, n)^{7}Be; the third main reaction ^{6}Li(n, T)^{4}He, which also occurs in the target, is similar to the breeding process in the blanket of a fusion reactor. The neutron spectra will be similar to the neutrons produced in a fusion reactor by the T(d, n)^{4}He reaction. Due to the fact that, depending on the location of the investigated materials, irradiation levels between > 20 dpa/fpy (displacement per atom/full power year) and about 1 dpa/fpy should be reached, the intensity of the 40 MeV cw proton beam will be about 250 mA at the target, which can only be achieved by combining two beams of 125 mA. A first prototype accelerator is under construction. The final prototype results will be available in 2013 and will be used for the final design.

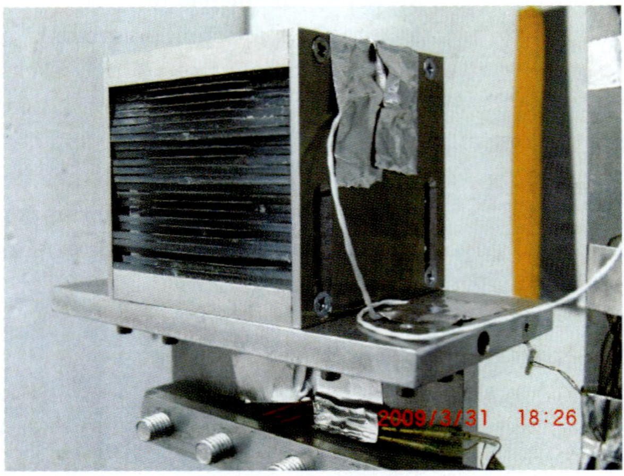

Fig. 5. Directionally focused moderators under tests. (Courtesy of D. Baxter, SNS and LENS.)

2.9. *Comparison of low to medium energy neutron accelerators*

Besides the existing facilities, it is of high interest to determine if low to medium energy accelerator neutron sources can be designed and constructed at the university level or in laboratories in emerging economies to boost neutron science and materials research. An IAEA report was devoted to this purpose [3]. The main characteristics of such a facility should be that the operational complexity should be low, which would be preferably true for turnkey systems commercially available. The pulsed beam option of a linear accelerator could deliver a substantially higher peak current in the pulse compared to a cyclotron with the same average current. The shielding requirement increases drastically with the beam energy. Neutrons with energies higher than about 20 MeV are very difficult to shield. However, prompt neutron and gamma radiation increases with increasing beam energy. An optimum for the required application has to be found. The same is true for the decommissioning requirements due to activation of materials, shielding and target. Careful accelerator engineering will be needed if no turnkey system is used. One of the most important parts for the neutron experiments is the design of the target and moderator structure. The target design will be finally crucial for experimental success and may require much more intense research, as can be seen by the liquid curved shape lithium target for the IFMIF neutron source.

D-T sources have only a limited lifetime, but due to their high flexibility are best used for many industrial applications, such as well-logging and NDT. The $Be(d, n)$ and $Be(p, n)$ sources have much higher intensities and a significantly longer lifetime through proper management of the targets. They are less flexible, but still can be designed in such a manner that the whole accelerator facility can be mounted on a truck or has the size of a typical container which can be moved. If an energy below 13.5 MeV is chosen, the buildup of additional radiation in the target can be avoided and this will lead to a cheaper design. Table 3 provides an overview of the recommendations made for future neutron sources. The approximate cost (at the 2005 level) is indicated. It has to be noted that a $Li(d, n)$ source can have much more moderate costs if a maximum beam power of about 100 kW at about 20 MeV is considered. The resulting neutron production rate is of course much lower than in the case for the IFMIF. The cost of a spallation source will always exceed the other costs by a factor of 10 or higher. Depending on the research

Table 3. Examples of neutron accelerator sources and estimated costs [3].

System	Reaction	Beam energy (MeV)	Beam power (kW)	Production rate (n/s)	Cost (approx.) (US$)
D-T	$T(d, n)^4He$	~ 0.3	0.05	10^9	100 k
AccSys DL1	$Be(d, n)$	1	0.12	10^{10}	0.5 M
AccSys PL11	$Be(p, n)$	11	11	10^{13}	3.5 M
LENS	$Be(p, n)$	13	30	10^{14}	20 M
Model A	$Li(d, n)$	20–30	100	10^{15}	> 50 M
Model B	Spallation	400–1000	100	10^{16}	> 500 M

and application opportunities, such a high cost may not be necessary.

3. Applications; Research and Development

It should be noted that many neutron applications which are usually done at a research reactor can be transferred to a neutron accelerator source. Especially for high energy spallation sources, the pulsed character of the neutron beam extends the possible research quite substantially. In this article we would like to outline briefly only applications and experimental setups, which are specific for low to medium energy accelerator sources, as mentioned in Sec. 2. Still, overlaps of activities exist and display the complementarity of the different sources.

Neutron production using ultrahigh intensity femtosecond lasers as demonstrated in pioneering work [39, 40] may lead to innovative neutron application at the newly designed European Extreme Light Infrastructure (ELI) facility [41] and will influence future neutron science.

There is a worldwide need for improved technologies for the efficient inspection of cargo containers, especially in the transportation sector. Neutron-based techniques for the detection of illicit materials and explosives can be used for this purpose [42], and are briefly mentioned in this article as detection of illicit materials includes many more topics, which are beyond the scope of this short review.

3.1. *Industrial application and nondestructive testing*

Well-Logging and nondestructive testing have already been mentioned as the main applications for neutron generators and neutron tubes commercially available. Even though as the techniques are mature and can be purchased through many companies, there is still the need for training in safe operation of radiation equipment. Especially in developing countries, such training is provided by the IAEA in completion of commercially available training.

3.2. *Radioisotope production*

In particular, lower energy accelerators utilizing (p, n) and (d, n) reactions can be employed as dual purpose machines. For higher energies above 13.5 MeV, the buildup of radioactivity in the target could be used for ^{57}Co and ^{201}Tl production in addition to an increased neutron flux. It has to be noted that additional facilities for the purpose of radioisotope production have to be included in the facility.

3.3. *Data mining and code testing*

Lower energy systems offer the possibility of modifying or changing the target or target structure more often. Thus, neutron codes such as MCNPx [43] can be tested for geometries similar to those of bigger and more expensive spallation sources. Scattering and absorption cross sections can be measured (see e.g. Ref. 44). Even neutron tubes have been used for testing the European Spallation Source target system [23]. Mixtures of light and heavy water and different moderators, such as methane, can be tested avoiding the risk of damage to the target by so-called "burps" [45].

The IAEA include data from low to high energy neutron sources for their benchmarking exercise of spallation models [46]. It is of great importance to validate on selected experimental data the abilities of the various codes to predict reliably the different quantities relevant for applications. The comparison has been made by different groups worldwide based on a variety of models as generally Monte Carlo implementations of intranuclear cascade (INC) or quantum molecular dynamics (QMD) models followed by de-excitation (principally evaporation/fission) models.

3.4. *Demining*

Portable neutron sources can be applied in the field for the detection of explosives as used in mines or improvised explosive devices (IEDs). The use of most neutron-based techniques relies on the detection of the characteristic gamma rays emitted by the excited nucleus. Depending on the elemental characteristics, matrices have been building up for the different constituents. Employed together with metal detectors and other demining techniques, thermal neutron analysis can be made using a radioisotope source, such as ^{252}Cf or pulsed fast and thermal neutron analysis (PFTNA) using a small sealed tube neutron source. The first tests were promising but turned out to be not sufficiently reliable.

Later experiments using fast neutrons and neutron backscattering showed more reliable results and detected certain main in dry soil at a depth of 15 cm. The fast neutrons were obtained from a variety of sources [47]: a radioactive ^{252}Cf source, 5×10^5 n/s; a D-D neutron generator, 2×10^6 n/s; and a radioactive PuBe source with a strength of about 5×10^7 n/s. The source was placed in the detector encasing, the point of neutron emission being around 12 cm from the ground. The standoff distance between the bottom of the detector encasing and the sand surface was roughly 10 cm.

The depth of the buried mines is the distance between the soil surface and the top of the mine. The Egypt SCAnning LAnd mine Detector (ESCALAD) consists of 16 ^{3}He proportional counter tubes with resistive wires (see Fig. 6). The system is being constructed at the Delft University [48].

Fig. 6. The Egypt SCAnning LAnd mine Detector (ESCALAD). (Courtesy of R. Megahid.)

3.5. *Cargo and air cargo inspection*

The application of neutron-based techniques to security operations has been intensively studied and at least provisionally implemented over the last few decades. Detection of explosive materials, illegal drugs, chemical and nuclear weapons is all potentially possible using generator-produced neutrons. Interrogation of luggage or cargo containers for hidden explosives or illegal drugs is an area of extreme interest. In addition, monitoring and identification of high explosive weapons, sometimes buried or underwater, is an application which requires nonintrusive methods of development. Several approaches have been developed in which the oxygen, nitrogen and carbon concentrations and ratios are correlated with known ratios in explosive materials as well as contraband [49–53]. Thermal neutron analysis (TNA) uses thermal neutrons as the analytical probe, so the method can make use of isotopic sources, as well as 14 or 2.5 MeV generators. This can also be done using a pulsed generator (PTNA), which allows the measurement of both prompt capture and decay gammas. This is sometimes combined with fast neutron analysis (FNA), which adds the capability of using inelastic scattering and other particle-emitting reactions. FNA requires neutron energies of > 6.5 MeV, so only the D-T neutron generators are applicable. Several other variations of these methods are possible, including pulsed fast neutron analysis (PFNA), associated particle imaging (API) and specialized detection methods like time of flight and gamma resonance attenuation. While a great deal of development has occurred in the application of accelerators and neutron generators to security operations such as airport locations and sea ports, rather few implementations are actually in use. This is an area of great potential utilization.

Cargo and air cargo inspection for detection of explosives or illegal materials can be done using PFNA [29]. Commercial systems are available, as mentioned in the previous section. However, there is still much room for further development. In particular, the scanning time has to be shortened, and the safety and security of such a setup in the environment of an airport or harbor has to be improved. Thus, most of the commercial systems still rely on X-ray or hard X-ray detection.

3.6. *Neutron capture therapy*

Boron neutron capture therapy (BNCT) stands as a second-generation promising alternative technique for treating cancerous tumors which are diffuse and infiltrating. Several clinical programs are well underway, most notably in Finland and Japan but also on a smaller scale in other countries like Italy and Argentina, and significant progress has been achieved in the control of different tumors, with an expanding scope as far as other tumor types are concerned [54, 55]. While a great deal of progress has been made using nuclear reactors, the advancement of BNCT requires neutron sources suitable for installation in hospital environments, such as low energy particle accelerators which are under development.

The direct use of proton and heavy ion beams (generally known as hadron therapy) for radiotherapy is a well-established cancer treatment technique, due to its clear advantages over conventional photon-based treatments. It requires accelerator facilities which are using essentially synchrotrons or cyclotrons. The efficacy of hadron therapy is accepted, but these facilities are expensive, which is the best and the worst for medicine in general. This is an important field of applications of accelerators but is not directly related to neutron production. Numerous publications and much information can be found at the website of the Particle Therapy Co-operative Group (PTCOG) [56].

Several centers around the world have used fast neutrons from accelerator sources for treating cancer, e.g. iThemba (South Africa) and CERCYL (Belgium) [57]. It has to be noted that only very specialized equipment can be used and very few cancer types can be treated. Thus, the number of such centers will not increase much. Still, there is place for improvement and research on utilizing the neutron beam and/or combining different types of treatment.

4. Instrumentation at Small Neutron Sources

Small and medium-sized neutron sources open about the same range of standard neutron diagnostics as for research reactors and big spallation sources. Of course, depending on the size and the neutron flux of the source, the layout and technical equipment have to be adjusted to fit into the experimental and often financial possibilities of the neutron source operator. It is advisable that the design of the source be made with respect to the most important application, and thus the instrumentation has to be tailored to the needs of the engineering and scientific community. Further, the moderator and the respective beamlines have to be adjusted to the needed neutron spectra and beam properties.

In principle, all neutron sources can be used for radiography and basic neutron experiments. Even very small sources, such as sealed neutron tubes or low capacity dense plasma focus, show results when used for radiography of simple objects, but these are more proof-of-principle experiments. The instrumentation described in this article is designed to be employed for accelerators with neutron production rates higher than about 10^{13} n/s. The expected neutron flux for these sources will be sufficiently high and the cost of the instrumentation does not exceed the cost of the source. It should be noted that the IAEA devoted a technical document especially to this kind of instrumentation at smaller neutron sources [3] and gave advice to researchers at smaller research reactors with similar instrumentation [59].

4.1. *Radiography*

Most applications of neutron radiography/tomography do not depend on a particularly high neutron flux. More important is a well-collimated, widely open neutron beam in a low background environment. On the other hand, to be competitive, neutron radiography stations must be equipped with modern, state-of-the-art imaging techniques and computing power.

The neutron radiography and tomography is complementary to X-ray radiography and tomography. This makes small sources attractive for radiography, as often X-ray sources are available at the same laboratory. For the Sec. 3 applications, we noted that for cargo inspection and for demining purposes this complementarity increases the value of the neutron measurements. Thus, neutron measurements can be used for verification or extension of radiographic results achieved with X-rays.

The collimated neutron beam passes through an object, and the "shade" picture is monitored behind using camera systems. The object may be static or rotated stepwise, the latter allowing three-dimensional reconstruction (tomography). Although the principle is very simple, well known, and has been widely applied for decades, the rapid development of new imaging technologies in the past few years, combined with the availability of specially tailored neutron beams (e.g. at SINQ), has opened a variety of new fields of application. This has been a very interesting development for both science and technology. The particularly strong interaction of neutrons with light elements (H, Li, B and others) can make devices in the interior of an object clearly visible, which would disappear into a fuzzy background if one was using X-rays. In many cases, it is such devices that are of particular interest for detailed inspection. Here again, the complementarity of X-rays and neutrons must be emphasized.

Further, with the computer power presently available in combination with the high penetration potential, neutron tomography allows destruction-free three-dimensional analysis of complicated structures, or devices, with high resolution and separating selected inner components. Besides identifying such internal components, or qualifying their integrity (like seals in valves, igniters and explosives in pyrotechnical devices), neutron tomography can be applied for "reverse engineering," i.e. analyzing the design of components, which are physically not, or not easily, accessible. This feature is very valuable for analysis of processes in hydrogen storage and hydrogen fuel cells [58].

The strong interaction with hydrogen makes neutron radiography very useful in biological applications (see Fig. 7). For example, it allows following in vivo plant root growth in soil, where the influence of soil contamination, or poisoning, on the root growth can directly be imaged. Other applications include the study of water ingress, or water

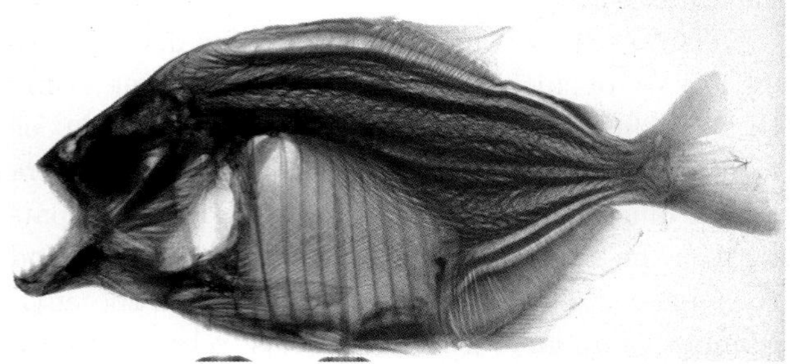

Fig. 7. This dried fish (a piranha) can be studied in great detail, because a high resolution has been obtained for the image. (Courtesy of IAEA [58].)

diffusion in construction materials, such as wood and wood products, or visualizing the flow of fluids in fuel cells and conventional engines.

The nondestructive nature of neutron radiography combined with the strong interaction between neutrons and light elements is a valuable prerequisite for physical investigations relevant to cultural heritage. Examples are proof of originality (or discovering fakes), analyzing ancient manufacturing techniques, inner signatures, hidden devices, etc.

4.2. *Activation analysis*

Activation analysis is a well-established tool for quantifying trace elements, and materials, with relatively high precision and sensitivity. To make a neutron source that is suitable for this kind of application, one has to choose the design of the target moderator system such that it provides sufficient and suitable positions to insert samples to be analyzed. The neutron energy spectrum at this location must be well characterized and stable, for quantitative analysis [59]. Possible applications are widely spread, including environmental and biological research on trace elements in blood and tissue.

4.3. *Small angle neutron scattering*

Small angle neutron scattering (SANS), reflectometry and powder diffraction instruments are usually attached to reactors or spallation systems. There are ambitious efforts to improve and adapt these systems to be used at smaller and lower intensity neutron sources [3] as supported by an IAEA-coordinated research project. First results involving a minifocusing small angle neutron scattering (mfSANS) instrument with an ellipse [60] are encouraging. Therefore, these methods are included to display the future potential use.

Typical SANS instruments probe the atomic structure of materials over length scales from one nanometer to a few hundred nanometers. By this technique, the nanoscale structures of various materials can be measured, such as the shape and size of a polymer chain inside plastic materials, supermolecular structures, precipitation structure in alloys, microdomain structures in magnetic materials, etc. Compared to X-rays, neutrons have relatively large cross sections for light atoms, such as hydrogen, oxygen, carbon and nitrogen, the constituent atoms

of organic materials, or soft matter. In X-ray scattering, such light elements can be very difficult to detect in the presence of atoms with a greater atomic number. For neutrons, the macroscopic cross section can in many cases be dominated by these light elements. Moreover, the scattering lengths for hydrogen and deuterium are quite different — hydrogen has a negative value and deuterium has positive one — giving opportunities to use the so-called contrast variation technique to provide element-specific structural information through isotope substitution. Neutrons have another unique feature: their "spin," associated with a magnetic moment. Neutron spins interact with electron spins, thus providing unique information about magnetic structure, or spin configuration, in magnetic materials.

There are many applications using the SANS technique, especially in the industrial area. Fuel cell and battery development are good examples. The nanopore structure of electrolytes can be investigated in various conditions so as to develop better materials. This is also applicable to many types of food, like chocolate and milk. These kinds of food are complex nanosize emulsions, whose details can be related to the quality of such foodstuffs. Many fundamental research studies with SANS can be performed at small and medium-size ADNSs, such as polymers, micelles, emulsions, protein solutions and magnetic nanophase. A SANS instrument is a perfect tool for a small or medium-size ADNS that is equipped with a cold source with wavelengths on the order of $1\,nm$ or greater. It could also be installed at an ambient temperature thermal moderator beamline, but in this case it would be difficult to perform measurements below $0.01\,\text{Å}^{-1}$. With a flux at the sample of as little as 10^4–10^5 n/cm^2/s and with a reasonable collimation of, say, 1–$3\,mrad$, useful research can be performed in a variety of fields.

4.4. *Reflectometry*

When neutrons are incident on a surface at a very shallow angle of incidence, they are reflected. By measuring the reflectivity as a function of the angle, and wavelength, the nanoscale scattering-length density variation perpendicular to the surface may be determined as a function of depth. Using this technique, one can measure the thickness of single-layer surfactant on liquid surfaces, interface structure of multilayer metallic coatings, polymers, or magnetic materials, etc. [61]. There are many industrial applications that can benefit from using small/medium-size ADNSs. Catalysts would be a good example, as would be adhesives. The flux requirement varies from application to application, depending on the information required. If the critical angle is the only information needed, a neutron flux of 10^4 n/cm^2/s would suffice, with a few-mrad collimation in one dimension, and ten times more beam divergence perpendicular to it. Reflectivities down to 10^{-6} are accessible with about 10^5–10^6 n/cm^2/s peak flux with the above collimation. There are many applications that can be served by using this kind of instrument. Again, a cold neutron moderator is desirable, but a thermal neutron beam can also be used for this kind of instrument.

4.5. *Powder/polycrystalline diffraction and strain measurements*

One special application of a powder diffraction instrument attached to smaller neutron sources is strain/stress analysis in structural components, e.g. at and around welds, in machine parts as fabricated, and after prolonged service. This special application is in high demand among engineers for guidance in design, structural optimization, manufacturing quality control and lifetime predictions. A concept for stress/strain measurements at low to medium energy neutron sources may be based on a time-of-flight diffraction with multiple pulse overlap [62].

Texture measurements fall into the same category of application as strain measurements. The demand for this application comes from the same industry groups, including applied mechanical fabrication technology, and it imposes similar requirements on the instrument design.

4.6. *TOF instrument for incoherent inelastic measurements*

Inelastic scattering studies are difficult to perform at small and medium-size flux neutron sources, especially for coherent inelastic scattering. However, there are broad possibilities of using incoherent inelastic scattering techniques. Since hydrogen has an extraordinarily large incoherent cross section of 80 barns in the thermal energy regime, it is possible to have a spectrometer that can access the

neutron energy range from sub-meV to a few hundred meV [16].

An inverted geometry crystal-analyzer spectrometer would be a candidate. Chemical vibration spectroscopy is one of the available research fields. A neutron flux of at least 10^6 n/cm^2/s, with a 3–10 mrad collimation in the horizontal and vertical directions, is needed to undertake this kind of measurement (Ref. 3, p. 18).

4.7. *Neutronic engineering*

Small neutron sources provide only a low neutron flux. Due to limited activation they can give very good opportunities for neutronic engineering research, neutron instrument component development, and testing of new instrument ideas and new devices. The electron linac facility at Hokkaido University is a very good example of this kind of application. With only 45 MeV of energy and 3 kW of power, which provides about 6×10^{12} n$_\mathrm{f}$/s (n$_\mathrm{f}$: fast neutrons), and about 6×10^{10} n$_\mathrm{th}$/s (n$_\mathrm{th}$: thermal neutrons), many new developments can be tested, such as the study of spatial and time distribution of neutrons leaking from an accelerator facility [63]. Compared to the high flux sources (see Fig. 1), the thermal neutron flux is orders of magnitude lower. For detector development, about 10^2–10^4 n/cm^2/s at the detector position is desirable. For testing other neutron optical devices, a relatively good collimation of the neutron flux is needed, of the order of 1–3 mrad in one direction. Neutron flux requirements for testing new concepts or new instrument ideas vary from application to application. Many studies can be made with relatively low fluxes, 10^2–10^4 n/cm^2/s neutrons, and with about 3 mrad beam divergence in one direction.

Even neutron tubes can be used, such as when testing the moderator setup of the proposed ESS neutron sources using the experiment JESSICA [64].

5. Cooperation in Neutron Research

Since about 1975, the International Collaboration of Advanced Neutron Sources (ICANS) has existed as a forum for neutron research and innovative neutron systems. The collaboration has supported and was in many cases even instrumental in the design of the high power neutron facilities. It is a forum where smaller laboratories can get access to information and data needed for the design of new facilities [65]. A further example of international collaboration on accelerators for neutron research is the IFMIF cooperation.

Cooperation in neutron research helps to optimize the use of experimental time at research reactors and intense neutron sources such as ADNSs, which are often heavily overbooked. Contrary to this, many reactors in the developing world are under-utilized [66]. The use of small to medium-size neutron sources based on accelerators opens up further opportunities, as outlined for example in research work supported by the IAEA. Specialized Coordinated Research Projects (CRPs) comprising research in developing and developed countries have been initiated to develop better detectors and neutron beams, which can be used for applications in science and research. Based on the experience with programs on accelerators [67, 68], cooperation in neutron research has been initiated. The general aims of the common research are adoptable for other kinds of cooperation [3]:

- Identification of possible stakeholders in developing and developed countries;
- Analysis of the stakeholders' research goals and priorities relevant to ADNSs;
- Training of scientists and engineers in existing facilities ("awareness of needs" and deployment of skills);
- Selection of a source concept based on the stakeholders' needs;
- Support for interested parties on the scientific and technical level;
- Supporting the building of a suitable infrastructure and facility management culture;
- Planning a "baseline" suite of research opportunities;
- Making funding agreements and site selection;
- Construction of "network nodes," if possible anchored to existing facilities;
- Expansion of the network.

Thus, a synergy between the operations of high and low to medium-size power sources is beneficial to both types.

6. Summary

Neutron research utilizing accelerators started in 1937 with the construction of the 37-inch cyclotron by Ernest Lawrence in Berkeley, which produced radioisotopes. Soon, research reactors took over and were the leading source of neutron production, but reached their limit in the 1970s with a thermal neutron flux of about $10^{15}\,\mathrm{n/cm^2/s}$. Based on the experiences with accelerators, specialized high power spallation neutron sources have been designed to overcome the limited flux in one pulse. These neutron sources are capable of reaching $10^{17}\,\mathrm{n/cm^2/s}$. They require high investments and are not flexible regarding design changes. Thus, the whole variety of smaller neutron sources are still needed and will continue to be used in research and industry. Due to new detector developments and more sensitive and improved methods, new applications in radiography and stress/strain measurements are possible. The smaller and less expensive sources can take over many important tasks from the bigger sources and save on the budget and beam time. Small accelerator-driven sources are capable of replacing underutilized research reactors while offering new possibilities in research and development.

Strong neutron research as a complementary method to synchrotron research and other methods needs international cooperation. One of the possible partners is the International Atomic Energy Agency, where laboratories from developing and emerging economies can find appropriate cooperation. New, bigger facilities, such as the IFMIF, need smaller devices for proof-of-principle experiments. The use of accelerator-driven neutron sources will continue and will pave the way for new neutron physics.

Acknowledgment

The authors would like to thank the partners in the IAEA-coordinated research projects on neutron research for providing their data and figures.

References

[1] R. Hamm, paper AP/IA-12, in *IAEA Proceedings Series*, STI/PUB/1433 (2010).
[2] R. Hamm, *RAST*, **1**, 163 (2008) [doi:10.1142/S1793626808000095].
[3] IAEA-TECDOC-1439 (2005).
[4] IAEA-TCS-34 (2009).
[5] IAEA-TECDOC-628 Rev. 2 (2009).
[6] J. G. Fantidis, C. Potolias and D. V. Bandekas, *Science and Technology of Nuclear Installations*, Vol. 2011 (2011), article ID 347320 [doi: 10.1155/2011/347320].

[7] http://nucleus.iaea.org/RRDB

[8] www.sns.gov

[9] j-parc.jp

[10] D. Filges and F. Goldenbaum, *Handbook of Spallation Research* (Wiley-VCH Verlag, 2009), and references therein.

[11] M. R. Hawkesworth, *At. Energy Rev.* **15**(2), 169 (1977).

[12] M. Drosg, *IAEA rep.* IAEA-NDS-87 Rev. 9 (2005).

[13] http://homepage.univie.ac.at/manfred.drosg/drosg99.PDF

[14] D. Filges and F. Goldenbaum, *Handbook of Spallation Research* (Wiley-VCH Verlag, Weinheim) p. 497.

[15] M. U. Khandaker *et al.*, paper SM/SR-12, in *IAEA Proceedings Series*, STI/PUB/1433 (2010).

[16] G. Bauer, lecture notes, ICTP School on Pulsed Neutrons (2005).

[17] K. N. Clausen, *Pramana* **71**(4), 623 (2008).

[18] www.accsys.com

[19] Wikipedia, http://en.wikipedia.org/wiki/Neutron_generator

[20] K. Nünighoff, Ch. Pohl, S. Koulikov *et al.*, *Eur. Phys. J. A* **38**, 115 (2008).

[21] V. Raspa, F. Di Lorenzo, P. Knoblauch *et al.*, *PMC Phys. A* (2008), 2:5.

[22] L. Soto, C. Pavez, A. Tarifeno *et al.*, *Plasma Sources Sci. Technol.* **19**, 055017 (2010) [doi: 10.1088/0963-0252/19/5/055017].

[23] V. A. Gribkov and A. Malaquias, *Nukleonika* **51**(1), 5 (2006).

[24] http://iec.neep.wisc.edu/index.php

[25] C. M. Lavelle, D. V. Baxter, A. Bogdanov *et al.*, arXiv:0803.4170v1.

[26] D. V. Baxter, J. M. Cameron, V. P. Derenchuk *et al.*, *Nucl. Instrum. Methods B* **241**, 209 (2005).

[27] K. Nueninghoff, W. Bernnat, V. Bollini *et al.*, Experimental investigation of advanced cold moderators at JESSICA at COSY-Juelich and comparison with MCNPX. Private communication.

[28] C. M. Lavelle, D. V. Baxter, A. Bogdanov *et al.*, arXiv:0803.4170v1 (p. 5).

[29] D. Strellis, T. Gozani and J. Stevenson, paper SM/EN-05, in *IAEA Proceedings Series*, STI/PUB/1433 (2010).

[30] http://www.pelletron.com/pellet.htm

[31] J. R. Granada, J. R. Santisteban, J. Dawidowski and R. E. Mayer, paper AP/IE-02, in *IAEA Proceedings Series*, STI/PUB/1433 (2010).

[32] Y. K. Kyanagi *et al.*, paper AP/AM-07, in *IAEA Proceedings Series*, STI/PUB/1433 (2010).

[33] G. N. Kim, H. S. Kang, Y. S. Lee *et al.*, arXiv:0008068v1.pdf.

[34] W. P. Swanson, *IAEA Tech. Rep.* 188 (1979).

[35] C. M. Lavelle, D. V. Baxter, A. Bogdanov *et al.*, *Nucl. Instrum. Methods A* **587**, 324 (2008).

[36] D. V. Baxter, J. Leung, H. Kaiser *et al.*, *Nucl. Instrum. Methods A* (2011), in press.

[37] www.frascati.enea.it/ifmif

[38] P. Garin and M. Sugimoto, paper FTP3-1, *IAEA Fusion Energy Conference*, 2010. To be published.

[39] T. Ditmire, J. Zweiback, V. P. Yanovsky *et al.*, *Nature* **398**, 489 (1999).

[40] G. Pretzler, A. Saemann, A. Pukhov *et al.*, *Phys. Rev. E* **58**, 1165 (1998).

[41] http://www.extreme-light-infrastructure.eu

[42] B. Sowerby, C. Franklyn and F. Mulhauser, in *Proc. Int. Topical Meeting on Nuclear Research Applications and Utilization of Accelerators* (4–8 May 2009; Vienna, Austria), IAEA-I3-CN-173 and papers referenced therein.

[43] The MCNPX radiation transport code, LA-UR-11-01502 (Los Alamos, 2011).

[44] F. Cantargi *et al.*, *Nucl. Instrum. Methods B* **248**, 340 (2006).

[45] E. Shabalin, E. Kulagin, S. Kulikov and V. Melikhov, *Radiat. Phys. Chem.* **67**, 315 (2003).

[46] D. Filges, S. Leray, G. Mank *et al.*, http://www.nds.iaea.org/spallations

[47] V. Bom, IAEA, TECDOC, STI/PUB/1300 (IAEA, 2007).

[48] V. R. Bom and R. Megahid, in *IAEA Proceedings Series*, STI/PUB/1441 (2010).

[49] L. Grodzins, *Nucl. Instrum. Methods B* **56/57**, 829 (1991).

[50] T. Gozani, *Nucl. Instrum. Methods B* **79**, 601 (1993).

[51] G. Vourvopoulos and F. J. Schultz, *Nucl. Instrum. Methods Phys. Res. B* **79**, 585 (1993).

[52] R. C. Smith, M. J. Hurwitz and K. C. Tran, *Nucl. Instrum. Methods Phys. Res. B* **99**, 733 (1995).

[53] J. C. Overley *et al.*, *Nucl. Instrum. Methods B* **99**, 728 (1995).

[54] A. Zonta *et al.*, (eds.), *Proc. 13th Int. Conf. Neutron Capture Therapy* (ENEA, 2008).

[55] S. Liberman, A. J. Kreiner *et al.*, (eds.), *Proc. 14th Int. Conf. Neutron Capture Therapy* (CNEA, 2010), www.14icnct.com.ar

[56] http://ptcog.web.psi.ch

[57] J. Gueulette, J. P. Slabbert, P. Bischoff, J. M. Denis, A. Wambersie and D. Jones, *Radiat. Meas.* **45**(10), 1414 (2010).

[58] IAEA-TECDOC-1604 (2008).

[59] IAEA-TECDOC-1234 (2001).

[60] M. Furusaka, K. Kamada, Y. Kiyanagi *et al.*, in *Proc. ICANS-XVIII* 25–29 Apr. 2007; Dongguan, China.

[61] *IAEA Proceedings Series*, STI/PUB/1246 (2006).

[62] U. Stuhr, *Nucl. Instrum. Methods A* **545**, 319 (2005).

[63] M. Kitaichi, S. Sawamura, M. Wakisaka *et al.*, *Radiat. Prot. Dosim.* **110**(1–4), 731 (2004).

[64] H. Stelzer, N. Bayer, H.-K. Hinssen *et al.*, *ICANS-XVI: Proceedings of the ICANS-XVI*, Jülich,

ESS, 2003.-(ESS; 03-136M1).-1433-559X.-S.873, M03.

[65] http://icans.web.psi.ch

[66] IAEA-TECDOC-1625 (2009).

[67] N. Dytlewski, G. Mank, U. Rosengard *et al.*, *Nucl. Instrum. Methods A* **652**(2), 650 (2006).

[68] G. Mank, A. Stanculescu, N. Dytlewski *et al.*, in *Proc. 8th Int. Topical Meeting on Nuclear Applications and Utilization of Accelerators, AccApp07*, pp. 53–57 (2007).

Guenter Mank is an atomic physicist who has worked on ECR ion sources, electron accelerators and particle accelerators during his career. After obtaining his Ph.D. in Giessen, Germany, he was the project leader for Cyclotron Injection at the National Superconducting Cyclotron Laboratory (NSCL) in East Lansing, Michigan, USA. He returned to Germany in 1990 to do research on particle fueling and exhaust at the fusion tokamak TEXTOR. From 2001 to 2003, he worked as the project coordinator for the European Spallation Source (ESS) project in Juelich, Germany. From 2003 to 2010, he was section head of the physics section of the International Atomic Energy Agency (IAEA), where he supported neutron and fusion research and was responsible for several schools and workshops on these topics at the International Centre for Theoretical Physics (ICTP) in Trieste, Italy. Since the end of 2010, Dr. Mank has been responsible for the German fusion project management and projects on renewable energies at the Projektraeger Juelich (PtJ), Germany.

Günter Bauer graduated in physics engineering from the Technical University in München and obtained a PhD in Physics at the University of Bochum. He devoted most of his career to the production of neutrons and their application to condensed matter physics. After developing the technique of diffuse elastic neutron scattering for the investigation of point defects in solids he became technical project leader for the target system of the SNQ project at the Jülich research centre, where he lead the development of a rotating target for 5 MW of beam power. In 1986 he moved to the Swiss Institute of Nuclear Research (now PSI) to lead the development efforts for the target systems of SINQ,

a continuous 1 MW spallation source which went into operation in 1996. As director of the Spallation Source Division at PSI he was responsible for the operations and improvement efforts of the facility and was one of the initiators of the MEGAPIE project, the first successful demonstration of a liquid lead-bismuth target for a beam power in the MW range. From 1995 on he also headed the study of a mercury target for a 5 MW spallation source in a collaborative effort with the ESS project. He went back to Jülich in 2001 as a Project Leader for the ESS target station design, from where he retired in 2005. He has served as an advisor and consultant for most of the spallation source efforts worldwide.

Françoise Mulhauser is a nuclear physicist. She obtained her Ph.D. in experimental nuclear/ atomic/particle physics at the University of Fribourg, Switzerland. She continued her career in fundamental research at various laboratories around the world, including TRIUMF in Canada, PSI in Switzerland, and Dubna in Russia. She was spokesperson for multiple experiments at those facilities, coordinating international research in the field of muon-catalyzed fusion for many years. Dr. Mulhauser is one of the authors of the publication "The Size of the Proton" (in the July 2010 issue of *Nature*), which was named one of the ten most important science results of the year by the IOP. Since 2005, she has been working for the International Atomic Energy Agency (IAEA) in Vienna, Austria, as a nuclear physicist in the department of nuclear science and applications. She is in charge of activities related to accelerator applications and education in nuclear physics. Her role is to promote these applications in developing countries when benefits are tangible and sustainable for the developing countries.

Reviews of Accelerator Science and Technology
Vol. 4 (2011) 235–255
© World Scientific Publishing Company
DOI: 10.1142/S1793626811000604

Prospects for Accelerator Technology

Alan Todd

*Advanced Energy Systems (AES), 100 Forrestal Road, Suite E,
Princeton, NJ 08540-6639, USA
alan_todd@mail.aesys.net*

Accelerator technology today is a greater than US$5 billion per annum business. Development of higher-performance technology with improved reliability that delivers reduced system size and life cycle cost is expected to significantly increase the total accelerator technology market and open up new application sales. Potential future directions are identified and pitfalls in new market penetration are considered. Both of the present big market segments, medical radiation therapy units and semiconductor ion implanters, are approaching the "maturity" phase of their product cycles, where incremental development rather than paradigm shifts is the norm, but they should continue to dominate commercial sales for some time. It is anticipated that large discovery-science accelerators will continue to provide a specialty market beset by the unpredictable cycles resulting from the scale of the projects themselves, coupled with external political and economic drivers. Although fraught with differing market entry difficulties, the security and environmental markets, together with new, as yet unrealized, industrial material processing applications, are expected to provide the bulk of future commercial accelerator technology growth.

Keywords: Accelerator technology; superconducting RF (SRF); normal-conducting; auxiliary systems; advanced accelerators.

1. Introduction

Reference 1 reported that more than 30,000 accelerators worldwide are producing particle beams today for research or commercial applications. The annual market for commercial accelerator products, dominated by the medical and semiconductor industries, now exceeds US$5 billion, with an estimated over US$500 billion worth of products processed by accelerator-based systems [1]. How will this evolve in the coming years? What new profitable applications will break through to the marketplace and what new technology will they apply? Which large-scale government R&D accelerator projects will be funded and built? It takes a degree of temerity to project any aspects of the future, let alone the prospects for accelerator technology. The perspective used in this article is principally that of an accelerator supplier based in the USA. Obviously, many of the opinions expressed are personal and open to challenge.

Accelerator projects fall into two principal categories distinguished by their funding source, government-sponsored devices for discovery science or future energy production, and commercial applications. This distinction is slightly clouded by "one-off" or "few-off" turnkey systems, produced by commercial companies for discovery science at universities or institutes. These are considered to be commercial systems where the project is assembled and commissioned by a commercial company rather than the purchasing organization. In addition, the possibility of government-assisted private development sometimes exists which is often needed for smaller-return markets, though less frequently in the USA than in Europe or Asia. An example of this market is the production of the radioisotope ^{99}Mo, where a critical shortfall is presently occurring but the total market is far too small to induce commercial development.

The recent US Department of Energy (DoE) workshop report on "Accelerators for America's Future" [1] provides an excellent backdrop to the present topic and goes into much greater detail on applications than can be covered here. Following a brief consideration of the key technologies advances that may be available in the future, the discussion below has been organized by the application area

 A. Todd

classification structure of that workshop. Specifically, accelerator applications to future discovery science, energy and the environment, medicine, security and defense, and material processing/industrial applications are addressed. The other chapters of this book provide excellent contemporary discussions on many of the applications in far greater depth than can be presented here. We can also build on the work of other authors [2–4] who have provided excellent and extensive summaries of the accelerator application marketplace that include useful bibliographies. The proceedings of several biannual conferences are also excellent sources of accelerator applications [5–7]. Hence, we focus the present discussion on where we think the different technologies will be heading and what new doors potential improvements may open.

The difficulty faced in gathering data and making projections is, that on the one hand, government-sponsored research is always subject to the unpredictable vagaries of the political process and macroeconomic changes, while on the other hand, evolving commercial directions are highly proprietary and closely held by the companies in the field. Accelerator technology sales must usually be inferred from spotty data [2], and market predictions are equally speculative and frequently inflated by the advocates of specific technologies. A further pitfall for commercial applications is the issue of "technology push" versus "market pull." Since almost all accelerator technology begins in a university or national laboratory, that technology must be successfully transferred to the commercial marketplace. All too often, the technical owners in the laboratory are enamored of their technology and may "push" it toward markets where technical viability but no clear economic "value added" or "market pull" exists. Where viable economic markets do exist, the problem of transitioning the technology to the marketplace has long been appreciated to be fraught with difficulty and has been the subject of much discussion and hand-wringing. As a result, the path from proof of concept to the marketplace and profitable sales has been dubbed the "valley of death" [8].

Partially to address these problems within the USA, the government sponsors the Small Business Innovative Research (SBIR) program [9] through the Small Business Administration, and technology transfer centers such as the National Center

at Wheeling Jesuit University [10]. The European Commission organizes sponsored collaborations in specific technology areas, such as the Future and Emerging Technologies (FET) program for information and communications technology [11]. Technology transfer is sponsored by organizations like the European Commission Enterprise Europe Network [12] and the Association of European Science and Technology Transfer Professionals (ASTP) [13].

In contrast to Europe and the USA, where national laboratories and universities typically develop accelerator technology prototypes, in Japan the Ministry of Economy, Trade and Industry (METI) has, in the past, encouraged Japanese companies to perform a significant fraction of the fabrication of prototypes in collaboration with their laboratory and academic infrastructure, thereby, in principle, spanning the valley of death. In China, India, Korea and other Asian countries, commercial accelerator production is embryonic and transfer procedures are yet to fully evolve. However, Zhang [14] recently indicated that transfer to industry in China has progressed significantly and that several companies are ready to compete strongly in the international accelerator technology marketplace.

Despite the acknowledgement of the problem in each of the three accelerator technology zones, it is not clear how successful these various technology-transfer-fostering approaches are in delivering academic technology to market.

As an example, ion therapy poses an interesting technology transfer question for the USA. The concept was conceived by Robert Wilson [15] and the original R&D was pioneered at the Harvard Cyclotron Laboratory in collaboration with the Massachusetts General Hospital and at other institutions, mainly in the USA. The conceptual design of the first dedicated hospital proton therapy facility to come on line at the Loma Linda University Medical Center (LLUMC) was produced by Fermilab [16]. Despite this history, today there are only two indigenous US suppliers of proton therapy systems. One is Optivus [17], which was spun off from LLUMC nearly 20 years ago, but which, to the best of our knowledge, has yet to register a sale of their Conforma 3000 synchrotron system. The other is a newcomer, Protom International, which was founded in 2008 and has announced three sales of its Radiance 330, a compact synchrotron solution [18]. The field is expanding quite rapidly but, with the exception of newcomer Protom, manufacturing expertise resides in Europe and Japan. Varian recently purchased the German company Accel GmbH in order to acquire proton therapy technology and market share, implying a failure of the US technology transfer system.

2. Enabling Technologies

There are four top-level metrics for adoption of new or improved accelerator technology. These are improved performance, improved reliability, reduced capital or life cycle cost, and reduced footprint or size/weight. Higher performance can manifest itself in many forms, like higher gradient, higher efficiency or higher beam brightness. Life cycle is the more usual cost metric, but some applications also have capital cost entry barriers. These metrics are not mutually exclusive, since, for instance, higher reliability sometimes, but not in medical applications, simply manifests itself as reduced operating and life cycle cost.

An important consideration is that accelerator technologies include not only the particle acceleration structures but also the technologies of the auxiliary systems that together define the complete accelerator package. These include the RF and DC power sources needed to power the accelerator components, their instrumentation and control systems (I&C), cryogenic and conventional cooling systems, diagnostics, magnets and so on. Often, the auxiliary systems dominate the footprint in applications where size is important, so we must consider all these components as a unit. In the medical and inspection fields that utilize RF linacs, smaller accelerators are desirable, implying higher-frequency structures. However, since there is often not a commensurate reduction in RF power system size with increasing frequency, when the complete system is considered the cost and size advantage achieved at higher frequency may be less obvious.

Table 1 (from Ref. 1) lists the importance to the participants in that workshop of R&D needs by application area. These R&D topics are organized somewhat differently from those we consider but the same metrics are covered, with the possible exception that only production cost, which is effectively capital cost, rather than life cycle cost is called out. While capital cost and size may not be so important to some large defense programs as to commercial ventures, in our estimation they are crucial to many security projects where commercial off-the-shelf (COTS) technology, cost and time to deployment

Table 1. Signal chart of technology R&D needs, indicating relative importance by application area. (Courtesy of Ref. 1.)

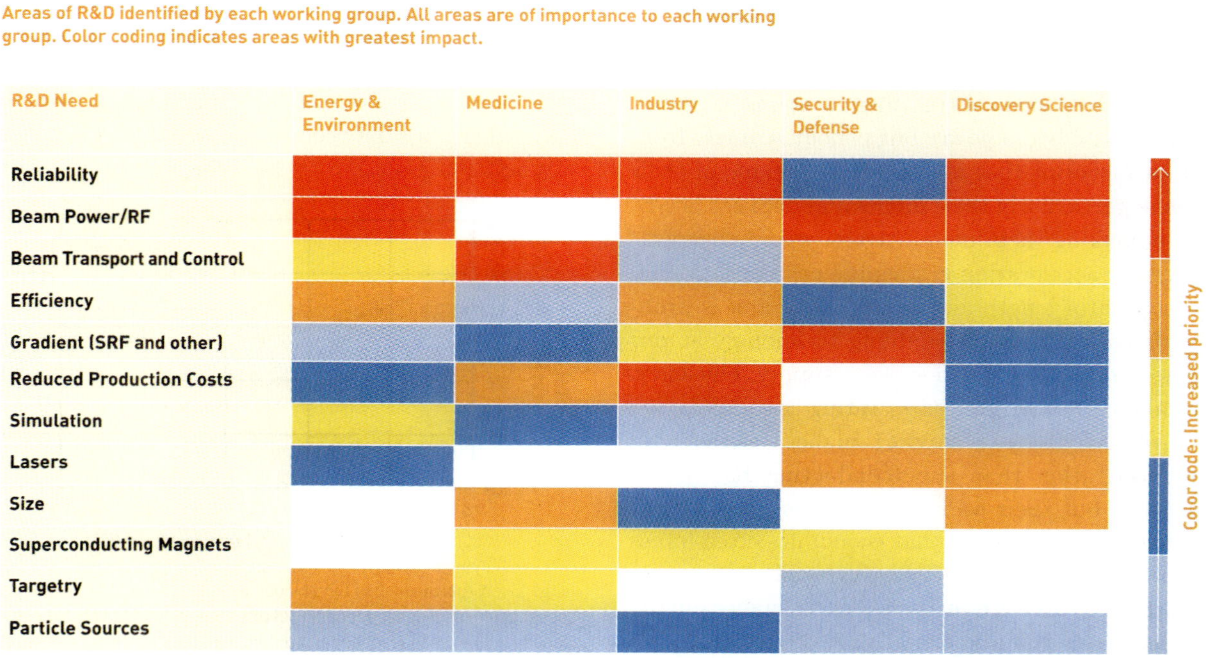

Areas of R&D identified by each working group. All areas are of importance to each working group. Color coding indicates areas with greatest impact.

R&D Need	Energy & Environment	Medicine	Industry	Security & Defense	Discovery Science
Reliability					
Beam Power/RF					
Beam Transport and Control					
Efficiency					
Gradient (SRF and other)					
Reduced Production Costs					
Simulation					
Lasers					
Size					
Superconducting Magnets					
Targetry					
Particle Sources					

Color code: Increased priority

often dominate procurements over performance characteristics. Also, size is a pre-eminent concern for the Navy Free Electron Laser (FEL) program [19, 20], where the available space on a ship is limited. The emergence as competitive devices of plasma accelerator systems, which could find application in proton medical therapy and other radiation generation applications, may require progress in laser technology, that is also not reflected in the table. Hence, we have some, generally minor, differences with the findings of the workshop.

Supporting technology advances listed in Table 1, like power technology, beam transport and control, lasers, simulation, superconducting magnets, targetry and particle sources will always be incorporated in accelerator systems as they become available, in order to deliver improved products. We do not specifically call out these areas in this section, though we note specific deficiencies in passing when applications are discussed. Improving the efficiency of power supplies of all types, in order to maximize system wall plug efficiency, is perhaps the single most enabling technology advancement that can be achieved from auxiliary systems. This is universally true but particularly important for high-power beam applications.

In the end, it is almost always about cost. Higher particle beam power per unit accelerator system volume is a driving goal in almost all applications. Some applications, like many in the environmental or testing areas, seek ever-increasing beam power to achieve economies of scale from throughput increase. Others, such as medical treatment systems, typically have adequate beam power but benefit strongly from reduced footprint that shrinks the envelope of their systems, particularly the gantries and delivery heads, thereby making them more attractive and cheaper.

A high accelerating gradient is critical in reducing accelerator length and capital cost, and is probably the single greatest enabling technology at our disposal, on a tier with power supply efficiency. Normal-conducting RF accelerating structures are more robust and can support higher gradients in pulsed operation than superconducting RF (SRF) structures, but their wall plug efficiency is impaired by resistive wall losses so that economics often forces operation at a reduced field. Typically, superconducting structures are utilized only in higher-energy systems where the additional cost of the cryoplant

required to cool the structures, as opposed to the significantly cheaper normal-conducting water cooling plant, is paid for by the increase in system wall plug efficiency. In time, plasma acceleration is expected to displace RF acceleration in virtually all high-energy accelerator applications because of the astonishing accelerating gradients that can be achieved.

For superconducting accelerators, there is an optimum operating temperature that minimizes life cycle costs. At the lowest temperatures, the heat lift is smallest but the cryoplant efficiency is low, while at higher temperatures the heat lift and cooling capacity increase, causing a minimum to occur at some temperature less than 10 K. Because of the cryogenic transition at around 4 K that drives cryoplant complexity and cost, there is fairly strong motivation for some applications to operate at or above 4 K. Since the optimum operating temperature is found to rise with decreasing frequency, these applications seek to utilize lower-RF-frequency (< 500 MHz) structures. For the Accelerator Production of Tritium (APT) program [21], the projected optimum life cycle cost was around 2.2 K, as shown in Fig. 1 [22].

For normal-conducting cavities, the optimum life cycle operating point is a function of the accelerating gradient. Wall losses and both the capital and operating cost of the associated RF power system scale as the square of the electric field dominating the economics at high fields. On the other hand, for given delivered beam energy the length of the accelerator and its associated capital cost scale linearly with the gradient and therefore dominate at a low gradient, yielding an optimum somewhere in between. The International Fusion

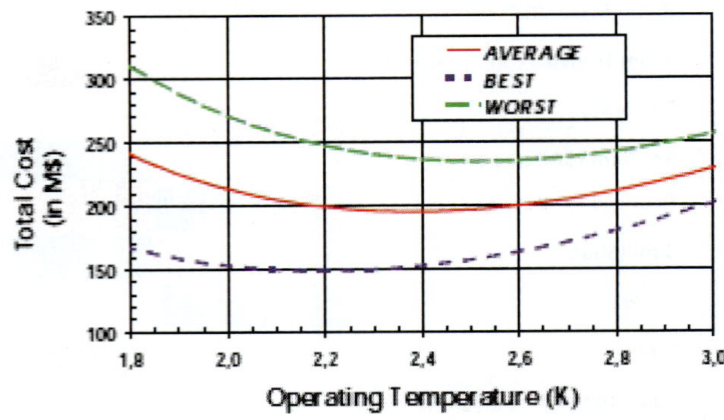

Fig. 1. Optimum life cycle cost as a function of cryogen temperature for the superconducting cavities of the Accelerator Production of Tritium. (Courtesy of Ref. 22.)

Materials Irradiation Facility (IFMIF) [23] and APT cases are particularly interesting, since the two accelerating options were considered by both projects and optimized. The accelerators were originally conceived of as normal-conducting, but systems trade studies [24, 25] demonstrated that the SRF option was cheaper with respect to total lifecycle cost. Since then, for similar reasons, all new large accelerator projects have evolved to SRF technology, including the Accelerator Transmutation of Waste (ATW) [26], the Spallation Neutron Source (SNS) [27], and the future International Linear Collider (ILC), which is described below [28]. Originally, the predecessor device, the Next Linear Collider (NLC), had embraced X-band copper technology [29], but once again system trade studies determined the life cycle cost advantage of SRF.

Nevertheless, the robust character of normal-conducting structures and the achievable gradients at a lower duty factor mean that they find application in many areas outside discovery science. Many normal-conducting S-band linacs are in use for medical therapy and nondestructive evaluation (NDE). Additionally, X-band normal-conducting structures are increasingly being used for these applications and in other areas. CW normal-conducting structures, such as the photoinjector in Fig. 2, have also been designed and fabricated [30] for specialized applications.

Driven to a large extent by ILC R&D, the recent advances in consistently achieved SRF gradients in $\beta = 1$ elliptical cavities have been rather remarkable. For pure niobium, the theoretical gradient limit is about 60 MV/m. The percentage of ILC R&D cavities reaching the vendor qualification value of

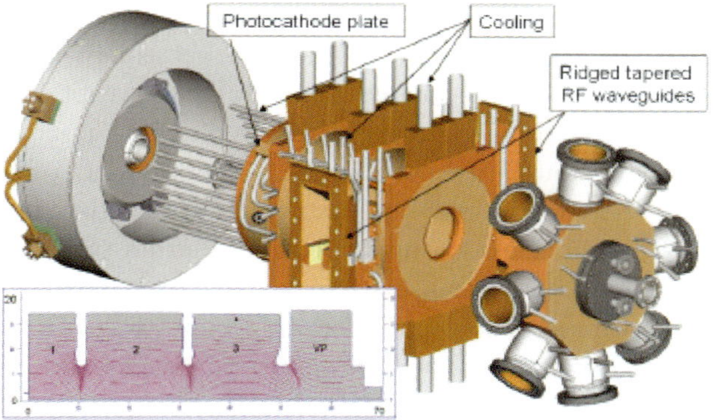

Fig. 2. The CW normal-conducting RF gun fabricated by Advanced Energy Systems (AES) for Los Alamos.

Fig. 3. Two AES-fabricated SRF cavities that exceeded the 35 MV/m ILC specification. The front structure reached 37.7 MV/m and the rear 41.1 MV/m.

35 MV/m with $Q_0 \geq 8 \times 10^9$ at 2 K in a vertical test assembly (VTA) is now around 70%, and it is believed that by 2012 this number will exceed 90%. About 10% of the cavities that have been produced to date under this program have exceeded 40 MV/m.

Figure 3 shows two cavities (manufactured by AES) that exceeded the ILC specification, while Fig. 4 shows the Q_0 versus gradient achieved for several Research Instruments (RI) and AES cavities. These firms are two of the ILC suppliers that have been qualified to date. Zanon is another qualified vendor and Mitsubishi Heavy Industries (MHI) is expected to be added to the list soon. The size of the ILC is such that no one company can begin to deliver the needed number of cavities to the project schedule, and thus it is essential to have multiple qualified vendors.

Meanwhile, single-cell R&D cavities of different shapes have exceeded 50 MV/m at Cornell University and KEK [31, 32]. New surface treatment processes such as centrifugal barrel polishing [33] promise an even closer approach to the absolute gradient limit for multicell cavities in the future.

Development of low-beta SRF spoke, half- and quarter-wave cavities [34], such as the spoke cavity designed by Argonne National Laboratory (ANL), shown in Fig. 5, is also progressing. Such cavities have reached surface magnetic and electric field values similar to those of high-beta elliptical cavities, though their geometry dictates different accelerating gradients. Cavities like these will likely be used for the ion beams of the Project-X injector [35] and IFMIF. High-beta spoke cavities are also of interest in high-power electron accelerators, because they typically operate at a lower frequency for a given radial footprint and hold out the promise of robust

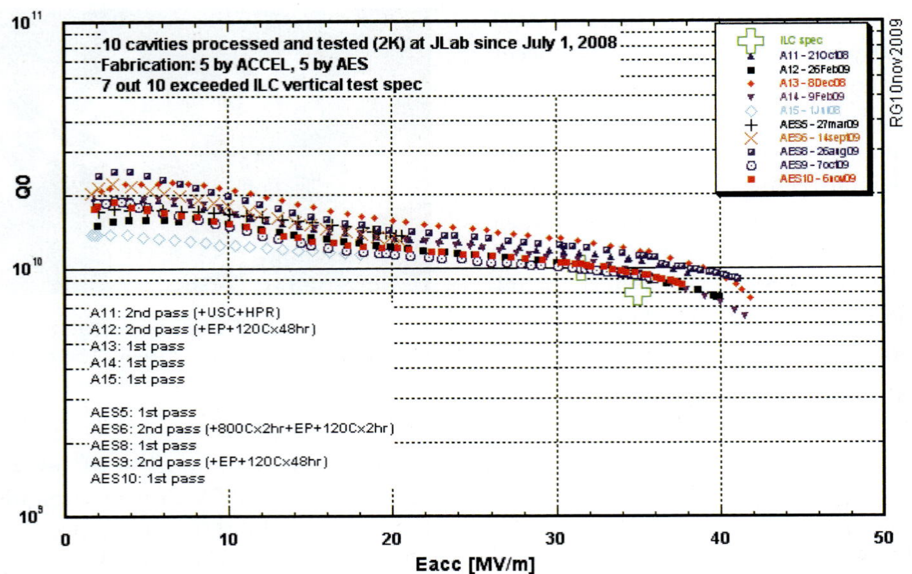

Fig. 4. Summary of ten nine-cell cavities processed and tested at Jefferson Laboratory in 2008 and 2009. Research Instruments GmbH and AES supplied five cavities each to this testing program. (Courtesy of R. L. Geng, JLAB.)

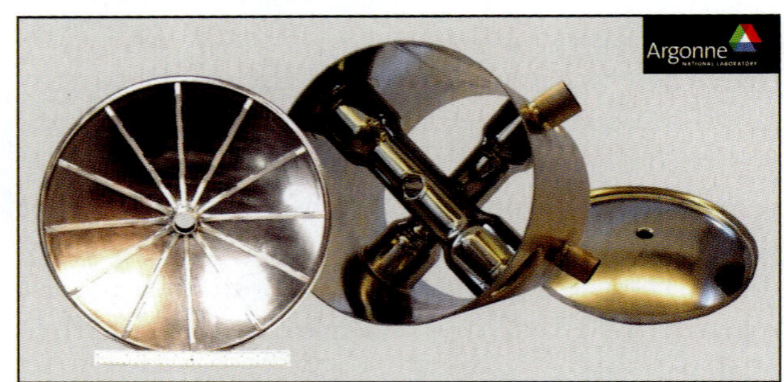

Fig. 5. The double-spoke resonator cavity fabricated by AES for ANL. (Courtesy of K. W. Shepard, ANL.)

operation above 4 K, which is economically attractive for defense and other applications.

The future holy grail of SRF clearly lies with structures that employ high-temperature superconducting (HTSC) material [36]. R&D is ongoing at many facilities but it will still be some years before such cavities are ready for production systems. Concepts show the material being deposited by a variety of processes on a cavity made of copper or some other metal. Difficulties presently exist with the methodology and uniformity of large-scale deposition inside the contoured structures. The simplified cooling required of these devices will significantly reduce the life cycle cost of SRF systems and promises to finally open the door for SRF to enter the commercial application marketplace. Right now, the only suggested commercial SRF application that might be economically viable is an energy recovery linac (ERL) for material processing [37–39], and little progress has occurred in the last 15 years to bring this to fruition. In contrast, Ref. 1 seems to suggest that SRF has near-term application beyond discovery science.

High-power FEL applications exhibit some of the pitfalls of transition to the market. The unique features of FELs are their wide wavelength tunability and the potential, which has yet to be fully realized, of very high average photon processing power. A drawback is the fact that for ERL FELs, the footprint is not a strong function of delivered photon power so that the cost per photon only becomes competitive with commercial lasers at extremely high power. Hence, the device is generally unacceptable for a low- or moderate-power commercial application. As fourth generation light sources with tunable wavelength, they are wonderful R&D tools for developing applications involving radiation processing. However, because there is usually a particular optimum wavelength for a specific application, tunability is not needed in the commercial system. There are several examples where the viability of an application has been proven using FEL radiation, only to have the actual commercial tool use a much

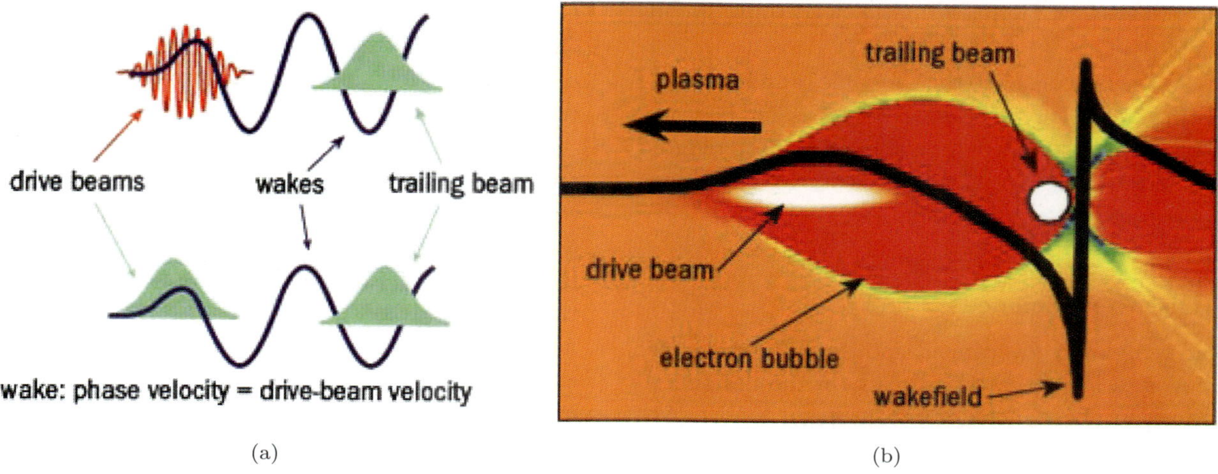

Fig. 6. (a) Simple 1D schematic of how wakefields are excited by a short-laser (top) or particle-beam (bottom) driver in a plasma. (b) 3D computer simulation of an extremely nonlinear wakefield excited by the drive beam in the "bubble" regime. The wakefield can accelerate an appropriately phased trailing beam at ultrahigh gradients. (Courtesy of Ref. 41 and CERN Courier.)

more economical, conventional laser tailored to the application. One such example is the use of lasers for sealing tooth enamel, where the original R&D was performed using an FEL [40], but in the end the commercial application utilized a CO_2 laser modified to deliver the optimum 9.4 μm wavelength.

Plasma accelerators that promise orders-of-magnitude increases in the accelerating gradient to greater than 100 GeV/m could result in compact systems for applications that require high particle energies. They will need to deliver the necessary beam quality and performance, and the drive systems they require must become robust and cost-competitive.

Figure 6 [41] schematically shows the basic concept of laser wakefield acceleration (LWFA) [42] and particle beam-driven plasma wakefield acceleration (PWFA) [43] drivers. Laser beat-wave (LBWA) and self-modulated laser (SMLWFA) concepts have also been proposed. Although the idea of using plasma acceleration is quite old and has been attributed to Budker and Veksler, the use of wakefields was first put forward by Tajima and Dawson in 1979 [44].

The LOASIS group at LBNL [45] and others have shown that these systems offer the promise of tabletop, high-energy accelerators. Strong focusing and accelerating gradients up to 50 GV/m have been demonstrated in plasmas [43]. Thus, electrons, positrons, protons and ions can be accelerated to very high energies in short distances with ultrashort, high-brightness bunches. Any

high-energy accelerator application stands to benefit from the significant cost and footprint gain of plasma acceleration. Medical therapy has been suggested by several authors [46] as a likely entry market for the technology. Another market might be radiation sources for institutions and companies that can now install their own on-site facilities rather than utilizing shared central resources such as light sources at national laboratories.

Plasma acceleration has also been proposed for future discovery-science colliders because of the potential major cost reductions resulting from the significantly shorter accelerating sections. However, much remains to be demonstrated. Muggli and Hogan note that key issues to be resolved are trailing bunch acceleration, preservation of emittance, drive beam or laser power, and the system concept of the plasma accelerator configuration [43].

In the nearer term, plasma acceleration has already been proposed as part of a new light source, LUNEX5 [47]. (LUNEX5 stands for "free electron Laser Using a New accelerator for the Exploitation of X-ray radiation of the 5th generation.") Thus, this French project would become the world's first so-called "fifth generation" light source. Plasma accelerators are not the only advanced-accelerator systems. Other concepts include the inverse free electron laser (IFEL) [48], inverse Cerenkov acceleration (ICA) [48, 49], the dielectric waveguide accelerator (DWA) [50] and vacuum laser acceleration (VLA)

[49], but we are unable to cover everything in this article.

We next consider the possible application of existing technology in new market areas and how new enabling technologies can be applied to both old and new markets to create new accelerator technology opportunities.

3. Discovery Science

Near-term large government-sponsored programs for discovery science that are already in the design and production phase include the European X-FEL [51], J-PARC Upgrade in Japan [52] and FRIB [53] in the USA. These, among others, are the projects of the present, not the projects of the future. They do, however, have one thing in common: they have all adopted SRF acceleration to one degree or another. In the case of J-PARC, whose main acceleration is provided by synchrotrons, only the linac from 400–600 MeV uses SRF technology. The adoption of SRF acceleration for large-scale accelerator projects was made for life cycle cost reasons, as noted above, and follows the trend that began with the IFMIF, APT, ATW and SNS.

As to the future, SRF technology will power the 29-km-long International Linear Collider (ILC) [54], shown schematically in Fig. 7. The high projected cost of the ILC forces international collaboration, thereby providing a degree of stability to the project due to the collective effect of the international partners. With a price tag that remains unclear, but is certainly more than US$10 billion, one has to be concerned that maintaining the schedule may be difficult given the present weak world economic situation.

As noted, ILC R&D has made great strides recently in pushing achieved accelerating gradients to consistently greater than 35 MV/m and often greater than 40 MV/m. The ILC linacs are made up of RF units consisting of three cryomodules and all the associated RF power. Two of the cryomodules have nine 1.3 GHz cavities, while the other one has eight cavities plus a quadrupole magnet, resulting in 26 cavities per RF unit. There are 610 RF units with 15,860 cavities planned for the two accelerators. The required real estate gradient for these units is 23.2 MV/m. ILC management is working to hold the final cost close to the US$6.7 billion (in 2007 US dollars) of the original report but this excludes nearly 15,000 person years of contributed institutional support. To do this, the cost of the ILC has been contained by switching to a single tunnel, reducing the estimates for SRF cavity production, and adopting a new RF distribution system and other improvements.

This represents a substantial opportunity for the accelerator fabricators, but of course it is only a one-time event and is not sustainable. Herein lies the problem for these suppliers, who must ride the roller coaster of occasional very large projects. It requires the sponsoring governments to invest in the private company infrastructure, else the return on investment (ROI) is not attractive. The methods employed

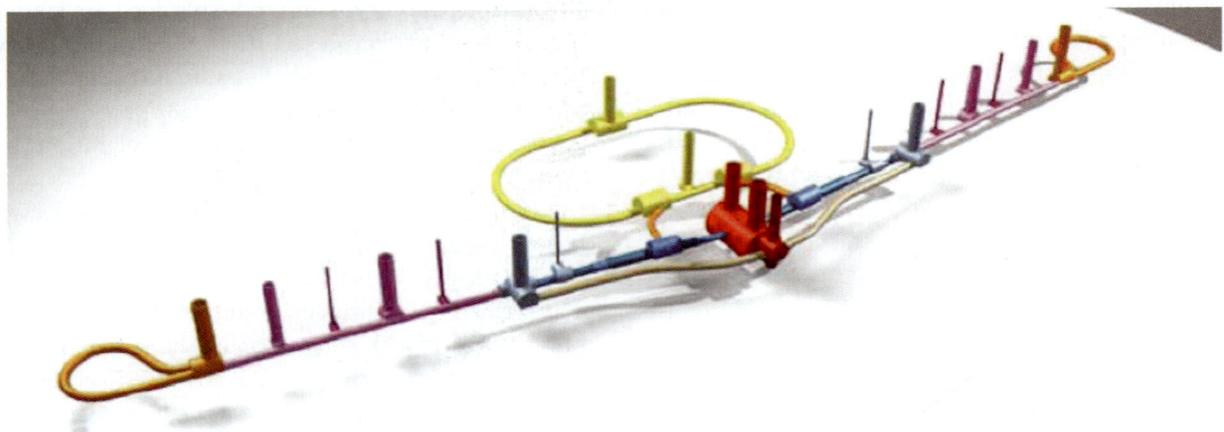

Fig. 7. Illustration showing the present 29 km end-to-end ILC design — the source section is in blue, the electron linac is to the left, the positron linac to the right, the central damping rings are in yellow and the detector area is in red. (Courtesy of Ref. 54.)

and the degree of investment differ in different parts of the world.

Other major discovery-science projects are evolving worldwide. Muon colliders [55], "extreme beams" such as the FermiLab Project-X to probe the intensity frontier [35], and electron–hadron colliders such as eRHIC [56] are to be anticipated. Hence, although cyclical, a healthy stream of large projects can be anticipated over the coming years though international collaboration is essential because of the high cost of these devices. We also anticipate that some time in the future, plasma acceleration may replace SRF acceleration for some collider devices.

In addition, a large number of smaller-scale discovery-science projects are being proposed, designed and built all over the world. These are typically second through fifth generation light sources. At the high end, we have the LUNEX5 proposal for a fifth generation light source using an advanced plasma accelerator [47], and a veritable plethora of fourth generation X-ray light source FELs either proposed or already in design and fabrication [57].

At the lower end of the light source spectrum, the availability from commercial concerns of more economical turnkey radiation sources for R&D is leading many individual research institutions to install their own light sources on site. Two recent examples are the THz FEL FLARE project [58] at Radboud University in The Netherlands, and the IR and THz FEL at the Fritz-Haber-Institut der Max-Planck-Gesellschaft [59] in Berlin. Figure 8 shows the latter device near the end of installation. These smaller turnkey devices appear to be an increasing but modest future market.

Fig. 8.   The IR and THZ FEL fabricated by AES nearing completion at the Fritz Haber Institute, Berlin.

Finally, we note that there will be new accelerator technology developed for discovery science that is presently difficult to predict and that will in time permeate down to the various commercial applications.

4. Energy and the Environment

This is the market segment where the author anticipates the greatest upcoming growth. With the exception of Heavy Ion Fusion (HIF) [60], IFMIF and Accelerator-Driven Transmutation (ADT) [61], the applications identified could just as well have been covered under "industrial applications." Here they are pulled out separately because of a common environmental thread.

At this time, HIF and ADT would closely resemble large government-funded discovery-science projects were they to transition beyond the present laboratory R&D programs they are today. This is not expected to happen for some time, although this may not prove to be the case with ADT, where activity in Belgium, India and China is proceeding quite seriously.

HIF is an inertial confinement fusion (ICF) approach with most R&D occurring in the USA and Germany, where it is funded at relatively low levels compared to the better-known magnetic confinement fusion (MCF) effort. MCF requires large, extremely powerful neutral particle beams of deuterium or tritium of approximately tens of MW to drive heating, rotation or current in the confined plasma. The International Thermonuclear Experimental Reactor (ITER) [62] is the flagship program of MCF and the present leading customer for these neutral beams. An additional ITER-related project is IFMIF, an accelerator-based 14 MeV neutron generating facility for the study of the plasma-facing and blanket materials that will be needed for the fusion reactors of the future. The IFMIF device can produce high-power deuteron beams up to 40 MeV.

ADT comes in two forms, each utilizing a different fuel cycle. The first seeks to render conventional nuclear power more palatable to the public by greatly reducing the requirements of permanent repository storage [26]. Originally called ATW, it proposes to induce fission in order to transmute long-lived radioactive byproducts in spent conventional nuclear reactor fuel. The resultant waste would require managed storage for about 500 years, as opposed to the 300,000 years needed for present unprocessed waste fuel. As a bonus, electricity is generated to improve the overall economics of the system.

The second form, which utilizes a fuel cycle based on thorium and uranium 233, was originally proposed by Carlo Rubbia [63]. Here the fission reactor is subcritical and is driven by an accelerator. There are several other advantages; namely, thorium is relatively plentiful, proliferation concerns are reduced because elements like plutonium are not produced, and managed storage of the end products is comparatively short and simple. India is seriously pursuing this option and the Chinese are not far behind.

For both ADT cycles, the accelerator technology required exists to a large extent, although further improvements in wall plug efficiency would be useful. There are areas of the processing chemistry where progress needs to be made. Given the present US attitude toward nuclear power, now further polarized by the aftermath of the Japanese earthquake and tsunami, it seems likely that progress in ADT will occur first in Europe or Asia. The Belgian government has already funded a US$1.3 billion program called MYRRHA (Multipurpose, hYbrid Research Reactor for High-end Applications) [64].

The real growth area in this market segment is anticipated to be in environmental remediation. The accelerators required are typically very-high-power for maximum throughput, place few requirements upon beam quality, but demand high efficiency to maximize the economies of scale. Many different remediation applications have been proposed. The two we see with the greatest near-term prospects are flue gas scrubbing to decrease acid rain and smog, and water treatment.

Full-scale electric power plants using electron beam treatment of the generated flue gases are in operation in Poland and China today. Fertilizer is produced as a marketable byproduct of eliminating the NO_x and SO_x from the exhaust gases. There is considerable interest in this application around the world, with several companies working toward efficient systems [65,66], as shown by Fig. 9. The efficacy of the concept has been proven and a strong market pull exists, so it is thought to be just a matter of getting the economics right for it to experience rapid growth.

Another proven concept with a strong market pull is water treatment. In this case, the pilot plant

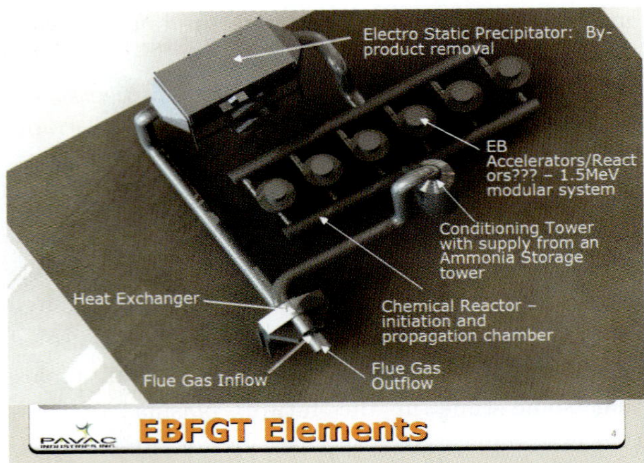

Fig. 9. Pavac industries electron beam flue gas treatment (EBFGT) concept. (Courtesy of Pavac Industries [65].)

shown in Fig. 10 is already operating in South Korea [67,68], at the Daegu textile dying Industrial complex. The treatment can be for biological and chemical control of drinking water or remediation of waste and certain toxically contaminated water.

Finally, we include the production of biofuels as an environmental application because of the potential to reduce the negative effects of oil exploration and the use of nonedible starch and cellulose for the fuel base, as opposed to edible crops that could otherwise be used as food. Conventional chemical processing of corn and other edible materials into ethanol produces toxins and has other less-than-ideal aspects. This is thus a potentially huge market whose economics has yet to be properly developed [69].

The only perceived impediment to the deployment of water and flue gas processing systems is reliability and wall plug efficiency, since uptime and cost per processed volume are the defining metrics. Commercial manufacturers are working to meet these requirements and the arrival of further prototype plants is expected soon. Biofuel production is another area with huge potential that requires study to determine the true economic tipping point.

5. Medicine

After processing tools for the semiconductor industry, medical devices are the largest present commercial market. For many years, S-band X-ray generators, typically using side-coupled standing

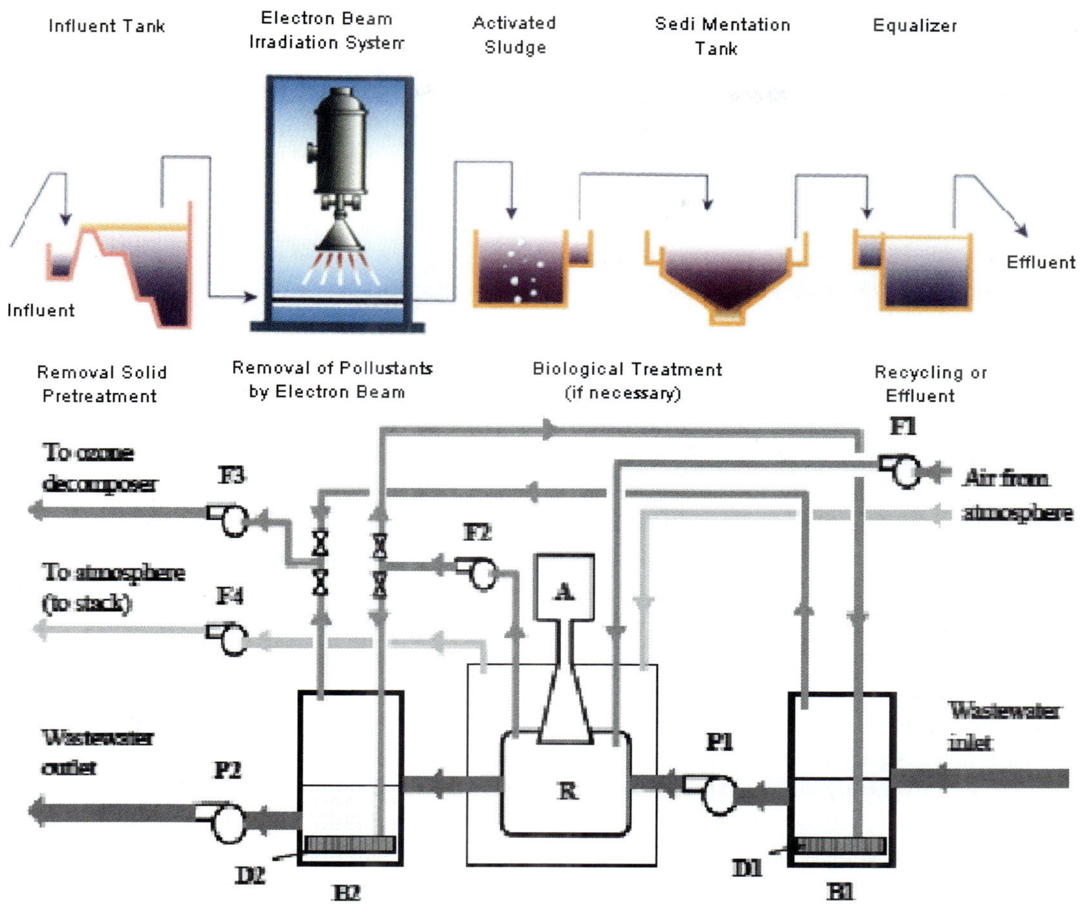

Fig. 10. Operating wastewater treatment plant in Daegu, South Korea. (Courtesy of EB-Tech [67] and Ref. 68.)

wave linac structures first developed by Knapp and Nagle at Los Alamos over half a century ago [70], have been in use for radiation therapy systems. Companies in Europe, the USA, Japan and China, such as Varian, Siemens or Elekta, have delivered more than 10,000 units to date, according to data from Varian Medical. Of these, approximately 5000 are believed to be presently functional worldwide. This market appears to have reached the "maturity" phase of its product cycle but treatment demand continues to grow as cancer occurrence is not diminishing as people live longer and world population increases rapidly. Hence, the possibility for innovation and the need to develop improved tools remain since improvements drive the replacement market in the developed countries.

Accuray has recently made inroads into the market with its CyberKnife system [71], which is illustrated in Fig. 11. Their product description notes

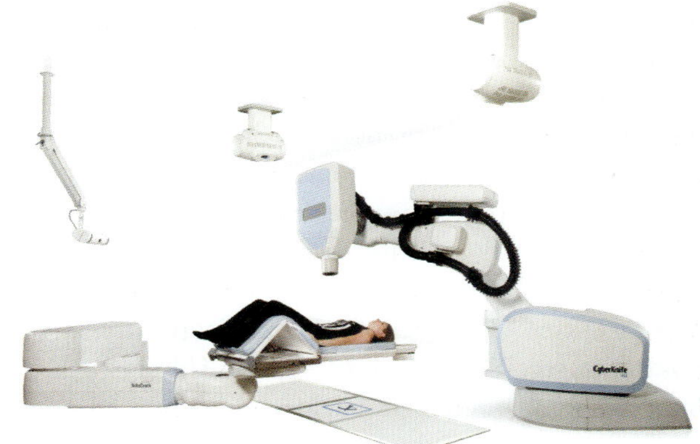

Fig. 11. Accuray X-band CyberKnife system. (Courtesy of Accuray [71].)

that "the compact, 1000 MU/min, 6 MV, X-band linear accelerator is capable of being positioned in virtually any direction by a high precision robotic manipulator with repeatable sub-millimeter accuracy." The shift to X-band yields the advantage of a smaller and lighter source for the moving radiation head but the deliverable dose is reduced. It is interesting to note that other major manufacturers, such as Varian, although they have studied X-band systems, do not appear to be moving in that direction at this time but rather are staying with S-band [72]. The author speculates that in addition to the delivered dose reduction, the RF power supply does not scale commensurately with frequency and hence the value added today by X-band is not overwhelming enough to redo an entire product line. Of course, a smaller improved X-band RF power supply somewhat changes the situation and highlights the critical importance of progress in auxiliary support systems as the dominant supporting technology across the entire board of accelerator applications.

X-ray medical therapy is predicted to continue to dominate the market for the foreseeable future and it will be interesting to see how the systems evolve to satisfy the economics of the marketplace.

At the same time, we are seeing major growth in the number of proton therapy centers that are coming into use worldwide, but in virtually all cases the base of the equipment suppliers is Japan or Europe. At present, there are 30 operating proton therapy centers and a further 14 under construction; 11 of the 44 are in the USA. When we consider heavier ion beams, the disparity widens dramatically. There are four operating facilities worldwide and five more are under construction, with none of the nine in the USA.

Reference 1 describes the problem well: "A careful look at government policies that affect development of accelerator-based medical technology could overcome roadblocks to progress in re-establishing US laboratories and industries as global leaders in the field. Public–private partnerships offer great opportunities for development. There are some notable successes, but in general the US lags behind international progress in such partnerships in these areas, hampered by legal restrictions, risk aversion within the funding agencies and in industry, and by intellectual property issues."

Due to the use of the Bragg peak, proton and heavy ion therapy promise the ability to more precisely deposit the dose in three dimensions. Some of this advantage is lost by the need to deliver a penumbra around the cancer site to insure complete irradiation. Also, treatments are often not attempted if there is body activity, such as breathing, that causes the tumor to move during the treatment, though both MELCO and Hitachi have developed "breathing mode" treatments where the beam delivered is timed to always be at the same place in the respiratory cycle. Nevertheless, proton and heavy ion therapy are procedures for specific cancers that are superior to stereotactic X-ray treatments [73]. The problem once again is cost due to the large size of the systems and the complexity of their gantries and accelerators, compared to an X-ray treatment procedure. This can lead to difficulties for patients with medical insurance plans. Future therapy machines of all kinds will have to deliver reduced cost or be squeezed from the market, at least under the present US medical insurance system.

Cyclotrons such as those which IBA delivers with its proton therapy systems are cheaper than the synchrotron that was used at Loma Linda, but they require an energy selection system. Varian has recently reduced the accelerator footprint by introducing the 250 MeV superconducting cyclotron shown in Fig. 12 for its proton therapy systems [74]. Higher-field magnets are also desirable, so as to enable smaller, lighter gantries for all ion beam therapy accelerator systems.

Another innovative accelerator concept that holds promise for reducing cost and footprint is the Dielectric Wall Accelerator (DWA) device being developed by CPAC [75]. By employing

Fig. 12. Varian superconducting proton therapy cyclotron. (Courtesy of Varian [74].)

induction accelerator technology with improvements in high-gradient insulators, a prototype system, using a radio frequency quadrupole (RFQ) accelerator as the proton injector, has been successfully tested. The concept promises to be highly reliable, which is essential for medical devices. CPAC plans to deliver a 150 MeV fixed beam DWA accelerator suitable for integration into a proton therapy system by late 2013.

In addition to therapy, accelerators are key to radioisotope production. The light isotopes used in positron emission tomography (PET) are made by lower-energy cyclotrons and linacs (produced by several vendors), up to around 15 MeV. Recently, Antaya has been developing the "Isotron," the highly compact superconducting high-field cyclotron prototype shown in Fig. 13, specifically for this market [76]. A similar device could be produced for proton therapy. The use of PET in oncology to both diagnose tumors and monitor treatment has caused the sales of PET accelerators to increase dramatically, with a 10% annual compounded growth.

A class of medium-energy cyclotrons can also deliver certain longer-lived isotopes using energies up to ~ 40 MeV. There are, however, some isotopes whose production requires beam energies of ~ 100 MeV, which are provided today by national laboratory linear accelerators as a subordinate mission. The one critical isotope whose production is in jeopardy at this time is the highly enriched uranium (HEU) fission product ^{99}Mo and its daughter ^{99m}Tc.

Fig. 13. The Isotron, a superconducting cyclotron for radioisotope production, showing the extremely compact footprint. (Courtesy of Antaya [76].)

The only present sources — in Canada, France, Belgium, The Netherlands and South Africa — are experiencing operational uncertainties.

In terms of prospects, cyclotrons and linacs have absorbed the low- and medium-energy parts of the market and will doubtless continue to fulfill that demand with modest growth. There is a need for a few dedicated facilities to produce isotopes that require higher energy or to replace HEU for their production. In particular, the US DoE is soliciting proposals for a new method of ^{99}Mo/^{99m}Tc production.

6. Security and Defense

There are many uses of accelerators in the defense industry. Most of them are the same industrial material processing, such as cutting, welding, hardening and nondestructive evaluation (NDE), that is found commercially and is part of the Sec. 7 discussion. Material characterization employing existing accelerator user facilities is also widespread. These activities are not addressed here.

At Los Alamos, the LANSCE proton accelerator facility [77] has been supporting "stockpile stewardship" and various other Department of Defense (DOD) needs for more than 30 years. Flash X-ray systems are used in nuclear weapon programs by the USA, France and other nations. The most advanced of these systems is the induction-acceleration-based Dual-Axis Radiographic Hydrodynamic Test Facility (DARHT) [78]. Los Alamos is now proposing the production of a 50 kV X-ray FEL device called the Matter–Radiation Interactions in Extremes (MaRIE) facility [79]. It is to be expected that a few of these types of accelerator-based devices will be constructed going forward.

During the Cold War, there were several major Strategic Defense Initiative (SDI) or "Star Wars" accelerator Directed Energy Weapon (DEW) programs. These included Neutral Particle Beams (NPB) for nuclear warhead discrimination and the Ground-Based Laser (GBL), which was conceived as a very large FEL operating with battle mirrors in space [80]. Today, there is known to be one active major accelerator-based DEW program. This is the $\sim$ US\$150 million Office of Naval Research (ONR) FEL Innovative Navy Prototype (INP) [20, 81, 82] contract, which was awarded to the Boeing Corporation in 2010. The Navy concept of future deployed electromagnetic (EM) weapons for defense against

Fig. 14.	Illustration depicting future Navy FEL DEW and other EM weapons, such as a railgun.

Fig. 15.	The PITAS prototype standoff detection accelerator at INL. (Courtesy of INL [87].)

cruise missiles and asymmetric threats is shown in Fig. 14.

High-power FELs have also been promoted in the past for other futuristic applications such as power beaming [83]. Recently, Colson [84] has proposed the use of a high-power FEL as the optimum means for communication with exoplanets and as a "flashlight" for science that can be employed to study the surfaces and atmospheres of moons and planets. On the negative side, it has transpired that a high-power FEL is apparently not useful for the elimination of space debris.

The other specialized area of accelerator use is in security applications. In addition to the large number of X-ray and now millimeter-wave systems deployed around the world for people, baggage and cargo screening, there are accelerators that seek to detect and/or neutralize chemical, biological, radiological, nuclear and explosive (CBRNE) threats [85]. Information on most of the work in this area is not available in the public domain but many of the thrusts are known. Ever-improving X-ray systems, some involving accelerator production of radiation, will continue to be a fertile and profitable market for the foreseeable future.

After X-ray systems, the best-known security accelerator application is port and border screening of cargo for standoff detection of special nuclear materials (SNMs) and other contraband [86]. X-rays are limited to discriminating density variation, and passive detection systems can be circumvented by

shielding, whereas other active radiation can stimulate a telltale nuclear fingerprint.

The Photonuclear Inspection and Threat Assessment System (PITAS) accelerator at Idaho National Laboratory [87], shown in Fig. 15, is a prototype for such a transportable standoff SNM detection electron accelerator-driven device. The concept of operation can be seen in the upper part of Fig. 16. The lower section of the figure shows a PITAS-like accelerator deployed in the container of the upper plot. Subsequent programs are under development within the US DOD [86, 88]. A large muon accelerator has even been suggested for SNM detection but the device footprint presents issues.

A problem that is being addressed in parallel is the development of efficient detectors for these SNM

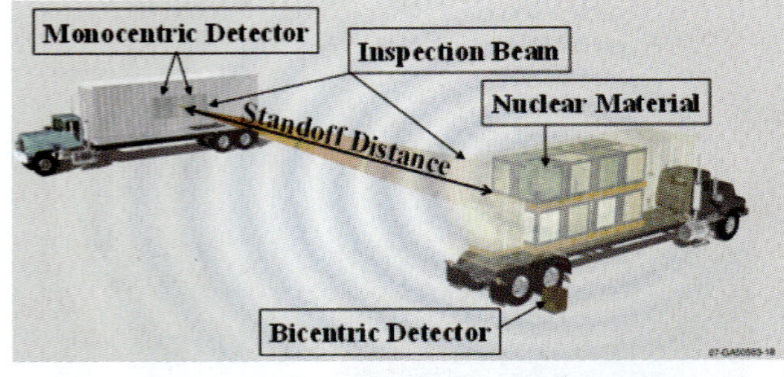

Fig. 16.	The INL PITAS prototype standoff detection accelerator operating concept. (Courtesy of Ref. 86.)

systems that do not require helium 3, due to the supply shortage caused by reductions in nuclear weapon programs.

In addition to the cancelation of the Secure Border Initiative Net, DHS has recently canceled the Advanced Spectroscopic Portal (ASP) effort after a negative report from the National Academy of Science [89]. ASP was to have been used in conjunction with accelerator-based CAARS 6–9 MeV X-ray systems [86]. Consequently, the future opportunity in this area is presently somewhat unclear.

The standoff detection of chemical and biological threats, using for instance THz LIDAR, is an active area of security R&D [90]. Typically, the desire of agencies like DARPA that are studying this is to field small man-portable devices with quantum cascade lasers leading the pack of prospective THz source technologies. This tends to preclude accelerator-based systems. However, standoff detection always benefits from power, and conventional accelerator-based systems have already demonstrated the generation of copious amounts of THz radiation, albeit in a larger footprint than QCL-based devices [91].

Smith-Purcell radiators are small accelerators that are being actively developed as THz sources by several groups [92]. They tend to be intermediate in size and performance and are related to backward wave oscillators (BWOs). These electronic devices provide more power than their photonic competitors but are too large for an individual to carry. However, the size and weight typically lies in the power supply and not the radiation production part of the system, so one basically gets what one pays for.

The push from below by the electronics advocates and from above by the photonics mavens has created a healthy competition in this transition region of the electromagnetic spectrum, where the efficiency of both types of sources tends to fall off drastically as one approaches 1 THz, as is shown in Fig. 17 [93]. We seem to stand at a point where this future market is not yet fully defined and it will therefore take time before the winners are identified.

THz radiation has been promoted for many other CBRNE missions, for instance to replace millimeter-wave scanners at airports because of the better resolution for explosive and contraband detection and because of the existence of THz-excitable

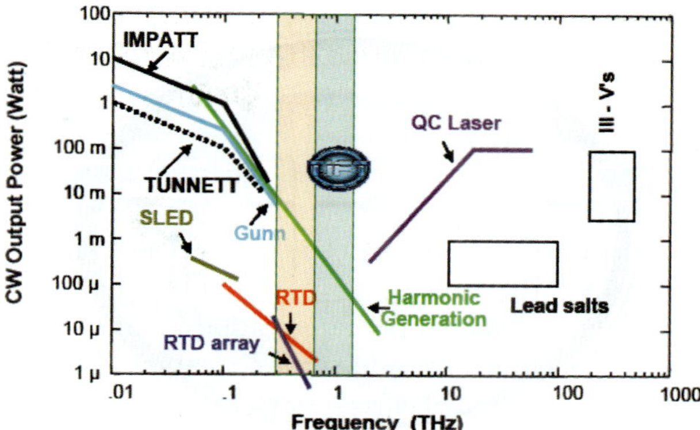

Fig. 17. Performance of various electronic and photonic devices plotted as maximum deliverable power versus frequency. The DARPA TIFT program sought the performance at the location of the logo, which is greater than either electronics or photonics could provide. Although the figure is dated, a performance gap still exists. (Courtesy of Mark Rosker [93].)

characteristic frequency fingerprints. It also offers promise as a method for secure communication and for sharp imaging of concealed metal objects. Through-wall imaging and other benefits have been cited for first responders, and scanning of mail for chem–bio agents and drugs has been proposed [91]. There are several hurdles to overcome, in that THz propagation in air and penetration in solids is limited.

It has been intimated that THz applications have been oversold even though there are several evolving and new photonics-based commercial applications reaching the market now. Regardless, where throughput is required, as in large-scale screening of contraband items, power seriously matters and accelerator solutions have an exploitable edge. It remains to be seen if accelerator-based systems can claim THz radiation generation market share in the future.

7. Material Processing/Industrial Applications

Even without the industrial applications described in the energy-and-environment sector, this market segment produces the highest annual commercial sales, estimated at $> US\$3.5$ billion ($\sim 70\%$ of the total current accelerator market), the largest number of annual accelerator unit sales, and has by far the largest installed base. It is also the area where the least information is available about where the suppliers are headed, since they are not about to tip their hands to competitors. Excellent summaries of the many applications can be found in Refs. 1–7 and little more will be addressed here. New directions are hard to predict but it is expected that industrial applications will be very profitable for the device suppliers and will continue to lead the use of accelerators throughout the world.

Electron systems find application in all sorts of material modification and surface treatment, generally through thermal effects, sterilization in medical, pharmaceutical and other industries, cross-linking of polymers and food irradiation. It is estimated that the installed base is ~ 1700 machines, led by low-energy polymer cross-linking devices, most notably for curing during the production of shrink-wrap materials.

The predicted applications of the future for electrons generally involve the "greening" and/or energy savings of existing thermal processes such as the curing of inks and coatings where significant added value, in this case, a 90% power reduction over current thermal processes is claimed for accelerator-based systems [1]. Wood preservation and composite material curing are two other green-tinged applications that have been identified as growth markets. Some years back, this author tried to develop a composite electron beam curing prototype system. In the end, the device did not provide sufficient added value to induce the large defense contractor customer to change its existing thermal approach to the curing of aircraft parts. As noted before, it is always about cost in the end. Not only is the cost of the processing itself important but the cost of transitioning to a new technology, coupled with the "comfort cost" of the existing technology, also has to be factored in.

The use of electrons to destroy pathogens in the food industry has a spotty history, owing to the reluctance of the public to accept the process with the FDA requirement to label the product as having been irradiated. In contrast, irradiation of spices has been occurring for many years with little publicity and no outcry. In 1998, this author picked food irradiation as the most likely to succeed in future industrial application of accelerators [94]. Unfortunately, this did not come to pass, and a better crystal ball is hoped for this time around. The potential value added of utilizing food irradiation, in terms of lost time due to illness and the reduction in proliferation concerns

from replacing the present use of conventional ^{60}Co sources, is enormous. It is hoped that public education can be brought to bear so as to allow this market to blossom to its potential.

If food irradiation does not take off, then the largest future industrial applications for electron beams are probably the systems for biofuel production, water remediation, and flue gas emission processing which were discussed in Sec. 4.

Ion systems, at $\sim 30\%$ of the total annual accelerator sales and an estimated 10,000-unit installed base, are the single largest market segment through semiconductor, medical and other implanters. Presumably, incremental development will continue until there is a paradigm shift in chip fabrication. Hardening of artificial joints, optical material development and similar applications should continue to provide a niche business. Of all the potential applications of ion beam technology, solar panel production would seem the most likely to experience explosive growth. Additionally, new ion beam applications will doubtless emerge in the future but are unclear at this time.

There are also a variety of industrial instruments utilizing accelerators that provide a growing niche market for some suppliers (see for instance Refs. 2 and 6). Accelerator mass spectrometry (AMS) deserves particular call out as a growing area with many diverse applications.

8. Discussion

The material presented indicates that the prospects are bright for accelerator technology development and sales. In the largest industrial-applications market, continued growth seems likely through incremental product enhancement, led by the big ticket items: semiconductor industry ion implanters and low-energy electron beams for polymer processing. The implanter market does experience the fluctuations of the semiconductor business as a whole but is a well-established market. At some point in the future, there will probably be a paradigm shift in semiconductor fabrication that displaces the implanters. But it is just as likely that some other accelerator application, about which we can at best speculate today, will be needed and developed to fill the void created.

Whether food irradiation takes off will depend on changing the public perception and acceptance of irradiated products. Meanwhile, the alternate "greener" electron beam processes, for instance the ink curing example, should reach the market if their economic advantage and reliability is as great as their proponents claim. If not, they will simply fall by the wayside.

Environmental applications are felt to have the greatest potential for growth in commercial accelerator sales. The next few years will tell if water remediation and flue gas treatment can make substantial market penetration. Conventional ethanol production has never made real inroads into fossil fuel consumption, so the jury is still out on the reality of the biofuel market. If the economics are really as favorable as advertised, one would expect this to become a very big market indeed.

The medical market should be mostly evolutional with growth to match population increases and as improved, smaller-footprint X-ray systems reach the market and more ion therapy facilities are commissioned. The radioisotope market for cyclotrons should be stable, with growth coming from improved medical care, particularly throughout the Third World. We noted above that new technology is driving replacement of X-ray systems in the existing developed world market and that PET is driving annual compounded growth in the low-energy cyclotron market.

The defense-and-security market remains fickle for accelerator suppliers. There is only one large accelerator DEW program at this time, the ONR FEL INP, and as with all large defense R&D contracts, funding to completion requires continuous attention. The defense market does use the same industrial applications as commercial ventures and has a need for a few R&D accelerators, though generally these are not cutting-edge. However, the bottom line is that there is very little unique accelerator activity in the defense segment at this time.

The security market is a little different but also quite fickle. There has been a tendency for some, but not all, procurement agencies to demand commercial off-the-shelf (COTS) technology, low-cost equipment with rapid delivery, and to resist investment in longer-term R&D. Consequently, the performance of some of these systems can be limited. Sales for border and container inspection have occurred, although recent government action canceled further passive procurements for SNM detection based on the National Academy of Science report that the systems did not work as advertised. Other active SNM systems are under development and it remains to be seen if they are eventually deployed.

Some companies are making a good business out of providing X-ray scanners for containers and luggage, as well as millimeter-wave imagers for the scanning of people. However, the accelerator-based market share, which is confined to the higher-energy border security market, is not substantial. Accelerator-based THz source systems have been proposed in competition with QCL and THz laser devices but have not achieved any significant market penetration to date despite the potential higher-power advantage they bring. There is a sense that there may eventually be good new opportunities in the security market but progress in market penetration remains weak.

Finally, we come to discovery science. Here, there is a niche market for smaller radiation source device sales to universities, institutes and also companies that need to perform internal R&D using radiation. Small light sources, based on IR and THz FELs, are an example. Another is electron sources for radiolysis in chemistry R&D. Soon perhaps, we will be able to offer compact UV and X-ray sources as well. Were the economics in place, many companies would love to have their own internal UV or X-ray source rather than making the trip to central user light source facilities to do their R&D.

In addition to ongoing construction projects like X-FEL, J-PARC and FRIB, several new large-scale accelerator programs are in the works, with the ILC being the largest and most visible. These programs bring substantial revenue to accelerator technology suppliers, but are cyclic and hence often not amenable to private financing of the infrastructure required to deliver the hardware. Welding, processing and testing equipment required for the ILC SRF cavities is an area where substantial investment is required to meet the volume and delivery schedule.

The USA is historically less nimble than Europe and Asia in providing needed private–public partnership support like this for its indigenous industry. We also noted other opportunities in different market segments with very significant social value added, but where the economic return is insufficient to spawn private investment. These are other areas requiring the creation of more innovative cooperative financing options.

Because of the nature of their government rather than private funding, the HIF, IFMIF and ADT programs look more like large discovery-science projects. Were ADT and/or HIF to prove successful after the first prototypical devices, the funding mechanisms could possibly change and become increasingly privatized.

To provide perspective, it is also interesting to note that the total budget for the ILC cryomodules is on the order of US$1 billion, which is considerably less than the annual sales of ion implanters to the semiconductor industry. There appears to be an element of a technology push in the Ref. 1 promotion of the use of SRF technology throughout many of the different market segments. Outside of the large discovery-science accelerators and their relatives like ADT and IFMIF in the energy segment, the SRF economics are questionable for commercial products until we are successful in developing HTSC cavities.

When plasma acceleration comes of age, it is expected to revolutionize accelerator applications. In the meantime, advances in power supply efficiency are very important for improving the economics of almost all applications.

9. Conclusions

The accelerator technology field is vibrant and growing, with about US$5 billion in annual sales and an estimated US$500 billion worth of processed products. In addition, it generates invaluable science through discovery science R&D using accelerator drivers, and also through private and public R&D at light sources and other user facilities.

The discovery-science market, led by the ILC, should continue to support considerable though cyclic growth. New fourth and fifth generation light sources are also a significant market area. SRF technology, while appropriate for large accelerators, will likely find widespread application only in commercial programs when we succeed in building HTSC cavities.

Present established industrial-application market sales drivers for the semiconductor and polymer processing industries look to contining their domination of market share for some time.

In the medical field, we anticipate incremental product improvement and growth due to population increases. Although small in terms of economic returns, the need for a new reliable non-fission-reactor source of ^{99}Mo is critical.

High-power FELs are the only major thrust in the defense market today, and the security market is highly volatile, making predictions uncertain. However, significant future accelerator opportunities should exist in security.

Within the energy segment, there appears to be considerable momentum for ADT in Europe and Asia. In addition to the Belgian MYRRHA effort, India and China seem committed to a thorium cycle reactor economy that will utilize ADT.

Meanwhile, we anticipate the environmental-remediation market to be the major growth area for near-term accelerator technology sales. The technology is proven for flue gas and water remediation, there is a market pull and the economics appear favorable. Food irradiation would be another major sales segment but overcoming public distrust first needs to be accomplished. Although a critical economic driver for all accelerators, increasing the efficiency of the needed power supplies provides enormous leverage to high-power accelerator applications such as these.

In principle, advanced accelerators, such as plasma accelerators that deliver significantly higher gradients in a greatly reduced total system footprint, could displace existing technologies all across the marketplace as soon as they are ready for prime time. They would also provide sufficient economic gain such that new applications that cannot be addressed effectively today could be undertaken.

Finally, we note the need for innovative political and financial strategies, and technology transfer approaches, to spur market penetration in those areas where the economic risk or return on investment is too low to attract private ventures. This is often the case in small markets and for those many, often "green," applications where the value added is partially social and not just economic.

Acknowledgments

The author gratefully acknowledges discussions with Bob Hamm of R&M Technical Enterprises on market size and industrial applications, Patric Muggli of the Max Planck Institute on plasma accelerators, Joe Lidestri of Columbia University on medical applications, Tony Favale of AES on ADT and the ILC, Brandon Blackburn of Raytheon IDS on security applications and Ed Pogue of Boeing on DEW.

References

[1] DOE October 2009 Workshop on Accelerators for America's Future. http://www.acceleratorsamerica.org/files/Report.pdf

[2] R. W. Hamm, *Reviews of Accelerator Science and Technology*, eds. A. Chao and W. Chou (World Scientific, Singapore, 2008), Vol. 1, Chap. 8, pp. 163–184.

[3] A. Sessler and E. Wilson, *Engines of Discovery* (World Scientific, Singapore, 2007).

[4] R. W. Hamm and M. E. Hamm, *Phys. Today*, June 2011, pp. 46–51.

[5] Biannual European Conference on Accelerators in Applied Research and Technology (ECAART), available in Elsevier publication *NIMB*.

[6] Biannual Conference on Applications of Accelerators in Research and Industry (CAARI), available in *AIP Conf. Proc.*

[7] Biannual Nuclear Applications and Utilization of Accelerators (AccApp), available in *AIP Conf. Proc.*

[8] E.g. www.ntis.gov/pdf/ValleyofDeathFinal.pdf

[9] US Small Business Innovative Research Program (SBIR). http://archive.sba.gov/aboutsba/sbaprograms/sbir/index.html

[10] US National Technology Transfer Center. http://www.nttc.edu

[11] research*eu focus, Vol. 9, May 2011.

[12] EC Enterprise Europe Network. http://www.enterprise-europenetwork.ec.europa.eu/services/technology-transfer

[13] European Science & Technology Transfer Professionals. http://en.wikipedia.org/wiki/Association_of_European_Science_and_Technology_Transfer_Professionals

[14] C. Zhang, Mega-science projects in China and their economic impact. http://accelconf.web.cern.ch/AccelConf/IPAC2011/PAPERS/WEIB04.PDF

[15] R. R. Wilson, Radiological use of fast protons, *Radiology* **47**, 487 (1946).

[16] *Conceptual Design of a Proton Therapy Synchrotron for Loma Linda University Medical Center* (Fermi National Accelerator Laboratory, June 1986).

[17] http://www.optivus.com/index.html

[18] http://www.protominternational.com/index.html

[19] ONR FEL INP. http://optics.org/news/2/2/19

[20] National Academy of Science Report, Scientific Assessment of FEL Technology for Naval Applications (2009). http://www.nap.edu/catalog.php?record_id=12484

[21] P. W. Lisowski, The accelerator production of tritium. http://accelconf.web.cern.ch/AccelConf/p97/PAPERS/9B003.PDF

[22] H. Safa, Optimum operating temperature of superconducting cavities. http://accelconf.web.cern.ch/AccelConf/l98/PAPERS/MO4095.PDF

[23] International Fusion Materials Irradiation Facility (IFMIF). http://www.frascati.enea.it/ifmif

[24] *Deuteron Linac Design Studies for ESNIT/IFMIF* (LA-CP-94-136, June 1994), ed. R. A. Jameson.

[25] T. J. Myers *et al.*, Assessment of Alternative RF Linac Structures for APT, Northrop Grumman Contract D69700016-8Y Report (26 Mar. 1997).

[26] F. Venneri *et al.*, Disposition of nuclear wastes using subcritical accelerator-driven systems (LA-UR 98-985). http://www.world-nuclear.org/sym/1999/venneri.htm

[27] Spallation Neutron Source (SNS). http://neutrons.ornl.gov

[28] International Linear Collider (ILC): a technical progress report. http://www.linearcollider.org

[29] C. Adolphsen, Advances in normal-conducting accelerating technology research from the X-band Linear Collider Program. http://accelconf.web.cern.ch/AccelConf/p05/PAPERS/TOPE002.PDF

[30] S. S. Kurennoy *et al.* Normal-conducting high current RF photoinjector for high power CW FEL. http://accelconf.web.cern.ch/AccelConf/p05/PAPERS/TPPE045.PDF

[31] F. Furuta *et al.*, Experimental comparison at KEK of high gradient performance of different single cell superconducting cavity designs. http://accelconf.web.cern.ch/AccelConf/e06/PAPERS/MOPLS084.PDF

[32] R. L. Geng *et al.* High gradient studies for ILC with single-cell re-entrant shape and elliptical shape cavities made of fine-grain and large-grain niobium. http://accelconf.web.cern.ch/AccelConf/p07/PAPERS/WEPMS006.PDF

[33] T. Higuchi *et al.*, Centrifugal barrel polishing of L-band niobium cavities. http://accelconf.web.cern.ch/AccelConf/s01/PAPERS/PR002.PDF

[34] M. P. Kelly, Recent developments in low and medium beta SRF cavities. http://accelconf.web.cern.ch/AccelConf/PAC2009/PAPERS/TU3RAI02.PDF

[35] Project X. http://projectx.fnal.gov

[36] T. Tajima, Would $> 50\,\mathrm{MV/m}$ be possible with superconducting RF cavities. http://accelconf.web.cern.ch/AccelConf/PAC2011/PAPERS/THOCS1.PDF

[37] G. R. Neil, Industrial applications of the Jefferson Lab high-power free electron laser, *Nucl. Instrum. Methods B* **144**, 40 (1998).

[38] M. J. Kelley, An FEL for the polymer processing industries, *Proc. SPIE* **2988**, 240 (1997).

[39] A. M. M. Todd, High power free-electron laser concepts and applications, *Proc. SPIE* **3931**, 234 (2000).

[40] S. Ogino and K. Awazu, *Nucl. Instrum. Methods B* **144**, 236 (1998).

[41] C. Joshi, Harnessing the power of the plasma wakefield, *Cern Courier*, June 2007. http://cerncourier.com/cws/article/cern/30148

[42] J. Faure *et al.*, *Nature* **431**, 541 (2004).

[43] P. Muggli and M. J. Hogan, *C. R. Physique* **10**, 116 (2009).

[44] T. Tajima and J. M. Dawson, *Phys. Rev. Lett.* **43**, 267 (1979).

[45] W. P. Leemans *et al.*, The BErkeley Lab Laser Accelerator (BELLA): A 10 GeV laser plasma accelerator, *AIP Conf. Proc.* **1299**, 3 (2010).

[46] T. E. Cowan *et al.*, Prospects for and progress towards laser-driven particle therapy accelerators, *AIP Conf. Proc.* **1299**, 721 (2010).

[47] M.-E. Couprie *et al.*, An FEL project towards the fifth generation in France. http://accelconf.web.cern.ch/AccelConf/FEL2011/PAPERS/TUPA09.PDF

[48] L. P. Campbell *et al.*, *IEEE Trans. Plasma Sci.* **28**(4), 1143 (2000).

[49] L. C. Steinhauer *et al.*, A new look at inverse Cerenkov acceleration and vacuum laser acceleration. http://accelconf.web.cern.ch/AccelConf/p91/PDF/PAC1991_0558.PDF

[50] M. E. Conde *et al.*, in *Advanced Accelerator Concepts: 8th Workshop*, ed. Wes Lawson, *AIP Conf. Proc. No. 472* (AIP, New York, 1999), pp. 626–634.

[51] X-FEL. http://www.xfel.eu

[52] J-PARC. http://j-parc.jp/index-e.html

[53] Facility for Rare Isotope Beams. http://frib.msu.edu

[54] International Linear Collider: a technical progress report. http://www.linearcollider.org/about/Publications/interim-report

[55] Muon Accelerator, Program. http://map.fnal.gov

[56] Electron Ion Collider Collaboration. http://web.mit.edu/eicc

[57] See for example the many papers on X-ray light source facilities to appear in *Proc. 2011 FEL Conference* (Shanghai, China, Aug. 2011). http://accelconf.web.cern.ch/AccelConf/FEL2011

[58] FLARE Facility at the University of Radboud. http://www.ru.nl/FLARE

[59] W. Schöllkopf *et al.*, Status of the Fritz Haber Institute THz FEL. http://accelconf.web.cern.ch/AccelConf/FEL2011/PAPERS/TUPB30.pdf

[60] Heavy Ion Fusion Science Virtual National Laboratory. http://hif.lbl.gov

[61] Accelerator Driven Transmutation. http://world-nuclear.org/info/inf35.html

[62] International Thermonuclear Experimental Reactor. http://www.iter.org

[63] C. Rubbia *et al.*, An energy amplifier for cleaner and inexhaustible nuclear energy production driven by a particle beam accelerator (CERN/AT/93-47, Nov. 1993).

[64] Multipurpose, hYbrid Research Reactor for High-end Applications. http://myrrha.sckcen.be

[65] Pavac Industries Electron Beam Flue Gas Treatment. http://www.ebfgt.com/index.html

[66] eScrub Systems Inc. http://escrub.com

[67] http://www.eb-tech.com

[68] B. Han *et al.*, Application of high power electron accelerator in wastewater treatment. http://accelconf.web.cern.ch/AccelConf/r06/PAPERS/THLO02.PDF

[69] K. Lee *et al.*, Cellulose modification study by e-beam irradiation and its application, in *Proc. ICOPS 2008: Plasma Science* (2008).

[70] D. E. Nagle, E. A. Knapp and B. C. Knapp, *Rev. Sci. Inst.* **38**, 1583 (1967).

[71] Accuray CyberKnife. http://accuray.com

[72] D. W. Whittum, priv. commun. (no archival paper). Presentation MOOBN4 in *Proc. 2011 Particle Accelerator Conference.*

[73] Stereotactic radiosurgery. http://www.nlm.nih.gov/medlineplus/ency/article/007274.htm

[74] Varian Proton Therapy Systems. http://www.varian.com/us/oncology/proton

[75] A. Zografos *et al.*, Engineering prototype for a compact medical dielectric wall accelerator. To appear in *AIP Conf. Proc. 11th Int. Conf. Applications on Nuclear Techniques* (Crete, June 2011).

[76] R. Lanza and T. Antaya, High energy protons for remote standoff detection of special nuclear materials. To appear in *AIP Conf. Proc. 11th Int. Conf. Applications on Nuclear Techniques* (Crete, June 2011).

[77] http://lansce.lanl.gov/about.shtml

[78] DAHRT delivers, in *Los Alamos National Laboratory Science and Technology Magazine* (Apr. 2007).

[79] B. Carlsten *et al.*, MaRIE X-ray free electron laser conceptual design. http://accelconf.web.cern.ch/AccelConf/PAC2011/PAPERS/TUODS1.PDF

[80] J. Schultz, M. Lavan, E. Pogue and T. Meyer, *Nucl. Instrum. Methods A* **318**, 9 (1992).

[81] http://www.fas.org/sgp/crs/weapons/R41526.pdf

[82] A. M. M. Todd, W. B. Colson and G. R. Neil, Megawatt-class free electron laser for shipboard self-defense, *Proc. SPIE* **2988**, 176 (1997).

[83] K.-J. Kim *et al.*, *Nucl. Instrum. Methods A* **407**, 380 (1998).

[84] W. B. Colson, Navy free electron laser, space applications. http://accelconf.web.cern.ch/AccelConf/FEL2011/PAPERS/MOPA01.PDF

[85] http://en.wikipedia.org/wiki/Chemical,_biological,_radiological,_and_nuclear

[86] http://www.fas.org/sgp/sgp/crs/nuke/R40154.pdf

[87] http://www.inl.gov/research/photonuclear-inspection-and-threat-assessment-system

[88] M. V. Hynes *et al.*, The Raytheon-SORDS trimodal imager, in *Non-intrusive Inspection Technologies II*, ed. B. W. Blackburn, *Proc. SPIE*, Vol. 7310 (731003 SPIE, Bellingham, 2009).

[89] National Academy of Science Report, Evaluating Testing, Costs and Benefits of Advanced Spectroscopic Portals (2011). http://www.nap.edu/openbook.php?record_id=13082&page=R1

[90] J. F. Federici *et al.*, THz standoff detection and imaging of explosives and weapons, *Proc. SPIE* **5781**, 75 (2005).

[91] *Annual Proceedings of the IRMMW-THz Conference Series* (IEEE).

[92] H. L. Andrews *et al.*, Observation of THz evanescent waves in a Smith–Purcell free-electron laser, *Phys. Rev. ST Accel. Beams* **12**(8). http://prst-ab.aps.org/abstract/PRSTAB/v12/i8/e080703

[93] M. Rosker, THz remote imaging: phenomenology and perspectives, in *Proc. SURA 2006 THz Applications Symposium* (Washington DC, June 2006).

[94] A. M. M. Todd, Emerging industrial applications of linacs. http://accelconf.web.cern.ch/AccelConf/l98/PAPERS/FR1004.PDF

Alan M. M. Todd was born in Scotland and grew up in Hong Kong. He received a first in Aeronautical Engineering from Bristol University, England in 1970. He performed his graduate work at Columbia University, New York, completing a Ph.D. in Plasma Physics in 1974. Accepting an appointment at Princeton University, he studied the magnetohydrodynamic stability of plasmas, before moving to Grumman Corporation in 1979. From 1985 on, his research interest evolved from plasma to accelerator physics. In September 1998, through a leveraged buyout, a group of former Northrop Grumman employees established Advanced Energy Systems (AES), Inc. Dr. Todd is the Vice President and a Director of the company. He is the author of more than 140 technical publications and a Fellow of the American Physical Society.

Reviews of Accelerator Science and Technology
Vol. 4 (2011) 257–277
© World Scientific Publishing Company
DOI: 10.1142/S1793626811000598

CERN: From Birth to Success

Herwig Schopper

CERN, CH-1211 Geneva 23, Switzerland
University of Hamburg, Edmund–Siemers–Allee 1,
20146 Hamburg, Germany
Herwig.schopper@cern.ch

A historical review is given of the development of CERN from its foundation to the present from the personal viewpoint of the author.

Keywords: International organization; accelerators; colliders; detectors; management.

1. CERN — A Unique Organization

In each volume of *RAST*, an article entitled "Person of the Volume" is published describing a large laboratory shaped mainly by that person. When I was asked to write an article about CERN, it was clear that this was impossible since so many personalities have played an essential role over the more than 50 years of CERN's history, from its establishment to the most recent successes. Indeed, there has always been a smooth transition from one director general to the next, while project leaders have successively passed their tasks on smoothly to their successors. This is the way of CERN, whose international nature has meant that the organization has had to accommodate a range of approaches, traditions and languages right from the start [1].

From its very beginning, CERN was a unique organization based on two quite different initiatives. European physicists started the first initiative as early as 1946. They realized that competition with the USA was possible only if European countries [2] joined forces. The first discussions were launched in the framework of UNESCO by Eduardo Amaldi from Italy [3], the two French physicists Pierre Auger and Lew Kowarski, and the American Isidor Rabi. A second, and rather independent, initiative which is much less well known was that of the Swiss writer Denis de Rougemont who had spent his time during the war at Princeton, where he had met and interviewed Einstein. After his return to Europe in 1948,

de Rougemont became, together with Raoul Dautry (*administrateur general* of the French Commissariat a l'Energie Atomique, CEA) and other far-sighted diplomats and administrators, one of the driving forces of the "European Movement," which resulted in the creation of the Centre Européen de la Culture at Lausanne in 1950. The objective was to build bridges between people who had been at war, and an international scientific laboratory was considered to be the best tool to bring scientists, administrators and politicians together for peaceful work — "science for peace".

The two initiatives were amalgamated into a proposal to the UNESCO General Conference in Florence in June 1950, and it was Isidor Rabi who, inspired by the foundation of the Brookhaven National Laboratory, formulated the decisive motion that was eventually submitted and accepted by UNESCO. When I invited Rabi to give a speech at the 30th anniversary of CERN in 1984, he said: "Europe had been the scene of violent wars... for 200 years. Now we have something new in the founding of CERN, namely Europe has gotten together, in the cause of science.... So I think it is most important for CERN to continue and be the symbol and the driving force of a possible unity of Europe.... I hope that the scientists at CERN will remember that they have other duties than exploring further into particle physics. They represent the combination of centuries of investigation and study... to show the power of

Fig. 1.   The start of excavation at the Meyrin site in 1954.

the human spirit. So I appeal to them not to consider themselves as technicians... but... as guardians of this flame of European unity so that Europe can help preserve the peace of the world" [4]. Thus CERN became the only international organization[a] that has two main objectives: to promote science and to contribute to peace.

2. Difficult Birth

After the principal decision of UNESCO, two groups — one of scientists and another of administrators and diplomats — started discussions on the specific structure and objectives of the new organization. After lengthy discussions reflecting serious differences of opinion concerning the future character of the laboratory, among the diplomats as well as the scientists, they proposed that a temporary organization be created with the aim of constructing the biggest proton synchrotron in the world, with an energy higher than that of the 6 GeV Bevatron in the USA.

In June 1953, a formal convention was agreed on for the preliminary organization, which was called the "Conseil Européen pour la Recherche Nucléaire" (European Council for Nuclear Research), or CERN, a name which became a mark of quality and is maintained today, although a subtitle was later added: "European Laboratory for Particle Physics." The CERN convention entered into force on 29 September 1954, after 12 countries (Belgium, Denmark,

Fig. 2.   F. Block, the first director–general, laying the foundation stone, watched by M. Petitpierre, president of the Swiss Confederation.

France, Germany, Greece, Italy, The Netherlands, Norway, Sweden, Switzerland, the United Kingdom and Yugoslavia, followed soon by Austria and later by Spain and Portugal) had sent documents of ratification to UNESCO. A group of historians that has described in detail the foundation of CERN came to the conclusion that "it succeeded because the entire project remained in the hands of scientists" [5].

A major controversial issue remained, however: the choice of a site for the new organization. Several sites were proposed: Geneva in Switzerland, Copenhagen in Denmark, Arnhem in The Netherlands and a site near Paris. After long bargaining, Geneva was selected, because it is in a small country and has an international environment. France had felt that it had a good chance in view of the strong involvement of French scientists in the early discussions. It finally agreed to the compromise, not least because Geneva is French speaking.

---

[a]The only other international organization with these two aims is the synchrotron radiation laboratory SESAME in Jordan. When I was involved in its establishment in 2004, also under the auspices of UNESCO in a similar way to CERN, I copied the CERN convention for that of SESAME.

The convention is a rather short document, wisely leaving a lot of room for interpretation. New member states required a unanimous vote in the council. The organization would provide for collaboration among European states, but the definition of "Europe" was left rather vague. Consequently, and following a sustained policy of openness, CERN has gradually developed from a European laboratory into a laboratory for the world. At present CERN has 20 member states, with Romania a candidate for succession and five others including Israel, Cyprus and Turkey, on their way to becoming members. Two conditions of the convention are still valid today: contributions are proportional to the GNP of the member states (with a cap of 25%) and they have to be paid in Swiss francs. The first makes it difficult for countries like the USA or Russia to become a member, since they would dominate the organization. Recently the status of associate member has been created, which will allow a stronger formal relationship with these and other countries. Scientific cooperation was never hindered by formal arguments. One of the founding principles was that any scientist of the world is welcome to participate in the CERN program provided that his or her participation improves the quality of the program.

3. The First Steps to Success

After the provisional foundation of CERN, Eduardo Amaldi, who had already played a major role as part of a group of experts, was appointed secretary general in 1952, a function that later became known as director general. At the same time, the Norwegian Odd Dahl became chairman of the Proton Synchrotron (PS) group — about a dozen of people with very little experience in accelerator technology, but sufficiently courageous to start the construction of the largest synchrotron in the world. In summer 1952, some of them visited the USA, and at Brookhaven they got to know to their big surprise a new idea for focusing particle beams; alternating gradient or strong focusing. Dahl convinced the CERN Council that the PS should be based on the strong focusing principle. However, he withdrew from CERN and was replaced by John B. Adams, who in the coming decades was to play a major role in the development of CERN. The strong focusing principle allowed the beam size to

be reduced by an order of magnitude, and consequently the scale of magnets providing the guiding field. Hence, a more ambitious machine could be envisaged for the original price tag of about 100 million Swiss francs, a lot of money in those days, and Adams proposed a PS with a maximum energy of 25 GeV, extendable to 28 GeV. The PS group, which grew to about 180 members, was faced with the enormous task of building a completely new machine on a green field site on the outskirts of Geneva.

After five years of construction and some difficult weeks of tests, the acceleration to 25 GeV was achieved for the first time on 24 November 1959. As planned, the PS had become the highest energy accelerator in the world, an achievement that was announced by Cornelis Bakker, who had succeeded the first director general of CERN, Felix Bloch. Thanks to the PS, the validity of the strong focusing principle was proven, although no real understanding of the beam behavior during acceleration had yet been achieved. The instruments for beam diagnostics and control were infinitely primitive compared to the sophisticated instrumentation of today's synchrotrons.

The PS has undergone various upgrade and extension programs which have continued to the present day. These include the acceleration of antiprotons, electrons, positrons and heavy ions. This is because the PS has served as injector for a string of later machines: the ISR, antiproton accumulators, the SPS, LEP and finally, the LHC. Thus, the PS has become the most versatile part of CERN's installations (Fig. 3).

But the PS was also the source of exciting physics using extracted beams. It was here that the principle was established that CERN facilities should be used mainly by outside users, not limited to those from member states but open to people from all over the world. The spirit of open international access was established right at the beginning and has continued through the whole history of CERN. Some of the early scientific directors — including Gilberto Bernardini and Wolfgang Gentner, to mention but two — were instrumental in setting the tone.

It cannot be the purpose of this report to cover all the physics results achieved with the PS [6]. However, one outstanding success was the discovery of

Fig. 3. The first hours of the startup of the PS. From left: J. Adams, H. Geibel, H. Blewett, Ch. Schmelzer, L. Smith, W. Schnell, P. Germain.

the neutral currents of the weak interaction with the heavy liquid bubble chamber Gargamelle, certainly worthy of a Nobel Prize had the leader of the collaboration, André Lagarrigue, not passed away before Stockholm could react.

Since it was expected that the construction of the PS would take several years, and since European physicists were eager to start physics experiments as soon as possible, another project was begun in parallel with the PS: a 600 MeV synchrocyclotron (SC) for protons. Although more modest as far as the energy was concerned, such a machine could be bought more or less from industry. It had a remarkably productive and versatile career in nuclear and particle physics for several decades, until it was finally closed down in 1991. The most original part of CERN's nuclear physics program was an on-line isotope separator, ISOLDE, which was attached to the SC. An upgraded version is still going strong, taking its primary beam from the PS booster. The very intense proton beams from the PS injection system are also used for a neutron time-of-flight facility, one of the most highly performing facilities of its kind in the world.

4. The Golden Years with the ISR and SPS

In the 1960s, a rapid expansion of CERN started with a considerable increase in staff and outside users and an almost exponential rise in the yearly budget (Fig. 8). This, and the increasing competence and self-assuredness of the staff, enabled not only an extension of the ongoing programs but also meant that several new projects could be envisaged for construction almost in parallel. All reasonable requests could be fulfilled. Victor Weisskopf was director general (1961–65) through this exciting phase of CERN's history.[b]

A unique "first" was achieved with the construction of the intersecting proton-proton storage rings (ISR), which remained the only p–p collider in the world until operation started with the LHC. The leader of this project was the Norwegian Kjell Johnson, who gathered around him a crew of excellent accelerator experts, some of whom went on to become pillars of the SPS, LEP and LHC. The ISR (Fig. 4) consisted of two separated, slightly distorted rings, each 300 m in diameter, intersecting horizontally eight times with a crossing angle of 14.8°. The rings were filled from the PS, with the highest energies at 26 GeV, by first stacking protons horizontally, which were stored unbunched after filling in the form of a ribbon several cm wide and 1–2 mm high. Then they could be accelerated to an energy of 31.4 GeV. The center-of-mass energy of the collisions of 62.3 GeV corresponded to a lab energy of about 2000 GeV and thus ISR was opening a window into an energy realm previously accessible only with cosmic rays. But, apart from its energy, this machine was breaking several other records. Beams of up to 40 A were stacked with a vacuum of less than 10^{-11} Torr, about four orders of magnitude lower than in other accelerators. New techniques for beam diagnostics and correction, and for luminosity measurements, were tried out, superconducting magnets were used for a low β-section, and stochastic cooling was tested with success. Finally, antiprotons were stored for days and weeks, proving directly that this particle is stable.

---

[b]Director generals are normally appointed by the CERN Council for five years. The only exceptions were the first DG, Felix Bloch, who resigned earlier; John Adams, who was acting DG in 1960–61 and DG Lab.II in 1971–75; and Herwig Schopper, whose mandate ran from 1981 to 88.

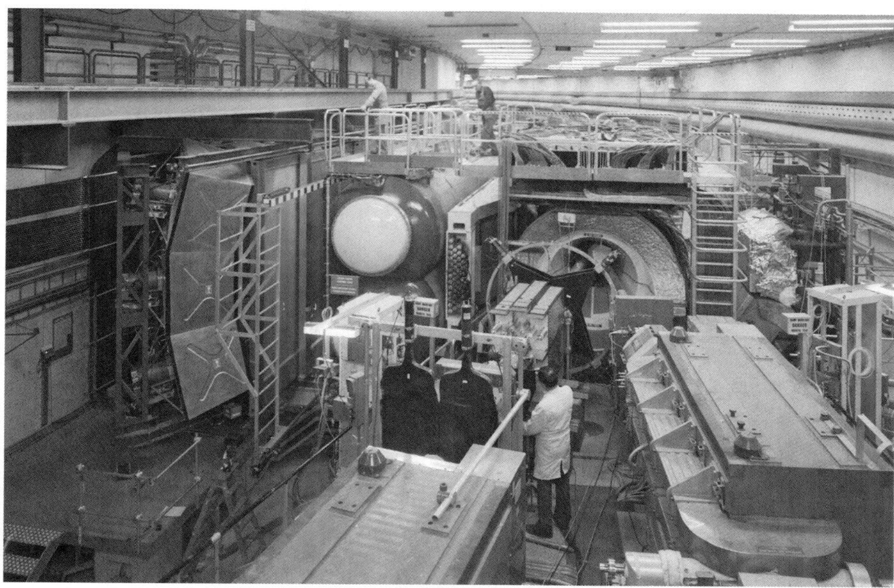

Fig. 4.   The intersection of the ISR.

A number of experiments permanently used six of the eight intersection regions, and many interesting results were obtained. For example, proton–proton total and differential cross-sections were measured, as were particle and jet production. Most of these experiments were observing hadron processes at small angles. They had been designed according to the mentality of fixed target experiments, allowing installation of several detectors at one interaction region. This conformed to the CERN tradition of accepting as many experiments as possible. Only at a later stage did the ISR experiments pay attention to observations with larger solid angles and to leptons which had been somewhat neglected with respect to hadron physics. One detector better adapted to new physics was the split field magnet detector SFM, which unfortunately came into operation rather late. Perhaps the most interesting result from the ISR was the discovery that the π^0 yield at 90° as a function of transverse momentum was much higher than an exponential decay: this was one of the first confirmations of the validity of QCD and the quark model. Unfortunately, the ISR missed some great discoveries: the J/ψ (discovered at BNL and SLAC) and the Y (found at Fermilab). The main reason was that measurements at large angles were not considered to be interesting for a long time. To my regret I had to decide to close down this great

machine in 1984 in order to find sufficient resources for the construction of the LEP.

In the 1970s, thanks to the favorable political climate, CERN was able to envisage a second new project besides the ISR. In order to harmonize the European high energy program, a European Committee for Future Accelerators (ECFA) had been established under the chairmanship of Eduardo Amaldi. As early as 1963, ECFA recommended the construction of a new proton accelerator with an energy of 300 GeV. A special study group at CERN worked out a proposal, which was submitted to the CERN Council in 1964. The Scientific Policy Committee supported this study and proposed the creation of a new laboratory in a different country. This posed two problems: the CERN convention had to be changed since only one laboratory was foreseen, and a new site had to be chosen. Discussions about how to solve these problems, and how to find the necessary funds, dragged on until 1971.

From the early 1960's, similar ideas were discussed in the USA, and in 1965 a study group at the Lawrence Radiation Laboratory at Berkeley published a design report for a 200 GeV proton machine. After some battles (site decisions are always delicate), it was decided to establish a new laboratory at Weston (Batavia), near Chicago. The new laboratory took the name "National Accelerator Laboratory,"

Fig. 5.   Directors-General until 1980 From left: J. Adams, W. Jentschke, F. Bloch, V. Weisskopf, L. van Hove.

which later became just "FermiLab," and Robert R. Wilson, who had been nominated as its director, proposed a machine based on a new magnet design leading to reduced cost and hence with an augmentable energy of 200–400 GeV.

The news spread quickly to Europe, and Willibald Jentschke, the director general of CERN, along with Amaldi, visited Wilson. They were so impressed with what they saw that discussion about the technical design of the European project hindered the CERN Council from taking any decision on the project for some time.

Politically, however, things were moving forward. A new convention was approved that allowed two laboratories to be established under the CERN umbrella. This was no trivial achievement, since it had to be approved by the governments and ratified by the parliaments of all of CERN's member states. Five years of legal work were needed before the convention could be amended in 1970. However, the other thorny issue, regarding where to site the new laboratory, had reached deadlock. Five member states had offered sites and, in spite of the many discussions, site visits and proposed bargains, it seemed that no decision could be reached.

In an attempt to cope at least with the technical problems, a steering group headed by the Nobel laureate Cecil Powell was nominated by director general Bernard Gregory (1966–70) in agreement with ECFA. They led the work until the end of 1969, when the council appointed John Adams as director of the 300 GeV program. Adams had left CERN after the construction of the PS and had gone back to plasma physics in the UK, where he was in charge of basic and industrial research for the British Atomic Energy Authority.

Adams immediately started to reconsider the main features of the project in his particular way, which can be characterized as pragmatic, confronting problems head-on, but also avoiding hasty decisions. After due consideration, he tactfully presented his decision to opt for the window-frame magnet proposed by Wilson, giving up the original choice of a C-type magnet, without pointing to any wrong conclusions from the steering group. The main argument was a reduction in the cost and higher possible energies. He suggested a ring with a 2 km diameter and a final energy of 400 GeV.

By the end of 1969, no agreement on the site had been achieved and only 6 member states had announced their willingness to support the 300 GeV project. No decision could be taken. In order to unblock the impasse, Adams presented a new version of the project, plan B, to the Council. In it, he showed that the project could be realized at the existing CERN site provided that an underground tunnel could be used. By employing the PS as injector and other infrastructure existing at CERN, the cost could also be substantially reduced. Gregory felt that his major task was to lead the council to agree to the project, and thanks to his talent of persuasion via well-crafted, rational arguments coupled with a great deal of patience, he was nearing a positive decision at the last session of the council he attended as director general in December 1970. The council agreed unanimously to the alternative version of the project, but asked for more studies. Hence, the meeting was adjourned for two months. When they met again in February 1971, the new proposal was approved by 10 of the member states, and the remaining two joined later.

However, the decision left CERN in a strange administrative situation. Jentschke had been appointed as successor to Gregory, while Adams had been nominated director-general for the laboratory of the new site before plan B had been proposed. These appointments could not be reversed and the only solution was to split CERN into two laboratories: CERN I and CERN II. At the end of Jentschke's mandate (end-1975), partial unification was decided by having two DGs with specific responsibilities: Leon van Hove as director general for research and John Adams as executive director-general. I consider it almost a miracle that this solution worked for five years, probably only thanks to the rational and balanced character of the two personalities. When van Hove's and Adams' mandates ended in December 1980, I was appointed as DG and the unification of the lab was formally completed by having only one director general. With hindsight, it is safe to say that the future development of CERN would have been impossible if a second site had been accepted for the SPS: the golden years of the 1960s and '70s were followed by lean years, and both the LEP and the LHC could not have been built without the resources concentrated at a single site.

Thanks to the enormous competence and outstanding leadership of John Adams, the SPS was built and became an example of European engineering capability. On 17 June 1976, at noon, the SPS achieved its nominal energy of 300 GeV, and after approval by the council at 15 : 30 the same day protons reached 400 GeV.

Apart from its excellent functioning, a few other features of the SPS should be remembered. It represented the first large accelerator based on a distributed computer control system with the hardware equipment locally connected to minicomputers that were integrated into a common network. Another innovation was the use of high-level languages for application programs that could be written by engineers and technicians. A few years later, this allowed the fast changeover to the proton–antiproton collider mode. A further technical first was the use of traveling wave accelerating structures instead of the conventional ferrite-loaded cavities.

With the SPS an essential step across the Swiss–French border was taken. The SPS ring lies mainly

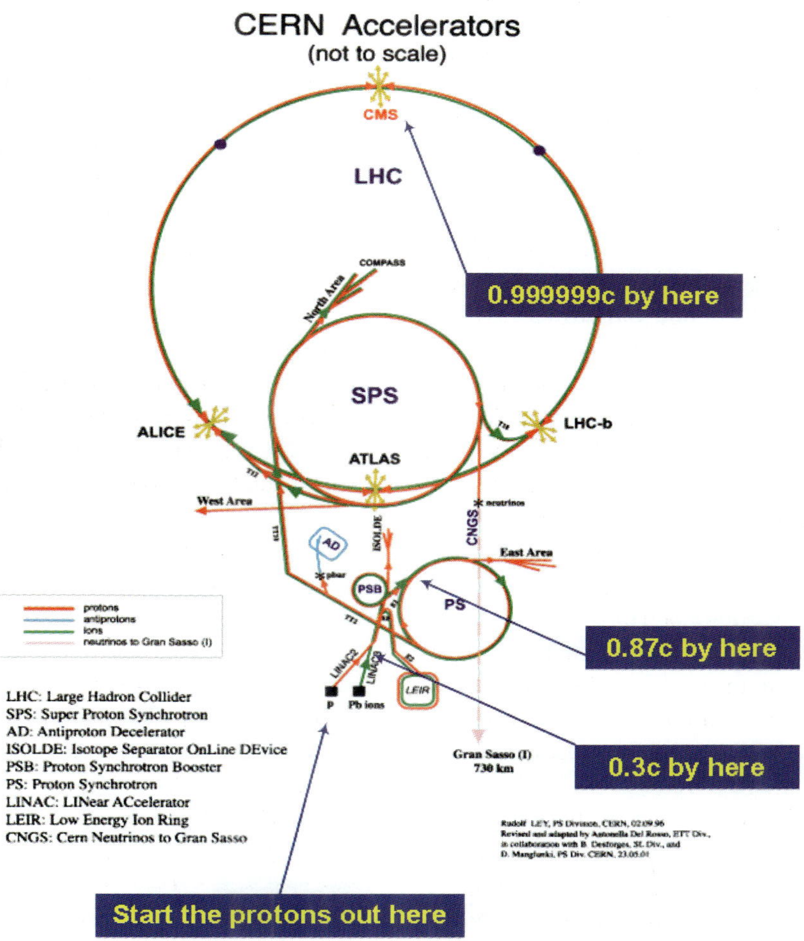

Fig. 6.   The present CERN system (c is the speed of light in vacuum).

in France and a whole new laboratory, the Prévessin site, with a large experimental hall was created for it. The confidence and pragmatism of the two host states turned out to be remarkable. Although large amounts of equipment and large numbers of personnel have been moved across the border over the last 50 years, CERN's host states confide management of this to CERN. It's a remarkable example of international cooperation!

The SPS provided a wide range of beams for physics, some of them (such as neutrino, muon and kaon beams) unique to the facility. This led to a rich harvest of exciting physics ranging from insights into the structure of the proton to clarification of the mechanism for CP violation [7]. Nevertheless, Carlo Rubbia and others thought that Europe should have a truly unique facility for elementary particle physics, something completely new and different. Rubbia proposed and lobbied hard for a new way of using the SPS: running it as a proton–antiproton collider. The basis for such a daring project had been work on the production and handling of antiprotons, which was well established at CERN.

Stochastic cooling as proposed by Simon van der Meer [8] had been tried at the ISR, but the results seemed doubtful since the effect appeared to be insignificant, requiring cooling times of days or even weeks. The situation was clarified by a little storage ring known as the Initial Cooling Experiment (ICE), which was built and operated in 1976 and 1977, when it demonstrated conclusively that stochastic cooling really worked. A special Antiproton Accumulator ring (AA) was constructed to deliver antiprotons to a Low Energy Antiproton storage Ring (LEAR), whose experiments produced many interesting results. These early efforts led to an extensive antiproton program that is still going strong today with various facilities trapping antiprotons, resulting in the first production of antihydrogen (consisting of an antiproton and a positron) and consequently enabling a study of the symmetries between matter and antimatter. For several years, CERN was the only place able to handle antiprotons in significant quantities, but it was later joined by Fermilab's Tevatron, which is unfortunately scheduled for closure this year after a remarkable 25-year career at the forefront of high energy physics. Meanwhile, at CERN, a new facility to produce antiprotons with very low energies has just been approved.

Going back to the 1970s, in spite of extensive experience with antiprotons, Rubbia's idea was not without risk. It was one thing to handle low energy antiprotons, quite another to convert the SPS into a high energy proton–antiproton collider. Eventually, it was thanks to the conservative and solid construction style of Adams that the SPS vacuum, beam stability and lifetimes were good enough for a collider. The two director generals proposed the conversion to the council in July 1978 and it was approved. Adams himself took on the responsibility as project leader and, as such, he had to make some crucial and controversial decisions, for example to use the PS as an injector instead of injecting antiprotons directly from the AA at an energy of 3.5 GeV.

The first antiproton–proton collisions at 2×270 GeV were achieved in 1981 and a fertile domain of physics was opened. Normal operation continued at 2×310 GeV and with a trick some collisions could be achieved at 2×450 GeV. Slowly, the number of collisions and the luminosity could also be increased by many orders of magnitude to allow the investigation of weak interactions. The culmination came with the discovery of the carriers of the weak force, the W in 1982 and the Z in 1983, which brought the Nobel Prize to Carlo Rubbia and Simon van der Meer in 1984 (Fig. 7).

Fig. 7. The Nobel Prize winners Carlo Rubbia and Simon van der Meer.

5. The Construction of the LEP — A New Era

As the 1970s gave way to the 1980s, CERN's golden era of growth came to an end. The SPS had been funded out of a special budget from the member states, above their previous annual contributions to CERN, but with emerging fields of science also demanding a bigger slice of the cake, it was clear that high energy physics would not have such an easy time in the 1980s.

In June 1980 the two acting DGs of CERN, along with me as their newly appointed successor, proposed the construction of the LEP, a large Electron–Positron collider with a circumference of 27 km. Long and difficult negotiations ensued, and were made particularly difficult by the fact that a unanimous vote of all member states was necessary. In June 1981, I submitted to the CERN council the final proposal, and the definitive approval was obtained in October 1981 — but with extremely difficult conditions attached. The greatest problem was that the LEP had to be built within a constant budget — no additional means and no additional personnel. This was aggravated by the fact that the constant budget was to be reduced. When I proposed a gentlemen's agreement to compensate for at least the material part of the budget for inflation while keeping the personnel expenditures constant, the answer I got was "You accept the budget level and we shall tell you each year how inflation will be compensated for," implying further budget reductions. Many future users of the LEP, represented by the Scientific Policy Committee, along with the CERN staff, thought that such a large project could not be achieved under such conditions, so they asked me to resign.

I, however, had full confidence in the ability of CERN staff and users to rise to the occasion and cope with the harsh new conditions, so I accepted the challenge and stayed on. The conditions which were imposed entailed some very difficult measures. The LEP had to be built as a "stripped-down" machine, implying that only the absolute minimum number of components would be installed in the first phase, just sufficient to produce Z particles in abundance. Running as a "Z-factory," the LEP's energy was 50 GeV. Upgrading the machine to reach the energy needed for W-pair production, about 80 GeV, would have to wait. No contingencies were foreseen. No provisions could be made for the experiments; only some cost for their infrastructure was taken into account. This was the first time that the experiments for a CERN machine were expected to be financed and staffed mainly by the users from outside CERN, with CERN taking responsibility for overall coordination. Furthermore, all of CERN's non-LEP programs had to be drastically reduced. One of the most difficult decisions was to stop the ISR, which was still in full swing and producing interesting data. On the other hand, we did not compromise on developments important for the future, such as cryogenic developments for accelerating cavities for the LEP and superconducting magnets for the LHC. A new program in heavy

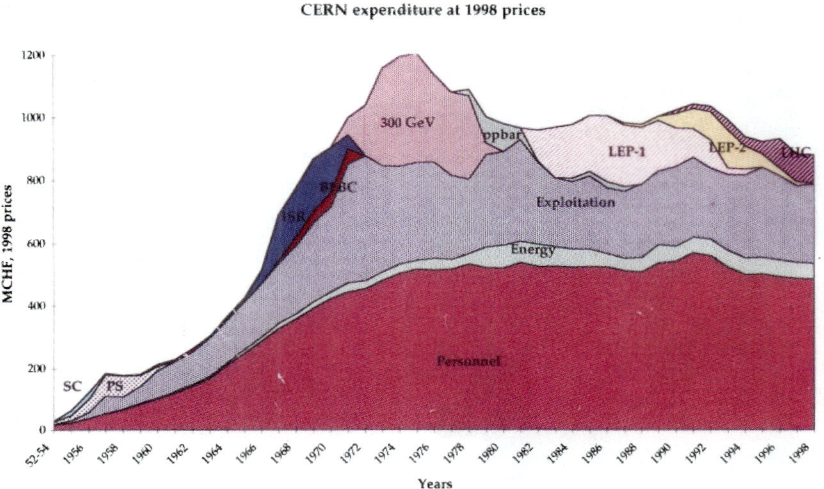

Fig. 8.   The CERN budget leveling off in 1980.

ion physics was even started against the advice of the scientific committees, going on to become an important part of the CERN program.

One major decision remained to be taken: the size and position of the tunnel. A tunnel with a circumference of 30 km had originally been proposed, but test borings under the Jura Mountains revealed that the ground was extremely bad for tunneling, being full of faults filled with water under high pressure. Geology experts therefore strongly advised avoiding the worst areas under the Jura, which would have had the consequence of reducing the tunnel circumference to about 20 km. This would have still allowed energies of about 80 GeV to be achieved, so it was tempting to follow the geologists' advice. Several prominent colleagues urged me to accept the reduction so as to avoid risking the whole project. However, discussions had already started about the possibility of adding a second magnet ring in the LEP tunnel for a hadron collider — ideas which came to fruition at a workshop in 1984 [9]. I believed that the most valuable part of the LEP was the tunnel, since everything could be changed except for the tunnel size, and so I resisted the suggestion to reduce its circumference to 20 km. This was partly motivated by competition with the SSC, the Superconducting Super Collider project that was getting underway at that time in the USA. After careful studies, we reached a compromise of 27 km, with 8 km in difficult terrain under the Jura. For geological reasons we also had to put the LEP on an inclined plane [10]. We paid the price for these decisions by having water break into the tunnel (Fig. 9) during the excavations

Fig. 9.   Water in the LEP tunnel.

under the Jura, delaying the project by one year. Looking back, most people would agree that these troubles were justified in view of the performance of the LEP, and the present performance of the LHC. Anyway, it is not generally known that the size of the LEP tunnel was chosen only in view of the LHC.

The council's decision not to approve additional resources for the LEP required a complete change in optimizing the management of CERN. Before the LEP the two labs CERN I and CERN II were in reality quite independent and the use of their resources was decided autonomously. This was even true to a large extent for individual CERN divisions. In order to find the necessary resources for the LEP, the unified CERN had to be optimized globally. This required the introduction of a new way of thinking and working, but to change people's habits is always difficult and takes time. Not only did the budget have to be distributed in a completely new way, but staff also had to become more internally mobile. This was particularly difficult since, in many groups, relationships of mutual trust and understanding had been built up between staff and their supervisors over many decades. Now it became unavoidable to assign to at least a third of the total staff new tasks in new environments.

In that context, the choice of an LEP project leader became crucial. I took the decision to appoint Emilio Picasso to that function, which created some surprise as he was an experimental physicist and not an accelerator expert. Since a considerable number of staff working for experiments had to be transferred to the LEP, and since Picasso also enjoyed the respect of the accelerator community, I thought that he would be the perfect choice to bridge the gap, and this proved to be the case.

With Picasso I agreed on a somewhat unorthodox method of running the project. We had made estimates for the different parts of the LEP but kept them secret. As per normal, we had appointed group leaders responsible for various parts of the project, such as magnets, RF, vacuum, cooling and ventilation, and power converters, but we did not tell them how much budget would be available for their specific tasks. We asked them to do their best to keep the price as low as possible. The outcome was surprising. Those parts that required considerable technological development and whose cost estimates were less certain came out less expensive

Fig. 10.   The first beam in the LEP; a group of physicists in the LEP control room on 14 July 1989. Director general C. Rubbia is in the middle and his predecessor, H. Schopper, is on his left, with Steve Myers on the right.

than expected, whereas the conventional equipment was more expensive — evidence of the ingenuity of CERN's technical experts. Giving responsibility to the group leaders and their teams also increased their motivation and engagement.[c] This made it possible to build the LEP within the foreseen budget.

One major problem concerned cash flow for investment. The yearly expenses of a large project resemble a Gaussian distribution; they start slowly, reach a peak and finally drop off. To accommodate such a cash flow within a constant low budget is almost impossible, since at the peak the expenses exceed the available constant funds. Of course, we tried to advance some expenses to early years and negotiate contracts with firms to delay some payments. But this did not solve the problem. When I asked the council to take a loan from a bank, it refused. We had to borrow money from the CERN pension fund. This lack of funds over many years created tensions, of course, and Picasso offered me several times his resignation, which I consistently rejected. Nevertheless we became and stayed as friends. Of course, I was not surprised that similar problems arose later with the LHC, which also had to be built within a constant budget, but the experience from the LEP had been almost forgotten. This time, however, the council agreed to a bank loan.

6. The LHC — CERN Became a World Laboratory

The idea that a proton collider in the same tunnel should succeed the LEP goes back to about 1977 and was studied in detail at a workshop in 1984. CERN, like most other high energy physics laboratories, has the tradition of looking far ahead, considering and planning for new projects while the previous one is still under construction. This brings the advantage that projects can be thoroughly designed and prepared so that they can be realized within foreseen schedules and budgets. At first a proton ring on top of the electron ring was planned with the possibility of alternating electron or proton operation, or even producing electron–proton collisions to extend the investigations of HERA at DESY to higher energies.

To prepare for the future of CERN, the council decided in 1985 to establish a Long Range Planning Committee chaired by Carlo Rubbia. The options were a proton collider, which was studied by a subpanel led by Giorgio Brianti, and an electron–proton machine. A recommendation made in 1987 came down in favor of a proton–proton collider — the LHC. A proton–antiproton collider was considered, but quickly rejected because it would not have been able to deliver sufficiently high luminosities: this decision has to be seen in the light of the SSC, which had been approved by President Ronald Reagan in the same year. In order to be competitive with the SSC, the beam energy of the LHC was proposed to be 8 TeV, considerably lower than the SSC's 20 TeV but partly compensated for by the LHC's higher luminosity. Rubbia, who had succeeded me as director general in 1989, formally proposed the LHC to the council with first beams foreseen for 1998 in parallel with LEP operation. One of his major achievements as DG was to keep the LHC alive in the face of the SSC.

It is always difficult to decide when an operational facility should take lower priority than a new project, as we had seen with the ISR and LEP. To some extent we were spared this when the competition between the LHC and the SSC came to an end with the US Congress' cancelation of the SSC project in 1993, but there was still interference between the LEP and the LHC. This continued until

---

[c]It is not possible to mention here all the people who made major contributions to the LEP. For a complete list see Ref. 8.

October 2000, when the LEP was finally shut down after a dramatic last few months of running. During the years 1994–1999, when Christopher Llewellyn-Smith was director general, the LEP was upgraded by replacing the copper cavities for the acceleration of the beams by superconducting cavities boosting the beam energies to close to 100 GeV. This not only allowed the abundant production of W particles, but also meant that the search for the Higgs particle could be extended to these energies. Toward the end of the LEP's scheduled running period, some experiments thought that they had seen an indication of this much-sought-after particle, and a last desperate effort was made to push the LEP to its absolute limit. By installing all the available copper and superconducting cavities, the LEP achieved a beam energy of 104.5 GeV in May 2000, initiating one final push. No Higgs was found, but a firm lower limit of 114 GeV for its mass was established, and that remains valid today. From all available data today, one can conclude that the Higgs mass must be quite low, and not far from the limit established by the LEP. With hindsight, one might argue that the LEP was shut down too soon. I had insisted that the LEP magnets, one of the cheaper parts of the machine, were able to go to 125 GeV. By adding more accelerating cavities, maybe the LEP's experiments would have found the Higgs in this region. The LHC data will very soon resolve this question!

In 1993, Lyn Evans was appointed project leader for the LHC and he proposed a cheaper "missing magnet machine" with one-third of the dipole magnets missing in a first phase, implying only two-thirds of the full energy for a number of years. LHC commissioning was planned for 2002. In the same year, a medium term plan was presented which foresaw an increase in the budget by 20%. An external review committee (ERC) had been established under the chairmanship of Robert Aymar, an expert on superconducting magnets, to check the estimates. The ERC confirmed toward the end of 1993 that the LHC parameters were "reasonable and realistic," and the proposed costing was also sound. After very difficult financial discussions, the CERN council approved the "missing magnet" LHC on 16 December 1994, with a plan to review the decision in 1997 in the light of expected non-member-state contributions. The essential prerequisite for approval of the LHC was the existence of the LEP tunnel, the other accelerators as

injectors, the overall infrastructure and, last but not least, the experience of the CERN staff. The lesson learned from the SSC was that it is very expensive to start a new laboratory on a green field site.

The LHC suffered many financial storms, the first coming in 1996, when Germany declared that because of the cost of reunification it was seeking reductions of about 10% in its contributions to all international scientific organizations. This move was immediately supported by the UK. Very difficult and complicated negotiations ensued. A new element became an agreement signed with the USA in December 1997, in which the country committed itself to providing in-kind contributions not only to the experiments but also to the collider itself. This became a major change in the policy of CERN, since in the past in-kind contributions had been made only for the experiments. Such a policy also required great efforts to convince the US authorities, in particular Congress, since it would be the first major US contribution to a project not located in the USA or in space. Other countries, including Japan, Russia, India, Canada, China and Israel, also declared that they would contribute to the building of the LHC.

While the political and financial negotiations went on, new decisions were taken concerning the construction of the machine. The original idea to have coexistence between the LEP and the LHC in the same tunnel was abandoned, and it was decided to remove the LEP after decommissioning. Furthermore, it was also agreed that a "one-stage project" was more reasonable than the missing magnet version, and this could be achieved by delaying commissioning to 2003 or 2004. However, at the end of his mandate in 1998 as director general, Llewellyn-Smith warned the council that the budget situation was very fragile, mainly because the LHC was approved in its R&D phase, with no definite design of the superconducting magnets and with no contingency.

When Luciano Maiani became director general in 1999, he took a rather optimistic view. The LEP was still in operation and some colleagues had hoped that it would continue into 2000. The real cost of the LHC was not yet known, nevertheless, Maiani got approval for a new project: CERN Neutrinos to Gran Sasso (CNGS). The objective was to send a neutrino beam to the Gran Sasso Laboratory in Italy to investigate the very interesting question of neutrino oscillations over long distances. This needed

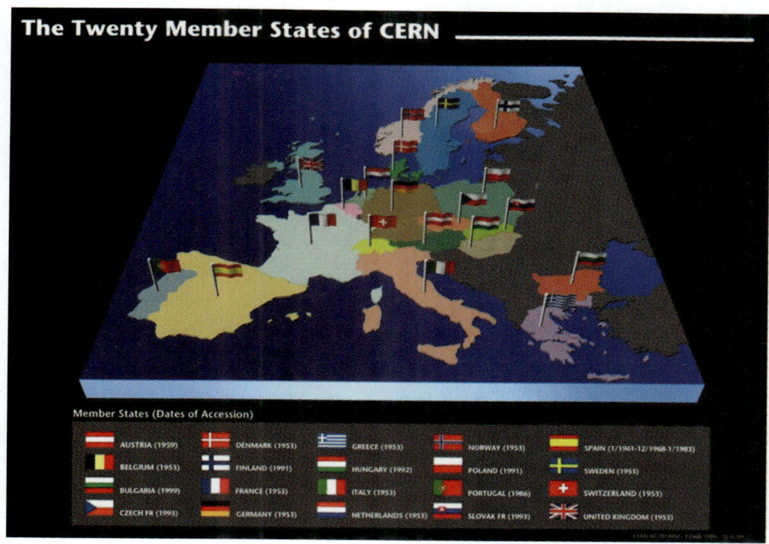

Fig. 11. The CERN member states.

extra manpower as well as funds for both construction and operation. In June 2001, Maiani reported to the council that everything was proceeding according to plan.

It then came as a big shock when the director general had to inform the council in December 2001 that the cost of the LHC would overrun by about 20%. For other high-tech projects in other organizations, such overspending is relatively normal; at CERN, however, this had never happened. The main reason was that the bids for the construction of more than 1200 magnetic dipoles were much higher than expected. A similar situation arose for the cryogenic distribution lines — another daring project, since superfluid helium at 1.8 K had never before been transported over a distance of kilometres. It became apparent that additional funds would be needed for personnel and the experiments. Since the council had not been warned early enough, hasty and misleading conclusions were drawn, and it was suggested that CERN's management had been economical with the truth in earlier reports, or was incompetent to manage such a large project. This was certainly not the case, since essentially the same team that had built the LEP was still in place for the LHC. Nevertheless, the council insisted that more formal procedures for cost monitoring and control should be set up, and the ERC was reactivated to review the cost estimate.

Serious budgetary restrictions on all CERN programs followed, with a crisis levy of 2% on salaries for a year being applied while further efficiencies and economies were sought. A bank loan was approved and strict control procedures adopted (CERN adopted Earned Value Management procedures), a bureaucratic measure not in line with the CERN tradition. The construction time was further stretched to spread the cost. When Maiani's term as DG came to an end in 2003, Robert Aymar was appointed as his successor. Having been the chairman of the ERC, he knew the project in all its details and he was considered to be tough enough to push through the necessary new procedures.

By autumn 2008, everything was ready to inject the beams into the machine for the first time, and after a good start another piece of bad luck hit the LHC on 19 September 2008. One electrical soldering among many thousands connecting superconducting cables failed, producing an electric arc. Two tonnes of superfluid helium were released with such force that some magnets broke their anchoring to the floor. This ultimately led to 53 of the LHC's 1600 superconducting magnet assemblies being damaged and a year-long repair and consolidation job before the LHC could run again. It was a painful learning experience, but led to systems being developed and put in place to ensure that a similar event could not happen again. In the meantime, with the LHC's construction phase over, Lyn Evans moved on to other things and Steve Myers, who had already played essential roles at the LEP and earlier machines at CERN,

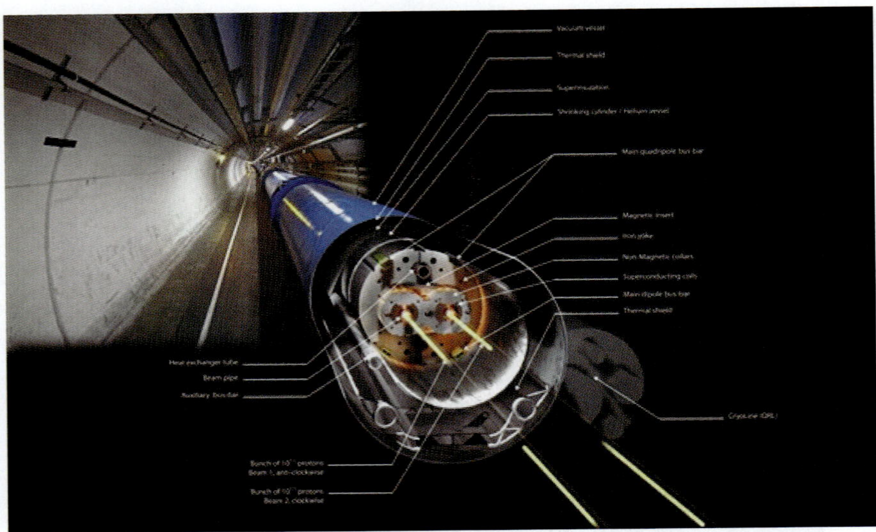

Fig. 12. A view of the LHC tunnel.

Fig. 13.   The LHC control room at the first beam in September 2008; happy directors general R. Aymar, L. Maiani, Ch. Lllewellyn-Smith, C. Rubbia, H. Schopper (from the left; from the right according to their appointment).

took responsibility for running the machine. At the beginning of 2009, Rolf Heuer, a former director of research at DESY and spokesperson of the OPAL experiment at LEP, began his mandate as director general and appointed Myers as director for accelerators and technology.

On 20 November 2009, beams were once again circulating in the LHC after an incredibly short time of adjusting the orbit parameters.[d]   A few days later, the LHC's first collisions were observed by the experiments at $2 \times 450\,\text{GeV}$, the LHC's injection energy, and on 30 November 2009 beams collided at $2 \times 1.18\,\text{TeV}$, only three hours after the first accelerated beam. With this success the LHC became the most powerful hadron collider in the world, overtaking the Tevatron. Since then, the LHC has gone from one record to the next. The beam energy was raised to 3.5 TeV in 2010, and the number of protons in each of the beams has gradually increased, producing ever-higher luminosities. Performance is improved with great care and attention, always giving priority to the protection of the machine. The power stored in each of the beams already corresponds to several kilos of explosive, and an uncontrolled loss of beams could damage the machine. For that reason, machine protection is given the utmost priority. The LHC will operate until the end of 2012 at below the design energy, after which the high current superconducting cable splices like the one that failed will be remade to a more robust design. This decision met with the full approval of the physicists, as in a proton collider lower energy can be partly compensated for by higher collision rates,[e] and

---

[d]At the LEP it took several days to achieve closed orbits. With the much better beam controls and programs, it took only hours for the LHC.

[e]The fundamental collisions occur between quarks and gluons. These particles move around inside the proton with energy distributions with high energy tails. These participate rarely in collisions but can be observed with sufficient data.

3.5 TeV per beam already opens up great prospects for new physics.

Since its restart in 2009, the LHC has been running like a Swiss clock, producing both proton–proton and lead–lead nuclei collisions. The data rates expected for the whole of 2011 were already achieved after just a few months. The experiments have already reported many interesting results; however, fundamental new discoveries, like the existence of the Higgs or supersymmetric particles, require more data and may be expected during 2012.

7. Experimental Techniques

The investigation of the microcosm requires ever more powerful accelerators or colliders. The simple reason for this is that the smaller the building blocks of matter are, the harder they are.[f] In order to break them up and look for smaller constituents, the somewhat paradoxical consequence of this is that the largest and most complicated machines have to be built to investigate the infinitesimally small. However, we may have reached a fundamental boundary: the smallest constituents of matter so far discovered, the quarks, cannot be produced as free particles (due to "confinement" of the strong nuclear force) and they seem to have no substructure. We see the quarks only indirectly. Therefore, the focus has moved from trying to break up the fundamental building blocks to producing new unknown particles that can help us to extend the Standard Model of particle physics and even answer questions about the universe. This also requires higher energies, since according to the Einstein formula $E = mc^2$ a particle with mass m needs an energy E to be produced (c is the speed of light in vacuum). The LHC collider at CERN now offers the most promise that a breakthrough in the understanding of the microcosm can be achieved, for example through the discovery of the long-sought-after Higgs particle. The reputation of CERN is to a large extent due to the successful development, construction and operation of accelerators and colliders, based on common efforts of CERN's applied physicists and engineers.

However, to produce powerful collisions would be useless without devices to record and analyze the data produced. To develop detectors and analyzing procedures to interpret the collisions has become an art in itself. Unlike the accelerators, the experimental techniques and detectors are to a large extent developed and constructed by external users at their home laboratories. CERN currently counts over 10,000 users from all over the world. Detector components are built at universities and institutes before being shipped to CERN to be assembled into the complete detector. This requires an immense coordination effort on the part of CERN and all the institutes involved, but it also provides a remarkable in-kind contribution to the CERN resources, which is particularly significant for non-member-states. This happened for the first time on a large scale with the LEP experiments and is now well established for the LHC experiments.

Bubble chambers were once an excellent tool for seeing directly what happened in collisions. However, bubble chamber pictures had to be taken blindly at arbitrary moments. As a result, millions of pictures had no interesting events and after scanning they had to be thrown away. Nevertheless, they produced some very important results. The heavy liquid bubble chamber Gargamelle has been mentioned already. BEBC, the European bubble chamber, constructed by a European collaboration and filled with liquid hydrogen, was one of the largest chambers ever built and produced many interesting results. Parts of both chambers can be seen today in one of the courtyards of CERN.

Bubble chambers were outdated by new technological developments. Various kinds of electronic tracking chambers were developed. These have the immense advantage that they can be triggered to record only when something interesting is going on. A fast electronic decision, within a few nano- or microseconds in a cascade of trigger levels, can be taken on whether a collision event contains interesting data that should be stored or whether it can be discarded.

A detector, which at the LHC is an enormous facility, consists of various kinds of individual detector components that allow the directions, energies and types of particles produced in a collision to be

---

[f]This is due to the Heisenberg uncertainty principle, which states that the product of the uncertainties of the momentum (and hence the largest possible momentum) of a particle and of its local position is given by Planck's constant. To restrict a particle to a small volume requires therefore strong binding forces.

determined. In some events, thousands of particles have to be recorded. Only a few of the many techniques that have been developed in connection with CERN experiments can be mentioned here.

The most outstanding achievement was the development of multiwire chambers by George Charpak, which brought him the Nobel Prize in 1992. He was an extraordinary personality. Born to a Jewish family in Poland, who moved to France, he joined the French resistance during the war and survived a concentration camp. Employed by CERN in 1959, he stayed there until his retirement, after which he initiated in the French schools the program "*mains dans la pâte*," a program similar to the "hands-on" program introduced by Leon Lederman in the USA.

In a multiwire chamber, each of many thousands of thin parallel wires is connected to an electronic channel and if a charged particle passes close to a wire an electric impulse is created. Combining several wire planes, some of them with wires at different angles, allows the tracks of particles to be reconstructed, providing pictures of an event almost as sharp as a bubble chamber picture. But these tracks are already in digital form and can immediately be used for further analysis. However, in some cases the local resolution, which is limited by the distance of the wires (fractions of mm), is not sufficient. This is the case, for example, if the decay vertex of a short-lived particle has to be determined. Silicon strip detectors have been developed for such purposes. Here, very fine strips of silicon giving a spatial resolution of about $10\,\mu m$ replace the wires. For the fabrication of such detectors, the production methods of computer chips are used. However, since such silicon strip detectors are needed to capture short-lived particles, they have to be positioned close to the collision points and therefore are exposed to strong radiation. Hence, considerable effort has to be devoted to making such detectors radiation-hard — an effort that has led to technological spinoffs. An enormous amount of ingenuity and hard work by many scientists in their home laboratories has gone into the invention, development and construction of such tracking detectors. These were used at a collider for the first time in the UA1 and UA2 experiments to detect the W and Z bosons. Later they played a vital role in the LEP experiments, and they are now ubiquitous in the LHC experiments — notably CMS,

which has the largest silicon detector ever built — and in experiments at other colliders.

A different method to register charged particle tracks in a large volume is the so-called time projection chamber (TPC). A cylindrical container is filled with an appropriate gas mixture such that a charged particle traversing it leaves a track of ionized molecules or atoms. Parallel to the cylinder axis, an electric field is provided, along which the ionization charges start to drift. After some time, they arrive at the endplates of the cylinder, which is subdivided into pads, and the impact point is registered. The drift time along the electric field provides the other coordinate. If properly designed, a TPC can register particle tracks with about the same precision as a wire chamber. The first TPC was employed at the PEP collider at SLAC, a quite large TPC was used in the ALEPH experiment at the LEP, and one is now implemented in the ALICE detector at the LHC.

Tracking detectors are not only used to determine the direction of the particles emitted at the collisions but, when combined with a magnetic field, can also be used to measure the momentum of the particles. For this purpose, the tracking detectors are imbedded in magnetic fields, necessarily covering large volumes. In order to measure high momenta with sufficient precision, strong magnetic fields are needed. To achieve this, superconducting coils are often used. In many cases, superconducting solenoids have been employed, and the largest such coils ever built are used in the CMS experiment at the LHC. The ATLAS experiment at the LHC uses an immense air-cored torroidal magnet system (Fig. 14).

All the methods mentioned so far do not work for electrically neutral particles such as photons and neutrons. A different kind of detector has to be used to measure their energy, and these detectors are called calorimeters — a very misleading name, since they have nothing to do with measuring temperatures. The incident particles are made to interact with matter, producing a cascade of new particles (an electromagnetic cascade in the case of photons or a hadronic cascade for particles subject to the strong interaction). In some calorimeters, these cascades emit scintillation light which is detected by photomultipliers. If they are properly constructed, the total light output will be proportional to the energy of the incident particle.

Fig. 14.   The ATLAS detector during installation. Six superconducting torroidal coils surround the calorimeter.

Electromagnetic cascades are relatively short and the whole cascade can be confined in a single crystal, provided that they contain a heavy element. Crystals of BGO containing bismuth, germanium and oxygen, for example, must be about 30 cm deep to capture the whole cascade. Since it is also important to know the direction in which the photons were emitted, one has to arrange many crystals around a collision point. In the L3 experiment at the LEP, such a BGO detector was used, containing about 30,000 crystals of ultrapure material.

Hadronic cascades are much longer and require at least half a meter of steel to be fully absorbed. To collect the scintillation light, one has to sandwich the detector, consisting of layers of absorber (iron, lead or uranium) interspersed by scintillators.[g] The total light output is collected and is proportional to the total energy of the incident particle. The first such "hadron calorimeter" was used at CERN to measure neutron scattering.

The precision for determining the energy of particles is much better for magnetic spectrometers than for calorimeters (except for very high momenta) but the solid angle which one can cover is much larger for the latter. In addition, magnetic spectrometers have to increase in size in proportion to the maximum momentum one wants to measure,[h] whereas calorimeters have to grow only logarithmically and hence require less space. These are the reasons why calorimeters have become so important for collider experiments, where the space around the collision point is restricted.

Many other technical developments for detectors at CERN cannot be discussed here. Looking back at the past 50 years, one can state that the progress for detectors has been as breathtaking as for accelerators and colliders. Of the many interesting results produced by the experimentalists at CERN only some could be mentioned since it is not the place here to give a comprehensive review of the scientific success of CERN [11].

8. Cooperation or Competition? The Sociology of Large Groups

The origin for such common efforts was not a few strong personalities who wanted to build their empires but rather a need for the talents and the competences of many people, all motivated by achieving the same complicated and demanding projects. This trend, originating in particle physics, can now be found in other fields, including astronomy, space research and even biology, as the frontier of knowledge is pushed back.

Particle physics has been in many respects the forerunner of other fields of physics or even other

---

[g]In some cases, the ionisation in a gas or a liquid is collected instead of the scintillation light.
[h]This is the main reason why experiments in particle physics have to be so big.

sciences, but one particularly interesting aspect is the formation of large collaborations where CERN played a leading role. The first large international cooperation involving several institutes outside the main laboratory was probably the UA1 experiment under the leadership of Carlo Rubbia. It was built to observe proton–antiproton collisions at the SPS and could detect, along with the smaller experiment UA2, the carriers of the weak force, i.e. the W and Z bosons. The UA1 collaboration became the prototype for the later and bigger LEP collaborations, and is now culminating in the big LHC collaborations. These collaborations are, in a very real way, institutions in their own right, dependent on, yet independent of, their host laboratory. This development is now culminating in the LHC collaborations ATLAS, CMS and ALICE, each involving more than 1000 scientists.

The surprising aspect is how well and efficiently these large collaborations work in spite of the strong individual ambitions of the participating scientists. The overriding wish to succeed with common objectives supersedes individual ambitions and allows openness of communication. This is even true to a certain extent concerning the exchange of information between competing experiments or even laboratories. Of course, the most exciting recent data are kept confidential until publication, but technical tools and technologies like software codes for the analysis of data are freely exchanged.

At the time of the LEP experiments, I made an interesting observation concerning the sociology of big collaborations. Two of the experiments, ALEPH and L3, had very strong personalities as leaders, namely Jack Steinberger and Sam Ting, whereas the other two, OPAL and DELPHI, were guided by rather democratically minded colleagues, namely Aldo Michelini followed by Rolf Heuer at OPAL and Ugo Amaldi at DELPHI. At a certain moment, I thought that these complicated collaborations involving many institutes and countries would work efficiently only with a very strong leadership. To my amazement, it turned out in the end that both ways worked very successfully. However, in the next step of the historical evolution, the democratic line won. All the LHC experiments have decided to have a "spokesperson" elected in a very democratic way for a given period of time. They adopted management structures similar to those of OPAL and DELPHI, and they have all become great successes.

9. Bringing Nations Together

A history of CERN would not be complete if the issue of science for peace was omitted. One of the original objectives of CERN was to bring together the Western European countries that had been adversaries in the Second World War. This was achieved in an outstanding way by insisting from the beginning that all experiments be done by international collaborations. However, these were not limited to member states of CERN. One of the secrets to the success of CERN was that political or bureaucratic arguments were avoided as much as possible and all scientists were welcome provided that their scientific quality was exceptional and they could make relevant contributions to the program.

When the Joint Institute for Nuclear Research (JINR) was founded at Dubna in the USSR in 1956, emulating the CERN model for the Warsaw pact states, cooperation with this institute provided one of the rare bridges for physicists between the West and the East during the Cold War. The cooperation became a new trend after the dissolution of the Soviet Union, since the cooperation now involves new countries that had been part of the USSR. In particular, CERN has benefitted from the considerable technical competence of the JINR by receiving components for experiments and accelerators. CERN was also the first organization to conclude an agreement with the Soviet Union (in 1968), establishing cooperation with the national Institute for High Energy Physics (IHEP) at Protvino, which at that time had the largest proton accelerator in the world.

Over recent years, CERN, still formally a European laboratory, has become an institution for the world. The LEP experiments were a first step in that direction, and the trend has continued strongly with the LHC. An extraordinary success in bringing people together was achieved when a group from China joined the L3 experiment at the LEP and worked there together with a group from Taiwan. This no doubt had political implications, requiring approval from the highest levels. Many other examples can be given now for the LHC experiments, where again political authorities are involved to

Fig. 15. Pope John Paul II visiting CERN.

Fig. 16. DG R. Heuer with C. Fernndez, president of Argentina.

permit cooperation, indicating that such collaborations have benefits far beyond the scientific domain. One event in particular serves well to underline this point. When disarmament negotiations in Geneva took place in the wake of the Reagan–Gorbachev summit, they ran into a deadlock. One day, the head of the US delegation, Alvin Trivelpiece, whom I knew from earlier collaborations, called me and suggested that I invite the heads of the two delegations to dinner at CERN, where in a neutral atmosphere respected by both parties they might find a solution. And indeed it worked. Later the ambassadors of the Disarmament Conference visited CERN and, at the end of their visit, stated they had learned that one of the objectives of CERN was to collide particles, whereas it was their task to avoid collisions between countries. But they added that CERN is probably doing better in both respects.

CERN has also become a meeting point between science, culture and religion. One example is the visit by Pope John Paul II in 1982. When we discussed the relation between religion and science, we agreed that there cannot be a conflict since the natural sciences are based on reproducible experiments whereas religion is inspired by transcendental revelation. Thus, they perceive different aspects of reality. The same agreement was achieved when the Dalai Lama visited CERN somewhat later.

10. The Future

For CERN, and indeed for the whole of particle physics based on accelerators, the physics output from the LHC will be decisive. Only the results of the LHC experiments will give an indication as to what the next steps should be, if any. If new phenomena are found, then an electron–positron collider could be the choice, since such collisions give much cleaner events than a hadron collider. If the Higgs and other new phenomena are found below 500 GeV, then a next generation of colliders reaching such an energy would be sufficient. Otherwise higher energies have to be aimed for.

Two electron–positron colliders, the International Linear Collider (ILC) and the Compact LInear Collider (CLIC), are being developed by international collaborations using different technologies. The ILC uses klystron-driven superconducting cavities and would probably be adequate if energies up to 500 GeV are sufficient. For higher energies, the CLIC, using room temperature cavities at very high frequencies powered by an intense drive beam, would be better suited and could go to 1000 GeV in an initial phase. Both machines could be extended to higher energies in a later phase. Only further technical developments will show which of the two possibilities is more advantageous as far as the physics goals and the cost are concerned. In any case, they could only be realized as an international project with worldwide participation, since the cost will be in the multi-billion-dollar range. In view of the present discussions on energy policies, the approval of such a project might hit a major obstacle because of the large electrical energy consumption, which would be higher than 200 MW already in a first phase.

The discussion concerning a future linear collider may appear in a completely new light when the LHC

Fig. 17.   Arial view of CERN, with Lake Geneva and Mont Blanc at the back.

has obtained more results. Already now it is surprising how clean the events at LHC experiments are and how well they can be analyzed. The big progress lies in a very fine granularization of the detectors which allow an excellent spatial resolution of events with many thousands of tracks in the collisions of lead nuclei. It has even become possible to analyze several events produced simultaneously in one particular beam crossing. The cleanliness of the events in hadron collisions has reached a quality that is comparable to that of electron–positron collisions and was not expected for the LHC. The advantage of a well-defined total energy in the case of electron collisions also loses some of its strength because of the so-called bremsstrahlung that is emitted at very high energies.

Meanwhile, the LHC has many years of running ahead of it and offers a number of possibilities for upgrading. In 2013, a major shutdown is scheduled to bring the energy to its design value of 7 TeV per beam. The first upgrading beyond the original design could start during long shutdowns in 2015 and 2018. By increasing the beam currents from 0.6 to 1 A and by squeezing the beam more strongly at the intersection points, the luminosity could be increased by a factor of about 7 compared to the design value of $1 \times 10^{34} \, \mathrm{cm}^{-2} \, \mathrm{s}^{-1}$.

In the more distant future, a major upgrade could be envisaged by changing the superconducting dipoles. If the present coil, made of NbTi, could be replaced by Nb_3Sn coils, the magnetic fields could be raised to 20 T, corresponding to a beam energy of 16.5 TeV. No other facility presently under consideration could reach similar energies. A working group at CERN is studying this high energy upgrade HE-LHC, which requires an aggressive program of R&D.

The old idea of electron–proton collisions, which had been considered already in LEP times, is also enjoying a renaissance. This LHeC option would, however, require more substantial changes. A new injector system (two different versions are under consideration) for the electrons with an energy of 10 GeV would have to be constructed and the actual experimental areas would require bypass tunnels for the new electron ring. Such a project would require 15–20 years.

Whatever the LHC results may bring, CERN can react according to the needs of physics, and a long bright future seems assured.

Acknowledgment

My warm thanks go to James Gillies, head of CERN Communication, who made essential contributions to this article and improved its form in many ways.

References

[1] This article is a "historical" review of CERN and cannot give the complete picture of accelerator or detector development. For complete reviews see: Landoldt-Boernstein, Springer Materials, Vol. I/21B, *Detectors for Particles and Radiation*, eds. C. W. Fabjan and H. Schopper (2011); and Vol. I/21C, *Accelerators and Colliders*, eds. S. Myers and H. Schopper, to be published in 2012.

[2] A. Hermann, J. Krige, U. Mersits and D. Pestre, *History of CERN*, Vol. 1 (North-Holland, 1987).

[3] H. Schopper, in *Italia at CERN, le ragioni di un successo*, ed. F. Menzinger (INFN, 1995).

[4] Allocutions prononcées à l'occasion du 30e Anniversaire du CERN, 21 Sep. 1984, brochure, CERN; J. Krige, *Phys. Today*, Sept. 2004, www.physicstoday.org/vol-57/iss-0/p44.html.

[5] A. Hermann, J. Krige, U. Mersits and D. Pestre, *History of CERN*, Vol. 1 (North-Holland, 1987), p. 199.

[6] A full description can be found in *CERN — 25 Years of Physics*, Physics Reports Series, Vol. 4 (North-Holland, 1981).

[7] *CERN — 25 Years of Physics*, ed. by M. Jacob (North-Holland, 1981).

[8] C. Chohan, Simon van der Meer (1925–2011): A modest genius of accelerator science, in this volume.

[9] Large Hadron Collider in the LEP tunnel, ECFA — CERN Workshop science (Lausanne, Geneva; Mar. 1984), ECFA 84/85, CERN 84-102.

[10] H. Schopper, *LEP — The Lord of the Collider Rings at CERN 1980–2000* (Springer, 2009).

[11] A comprehensive review of the state of the physics, the detectors and accelerators and colliders can be found in: Landolt-Boernstein, Springer Materials, *Elementary Particles*, Vol. I/21A, B, C (2010).

Herwig Schopper was born in Bohemia, Germany. He received his physics diploma and doctorate from the University of Hamburg and is Professor Emeritus there. He was Research Assistant at the Technical University, Stockholm (Lise Meitner); Cavendish Laboratory, UK (O. R. Frisch); and Cornell University (R. R. Wilson). His fields of work include optics, nuclear physics and elementary particle physics. Dr. Schopper held various professorships in Germany. He was Division Leader at CERN, Director at DESY and Director General at CERN. He was President of the Association of German Research Centers (now the Helmholtz Association), the German and European Physical Society, and the SESAME Council in Jordan. He is a member of various academies, and a fellow of the Institute of Physics, London and APS. Dr. Schopper has received several honorary doctorates, the Tate Medal of AIP, the UNESCO Albert Einstein Gold Medal, the Niels Bohr Gold Medal, (Denmark) and various state medals (Germany, Russia, Jordan), as well as other distinctions.

Reviews of Accelerator Science and Technology
Vol. 4 (2011) 279–291
© World Scientific Publishing Company
DOI: 10.1142/S1793626811000550

Simon van der Meer (1925–2011):
A Modest Genius of Accelerator Science

Vinod C. Chohan

CERN, CH-1211, Geneva 23, Switzerland
vinod.chohan@cern.ch

Simon van der Meer was a brilliant scientist and a true giant of accelerator science. His seminal contributions to accelerator science have been essential to this day in our quest for satisfying the demands of modern particle physics. Whether we talk of long base-line neutrino physics or antiproton–proton physics at Fermilab or proton–proton physics at LHC, his techniques and inventions have been a vital part of the modern day successes. Simon van der Meer and Carlo Rubbia were the first CERN scientists to become Nobel laureates in Physics, in 1984. Van der Meer's lesser-known contributions spanned a whole range of subjects in accelerator science, from magnet design to power supply design, beam measurements, slow beam extraction, sophisticated programs and controls.

Keywords: Accelerator physics; stochastic cooling; magnetic horn; antiproton accumulator; Nobel Prize; Simon van der Meer; CERN; antiproton–proton physics; hadron colliders; SPS; ISR; AA; LEAR.

Whilst writing homage to Simon van der Meer for this review, I soon realized that I had to start with a caveat. He was a practical genius often unjustifiably feared, particularly because he could seldom be equalled, was rarely challenged and, invariably, he would be right in any scientific discourse. Often, when approached with a novel suggestion, he would have had the same idea already and knew if it would work or not, restraining discussion. One would easily get short shrift if one argued with him in mistaken belief or disturbed him during his contemplative moments. Nevertheless, he was a kind and honest person with a brilliant, fast-thinking brain. Many at CERN since the 1960s knew about Simon's temperament, his shyness and perhaps his taciturn nature. However, those who came to know him closely realized quickly that his modesty and kindness never excluded helping those who sought his counsel or help. So, while he did not suffer fools gladly, a fact even acknowledged by his progeny, Esther and Mathijs, he was indeed very practical in his approach, helpful to those acceptable as well as being modest. The latter is best illustrated by his statement during the Panel Discussion on "Future of Particle Physics" chaired by Carlo Rubbia in September 1993, where he was amongst ten eminent personalities and Nobel Prize winners. This statement followed a day-long symposium on "Prestigious Discoveries at CERN, 1973 & 1983" [1]:

"Pierre [Darriulat] just said that he was embarrassed. I think that I have the right to be even more embarrassed because I feel a complete outsider in this company. I have never done any particle physics and even in the range of machine physics I always felt like an amateur. So I will not say anything about the future of CERN, this is really beyond me. But what I want to say to all people who are working on machine physics is to think of two things: first of all, do not believe it when people tell you that something is impossible. Always try to follow up crazy ideas. And don't forget that all the experts in machine physics sometimes forget things which you can do by making some kind of 'bricolage,' those things, which people thought could not work, still work, if you work on it long enough. I think that's all I want to say. Thank you."

To lend weight to my personal experience and belief, I can't help but cite Martinus Veltmann's statement, just following Simon's:

"I think that I have seldom seen such a spectacle of modesty as in Simon's view on these things. Few people have contributed so much to this Laboratory as he has. Thank you."

1. Introduction

Simon's practical approaches and versatility in all aspects of work and life are legendary. This, coupled with his prolific inventiveness, meant that he was arguably the cleverest person that I have come across in my 30-odd years at CERN. This statement also reflects the views of Hans-Karl Kuhn from his SPS power supplies days. Simon's contributions to CERN and accelerator physics speak for themselves; they ranged from magnet design in the 28 GeV PS era in the fifties to the 1961 invention of the pulsed focusing device (known as the van der Meer horn). This was followed in the sixties by the design of a small storage ring for a physics experiment studying the anomalous magnetic moment of the muon. The following decade saw him doing some very innovative work on power supplies regulation and control for the ISR and later the SPS. His ISR days in the seventies led to his technique for luminosity calibration of colliding beams, first used in ISR and still used today in the LHC. Last but not least, there came the Nobel-Prize-winning ideas behind stochastic cooling and application at CERN in the late seventies and in the eighties. Simon's prolific inventiveness means that the whole park of CERN's accelerators running so well today for physics, whether it is neutrinos to Gran Sasso, colliding proton beams in the LHC or the antiproton physics at the Antiproton Decelerator, owe him a immense gratitude. Likewise, the Fermilab antiproton program since 1983–85 on the other side of the Atlantic, the successes of the p–pbar Tevatron Collider to date and its discovery of the top quark, owe him considerable gratitude.

2. The Early Years

During his formative years in Holland, Simon was known to have frequented the local flea markets, taking home his purchases to invent and repair all things electromechanical at his parental home. When he was once asked what he was doing when the war broke out and everyone was glued to their radios, Simon's typically nonchalant answer was that he was probably repairing a radio instead! When refrigerators were still new in Holland and lacked a light coming on when one opened the door, he invented one for the home which did exactly that.

Simon was born in The Hague (The Netherlands) in 1925, the third child of Pieter van der Meer and Jetske Groeneveld. His father was a schoolteacher and his mother came from a teacher's family. Hence, it was hardly surprising that education and learning was cherished in the van der Meer family; the parents made sacrifices to provide this to Simon and his three sisters. Having attended the high school (gymnasium-science section) in The Hague, he passed his final examination in 1943, during the German occupation of Holland during the war. Although qualified to pursue further education, he could not attend the university because the German occupation had closed the establishments. So he stayed on in the high school, attending classes in the humanities section but at the same time assisting his former physics teacher with the preparation of numerous school physics demonstrations. This was a highly formative period for Simon; his interest in physics and technology knew no bounds and the physics teacher provided much inspiration and encouragement. Simon avidly dabbled in electronics — as he would say, equipping the family home with various gadgets, including the fridge light!

In 1945, Simon began studying "technical physics" at Delft University, where he specialized in measurement and regulation technology. This was the precursor to his influential work on the ISR and SPS voltage-controlled power supplies and closed loop control. The technical physics knowledge that Simon acquired at Delft was excellent, albeit "restricted out of necessity" (using his words). He regretted not having the intensive physics training that his contemporaries had undergone; however, he felt that his highly practical experience coupled with what he called a "slightly amateur" approach in physics was an asset enabling him to make original contributions to the field of accelerators. In a similar vein to his decisive power supplies work, his training in measurement and feedback at Delft led him to his well-acknowledged invention, stochastic cooling, which is also a combination of measurement (of the position of the particles) and feedback.

After obtaining his degree in 1952, Simon joined the Philips research laboratory in Eindhoven, working mainly on high-voltage equipment and electronics for electron microscopes. In 1956, he decided to move to the newly founded European Organization

Fig. 1. Simon van der Meer, 37 years old, explaining the principles of a magnetic horn (June 1962).

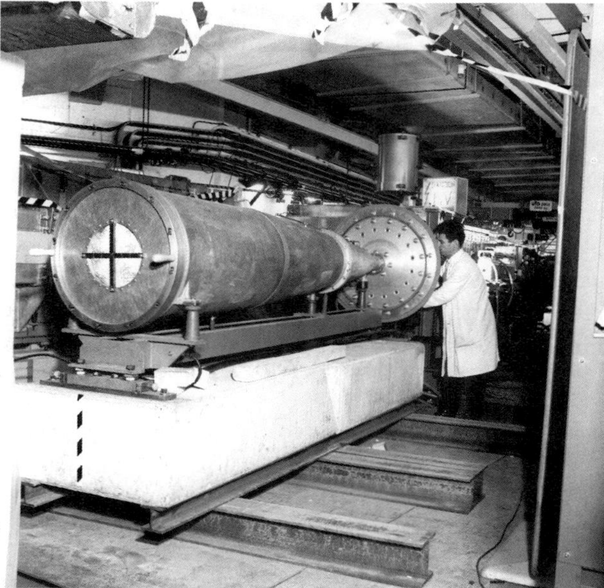

Fig. 2. The 1963 neutrino experiment beams from the PS.

of Nuclear Research (CERN), in Geneva. At CERN, under the leadership of John Adams and Colin Ramm, Simon became involved in the technical design of the pole-face windings, multipole correction lenses and their power supplies for the 28 GeV Proton Synchrotron (PS), under construction at that time. Fifty years on, the PS is still in operation, as the heart of CERN's accelerator complex and feeding the SPS and the LHC. Simon had a growing interest in the handling of particles. After working for a year (in 1960) on a separated antiproton beam, he proposed the idea of a high-current, pulsed focusing device, the magnetic horn [2]. This was his first CERN invention, at the tender age of 37 (Fig. 1); here, the charged particles traverse a few-mm-thick metal wall in which a pulsed high current flows. The original application of the magnetic horn was in the context of neutrino physics, where beams of pions have to be tightly focused. When the pions then decay into muons and neutrinos, an equally well-focused neutrino beam is obtained. The first horn [3, 4], for the PS neutrino beams in the early 1960s, was of monstrous size (Fig. 2), and today's horns for SPS beams to Gran Sasso, 732 km away, are equally huge. The magnetic horn of much smaller size was also essential in the collection of antiprotons from a production target and the subsequent accumulation and storage in the Antiproton Accumulator in 1980; an improved 1993 version is still in use today, at CERN's Antiproton Decelerator (AD) facility. Simon's interest in handling particles even extended to calculating the neutrino flux, taking into account the focusing provided by the horn [4, 5].

In 1965, Simon joined a small group led by Francis Farley preparing for a second experiment for the measurement of the anomalous magnetic moment of the muon. There, he designed a small storage ring (the "g-2" ring) and participated in all phases of the experiment. As he stated later, this period was an invaluable experience not only for learning the principles of accelerator design but he also got acquainted with the workings and lifestyle of experimental high-energy physicists.

It was also in 1966 that Simon met his future wife, Catharina Koopman, during a skiing excursion in the Swiss mountains. He later described his decision to get married soon after as "one of the best decisions of my life." The couple had two children, Esther and Mathijs, born in 1968 and 1970, respectively.

In 1967, Simon went back to more technical work, as opposed to particle handling or accelerator design, by becoming responsible for magnet power supplies. This was for the Intersecting Storage Rings (ISR), under construction at that time. During the ISR period, he developed the two-dimensional magnetic field calculation program MARE [6], which was used in both the ISR and SPS magnet design. Here, too, Simon's modesty meant that he did not want to put his name first on the CERN Yellow Report, as recounted to me recently by Romeo Perin.

Soon after, the so-called 300 GeV project for the construction of the Super Proton Synchrotron (SPS) at CERN was approved under the leadership of John Adams; Simon was part of the SPS Design Committee, as head of the SPS power supplies group (Figs. 3 and 4). Here, he proposed the generation of the reference voltages for the bending and quadrupole supplies to be based on the measurement of the field along the cycle, and detailed the outlines of the correction algorithms [7]. His proposal resulted in the first-ever computer-controlled closed loop system for a geographically distributed system as the 7 km circumference SPS was; this was a no mean feat for the early seventies. Main magnet current measurements were introduced only later, when the SPS had to run as a storage ring for the SPS p–pbar collider. Simon's work on the SPS power supplies using DAC's and simple integrators and a dedicated minicomputer remained the mainstay of the SPS way up to 1996. He also left a large software legacy in the SPS control room consoles for generating main power supply ramps, trimming the main dipoles and adjusting the tune together with archiving of power supply settings. Many of these pioneering tools and ideas were taken up or integrated into other application programs for the SPS and inherited now for the LHC.

It is also important to note that Simon's work for the SPS power supplies [7] was written up by him in August 1972, the same month as the seminal note on "Stochastic Damping of Betatron Oscillations in the ISR" [8].

During his activities at the ISR, Simon developed the technique using steering magnets to vertically displace the two colliding beams with respect to each other; this permitted the evaluation of the effective beam height [9], leading to beam luminosity at an intersection point. The famous so-called "van der Meer scans" are indispensable even today for the LHC experiments; without them, the precision of calibration of luminosity at any intersection point in the Collider would be much lower.

It was in 1968, for the ISR, that a new and brilliant idea to increase luminosity was conceived: the concept of stochastic cooling. "The cooling of a single particle circulating in a ring is particularly simple," as Simon said in a qualitative description of betatron cooling for the Nobel Lecture in 1984 [10]. All that was needed was to measure the amount of deviation from the central orbit and correct it later with a kicker at a suitable location in the ring; the correction kicker signal had to arrive at a suitable location before the beam trajectory through the ring. However, the devil is in the details of such a system. In reality it is not possible to measure the position of just one particle, because there are many particles in the ring and there is the electronic noise of the system so that a single particle is impossible to resolve. So, groups of particles, referred to as beam "samples" or "slices," must be considered instead.

Fig. 3.   The SPS 300 GeV Project Group Leaders, with Simon on the extreme right at near end (Oct. 1971).

Fig. 4.   The first beams in the SPS (summer 1976), with Simon standing on the extreme left.

For such a beam slice, it is possible to measure the average position during its passage through a pickup and to correct for this when the same slice goes through a kicker. In addition, there is the need for "mixing": because there is a spread around the central momentum, some particles are faster and others are slower. This leads to an exchange of particles between adjacent beam slices and this is vital for stochastic cooling. With the combination of many thousands of observations (many thousands of turns), a sufficiently large bandwidth of the cooling system, low-noise or even cryogenically cooled electronics and good mixing, stochastic cooling works.

Simon only published the first internal note [8] on stochastic cooling in August 1972, apparently after much persuasion from W. Schnell, the ISR Instrumentation and RF group leader at that time. However, as Simon says in the final remark of this decisive note, it was also because "fluctuations upon which the system is based were experimentally observed recently," meaning early 1972; see e.g. Ref. 11 and later Ref. 12. The first experimental record of the stochastic damping (cooling) observation over $\sim 4\,\text{h}$ is in Ref. 13, from November 1973.

Figure 5 illustrates the first pickup to kicker stochastic cooling system between ISR points 5 and 6 with a 120 m coaxial line. The experiment, carried out using that system, led to the results shown in Fig. 6, with a clear indication of reduction in the rms amplitude. The effect of stochastic cooling on or off, spread over 13 h on a stored beam, is more discernible in a test carried out in 1974 [14].

3. The CERN Proton–Antiproton Collider

In the late 1960s, a major challenge for particle physicists worldwide was the search for the W and Z particles — the carriers of the weak force — but no accelerator at CERN, or elsewhere, provided enough energy to create these predicted particles.

In 1966, Budker from Novosibirsk suggested that high-luminosity proton–antiproton collisions in a single magnet ring might become feasible by means of electron cooling. This was followed by the development and experimentation of stochastic cooling in the ISR in the early 1970s. This then led to the bold and imaginative idea of Carlo Rubbia in 1976 to use the CERN 400 GeV SPS or the Fermilab Main Ring as a single magnet ring p–pbar collider.

Head-on collisions between two beams provide the highest collision energies, a principle exploited at CERN in the 31 GeV ISR, in which two proton beams circulated in two interlaced rings. The new idea was that if a dense-enough antiproton beam could be produced, an accelerator with higher energy than the ISR, filled with a proton beam traveling in one direction and an *antiproton* beam in the *opposite* direction, could be used as a collider.

To apply this idea to the SPS, antiprotons had to be provided in sufficiently high quantities, and this required a new, small storage ring — the Antiproton Accumulator (AA). For CERN, this project was considered as an "experiment" because of the multitude of challenges and unknowns that had to be overcome. The AA was indeed an adventure into uncharted territory. Never before had a project called for such imagination, involving the whole of CERN.

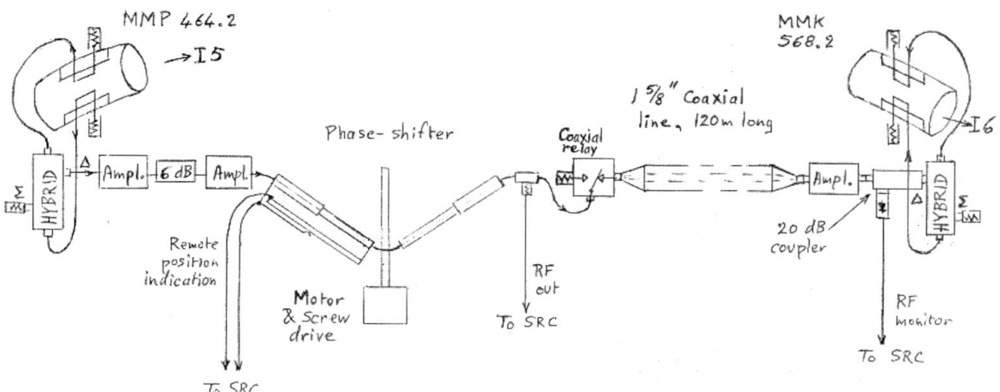

Fig. 5.　The first stochastic cooling system used in the ISR (November 1973).

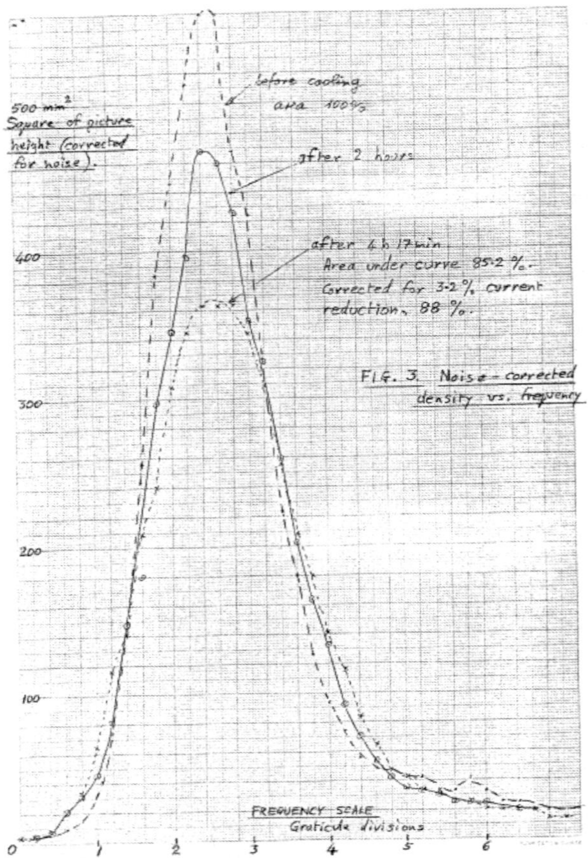

Fig. 6.   Treated data of vertical Schottky scans after subtraction of the noise floor, 27 November 1973.

Two CERN working groups examined in 1976 the technical aspects of such schemes and the physics potential. Finally, CERN decided to pursue two courses of action in parallel. One was to construct rapidly a small ring (Initial Cooling Experiment — ICE) to study both electron and stochastic cooling; the other was to set up a study group to prepare a design for the p–pbar facility using the SPS as a storage ring for collisions. Initially, the study group proposed using two separate rings for collecting and cooling antiprotons, because it was clear that the electron cooling scheme would only work at low energy for the large emittance antiproton beams. Hence, the second ring was to decelerate the antiprotons. Meanwhile, the many experimental tests in the ISR, further theoretical developments and, most importantly, the proposal of a faster and more efficient method of longitudinal cooling (the Thorndahl filter method, which uses the shorter, high signal-to-noise "sum" pickup, as opposed to "radial"

pickup), gave CERN the possibility of a solution based entirely on stochastic cooling and stacking. This then is how the AA was conceived and born — a fixed field, single DC-operated accumulator ring. The potential savings in cost and complexity compared to the two-ring idea were the ultimate criteria, despite the fact that it represented three orders of magnitude in extrapolation over the ISR stochastic cooling experiments.

Just as a side remark, electron cooling works by transferring energy from an antiproton beam circulating in a storage ring to a "cold" electron beam traveling in synchronism with the antiproton beam over part of its path. The electron beam has to be continuously refreshed from an electron gun. In other words, by precise matching of velocities of the two beams, the Coulomb interaction tends to equalize their "temperatures" so that an electron beam with very little transverse motion will reduce that of the antiprotons, whose density is thus increased. However, electron cooling works best at very low energy, whereas for the stochastic technique the cooling rate is independent of energy. As mentioned earlier, it was also the fast and efficient precooling of momentum spread by Lars Thorndahl's filter method that clinched the decision to exclusively use stochastic cooling for the p–pbar collider project. Electron Cooling also needs a rather large amount of well-functioning, reliable equipment.

The AA proposal called for an overall increase in antiproton density from the production target to the stack core of over 1E9. Meanwhile, the ICE ring in 1978 gave encouraging results in stochastic cooling [15], confirming cooling in all planes, though at timescales (longitudinally) of the order of 10 s, still about 30 times slower than what was required in the AA. However, the stochastic stacking process, which was prone to instabilities (as first pointed out by F. Sacherer [16]), was an essential feature in the accumulation scheme; this could not be tested in ICE. The process involved simultaneous cooling in both transverse planes and increasing the longitudinal density by four orders of magnitude whilst moving the particles into the dense core, using a combination of filter and radial-pickup-based Palmer cooling techniques to avoid instabilities. This then was the biggest gamble in the launch of the AA, because it could only be studied in detail by theoretical calculation. Fortunately, in this respect the AA performed as expected.

I first met Simon in the heady days of mid-1980, when the AA had just been constructed in a record time of two years, under the inspirational leadership of Roy Billinge, with Simon as a joint project leader. This was the era when the AA project team had meetings every Friday morning where no hierarchies were discernible (a far cry from today's CERN) and most junior technicians to senior accelerator physicists came together to hold discussions. If Simon spoke in dissent or agreement, his views would carry the day, and that was how things would be implemented; as a newcomer to this august crowd of accelerator builders and experts, it was quite an eye-opener, being part of this very vibrant and stimulating group of multiple nationalities working across many CERN divisions. The meeting would then adjourn to Tortella's, CERN's No. 2 restaurant, where R. Billinge had an astute arrangement, with food and drinks flowing for a modest fixed sum. Simon was a regular attendee, participating actively in over-the-table discussions in all matters of work or otherwise.

Those early AA days were also a period of prolonged controversies in the controls philosophy, such as simple, SPS style touch terminals directly connected to the AA controls computer or a two-layer model with a control room console computer connected to the front-end accelerator-related computer via a proprietary network. Simon's insistence on direct AA controls and facilitated interpreter (NODAL) programming carried the day. This was also the era when he gradually took over all the AA application programs, written initially by equipment builders and others; the stand-alone HP computers or other independent beam measurement instruments were frowned upon and he succeeded in eliminating all such devices by insisting on direct connection to the AA computer via the established CAMAC interfaces.

Issues of operation centered on the choice between using the local AA (Figs. 7 and 8) or the central Meyrin control room and how to man the delicate AA operation, with accelerator experts or operations group staff. After all, the AA was built as an "experiment" with the help of people from seven different CERN divisions! Simon's prolific programming ability meant that by 1981 he had implemented

Fig. 7.   First proton test beams for the AA (July 1980), with Simon seated on the chair closest — waiting for the beam flash on a scintillation screen!

Fig. 8.   Simon in the local AA control room with the SPS style touch terminal (Jan. 1984).

virtually all the user-friendly tools to man the delicate AA operation by the operations group team, put together by then. There were also the proxy wars about local control room devices and instruments — whether they were essential in the local control room or not. The final, *de facto* decision was Simon's in integrating the device's usage in his sophisticated application programs or not. In the case of the latter, the use of the instrument would fall into abeyance, much to the chagrin of the initial provider.

Simon was indeed a hands-on man. For machine improvements or experiments, he would often agree to work night shifts, particularly if he deemed it necessary to work quietly, without the usual perturbations of day shifts and more steady availability of test beams from the PS. A case in point was certain corrections on magnets where he was the unique expert. The AA and later the AC (Antiproton Collector) both had many packs of washers bolted to the quadrupole end shims, for the purpose of calibrating the gradient. Based on beam measurement, Simon would meticulously calculate a new configuration of washers. He and I worked on such night shifts on several occasions, where I would assist him with the tedious and time-consuming process of removing or adding washers on all the specific magnets and measurements.

We would make the beam tune measurements, recalculate the washer configuration, make the machine electrically safe by manually locking out the magnet circuit breakers, enter the machine and modify the washers; then it was back to the control room via the washroom to clean off all the graphite grease from our hands, unlock the circuit breakers, switch on the magnets and request the proton test beam from the PS. The injected beam would then be captured by the RF at injection and moved across the aperture to the desired momentum position and the tune remeasured, using transverse Schottky scans. Simon would then recalculate another washer configuration, and the whole process would start all over again. His hands-on versatility meant that he even showed me how to get the black grease off my hands most efficiently, without recourse to water!

The goal was, of course, to obtain the desired tune across the full momentum bite of the ring. Finally, Simon would fastidiously note all the results in the control room paper logbook. While others calculated and noted the AA or AC parameters to five or more decimal places, he knew that three would do and had it all in his head.

Simon had a vested interest in the AA machine being perfectly tuned and he wasn't going to leave this job to anyone else. This philosophy he also applied with equal vigour to the AC ring which we

commissioned in 1987. He would work at the highest computational level and then apply himself to the manual tasks with equal dedication. He never complained; it was a job which had to be done right and the only way to achieve that was to do it himself. So that is exactly what he did. Of course, he fully expected everyone around him to be equally committed. Not always an easy act to follow!

After Roy Billinge became the PS Division Head, he usurped the AA Group's Friday morning meeting slot and lunch, and we had to move to Mondays for weekly meetings and lunch, which also moved, to the local pizza restaurant across the road from CERN. The lunch gatherings had a much-reduced crowd, with an ebullient and charismatic group leader, Eifionydd Jones. The success of AA meant a series of Fermilab scientists partaking in these meetings and lunches. Particular names that come in mind are Rol Johnson, Gerry Dugan, John Marriner, Carlos Hojvat and, occasionally, John Peoples, Alvin Tollestrup and Fred Mills.

Fermilab too, went through the ideas of moving shutters inside the accumulator ring, similar to the CERN AA, moving in every 2.4 s to decouple the cooling systems, i.e. protecting the accumulated stack from the powerful and fast precooling systems. Electron cooling was also discussed at Fermilab; however, eventually, the Fermilab pbar source took the final shape of debuncher and accumulator rings with higher-frequency-band stochastic cooling systems only. Many from Fermilab sought Simon's counsel, as of others at CERN in those days. The then young physicist Jeff Hangst (recently, of antihydrogen at AD fame) visiting CERN would not have missed his first encounter with Simon then, while reporting on the Fermilab ideas of a Sem-Grid before the production target receiving the 120 GeV beam. As Jeff would recount about this later, Simon was not shy to castigate him, albeit discretely, calling it "fiction" due to the resolution of grid wires — much to Jeff's astonishment! The Fermilab pbar source technical notes and reports of that era reflected very much the reports and notes of the CERN AA, sometimes even down to the notation of q instead of ϑ for tunes.

That period was the beginning of the very fruitful CERN–Fermilab collaboration of the antiproton accelerator community. Ideas and people crossed the Atlantic and eventually ideas even came back, like CERN's construction of the larger acceptance Antiproton Collector (AC) ring for fast "precooling" systems, disuse of shutters in the AA and the advent of higher-frequency-band stochastic cooling systems at CERN. I believe that Simon did not like to travel much. On one occasion, I remember that he agreed to travel for some review or something similar and the meeting was held at Chicago O'Hare, and he took the flight straight back. This had to be via Amsterdam so that he could buy books at the airport in his mother tongue!

4. The Nobel Prize

The first SPS p–pbar collisions occurred in July 1981 and the first real period of physics runs took place in 1982. December 1982 saw the Collider arriving at an integrated luminosity of 28 inverse nanobarns and Carlo Rubbia offering a champagne-only party with 28 champagne bottles! Suffices to say that the first signs of the W boson were announced soon after, in January 1983. This was to be followed by the discovery of the Z (Fig. 9), announced in May 1983. I would not go into too many details of the W and Z discoveries and the Nobel Prize, because these have been amply recorded elsewhere then and since then. When the telex came on 17 October 1984, Simon was sitting in his office in Bldg. 19 and, characteristically, Rubbia was in a Milan cab on the way to Milano Linate airport *en route* to Trieste for a conference. The story goes that the taxi radio was interrupted by a news flash because of an Italian sharing the physics Nobel Prize; initially, Rubbia's excitement

was hardly believed by the taxi driver but, on persistent convincing, Rubbia got a free ride!

The Nobel citation read:

"The Nobel Prize in Physics for 1984 was awarded jointly to Simon van der Meer and Carlo Rubbia for their decisive contribution to the Large Project, which led to the discovery of the field particles W and Z, communicators of weak interactions."

There was euphoria in the control rooms, particularly the local AA control room, where we gloated over the telex copy, quickly stuck in the AA logbook, and champagne started flowing. There were similar scenes in the UA1 cavern/control room (Fig. 10). The antiproton accelerator community certainly liked the words "Large Project" in the citation, because it gave a collective recognition too! Makes one wonder what the Nobel committee would need to invent if prize-winning discoveries are at the LHC's fore.

The success of the CERN antiproton adventure meant that we got some consolidation money, managed parsimoniously by Ted Wilson. 100 MHz sampling transient recorders with sufficient memory and based in CAMAC were just about available then from LeCroy. It was Colin Johnson's idea to invest in these. Dollar-wise, they were very expensive, and even some UA2 detector physicists were envious of our purchases and borrowed them for trials. Simon used to be highly irritated with another commercial device (much pricier), connected via the GPIB interface to the controls system. I suspect it was purchased without his tacit benediction! Hence,

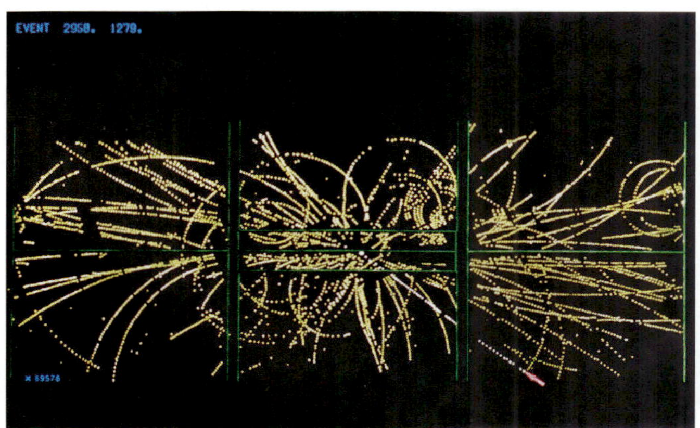

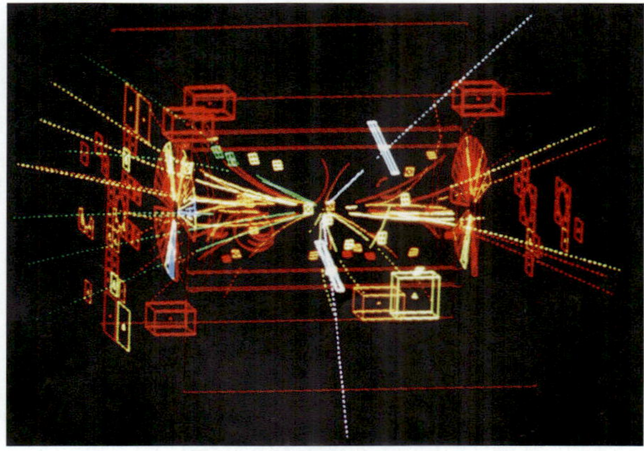

Fig. 9.   First W event, left (Dec. 1982) and first Z event in the UA1 experiment (Apr. 1983).

Fig. 10. C. Rubbia and Simon cheering in the UA1 cavern, celebrating the Nobel Prize (Oct. 1984).

my next strong collaboration with Simon began. We worked together for months over this cause till we got the Digitizer system fully operational, according to Simon's wishes, and to get the maximum out of the faster middleware software, I had to negotiate with the usual groups concerned. Subsequently, Simon would invent more and more applications using digitized signals from position pickups and the like, till he retired. The other GPIB device went into its natural obsolescence in usage soon after! Not much later, such digitizers provided the backbone for LeCroy's pioneering digital oscilloscopes.

Perhaps a bit different from today's LHC and several huge experimental teams, there was much more cohesion between the UA1, UA2 experimental physicists and the accelerator community. We used to have the daily, animated five o'clock meetings but Simon refrained from attending them. Even though he was very much involved in daily operation, he preferred others to take up the AA cause for collider operation.

The Nobel Prize apart, the AA was not without its gremlins that hounded us to the bitter end. The machine acceptance doggedly hovered around the $80\,\pi\,\mathrm{mm \cdot mrad}$ mark in both planes instead of the $100\,\pi$ foreseen, and the pbar yield obtained stayed frustratingly low compared to the design. This then led to innumerable studies and proposals

for yield improvement using ideas for better target assemblies, conducting targets, Li lenses for collection and eventually the design and construction of the larger acceptance AC Ring in 1986–87, which surrounded the AA Ring.

The ACOL project, as it was called, required all the usual services and equipment, particularly the new beam instrument devices and their integration into the existing controls system. Of course, it goes without saying that Simon would insist on everything following the same standards as the AA so that we could adapt or augment the necessary hardware and software easily. The Project Leader was Eifionydd Jones, who had a way of dealing with everyone, including Simon; his explicit instruction was for me to keep Simon happy and ensure all went well during all the planning, construction and implementation for all equipment integration, including the beam instrumentation and controls issues. We also had only 11 months for installation activities in 1986–87, with startup/commissioning in June 1987. We were late in full machine equipment installation by June 1987 but kept that date; the outstanding installation work was done in daytime and commissioning at night. Simon and I were the two fully excused from night shifts so that we could iron out all the problems that were exposed in the commissioning night shift with test beams. He and I had the tacit arrangement to go home in the evening as usual but came in later daily, to see the fruits of our daytime modifications with evening beams till fairly late in the evening.

The AA and AC pbar source complex from 1987 onward till 1996 remained the most highly automated set of machines [17] in CERN's repertoire of accelerators, again thanks to Simon.

The success of the p–pbar Collider program also had a spinoff in the form of antiproton physics at very low energies. A Low Energy Antiproton Ring (LEAR) came into being at CERN around 1983, using both stochastic and electron cooling systems. Antiprotons were extracted and sent to the experiments in the CERN South Hall. Simon's invention of noise-assisted slow extraction (so-called stochastic extraction) permitted antiproton physics spills for up to 24 h — a boon for feeble quantity antiproton trapping. Prior to that, low-ripple spills could last at best a few seconds. The CERN LEAR program ended in 1996, but not before culminating in the creation of the first antihydrogen atoms in 1995 and subsequent conversion of the AC Ring into the Antiproton Decelerator, a new facility to continue the low-energy antiproton physics experiments.

Stochastic cooling systems became operational at Fermilab in the early 1980s and in the early 1990s, at GSI Darmstadt and Forschungszentrum Jülich (FZJ). Today, there are nine or so other cooling rings, all over the world.

5. Retirement

A few months before Simon was due to retire in December 1990, the Division Head, Roy Billinge, organized a meeting together with Simon and me, to cater for Simon's software legacy for the future operation of the pbar complex. In the following months, Simon meticulously documented the programs which had his direct implication and knowledge. Those were early days for word processors or PCs and everything was done very clearly as text files in his favorite interpreter. This then was the legacy he left to me, up to the last days of the AA (and the AC) in 1996, when the pbar program in that form ended at CERN. On his last day at CERN, he took home the large output listings of all programs, and I was welcome to call him if I needed him so that he could look up something if needed.

Simon cleared out his office and insisted on not keeping a foot in CERN, unlike many others, then and subsequently. That showed the modesty and spirit that he had expressed throughout his career at CERN. He had done his bit, and now it was time to do other things — including his voracious reading!

All those years our Monday lunches had continued, till the early, regrettable passing away of Eifionydd Jones in early 1990. By then, the lunch participants were down to four or five; others had dispersed or disappeared to other activities. At the Monday lunches, Simon was always well-informed about current affairs — including, surprisingly, news from Britain; this I found out later because he bought *The Observer*, the English Sunday paper. Once, I chided him, saying that *The Sunday Times*, which I read on Sundays, was of much superior quality! Little did I know that the crossword puzzle in *The Observer* was *the reason*; he would usually succeed in completing it in no time on the Sunday afternoon, hence his excellent command of English.

After Simon's retirement, it was the AA group's veterans and antipodeans Colin Taylor (the Australian who built the first CERN linac as well as the AA core cooling systems) and Ray Sherwood (the New Zealander of linac and beam line fame), together with Simon and me, that held the fort for pizza lunches in the local establishment(s). Later, Flemming Pedersen, on his return from SLAC, joined us regularly, while Ted Wilson would join in occasionally, in between his CERN Accelerator School jaunts. Satellite dishes were in vogue those days, and both English and Dutch channels were just about receivable in Geneva with different satellites and orientation. Simon's first retirement project was to build an 80 cm dish with orientation controls available from the living room at home. I remember the related over-the-pizza discussion and, within a few days, the whole thing was functional. The dish was in the garden and he had made a flat surface with a cable trench neatly covered to avoid mowing problems. I found out later that he had built a motorized antenna holder system in his usual efficient but frugal style: using an idle wheel of a container, a barbecue grill motor and a Tupperware box to keep out humidity!

In the early part of his retirement, Simon agreed to assist in some worldly cause, if my memory is correct. This involved writing and posting letters under the auspices of the World Economic Forum, which is based in Geneva. However, it would be more appropriate to say that unlike many other Nobel laureates with exuberant personalities, he would rather not proffer counsel or take up causes, of whatever noble or laudable claim, because it would just be against his grain to do so. He was not a man in quest of self-serving platitudes.

Late in 2003, in preparations for CERN's 50th birthday in 2004, it was decided to compile and publish a 50th anniversary glossy book, with 50 chapters. Each chapter would denote a year and notable events in that year, in the history and life of CERN. I was asked to approach Simon about writing the 1980 chapter, on the beginnings of the antiproton adventure, just as Carlo Rubbia had agreed for the 1983 chapter, amongst other personalities of CERN, spanning 50 years. It goes without saying that Simon's characteristic response was negative, because it was all a long time ago and he did not remember much and so forth — his usual, modesty-driven reasons. After an iteration or two, he agreed that if I wrote the piece, he would read and correct it. This, then, is what is published in Ref. 18.

In May 2004, I suggested to Simon to come and visit the LHC Magnet Test Facility, where I was responsible for the 24/7 operation to test all the LHC cryomagnets. This he did after one of our lunches, and it was one of his last visits to CERN (Fig. 11). In the following years, some back problems meant that he was reluctant to come for lunches, and later he would refrain from driving and refused rides as well. Without Simon, the regularity of lunches too waned away.

Simon van der Meer left his mark in many ways in the nearly 60 years of CERN's existence and accelerators in general. While accelerator physicists continue to come up with new developments and add to our growing knowledge, Simon's practicality and approach provided dramatic added value not equalled in the history of CERN. It would be a moot point whether in large science projects anywhere such a personality could still shine through, despite the growth of Weberian hierarchies, in the name of increased oversight and scrutiny.

To conclude, I cannot but quote Mathijs van der Meer, in his epitaph to his father in March 2011:

"Simon van der Meer had grown up in a religious family but the war made him lose his faith. Still, he would not have minded a passage from Cardinal Newman's prayer, 'The Mission of my life,' which says, 'I am a link in a chain, a bond of connection between persons.' So, even though Simon did not believe in the soul or life after death, he would probably agree that there is such a thing as a spirit which would remain — in and with the people with whom he interacted."

Amongst others, I was one of those.

Acknowledgments

I am grateful to Mathijs and Esther van der Meer for confiding to me some details of Simon's early period in Holland. Apart from my own records, I have used the CERN document archives, including the CERN Courier article (June 2011 issue) compiled by F. Caspers, D. Mohl and H. Koziol.

References

[1] *Eur. Phys. J. C* **34**, 91 (2004).

[2] S. van der Meer, A directive device for charged particles and its use in an enhanced neutrino beam, *CERN Yellow Report* 61–7.

[3] M. Giesch, B. Kuiper, S. van der Meer *et al.*, *Nucl. Instrum. Methods* **20**, 58 (1963).

[4] M. Giesch, S. van der Meer *et al.*, Magnetic horn and neutrino flux calculations, NPA/Int. 63–26, 4.11.63.

[5] S. van der Meer and K. M. Vahlbruch, Neutrino flux calculations, NPA/Int. 63–11.

[6] R. Perin and S. van der Meer, The program MARE for the computation of two-dimensional static magnetic fields, *CERN Yellow Report 67-7*, Mar. 1967.

[7] S. van der Meer, Programming of the main magnet power supplies, Lab II/PS/Int/72-3, Aug. 1972. Also, see http://cdsweb.cern.ch/record/213191/files/p141.pdf.

[8] S. van der Meer, Stochastic damping of betatron oscillations in the ISR, CERN/ISR-PO/72-31, Aug. 1972.

[9] S. van der Meer, Calibration of the effective beam heights in the ISR, CERN Internal Report ISR-PO/68-31, June 1968.

[10] S. van der Meer, Nobel Foundation, Nobel Lecture, Dec. 1984. http://nobelprize.org/nobel_prizes/physics/laureates/1984/meer-lecture.pdf.

[11] P. Bramham, K. Hubner and W. Schnell, Observation of microwave signals, ISR Performance Report, 4 Apr. 1972.

[12] W. Schnell, About the feasibility of stochastic damping in the ISR, CERN-ISR-RF/72-46, Nov. 1972.

[13] P. Bramham and L. Thorndahl, Stochastic damping (cooling) experiment, ISR Performance Report, Run 390 — Ring 2 (27.11.73), 5 Dec. 1973.

Fig. 11. Simon visiting the SM18 LHC Magnets Facility (May 2004). This is one of the last photographs of him visiting CERN.

[14] P. Bramham, G. Carron, H. Hereward, K. Hubner, W. Schnell and L. Thorndahl, *Nucl. Instrum. Methods B* **125**, 201 (1975).

[15] G. Carron *et al.*, *Phys. Lett. B* **77**(3), 353 (1978).

[16] F. Sacherer, Stochastic cooling theory, CERN/ISR/TH/78-11 (1978).

[17] V. Chohan and S. van der Meer, *Nucl. Instrum. Methods Phys. Res. A* **293**, 98 (1990).

[18] *Infinitely CERN, 1954–2004: Memories of Fifty Years of Research* (Hurter, Geneva, 2004).

Vinod C. Chohan graduated in Wales (B.Sc.) and England (Ph.D.), and joined CERN's PS Division in January 1975 for three years. He also worked at the Swiss Institute of Nuclear Research before joining the Antiproton Accumulator team at CERN in 1980. He worked closely with Simon van der Meer on many beam measurements, automation and operational issues, and played key roles in the CERN antiproton construction projects through the 1980s and '90s. He has spent time at Fermilab, during the commissioning and operation of the antiproton source machines in 1985 and 1986, as well as being briefly at the PSR Ring in Los Alamos in 1989. In recent years, he has headed the team responsible for the operation and testing of all superconducting cryomagnets for the LHC. At present, he leads a team responsible for technical coordination activities in CERN's chain of injector machines and beam lines right up to the LHC.